Principles of Paleoclimatology

Principles of Paleoclimatology

Editor: Andrew Hyman

RCALLISTO
REFERENCE

www.callistoreference.com

Callisto Reference,
118-35 Queens Blvd., Suite 400,
Forest Hills, NY 11375, USA

Visit us on the World Wide Web at:
www.callistoreference.com

ISBN: 978-1-63239-853-6 (Hardback)

Cataloging-in-publication Data

Principles of paleoclimatology / edited by Andrew Hyman.
    p. cm.
Includes bibliographical references and index.
ISBN 978-1-63239-853-6
1. Paleoclimatology. 2. Climatic changes. 3. Climatology--History. I. Hyman, Andrew.
QC884 .P75 2017
551.60901--dc23

Table of Contents

Permissions

List of Contributors

Index

Preface

The study of the weather and weather patterns enable researchers to record and analyze climate change as it unfolds in a particular region over a period of time. This research helps in adding great detail to the evolutionary history of the earth when extreme climatic changes occurred. Paleoclimatology relies on fossils that are trapped in the earth's surface, such as sediments, ice sheets, corals and tree rings which reveal the climatic identity of the region through its inner structure. This book provides significant information of this discipline to help develop a good understanding of paleoclimatology and related fields. It includes some of the vital pieces of work being conducted across the world on various topics related to this field. This book is a resource guide for experts as well as students. Those in search of information to further their knowledge will be greatly assisted by this book.

This book is a comprehensive compilation of works of different researchers from varied parts of the world. It includes valuable experiences of the researchers with the sole objective of providing the readers (learners) with a proper knowledge of the concerned field. This book will be beneficial in evoking inspiration and enhancing the knowledge of the interested readers.

In the end, I would like to extend my heartiest thanks to the authors who worked with great determination on their chapters. I also appreciate the publisher's support in the course of the book. I would also like to deeply acknowledge my family who stood by me as a source of inspiration during the project.

Editor

Late Miocene–Pliocene Paleoclimatic Evolution Documented by Terrestrial Mollusk Populations in the Western Chinese Loess Plateau

Fengjiang Li[1]*, Naiqin Wu[1], Denis-Didier Rousseau[2,3], Yajie Dong[1], Dan Zhang[1], Yunpeng Pei[4]

1 Key Laboratory of Cenozoic Geology and Environment, Institute of Geology and Geophysics, Chinese Academy of Sciences, Beijing, China, **2** Laboratoire de Meteorologie Dynamique, UMR INSU-CNRS 8539 & CERES-ERTI, Ecole Normale Superieure, Paris, France, **3** Lamont-Doherty Earth Observatory of Columbia University, Palisades, New York, United States of America, **4** School of the Earth Sciences and Resources, China University of Geosciences, Beijing, China

Abstract

The Neogene eolian deposits in the Chinese Loess Plateau (CLP) are one of the most useful continental deposits for understanding climatic changes. To decipher Late Neogene paleoclimatic changes in the CLP, we present a terrestrial mollusk record spanning the time interval between 7.1 and 3.5 Ma from the western CLP. The results indicate four stages of paleoclimatic evolution: From 7.1 to 6.2 Ma, cold and dry climatic conditions prevailed as evidenced by high values of the total number of cold-aridiphilous (CA) mollusk species and by low values of all of the thermo-humidiphilous (TH) mollusk indices. From 6.2 to 5.4 Ma, the climate remained cold and dry but was not quite as dry as during the preceding phase, as indicated by the dominance of CA mollusks and more TH species and individuals. From 5.4 to 4.4 Ma, a warm and moist climate prevailed, as indicated by high values of the TH species and individuals and by the sparsity of CA species and individuals. From 4.4 to 3.5 Ma, all of the CA indices increased significantly and maintained high values; all of the TH indices exhibit high values from 4.4 to 4.0 Ma, an abrupt decrease from 4.0 Ma and a further increase from 3.7 Ma. The CA species of *Cathaica pulveraticula*, *Cathaica schensiensis*, and *Pupopsis retrodens* are only identified in this stage, indicating that the CA species were diversified and that the climate was becoming drier. Moreover, the CA mollusk group exhibits considerable diversity from 7.1 to 5.4 Ma when a cold, dry climate prevailed; whereas the diversity of the TH group was high during the relatively warm, wet interval from 5.4 to 4.4 Ma. This indicates that variations in the diversity of the CA and TH mollusk groups were closely related to climatic changes during the Late Miocene to Pliocene.

Editor: Lorenzo Rook, University of Florence, Italy

Funding: This study is supported by the Chinese Academy of Sciences (KZCX2-EW-QN107), the National Basic Research Program of China (No. 2010CB950204), the Strategic Priority Research Program: Climate Change, Carbon Budget and Relevant Issues (XDA05120203), the National Natural Science Foundation of China (Projects 41072130, 41272205, and 41230104), and part of the work was achieved during a stay in Paris through a CNRS-NSFC grant. This is Lamont-Doherty Earth Observatory contribution 7763. The funders had no role in study design, data collection and analysis, decision to publish, or preparation of the manuscript.

Competing Interests: The authors have declared that no competing interests exist.

* E-mail: fengjiangli@mail.iggcas.ac.cn

Introduction

As evidenced by the marine benthic foraminiferal $\delta^{18}O$ record, Earth's climate underwent a gradual cooling trend during the Late Miocene and Pliocene. This interval witnessed the progressive cooling of oceanic deep water, the expansion of permanent ice sheets in Antarctica, the occurrence of ice rafted detritus in Northern Hemisphere (mostly in north Atlantic), and both hemispheres covered by ice sheets after middle Pliocene [1]. Coincident with ice development in both Polar regions are significant ecological, climatic, and tectonic events elsewhere and especially around Asia [1–12], and which demonstrate that the Late Miocene and Pliocene was an important and complex interval that needs to be better understood.

In East Asia, one of the most important climatic changes is the evolution of the East Asian (EA) monsoon. Numerous sedimentological, geochemical, and paleontological studies, including of fossil mammals, mainly from the Chinese Loess Plateau (CLP) and the South China Sea (SCS), have contributed significantly to our understanding of the EA monsoon changes during the Late

Miocene to Pliocene. However, the results are inconsistent and some are even in conflict. In the CLP, sedimentological evidence of changes in sediment grain size and sedimentation rate from the Xifeng, Lingtai, Lantian, Xunyi, Luochuan, Jiaxian, Baode, and Jingle (Figure 1) Red Clay deposits indicate the occurrence of a strong winter monsoon and pronounced aridity in the Asian interior during the Late Miocene [8,13–16]. However, the geochemical record of $^{87}Sr/^{86}Sr$ ratios from the Lingtai Red Clay sequence implies a weak EA winter monsoon from 7 to 2.5 Ma [17]; and in addition, this interval can be sub-divided into a large amplitude and high frequency stage from 7–4.2 Ma and a weak, stable stage from 4.2–2.6 Ma, as evidenced by records of Zr/Rb ratio and mean quartz grain size [18]. Furthermore, grain-size records from fluvial deposits in the Linxia Basin (Figure 1) indicate that the EA winter monsoon intensified at 7.4 Ma and 5.3 Ma [19], the latter datum being significantly different to evidence from the Red Clay record.

In the case of the EA summer monsoon, the magnetic susceptibility record from the Xifeng Red Clay sequence indicates a weakened summer monsoon during the Late Miocene (6–5.4

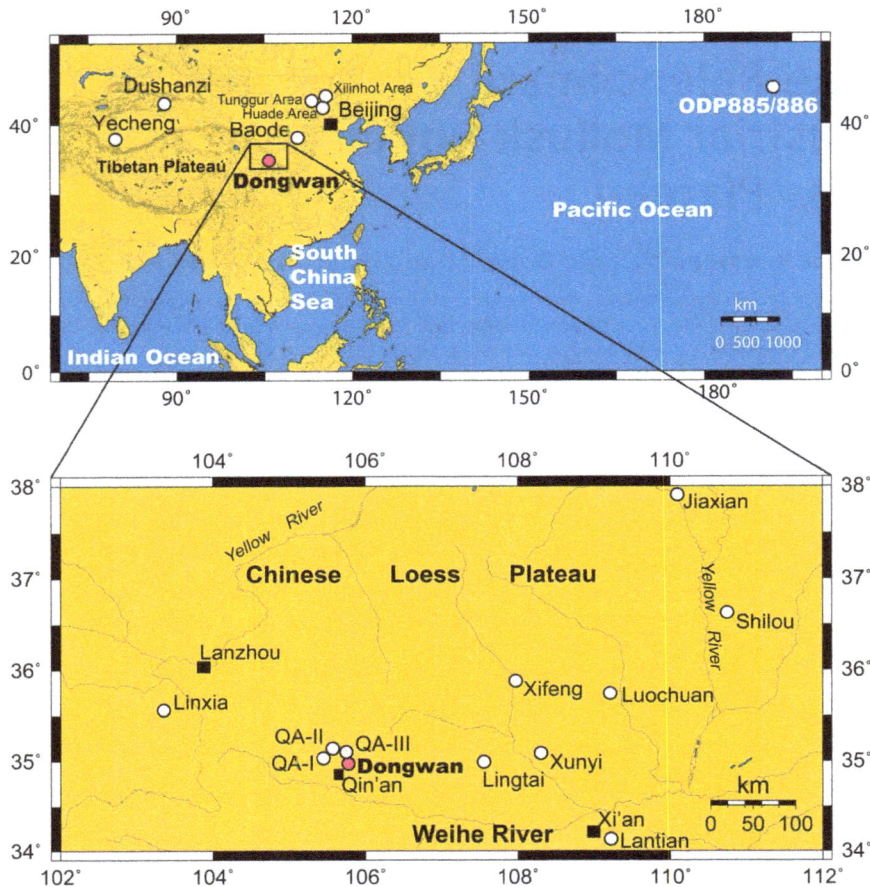

Figure 1. Location of the Chinese Loess Plateau (CLP), the studied Dongwan loess-paleosol sequence and other Neogene sections mentioned in the text. The Dongwan sequence is indicated by a red circle and other sections by white circles. Main cities are indicated by black squares.

Ma) and a strong summer monsoon during the early Pliocene [8]. Additional information about the EA summer monsoon is provided by records of pedostratigraphy and iron geochemistry from the Lingtai Red Clay deposits. These records indicate that the EA summer monsoon was relatively weak from 7.05 to 6.2 Ma, strengthened from 6.2 to 5.5 Ma, was very strong from 5.5 to 3.85 Ma corresponding to the interval of strongest soil development, and weakened significantly over the interval 3.85 to 3.15 Ma [20,21].

However, studies of mammalian fossils from the CLP have yielded very different results: Hypsodonty analysis indicates that northern China became more humid at 7–8 Ma, coincident with the onset of Red Clay deposition in the eastern CLP, and which was interpreted by the authors as representing the onset or intensification of summer monsoonal precipitation [22,23]. A general rise in $\delta^{18}O$ values of soil carbonate from the Lantian fluvial and Red Clay deposits reflects increased summer precipitation related to the onset and/or intensification of the EA monsoon during the Late Miocene to Pliocene [24,25], and which is supported by a mammalian faunal turnover event implying a marked change to more humid and closed habitats [26]. Recent isotopic evidence from fossil mammals and soil carbonates indicates a strengthened EA summer monsoon from 7–4 Ma [27]. However, the $\delta^{13}C$ values of fossil enamel from a diverse group of herbivores and of paleosol carbonate and organic matter

from the Linxia Basin indicate that C4 grasses were either absent or insignificant in the Linxia Basin prior to 2–3 Ma, suggesting that the EA summer monsoon was probably not strong enough to affect this part of China throughout much of the Neogene [28]. However, this result conflicts with the interpretations of Fortelius et al. (2002) [22], Kaakinen et al. (2006, 2013) [24,25], Liu et al. (2009) [23] and Passey et al. (2009) [27].

In the SCS, the EA winter monsoon developed progressively from 7 Ma as shown by an increasing trend in black carbon concentration and accumulation rate [29]. In addition the EA summer monsoon weakened after 7.5 Ma, as evidenced by combined planktonic foraminiferal Mg/Ca and $\delta^{18}O$ records [30], consistent with the Red Clay sequences in the CLP. The suggestion of a weakened EA summer monsoon is supported by geochemical records which indicate that chemical weathering intensity decreased from the early Miocene with a rapid decrease centered at 7.2 Ma [31]. This result is generally consistent with records of radiolarian species numbers and individuals, and diversity, from the southern SCS and which suggest a major decrease in summer monsoon intensity after 7.70 Ma [32]. In contrast, records of clay/feldspar ratio, kaolinite/chlorite ratio and biogenic opal MAR from the SCS suggest that the summer monsoon was strong from 7.1–6.2 Ma and remained relatively stable from 6.2–3.5 Ma [33]; however, the authors emphasize that their study is a schematic reconstruction which only outlines the principal stages but neglects the details.

A semi-quantitative reconstruction of the Neogene vegetation in China indicates that Miocene aridification associated with strengthening of the EA winter monsoon is consistent with Neogene global cooling, and that the EA summer monsoon did not weaken during the Pliocene [34], and this result agrees with most of the geological records from the CLP and SCS. However another quantitative reconstruction from plant fossil records yielded contrasting results, indicating that records of both temperature and precipitation from north China exhibit no significant difference between the western and eastern regions during the Miocene, suggesting that the monsoon climate did not commence or intensify at that time [35]. Regional climate model experiments also reveal that during the Late Miocene, from 11–7 Ma, the monsoonal climate may not have been fully established in various Asian regions, including northern China [36]; and this finding is contrary to that of numerous previous studies of the EA monsoon which suggest that it was initiated around the time of the Oligocene/Miocene boundary [37–39].

The foregoing review demonstrates that more work is needed to better understand the evolution of the EA monsoon during the Late Miocene and Pliocene. In particular, higher resolution studies using more sensitive monsoon proxies from key monsoon-dominated regions may be one of the best solutions for resolving the various inconsistencies and even conflicts regarding the process of monsoon development.

As mentioned above, the CLP (Figure 1), located to the northeast of the Tibetan Plateau, is a key continental region for the study of the EA monsoon. Deposition of eolian sediments commenced in the western CLP from 22 Ma, as observed in the QA-I, QA-II, and QA-III (Figure 1) Miocene loess-paleosol sequences which have a basal age of about 22 Ma [37,40], while the upper boundary is dated at about 3.5 Ma in the Dongwan late Miocene-Pliocene loess-paleosol sequence in the western CLP [41]. In contrast, in the eastern CLP, the age of the lower boundary of eolian deposits, the Red Clay sequences, is about 7–8 Ma as evidenced in the Lingtai, Xifeng, Liantian, and Baode sequences [15,20,42,43]. However, results from the recently reported Shilou Red Clay sequence indicate that eolian sediments were deposited from 11 Ma in the eastern CLP [44]. Despite the occurrence of a totally different lower boundary age between the east and west CLP, these deposits have great potential for deciphering in detail the processes of ecological and climatic evolution in the CLP, and by extension in East Asia, during the Miocene and Pliocene. As reviewed above, sedimentological and geochemical studies have so far contributed a great deal to our understanding of climatic changes during the interval of interest [6,8,9,13–18,20,21,24,27,37,40–45]; however, to date there has been only a limited application of a biological approach to analyzing these sequences [22,26,46,47].

Land snails are generally the most common and abundant fossils in Quaternary loess sequences, and this fact, together with the limited degree of success of most of the other paleontological studies of Quaternary loess, makes them especially important paleoenvionmental indicators for loess deposits. Their occurrence in Quaternary loess was first documented in Europe in the early 1820s [48], and since then they have contributed significantly to understanding the origin and paleoclimatic evolution of Quaternary loess deposits in Eurasia and especially in the CLP [46,49–57]. However, terrestrial mollusks preserved in the Neogene sequences in the CLP have not been investigated in detail until recently. Although they have provided crucial paleontological evidence for the wind-blown origin of the Neogene loess sequences

in the western CLP [58,59], the ecological and climatic information that terrestrial mollusks may provide has not been well deciphered, with the exception of the record from the Xifeng Red Clay sequence spanning the interval from 6.2 to 2.4 Ma in the eastern CLP [46]. Furthermore, it is unknown how terrestrial mollusk diversity varied in the CLP against the background of climate changes during the Neogene, since the necessary studies have not been performed. In this study, we studied the record of terrestrial mollusks preserved in the Late Miocene to Pliocene Dongwan section in order to investigate the evolution of paleoclimate and terrestrial mollusk diversity in the western CLP during the Late Miocene to Pliocene.

Materials and Methods

The Dongwan loess-paleosol sequence (105°47′E, 34°58′N) [41] is located in the northeastern part of Qin'an County in the western CLP (Figure 1). The current climate in Qin'an County is mainly controlled by the EA monsoon, with mean annual precipitation of about 400–500 mm and mean annual temperature of ~10.4°C. Mean July and January temperatures are ~22.7°C and −3.4°C, respectively. Vegetation in the region corresponds to a semi-arid temperate steppe [60].

The Dongwan sequence, located in the western CLP, is the first counterpart of the Red Clay sequences in the eastern CLP [41]. The section is about 73.7 m thick, and is composed of 84 distinguishable loess-paleosol couplets. The chronology of the sequence has been established by Hao and Guo (2004) [41] using magnetostratigraphic and micromammalian studies (Figure 2). First, Hao and Guo (2004) [41] ascribed the approximate age of the sequence to the Late Miocene to Pliocene using micromammalian fossil teeth sampled from 20 depths of the Dongwan sequence. Second, they defined a series of magnetozones based on 319 oriented samples collected at 20 or 25 cm intervals and correlated them with the Geomagnetic Polarity Timescale [61]. Finally, they established a chronology using paleomagnetic reversals as age controls and interpolated in between them using the magnetic susceptibility age model developed by Kukla et al. (1990) [62]. Their results yield an age duration of 7.1 to 3.5 Ma for the Dongwan loess-palaeosol sequence [41].

A total of 310 mollusk assemblages were collected from the Dongwan sequence using continuous 20-cm-thick samples; however, some intervals were sampled at intervals of 10–50 cm, according to lithological changes [58]. About 30 kg of sediment were obtained for each sample. In the field, we progressively broke each sediment sample into smaller pieces of about 0.5 mm in diameter, at the same time collecting all available shells and visible broken pieces. No necessary permits for the described field investigations were needed. In the laboratory, we attempted to restore any broken shells, and then identified and counted them under a binocular microscope. All of the identifiable mollusk remains were considered in the individual totals using the method of Puisségur (1976) [63]. All of the mollusk shells are stored in the Institute of Geology and Geophysics, Chinese Academy of Sciences.

For each mollusk assemblage, the numbers of species (S) and individuals were counted and a diversity index was calculated for all species, thermo-humidiphilous (TH) species, and cold-aridiphilous (CA) species in order to investigate changes in terrestrial mollusk populations and in different ecological groups. We used the most widely applied Shannon index [64], sometimes referred to as the 'Shannon–Weaver' index and sometimes as the 'Shannon–Wiener' index by researchers, in order to calculate the values of diversity, H (S), of the total, TH, and CA species, as follows:

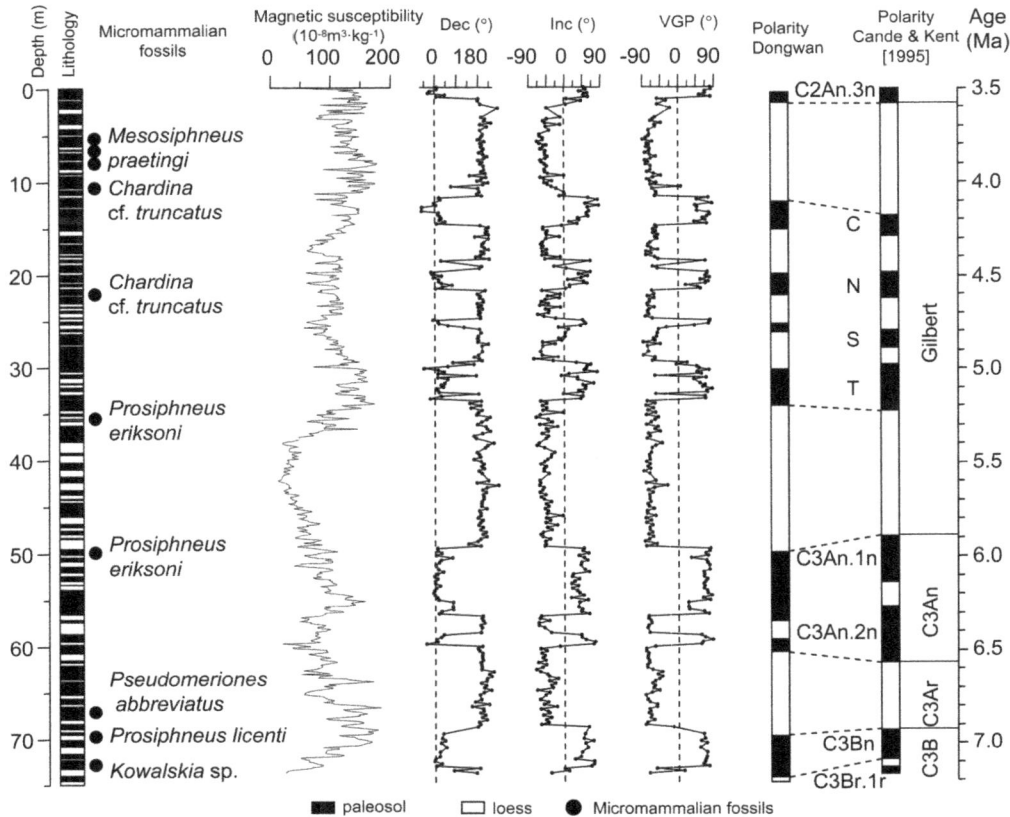

Figure 2. Chronology of the Dongwan loess-paleosol sequence (modified after Hao and Guo, 2004) [41].

$$H = -\sum_{i=1}^{S} \frac{ni}{N} \log_2 \frac{ni}{N}$$

where n_i is the density measure (in this case the mollusk individuals) of the i-th species (I varying between 1 and n); S is the number of species in the sample, and

$$N = -\sum_{i=1}^{S} n_i$$

The theoretical maximum (H_{max}) of diversity in any sample is expressed as

$$H_{max} = \log_2 S$$

Equitability (or evenness, E) is expressed as

$$E = \frac{H}{H_{max}}$$

These indices have been applied to European and North American Quaternary terrestrial mollusk assemblages [65–67].

Results

Mollusk fossils are relatively abundant in the Dongwan sequence with significant concentrations at ~2 m, 10 m, 30 m, 50 m, and 70 m depth. Amongst the total of 310 samples, 298 yielded 16439 mollusk individuals and 12 samples were barren. The maximum count reached 1121/30 kg at 54 m depth (Figure 3). Variations in total mollusk individuals parallel fluctuations in magnetic susceptibility [58], indicating that pedogenic processes such as carbonate dissolution did not affect the preserved assemblages [58].

A total of 24 mollusk species were identified in the Dongwan section. They are all terrestrial taxa and consist of *Gastrocopta armigerella* (Reinhardt, 1877), *Gastrocopta* sp., *Punctum orphana* (Heude, 1882), *Punctum* sp., *Metodontia huaiensis* (Crosse, 1882), *Metodontia yantaiensis* (Crosse et Debeaux, 1863), *Metodontia beresowskii* (Moellendorff, 1899), *Metodontia* cf. *huaiensis*, *Metodontia* cf. *yantaiensis*, *Metodontia* cf. *beresowskii*, *Metodontia* sp., *Kaliella* sp., *Macrochlamys* sp., *Opeas* sp., *Cathaica* sp., *Cathaica pulveratrix* (Martens, 1882), *Cathaica pulveraticula* (Martens, 1882), *Cathaica schensiensis* (Hilber, 1883), *Cathaica placenta* (Ping et Yen, 1933), *Pupilla aeoli* (Hilber, 1883), *Pupilla grabaui* (Ping, 1929), *Pupilla* sp., *Vallonia* sp., and *Pupopsis retrodens* (Martens, 1879).

All of these species, except *Pupilla grabaui* and *Pupopsis retrodens*, have been previously identified in the Chinese Quaternary loess-paleosol sequences and most of them have modern representatives. For example, *Cathaica pulveratrix*, *C. pulveraticula*, *C. schensiensis*, and *Pupilla aeoli* are the most common species that prefer living in relatively cold, dry environments, and are presently distributed in northwestern China. They have been regarded as indicative of a

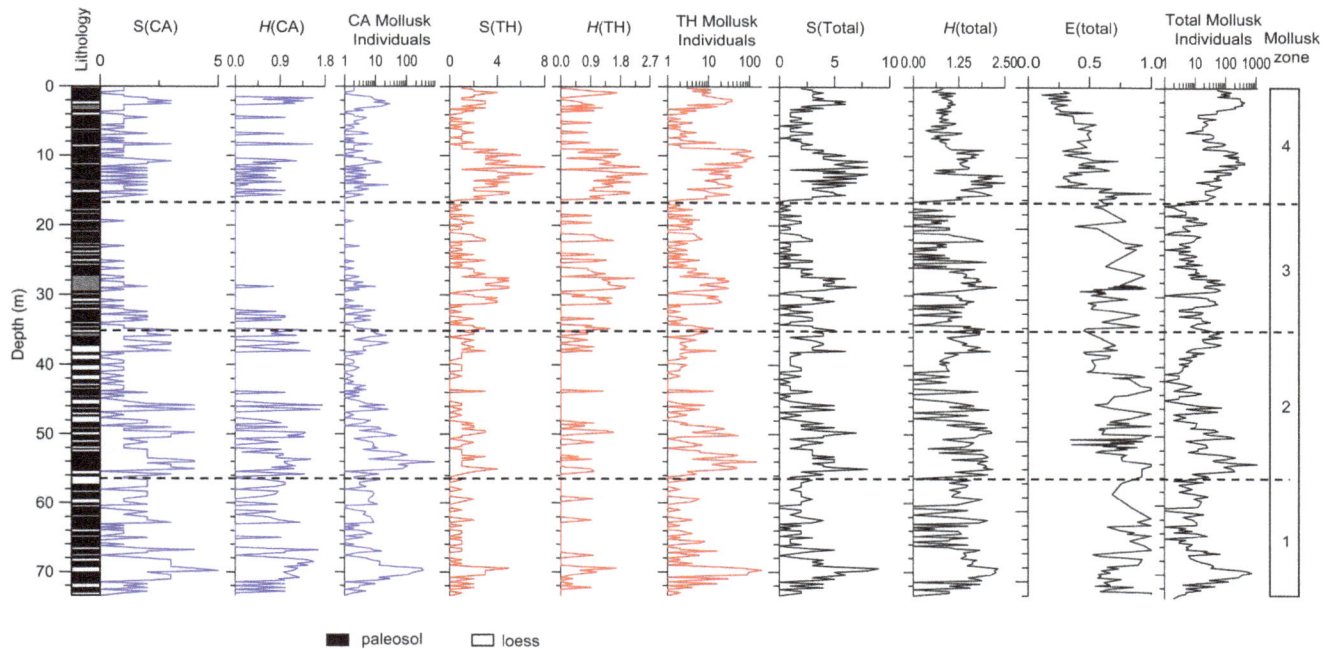

Figure 3. Variations in terrestrial mollusks versus depth in the Dongwan sequence. Lithology and total mollusk individuals are from Li et al., (2006a) [58] and Li et al. (2008) [47]. S(CA)–Total number of species of the cold-aridiphilous (CA) mollusk group. H(CA)–Diversity of the CA mollusk group. S(TH)–Total number of species of the thermo-humidiphilous (TH) mollusk group. H(TH)–Diversity of the TH mollusk group. S(total)–Total number of species of the total mollusk group. H(total)–Diversity of the total mollusk group. E(total)–Equitability of the total mollusk group.

strengthened winter monsoon [49–58]. Conversely, *Metodontia huaiensis*, *M. yantaiensis*, *M. beresowskii*, *Gastrocopta armigerella*, and *Punctum orphana* are species living in a warmer and more humid environment, and are distributed mainly in the southeastern part of the CLP, where the warm, moist summer monsoon brings sufficient precipitation [49–58]. Therefore based on their present requirements of moisture and temperature, as well as their modern geographical distribution, the Dongwan terrestrial mollusks can be grouped into CA and TH ecological groups, as have been previously defined in the Chinese Quaternary loess-paleosol sequences [49–57]. In the Dongwan section, the TH group comprises *Gastrocopta armigerella*, *Gastrocopta* sp., *Punctum orphana*, *Punctum* sp., *Metodontia beresowskii*, *Metodontia huaiensis*, *Metodontia yantaiensis*, *Metodontia* cf. *huaiensis*, *Metodontia* cf. *yantaiensis*, *Metodontia* cf. *beresowskii*, *Metodontia* sp., *Kaliella* sp., *Macrochlamys* sp., and *Opeas* sp.; and the CA group comprises *Cathaica* sp., *Cathaica pulveratrix*, *Cathaica pulveraticula*, *Cathaica schensiensis*, *Cathaica placenta*, *Pupilla aeoli*, *Pupilla grabaui*, *Pupilla* sp., *Vallonia* cf. *pulchella*, *Vallonia* sp., and *Pupopsis retrodens*.

The distributions of these terrestrial mollusks in the Dongwan section have been described previously [58]. *Cathaica* sp. is the most continuously distributed mollusk taxon in the Dongwan section, and *Gastrocopta* sp., *Pupilla* sp., and *Vallonia* sp. are other well represented taxa. The TH species of *Metodontia*, *Punctum*, *Macrochlamys*, and *Opeas* are concentrated in the upper part of the section. In general, the CA species are dominant in the loess layers, while the TH species mainly occur in the paleosols [58].

Variations in the number of species (S), diversity (H), equitability (E), and total individuals of all species, CA species, and TH species in the Dongwan sequence are plotted against depth in Figure 3. These data can be found in the supporting information data (Table S1). Variations in these parameters allow definition of four mollusk zones. In Zone 1, from the bottom of the sequence to

about 56.8 m depth (~7.1–6.2 Ma), high values of all the CA species (S(CA)) and individuals and diversity (H(CA)) dominate over low values of all of the TH species (S(TH)) and individuals and diversity (H(TH)). A prominent feature of this zone is that high values of S(CA), H(CA), CA mollusk individuals, S(TH) and H(TH), as well as total mollusk species and individuals, occur at around 70 m depth, at the very base of the sequence; however, these values then decrease upwards. H(CA) and H(total) remain at high levels throughout the zone, and the equitability (E) of total species is also higher than in the other zones.

In Zone 2, from 56.8–34.8 m depth (~6.2–5.4 Ma), the total number of CA species and individuals remains at a similarly high level as in the previous zone, except for the interval from 42 to 38 m. In contrast, the total number of TH species and individuals is slightly higher than in the previous zone. The total number of species and individuals of all of the species (CA and TH combined) are generally higher than in the previous zone but with a decreasing trend. The diversity of all species, H(total), does not exhibit significant changes compared to Zone 1, and E(total) is somewhat lower than in Zone 1.

Zone 3, from 34.8–16.2 m depth (~5.4–4.4 Ma), is characterized by low numbers of CA species and individuals being almost absent in the upper part, from 24 m to 16.2 m. This pattern of variation is paralleled by a clear increase of all the TH indices, including diversity. The total number of species and individuals of all of the species is lower than in Zone 2, except for the interval from about 32 to 28 m. There are no large magnitude changes in H(total) and E(total) within the zone.

Zone 4 corresponds to the depth from 16.2 m to the top of the sequence (~4.4 to 3.5 Ma). The number of CA species and individuals increases markedly and maintains relatively high values throughout the zone. The TH indices exhibit high values from about 16.2 to 9 m and then decrease significantly to very low values up to 4 m, and increase again above about 4 m depth. All of the indices for all species, i.e., S(total), H(total), and E(total), increase markedly; however, H(total) and E(total) exhibit a generally declining trend, implying that the diversity of the terrestrial mollusk populations decreased and that the distribution of individuals of different species was uneven. The total mollusk individuals in this zone is high and remains relatively stable except for the interval between 8 and 4 m, a pattern which differs from the other three zones where the total number of mollusk individuals is high at the beginning of the zone and then declines thereafter.

Moreover, the variation in diversity of the CA and TH ecological groups from the entire Dongwan sequence, as exhibited in Figure 3, can be differentiated into two major intervals based solely on the Shannon index of the CA and TH groups; and this result probably reflects different ecological population dynamics. First, from the base of the Dongwan sequence to 34.8 m depth, the CA group is dominant, exhibiting high diversity with H values varying between 0 and 1.75 (mean of 0.52). In contrast, the diversity of the TH populations exhibits significantly low values, ranging from 0 to 1.58 (mean of 0.18) within this depth interval, although the diversity of all mollusk species, i.e. the sum of CA and TH, does not exhibit any clear changes. Second, from 34.8 m depth to the top of the sequence, the diversity of the CA group declines significantly with H values varying between 0 and 1.56 (mean of 0.14); and in contrast the TH group becomes dominant in terms of diversity with H values ranging from 0 to 2.64 (mean of 0.64). The diversity and equitability of all mollusk species also exhibit a significant change at about 16.2 m depth within this interval.

Discussion

Late Miocene and Pliocene Paleoclimatic Evolution in the Western CLP

Previous studies of European and North American Quaternary terrestrial mollusks have shown that mollusk assemblages, based not only on the occurrence of characteristic species but also on the statistical dispersion of the assemblages (e.g., diversity, as described by the Shannon index (H)), can provide information about climatic conditions and mollusk populations [65–67]. For example, correlation of terrestrial mollusk groups with changes in dust flux, and thus climatic conditions, has previously been observed in the European Quaternary terrestrial mollusk diversity record [65]. Here we use the Dongwan terrestrial mollusk record to extend the statistical analysis of terrestrial mollusk populations and the interpretation of climatic changes to the Late Neogene. As shown in Figure 4, four stages can be recognized according to changes in the mollusk indices used in the present study, and which outline the evolutionary history of ecological and climatic conditions in the CLP from 7.1 to 3.5 Ma.

First, from 7.1 to 6.2 Ma (Zone 1), the total number of CA species (S(CA)) and diversity (H(CA)) index are high; however, the number of CA mollusk individuals is generally low apart from a large high peak at around 7 Ma and which indicates diversified CA species. Conversely, all of the indices of the TH species exhibit low values except for the peak at around 7 Ma, indicating low diversity. These features may be related to the occurrence of very cold, dry climatic conditions, the occurrence of which is roughly coincident with the expansion of C4 vegetation in the northeastern CLP [27], Central Inner Mongolia (Tunggur Area, Xilinhot Area, and Huade Area) [68] (Figure 1), and Pakistan [2]. They also indicate the occurrence of seasons with water stress for vegetation and terrestrial mollusks as well as the fact that relatively dry

climatic conditions had already appeared in northern China during the Late Miocene, coincident in age with the global transition from C3 to C4 vegetation [4]. The occurrence of C4 plants during the latest Miocene should have been restricted to limited areas within deserts, and the occurrence of these niches could be coincident with, or occurred after, major uplift of the Tibetan Plateau at about 8 Ma [3,8,10,11]. However, C4 vegetation was not yet to develop in the central and southern CLP, demonstrating that C3 plants were still dominant as indicated by $\delta^{13}C$ values of soil carbonate in the Lingtai, Xifeng and Liantian Red Clay sequences in the eastern CLP [6,9,24]. This is also evidenced by the thickness of loess layers in the studied Dongwan sequence and by dust mass accumulation rates in the western Pacific. High dust deposition rates with a gradually increasing trend culminated at about 6.2 Ma in the western Pacific [69], corresponding to thicker loess layers in the Dongwan sequence (Figure 4). Both features indicate drier climatic conditions in the Asian interior, the potential source region of dust deposits in the CLP and western Pacific. While dust accumulation in the western Pacific reached a maximum from about 6.5 to 6.2 Ma, the number of mollusk species and individuals at Dongwan was very low, implying very dry climatic conditions which were unsuitable even for the development of CA species. Pedostratigraphy and iron geochemistry of the Lingtai Red Clay deposits in the CLP indicate that the EA summer monsoon was relatively weak from 7.05 to 6.2 Ma [20,21]. The $\delta^{13}C$ values of fossil enamel from a diverse group of herbivores and of paleosol carbonate and organic matter from the Linxia Basin indicate that C4 grasses were either absent or insignificant in the Linxia Basin prior to ~2–3 Ma, suggesting that the East Asian summer monsoon was probably not strong enough to affect this part of China throughout much of the Neogene [28].

However, during our studied interval there is much evidence for a strong summer monsoon, evidenced mainly from mammal assemblages as well as analyses of mammal tooth and soil carbonate isotopes [22–25,27]. Hypsodonty analysis of fossil mammals indicates that northern China became more humid at 7–8 Ma, coincident with the previously recognized onset of Red Clay deposition in the eastern CLP, and which was interpreted by the authors as reflecting the onset or intensification of summer monsoonal precipitation [22,23]. Soil carbonate isotopes from the Lantian fluvial and Red Clay deposits indicate the absence of any marked change in plant photosynthetic pathway or climate, implying the occurrence of pure C3 vegetation with no indications of any C4 plants during the Late Miocene and Pliocene. A general rise in $\delta^{18}O$ values probably reflects increased summer precipitation related to the onset and/or intensification of the Asian monsoon system [24,25], which is supported by a mammalian faunal turnover event implying a marked change to more humid and closed habitats [26]. A strong summer monsoon at 7.1–6.2 Ma is also supported by the records of clay/feldspar ratio, kaolinite/chlorite ratio and biogenic opal mass accumulation rates (MAR) from the SCS, although the authors emphasize that their study is a schematic reconstruction which only outlines several principal stages and neglects the details [33].

There is also much evidence for a weakening of the summer monsoon and strengthening of the winter monsoon during the Late Miocene. The EA summer monsoon weakened after 7.5 Ma as evidenced by combined planktonic foraminifera Mg/Ca and $\delta^{18}O$ records from Ocean Drilling Program (ODP) Site 1146, northern SCS [30]. Radiolarian species numbers and individuals and diversity from ODP Site 1143, southern SCS, suggest a summer monsoon maximum at 8.24 Ma and a major decline after 7.70 Ma [32]. Geochemical records from the ODP Site 1148,

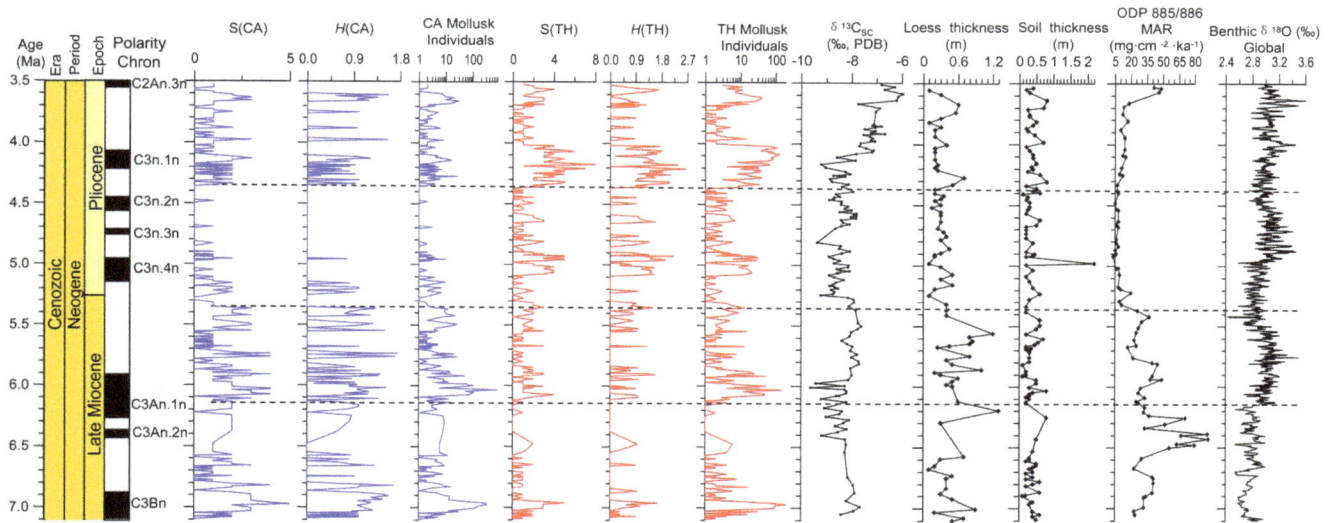

Figure 4. Variations in the terrestrial mollusks of the Dongwan sequence in the western CLP from 7.1 to 3.5 Ma, and comparison with other climate proxies. The proxies are $\delta^{13}C$ of soil carbonate ($\delta^{13}C_{SC}$) in the Lingtai Red Clay sequence [6], loess and soil thickness in the Dongwan sequence (this study), global cooling trend deduced from the marine benthic foraminiferal $\delta^{18}O$ record [1], and continental aridity inferred from the mass accumulation rate (MAR) of eolian dust at ODP Site 885/886 in the western Pacific [69]. Geomagnetic polarity chronology and age of the Dongwan sequence is from Hao and Guo (2004) [41]. S(CA)–Total species of the cold-aridiphilous (CA) mollusk group. H(CA)–Diversity of the CA mollusk group. S(TH)–Total number of species of the thermo-humidiphilous (TH) mollusk group. H(TH)–Diversity of the TH mollusk group.

northern SCS, indicate that chemical weathering intensity decreased after the early Miocene with a rapid decrease centering at 7.2 Ma, compatible with a weakening of the EA summer monsoon [31]. The trend of increasing black carbon concentration and accumulation rate at ODP Site 1148 suggests the progressive development of the East Asian winter monsoon after about 7 Ma [29]. In addition, grain-size evidence from the Linxia Basin of the CLP indicates that the EA winter monsoon intensified after 7.4 Ma and again at 5.3 Ma [19].

Thus it is noteworthy that the TH mollusks at Dongwan do exhibit peak values at around 7 Ma, suggesting warm, humid climatic conditions consistent with a mammalian faunal turnover event [26]; however, the peak does not extend to the subsequent period, implying that the interval of relatively warm and humid climate may have been a relatively brief event within the context of an overall very cold and arid climate. In addition, a recent synthesis of isotopic and sedimentological analyses, climate modeling and an extensive mesowear analysis of the Baode Red Clay sequence indicates that the climate at 7.5 Ma was more humid than during the youngest interval at 5.7 Ma and that variable climatic conditions occurred at ~6.5 Ma. Thus a significant decrease in the EA summer monsoon strength from 7–5.7 Ma in the Baode region, based on three samples dated at 7.0, 6.5, and 5.7 Ma [70], does not contradict the mollusk results. We suggest that much more definitive results would have been obtained if a higher sampling density had been used. In addition, numerous paleoclimatologists regard the occurrence of Red Clay sequences as reflecting aridity or desertification and an intensified winter monsoon [16,37,42,43,69,71], which contradicts the mammal and isotope record.

We suggest four possible reasons for the discrepancies amongst the extant research results. First, the sensitivity of the proxies used may be one of the main reasons; and indeed, it has long been found in Quaternary loess studies that different climatic proxies can respond differently to climatic changes [55]. Second, as summarized by Wang et al. (2005) [72] and Kaakinen et al. (2006)

[24], interpretations of expansions of C4 vegetation are contentious in terms of whether or not they are a summer monsoon proxy [2,4,45]. Third, regional difference in patterns of climate change may have occurred, as suggested by Passy et al. (2009) [27]. Finally, low stratigraphic resolution may have resulted in the failure to resolve environmental changes in sufficient detail, and high resolution mammalian faunal studies (for example using a 20-cm interval) may shed light on the observed differences. In summary, the conflicts cannot be resolved by the present study and future studies employing more sensitive proxies, higher resolution or continuous sampling combined with accurate dating may be necessary.

Second, from 6.2 to 5.4 Ma (Zone 2), the dominant CA terrestrial mollusks indicate that the climate during this interval remained cold and dry, but was not drier than in Zone 1, as evidenced by the slightly greater abundance of TH species and individuals. At the beginning of this zone, from about 6.2 to 5.8 Ma, the number of TH mollusk individuals was higher, and this corresponds to a declining rate of dust deposition as indicated by the thinness of the loess layers in the Dongwan sequence (Figure 4). This supports the finding that decreasing or increasing dust deposition affects the development of terrestrial mollusks by creating either more- or less- favorable environmental conditions [65]. After about 5.8 Ma the number of TH mollusk individuals was reduced, which is probably related to the increase of dust deposition in the CLP, as shown by the increased thickness of the loess represented by this interval (Figure 4). These cold, dry climatic conditions were not solely restricted to the western CLP; in the eastern CLP, mollusks in the Xifeng Red Clay sequence exhibit dominant percentages of the CA group associated with a few meso-xerophilous components and a lesser occurrence of TH species, also indicating a cold, dry climate [46]. This is also reflected by the coarse grain size of the Red Clay sequence in the CLP [16], and by the fact that dust deposition rates in the western Pacific remained high, albeit lower than before, and with a declining trend towards the subsequent time interval from 5.4 and

4.4 Ma [69]. The thickness of the Dongwan loess layers was not higher than before, except at about 6.2 and 5.6 Ma; however, the thickness of the soil layers does not change significantly. The $\delta^{13}C$ values of the soil carbonates in the Lingtai and Xifeng Red Clay sequences were similar to those of the previous time interval from 7.1 to 6.2 Ma, indicating again that C4 plants were not the dominant vegetation in the studied area of the CLP [6,9]. These lines of evidence indicate generally cold, dry climatic conditions in the EA continent but which were not drier than the lower stage. However, the global mean benthic foraminiferal $\delta^{18}O$ record demonstrates that global ice volume was increasing with the highest values occurring at about 5.8–5.7 Ma, followed by a decrease towards the upper stage [1].

There is much evidence supporting the conclusion from the Dongwan mollusk results that the winter monsoon may still have been strong from 6.2–5.4 Ma but that the summer monsoon strengthens than before although may be to a lesser extent, as confirmed by the pedostratigraphy and iron geochemistry of the Lingtai Red Clay deposits in the CLP [20,21]. The coeval pollen record from the Lingtai section exhibits a predominance of Chenopodiaceae and *Artemisia*, indicating a desert or desert-grassland landscape during the Late Miocene [73]. A higher dust accumulation rate and coarser eolian grain size in the CLP and in the North Pacific suggest stronger continental aridity in the Asian interior from ~6.2 to ~5 Ma [14,16,69]. The variation in the U-ratio of the grain size of the Red Clay, reflecting the changing strength of the winter monsoon, indicates a strong winter monsoon between 6.1 and 5.4 Ma [13]. The sedimentation rate across the CLP, including at the Baode Red Clay sequence, indicates that the EA winter monsoon strengthened between 6.26 and 5.4–5.25 Ma [14,15].

Third, from 5.4 to 4.4 Ma (Zone 3), the prominent feature is that all of the CA species and individuals are very few in number and in fact they decrease significantly with almost no occurrence after about 5 Ma. In contrast, there are high values of the TH species numbers and individuals with the maxima occurring at around 5–5.1 Ma when a thick soil layer formed, indicating warm, moist climatic conditions. In the eastern CLP, the mollusk fauna was characterized by maximum abundance of TH species, an absence of CA species, and relatively abundant meso-xerophilous taxa [46]. The Xifeng pollen record indicates a significant increase of temperate forest plants, also implying humid regional climatic conditions during this period [74]. The $\delta^{13}C$ values indicate a slight decrease and thus a shift towards more C3 plants in the CLP in agreement with the pollen results (Figure 4). Ding et al. (1999) identify extremely mature soils in the 5.5–3.85 Ma interval at Lingtai, and interpret these as indicating warm and wet climates, and possibly the strongest summer monsoons in the past 7 Ma [20]. Minimum grain size and sedimentation rate in the CLP [15,16], decreased thickness of loess layers in the Dongwan sequence (Figure 4), and very low dust deposition rates in the western Pacific [69] together suggest very warm, moist climatic conditions in the CLP and less dry conditions in the Asian interior, the potential source area of dust in the CLP and western Pacific. However, this warm climate is not clearly recorded by the global mean benthic foraminiferal $\delta^{18}O$ record compiled by Zachos et al. (2001) [1].

Last, from 4.4 to 3.5 Ma (Zone 4), all of the CA indices increase significantly and remain almost unchanged, while in contrast all of the TH indices exhibit a different pattern of variability with high values from 4.4 to 4.0 Ma, coincident with changes in the CA indices, an abrupt decrease from 4.0 Ma and increased values again from 3.7 Ma. The CA species during this time interval were different from those between 7.1 and 5.4 Ma. Indeed, the CA

species of *Cathaica pulveraticula*, *Cathaica schensiensis*, and *Pupopsis retrodens* have only been identified in this zone, as shown by Li et al. (2006a) [58], probably indicating that climatic conditions became drier. The coeval Xifeng land snail record from the eastern CLP is dominated by meso-xerophilous taxa, associated with a significantly reduced abundance of TH species and the paucity of CA taxa during the late period [46]. Pollen data from the Xifeng Red Clay sequence indicate a typical steppe ecosystem during this period [74]. These variations are in good agreement with the expansion of C4 plants in the central CLP as observed at the Lingtai and Xifeng Red Clay sequences [6,9]. In addition, the loess layers in the Dongwan section during this time interval are of moderate thickness, thinner than from 7.1 to 6.2 Ma but thicker than from 5.4 to 4.4 Ma. In contrast the paleosols are somewhat thicker than before (Figure 4), indicating longer pedogenesis under different climate conditions, and probably enhanced seasonality, corresponding to expansions of C4 plants in the central CLP [6,9]. Pedostratigraphy and iron geochemistry of the Lingtai Red Clay deposits in the CLP indicate that the EA summer monsoon weakened significantly over the interval 3.85 to 3.15 Ma [20,21]. Dust deposition increased again in the Pacific in parallel with increased aridity in the dust source areas [69] (Figure 4).

Moreover, the entire Dongwan terrestrial mollusk record provides information about the evolution of mollusk diversity and climatic changes from the Late Miocene to the Pliocene. Previous studies indicate that in most Quaternary loess sequences in Europe and North America generally fewer than 20 mollusk species can be identified and diversity varies with H values between 0 and 4, suggesting that the loess deposits provided few additional ecological niches for land snails to grow and develop [65–67]. In the Neogene Dongwan loess sequence in the CLP, East Asia, 24 mollusk species were identified and diversity varies between 0 and 2.5, which also suggests that fewer niches were provided. The response of the Neogene terrestrial mollusk populations to climatic changes depends on their ecological requirements. Different ecological groups, such as CA and TH, respond differently. During the time interval from 7.1 to 5.4 Ma when a cold, dry climate obtained, as indicated by a high flux of dust in the CLP and western Pacific, the diversity of the CA group was high, indicating that relatively cold, dry climatic conditions may be favorable for the development of CA terrestrial mollusk populations. In contrast, during the relatively warm, moist time interval from 5.4 to 4.4 Ma, corresponding to a reduced loess thickness in the CLP and low dust flux in the western Pacific, the diversity of the TH terrestrial mollusk populations was high.

Possible Causes of Paleoecological and Paleoclimatic Evolution in the Western CLP during the Late Miocene and Pliocene

Dust in the Asian interior, including northwestern China, was emitted and transported in two possible modes, corresponding to patterns of low and high level atmospheric circulation. One mode is that the Asian dust was transported eastwards by high atmospheric circulations (westerlies) and reached northwestern Pacific, as recorded by high values of dust flux to the northwestern Pacific. The other mode is that dust in the Asia interior was transported by low level atmospheric circulation, the EA winter monsoon, to the middle reaches of the Yellow River, leading to the formation of the CLP [37,49,75,76]. Thus loess sequences in the CLP and dust deposits in the western Pacific both relate to climatic changes that impacted the regions to the north of the Tibetan Plateau, i.e., the Asian interior, as has been indicated by the previous studies of Hovan et al., (1989) [77] and Rea et al., (1998) [69].

At approximately 8–7 Ma, the accumulation rate of eolian deposits in the CLP reached high values of about 4 cm/ka [37] and dust deposition in the western Pacific was maximal [69] (Figure 4), indicating that particularly dry climatic conditions must have prevailed in the Asian interior resulting in the mobilization and transport of large amounts of dust. Indeed, a palynological study of the Late Miocene–Pliocene sediments of the Dushanzi section from northwestern China (Figure 1) indicates that steppe taxa (*Artemisia* and Chenopodiaceae) were generally dominant, implying that a dry climate existed in the inland basins of northwestern China since 8.7 Ma, except for a warm and humid phase that lasted from 5.8 to 3.9 Ma [78]. In addition, significant desertification in northwestern China prevailed as early as 7.2–7 Ma, as shown by the development of eolian sand dunes in the Taklimakan Desert [79] and eolian Red Clay deposition around the Lanzhou region [80]. These climatic and environmental conditions probably provided a suitable environment for the growth and development of C4 plants in northern China and promoted the expansion of eolian deposits from the west towards the eastern part of the CLP [42,43] and which constituted suitable environments for terrestrial mollusks. Thus terrestrial mollusks grew and developed in the CLP and were able to record the patterns of environmental evolution as recorded by the Dongwan sequence.

Variations in the Dongwan terrestrial mollusk assemblages indicate that, during the time interval from 7.1 to 3.5 Ma, major paleoecological and paleoclimatic changes occurred at about 6.2, 5.4, and 4.4 Ma. The change at around 6.2 Ma, the boundary between Zones 1 and 2, is less significant than the one at 4.4 Ma, and therefore it is not clearly reflected in terms of a transition in periodicity recorded by the relative abundance of mollusks at this time [47]. Indeed, a relatively high number of CA mollusk species and individuals, consistent with the high relative abundance of CA mollusks [47], prevailed during Zone 1 and Zone 2, indicating that a cold, arid climate prevailed in the study area. These two zones exhibit the same dominant 100 kyr periodicity [47], again suggesting that the 6.2 Ma datum was not particularly significant at Dongwan. However, the TH mollusk individuals and species are somewhat different within these two zones, and in addition the appearance of mollusks in the Xifeng Red Clay sequence and the upper age of the QA-I section imply that 6.2 Ma may be an important datum in the paleoclimatic evolution of the CLP [37,46].

Changes in the CA mollusks of Zones 1 and 2 are paralleled by a global cooling trend [1] (Figure 4), increased ice-rafted detrital flux in the Northern Hemisphere [81–85], and the buildup of the Western Antarctic ice sheet [1]. The CLP is particularly sensitive to changes in high northern latitudes through the EA winter monsoon circulation [76,86]. Extended ice sheets in the Northern Hemisphere reinforce the southward movement of cold air and thereby enhance the Siberian High that controls the EA winter monsoon wind system [86]. Thus, global cooling, especially the expansion of ice deposits in high northern latitudes, could have affected the Siberian High and the EA winter monsoon, thereby expanding habitats for the CA mollusks in the CLP.

There is no evidence for major uplift of the Tibetan Plateau at about 6.2 Ma, and therefore this mechanism is excluded as a cause of the change at this time. However, if the Tibetan Plateau reached a significant height at about 8 Ma it would have accelerated climatic cooling and strengthened the EA winter monsoon [3,8,87–90], causing cold, dry climatic conditions in the western CLP during the Late Miocene, as discussed above.

The shift at about 5.3–5.4 Ma roughly corresponds to the onset of the Pliocene when global climate became warmer than before.

However, the climatic drivers contributing to the Pliocene global warming are still highly debated. As summarized by Haywood et al. (2009) [91], possible candidates include paleogeographic changes [92], altered atmospheric trace gas concentrations and water vapor content [93], changes in oceanic circulation [94,95], oceanic heat transport [5,96], thermal structure of the oceans [97–99], and feedbacks generated through altered land cover (including ice sheet extent), surface albedo, cloud cover and temperature [100].

Our results seem to support the suggestion that the warming interval may be related to changes in ocean circulation and ocean heat transport caused by the closures of the Panama and Indonesian seaways [5,94–96]. Closure of these two seaways may have caused changes in heat distribution between the Pacific and Atlantic, causing reorganization of global climatic patterns and changing the pattern of atmospheric moisture flux from latitudinal to meridional, resulting in increased moisture flux to high latitudes [5,95,101] and contributing to climate changes in the CLP. Moreover, both geological records and modeling studies show that closures of the Panama and Indonesian seaways likely played important roles in the strengthening and enlargement of the western Pacific warm pool [5,102–106], providing more moisture and heat to the CLP favorable for the abundant occurrence of TH mollusk species.

The 4.4 Ma shift, representing a climatic transition from early Pliocene warming to late Pliocene cooling, is observed in the Dongwan mollusk diversity and species numbers and individuals; however, it is not clearly shown in the relative abundance of mollusks [47]. It may be related to tectonic events such as the uplift of Tibetan Plateau, since climate models suggest that uplift played a particularly important role in the evolution of the global paleoenvironment [8,87–90], and the effects of which on the EA winter monsoon are more significant than on the summer monsoon [89]. Changes in depositional facies from distal alluvial plains to proximal alluvial fans and an increase in sedimentation rate near Yecheng (Figure 1) in the western Kunlun Mountains indicate the uplift of the northern Tibetan Plateau at about 4.5 Ma [7]. This tectonic activity is thought to trigger not only the enhancement of the EA monsoon, but is also considered to be a driver of general cooling through the consequent increase in the rate of chemical weathering, thus accelerating ice expansion in Northern Hemisphere high latitudes [88,107,108] and further strengthening the Siberian High and the EA winter monsoon and transportation of dust to the CLP. Furthermore, uplift of the Tibetan Plateau could have blocked moisture transport to the Asian interior, contributing to its aridification. Under these climatic conditions, the northern CLP would have commenced a drying trend earlier than the southern CLP, causing significant ecological changes in terrestrial ecosystems including terrestrial mollusks and C4 plants at roughly 4.4 Ma.

Conclusions

The Dongwan terrestrial mollusk record from the western CLP exhibits four stages during the time interval from 7.1 to 3.5 Ma, indicating the phased evolution of paleoecology and paleoclimate. From 7.1 to 6.2 Ma, very cold and dry climatic conditions prevailed. From 6.2 to 5.4 Ma, the climate remained cold and dry but was not as dry as the preceding interval, as evidenced by the dominance of CA mollusks and rather more TH species and individuals. From 5.4 to 4.4 Ma, very warm and moist climatic conditions prevailed, evidenced by high values of the TH species and individuals, as well as by the very small numbers of CA species and individuals and their almost complete absence after about

5 Ma. From 4.4 to 3.5 Ma, all of the CA indices increase significantly and remain at a high level, indicating a cooling climate; however, all of the TH indices exhibit relatively high valuesfrom 4.4 to 4.0 Ma, an abrupt decrease from 4.0 Ma and an increase again from 3.7 Ma. The three CA species, *Cathaica pulveraticula, Cathaica schensiensis, Pupopsis retrodens*, occurred solely during this period and were absent from 7.1 to 5.4 Ma, suggesting that the climate from 4.4 to 3.5 Ma was becoming colder and drier than previous stage.

The very cold and arid climatic conditions, with changes at about 6.2 Ma, are paralleled by a global cooling trend, increased in ice-rafted detrital flux in the Northern Hemisphere [81–85], and the buildup of the Western Antarctica ice sheet [1]. The shift at about 5.3–5.4 Ma roughly corresponds to the onset of the Pliocene when global climate became increasingly warmer than previously. Our results seem to support the suggestion that the warming interval may be related to changes in oceanic circulation and oceanic heat transport [5,94–96]. The 4.4 Ma shift may be related to tectonic events such as the uplift of Tibetan Plateau.

Variations in the diversity of the CA and TH mollusks, H(CA) and H(TH), are closely related to climatic changes during the Late Miocene to Pliocene. From 7.1 to 5.4 Ma when a cold, dry climate prevailed, the CA group was the more diverse. In contrast, during the relatively warm, wet time interval from 5.4 to 4.4 Ma, the TH terrestrial mollusk populations became more diverse. It should be pointed out that most of the Neogene terrestrial mollusks in the CLP have modern representatives and therefore they have the potential to estimate quantitatively Neogene changes in temperature and precipitation in the CLP. However, such estimates depend upon the development of a training set based on a large number of surface samples. Changes in fossil terrestrial mollusk diversity are significant for the prediction of terrestrial biodiversity changes, and Quaternary loess deposits in Europe and North America have been studied in this context. The 22-Ma loess deposits in China provide an excellent opportunity for understanding long term changes in terrestrial mollusk diversity; and although the present paper focuses on the Late Neogene terrestrial mollusk in the CLP, ongoing studies will focus on other time intervals.

Supporting Information

Table S1 The depth, age, and mollusk data from the Dongwan loess-paleosol sequence. S(CA)–Total number of species of the cold-aridiphilous (CA) mollusk group. H(CA)–Diversity of the CA mollusk group. CAMI–Mollusk individuals of the CA mollusk group. S(TH)–Total number of species of the thermo-humidiphilous (TH) mollusk group. H(TH)–Diversity of the TH mollusk group. THMI–Mollusk individuals of the TH mollusk group. S(total)–Total number of species of the total mollusk group. H(total)–Diversity of the total mollusk group. E(total)–Equitability of the total mollusk group. TMI–Total Mollusk individuals.

Acknowledgments

We thank all of the reviewers for valuable comments, criticisms and suggestions. We are grateful to Dr. Jan Bloemendal for editing the English and for suggestions, Prof. Qingzhen Hao for field assistance and Prof. Shiling Yang for providing us with δ^{13}C data of soil carbonates from the Lingtai Red Clay sequence.

Author Contributions

Conceived and designed the experiments: FL NW. Performed the experiments: FL NW YP. Analyzed the data: FL NW DDR YD DZ. Contributed reagents/materials/analysis tools: FL NW. Wrote the paper: FL NW DDR.

References

1. Zachos J, Pagani M, Sloan L, Thomas E, Billups K (2001) Trends, rhythms, and aberrations in global climate 65 Ma to present. Science 292: 686–693.
2. Quade J, Cerling TE, Bowaman JR (1989) Development of Asian monsoon revealed by marked ecological shift during the latest Miocene in northern Pakistan. Nature 342: 163–166.
3. Harrison TM, Copeland P, Kidd WSF, Yin A (1992) Raising Tibet. Science 255: 1663–1670.
4. Cerling TE, Harris JM, MacFadden BJ, Leakey MG, Quade J, et al. (1997) Global vegetation change through the Miocene/Pliocene boundary. Nature 389: 153–158.
5. Haug GH, Tiedemann R (1998) Effect of the formation of the Isthmus of Panama on Atlantic Ocean thermohaline circulation. Nature 393: 673–676.
6. Ding ZL, Yang SL (2000) C3/C4 vegetation evolution over the last 7.0 Myr in the Chinese Loess Plateau: evidence from pedogenic carbonate δ^{13}C. Palaeogeogr. Palaeoclimatol. Palaeoecol. 160: 291–299.
7. Zheng HB, Powell CM, An ZS, Zhou J, Dong GR (2000) Pliocene uplift of the northern Tibetan Plateau. Geology 28: 715–718.
8. An ZS, Kutzbach JE, Prell WL, Porter SC (2001) Evolution of Asian monsoons and phased uplift of the Himalayan-Tibetan plateau since Late Miocene times. Nature 411: 62–66.
9. Jiang WY, Peng SZ, Hao QZ, Liu TS (2002) Carbon isotopic records in paleosols over the Pliocene in Northern China: implication on vegetation development and Tibetan uplift. Chin. Sci. Bull. 47: 687–690.
10. Clark MK, House MA, Royden LH, Whipple KX, Burchfiel BC, et al. (2005) Late Cenozoic uplift of southeastern Tibet. Geology 33: 525–528.
11. Fang XM, Yan MD, Voo RV, Rea DK, Song CH, et al. (2005) Late Cenozoic deformation and uplift of the NE Tibetan Plateau: Evidence from high-resolution magnetostratigraphy of the Guide Basin, Qinghai Province, China. Geol. Soc. Am. Bull. 117: 1208–1225.
12. Molnar P (2005) Mio-Pliocene growth of the Tibetan Plateau and evolution of East Asian climate. Palaeontologia Electronica 8: 2A. 1–23.
13. Vandenberghe J, Lu HY, Sun DH, van Huissteden J (2004) The late Miocene and Pliocene climate in East Asia as recorded by grain size and magnetic susceptibility of the Red Clay deposits (Chinese Loess Plateau). Palaeogeogr. Palaeoclimatol. Palaeoecol. 204: 239–255.

14. Wen LJ, Lu HY, Qiang XK (2005) Changes in grain-size and sedimentation rate of the Neogene Red Clay deposits along the Chinese Loess Plateau and implications for the palaeowind system. Sci China Earth Sci 48: 1452–1462.
15. Zhu YM, Zhou LP, Mo DW, Kaakinen A, Zhang ZQ, et al. (2008) A new magnetostratigraphic framework for late Neogene Hipparion Red Clay in the eastern Loess Plateau of China. Palaeogeogr. Palaeoclimatol. Palaeoecol. 268: 47–57.
16. Guo ZT, Peng SZ, Hao QZ, Biscaye PE, An ZS, et al. (2004) Late Miocene-Pliocene development of Asian aridification as recorded in the Red-Earth Formation in the northern China. Glob. Planet. Change 41: 135–145.
17. Wang YX, Yang JD, Chen J, Zhang KJ, Rao WB (2007) The Sr and Nd isotopic variations of the Chinese Loess Plateau during the past 7 Ma: Implications for the East Asian winter monsoon and source areas of loess. Palaeogeogr. Palaeoclimatol. Palaeoecol. 249: 351–361.
18. Chen J, Chen Y, Liu LW, Ji JF, Balsam W, et al. (2006) Zr/Rb ratio in the Chinese loess sequences and its implication for changes in the East Asian winter monsoon strength. Geochim. Cosmochim. Acta 70: 1471–1482.
19. Fan MJ, Song CH, Dettman DL, Fang XM, Xu XH (2006) Intensification of the Asian winter monsoon after 7.4 Ma: Grain-size evidence from the Linxia Basin, northeastern Tibetan Plateau, 13.1 Ma to 4.3 Ma. Earth Planet. Sci. Lett. 248: 186–197.
20. Ding ZL, Xiong SF, Sun JM, Yang SL, Gu ZY, et al. (1999) Pedostratigraphy and paleomagnetism of a 7.0 Ma eolian loess-red clay sequence at Lingtai, Loess Plateau, north-central China and the implications for paleomonsoon evolution. Palaeogeogr. Palaeoclimatol. Palaeoecol. 152: 49–66.
21. Ding ZL, Yang SL, Sun JM, Liu TS (2001) Iron geochemistry of loess and red clay deposits in the Chinese Loess Plateau and implications for long-term Asian monsoon evolution in the last 7.0 Ma. Earth Planet. Sci. Lett. 185: 99–109.
22. Fortelius M, Eronen JT, Jernvall J, Liu L, Pushkina D, et al. (2002) Fossil mammals resolve regional patterns of Eurasian climate change during 20 million years. Evol. Ecol. Res. 4: 1005–1016.
23. Liu LP, Eronen JT, Fortelius M (2009) Significant Mid-Latitude Aridity in the Middle Miocene of East Asia. Palaeogeogr. Palaeoclimatol. Palaeoecol. 279: 201–206.
24. Kaakinen A, Sonninen E, Lunkka JP (2006) Stable isotope record in paleosol carbonates from the Chinese Loess Plateau: implications for late Neogene

paleoclimate and paleovegetation. Palaeogeogr. Palaeoclimatol. Palaeoecol. 237: 359–369.

25. Kaakinen A, Passey BH, Zhang Z, Liu L, Pesonen LJ, et al. (2013) Stratigraphy and Paleoecology of the classical dragon bone localities of Baode County, Shanxi Province. p. 203–217. In Wang X, et al. (eds): Fossil Mammals of Asia: Neogene Biostratigraphy and Chronology. Columbia University Press, New York.

26. Zhang ZQ, Gentry AW, Kaakinen A, Liu LP, Lunkka JP, et al. (2002) Land mammal faunal sequence of the late Miocene of China: new evidence from Lantian, Shaanxi Province. Vert. PalAsiat. 40: 165–176.

27. Passey BH, Ayliffe LK, Kaakinen A, Zhang ZQ, Eronen JT, et al. (2009) Strengthened East Asian summer monsoons during a period of high-latitude warmth? Isotopic evidence from Mio-Pliocene fossil mammals and soil carbonates from northern China. Earth Planet. Sci. Lett. 277: 443–452.

28. Wang Y, Deng T (2005) A 25 m.y. isotopic record of paleodiet and environmental change from fossil mammals and paleosols from the NE margin of the Tibetan Plateau. Earth Planet. Sci. Lett. 236: 322–338.

29. Jia GD, Peng PA, Zhao QH, Jian ZM (2003) Changes in terrestrial ecosystem since 30 Ma in East Asia: Stable isotope evidence from black carbon in the South China Sea. Geology 31: 1093–1096.

30. Steinke S, Groeneveld J, Johnstone H, Rendle-Bühring R (2010) East Asian summer monsoon weakening after 7.5 Ma: Evidence from combined planktonic foraminifera Mg/Ca and $\delta^{18}O$ (ODP Site 1146; northern South China Sea). Palaeogeogr. Palaeoclimatol. Palaeoecol. 289: 33–43.

31. Wei GJ, Li XH, Liu Y, Shao L, Liang X (2006) Geochemical record of chemical weathering and monsoon climate change since the early Miocene in the South China Sea. Paleoceanography 21: PA4214, doi:10.1029/2006PA001300.

32. Chen MH, Wang RJ, Yang LH, Han JX, Lu J (2003) Development of east Asian summer monsoon environments in the late Miocene: radiolarian evidence from Site 1143 of ODP Leg 184. Mar. Geol. 201: 169–177.

33. Wan SM, Li AC, Clift PD, Jiang H (2006) Development of the East Asian summer monsoon: Evidence from the sediment record in the South China Sea since 8.5 Ma. Palaeogeogr. Palaeoclimatol. Palaeoecol. 241: 139–159.

34. Jacques FMB, Shi G, Wang WM (2013) Neogene zonal vegetation of China and the evolution of the winter monsoon. Bull. Geosci. 88: 175–193.

35. Liu YS, Utescher T, Zhou ZK, Sun BN (2011) The evolution of Miocene climates in North China: Preliminary results of quantitative reconstructions from plant fossil records. Palaeogeogr. Palaeoclimatol. Palaeoecol. 304: 308–317.

36. Tang H, Micheels A, Eronen JT, Fortelius M (2011) A regional climate model experiment to investigate the Asian monsoon in the Late Miocene. Clim. Past 7: 847–868.

37. Guo ZT, Ruddiman WF, Hao QZ, Wu HB, Qiao YS, et al. (2002) Onset of Asian desertification by 22 Myr ago inferred from loess deposits in China. Nature 416: 159–163.

38. Sun XJ, Wang PX (2005) How old is the Asian monsoon system?– Palaeobotanical records from China. Palaeogeogr. Palaeoclimatol. Palaeoecol. 222: 181–222.

39. Qiang XK, An ZS, Song YG, Chang H, Sun YB, et al. (2011) New eolian red clay sequence on the western Chinese Loess Plateau linked to onset of Asian desertification about 25 Ma ago. Sci China Earth Sci 54: 136–144.

40. Hao QZ, Guo ZT (2007) Magnetostratigraphy of an early-middle Miocene loess-soil sequence in the western Loess Plateau of China. Geophys. Res. Lett. 34: L18305, doi:10.1029/2003GL031162.

41. Hao QZ, Guo ZT (2004) Magnetostratigraphy of a late Miocene-Pliocene loess-paleosol sequence in the western Loess Plateau in China. Geophys. Res. Lett. 31: L09209, doi:10.1029/2003GL019392.

42. Sun DH, Shaw J, An ZS, Chen MY, Yue LP (1998) Magnetostratigraphy and paleoclimatic interpretation of a continuous 7.2 Ma late Cenozoic eolian sediments from the Chinese Loess Plateau. Geophys. Res. Lett. 25: 85–88.

43. Ding ZL, Sun JM, Yang SL, Liu TS (1998) Preliminary magnetostratigraphy of a thick eolian red clay-Loess sequence at Lingtai, the Chinese Loess Plateau. Geophys. Res. Lett. 25: 1225–1228.

44. Xu Y, Yue LP, Li JX, Sun L, Sun B, et al. (2009) An 11-Ma-old red clay sequence on the Eastern Chinese Loess Plateau. Palaeogeogr. Palaeoclimatol. Palaeoecol. 284: 383–391.

45. An ZS, Huang YS, Liu WG, Guo ZT, Clemens SC, et al. (2005) Multiple expansions of C4 plant biomass in East Asia since 7 Ma coupled with strengthened monsoon circulation. Geology 33: 705–708.

46. Wu NQ, Pei YP, Lu HY, Guo ZT, Li FJ, et al. (2006) Marked ecological shifts during 6.2–2.4 Ma revealed by a terrestrial molluscan record from the Chinese Red Clay Formation and implication for paleoclimatic evolution. Palaeogeogr. Palaeoclimatol. Palaeoecol. 233: 287–299.

47. Li FJ, Rousseau DD, Wu NQ, Hao QZ, Pei YP (2008) Late Neogene evolution of the East Asian monsoon revealed by terrestrial mollusk record in western Chinese Loess Plateau: from winter to summer dominated sub-regime. Earth Planet. Sci. Lett. 274: 439–447.

48. Zoller L, Semmel A (2001) 175 years of loess research in Germany–long records and "unconformities". Earth-Sci. Rev. 54: 19–28.

49. Liu TS (1985) Loess and the Environment. China Ocean Press, Beijing, 251 p.

50. Rousseau DD, Wu NQ (1997) A new molluscan record of the monsoon variability over the past 130 000 yr in the Luochuan loess sequence, China. Geology 25: 275–278.

51. Rousseau DD, Wu NQ (1999) Mollusk record of monsoon variability during the L_2-S_2 cycle in the Luochuan loess sequence, China. Quat. Res. 52: 286–292.

52. Rousseau DD, Wu NQ, Guo ZT (2000) The terrestrial mollusks as new indices of the Asian paleomonsoons in the Chinese loess plateau. Glob. Planet. Change 26: 199–206.

53. Wu NQ, Rousseau DD, Liu TS (1996) Land mollusk records from the Luochuan loess sequence and their paleoenvironmental significance. Sci. China, Ser. D: Earth Sci. 39: 494–502.

54. Wu NQ, Rousseau DD, Liu XP (2000) Response of mollusk assemblages from the Luochuan loess section to orbital forcing since the last 250 ka. Chin. Sci. Bull. 45: 1617–1622.

55. Wu NQ, Rousseau DD, Liu TS, Lu HY, Gu ZY, et al. (2001) Orbital forcing of terrestrial mollusks and climatic changes from the Loess Plateau of China during the past 350 ka. J. Geophys. Res. 106: 20045–20054.

56. Wu NQ, Liu TS, Liu XP, Gu ZY (2002) Mollusk record of millennial climate variability in the Loess Plateau during the Last Glacial Maximum. Boreas 31: 20–27.

57. Wu NQ, Chen XY, Rousseau DD, Li FJ, Pei YP, et al. (2007) Climatic conditions recorded by terrestrial mollusc assemblages in the Chinese Loess Plateau during marine Oxygen Isotope Stages 12–10. Quat. Sci. Rev. 26: 1884–1896.

58. Li FJ, Wu NQ, Pei YP, Hao QZ, Rousseau DD (2006a) Wind-blown origin of Dongwan late Miocene-Pliocene dust sequence documented by land snail record in western Chinese Loess Plateau. Geology 34: 405–408.

59. Li FJ, Wu NQ, Rousseau DD (2006b) Preliminary study of mollusk fossils in the Qinan Miocene loess-soil sequence in western Chinese Loess Plateau. Sci. China, Ser. D: Earth Sci. 49: 724–730.

60. Hou XY (1983) Vegetation of China with reference to its geographical distribution. Annals of the Missouri Botanical Garden 70: 509–549.

61. Cande SC, Kent DV (1995) Revised calibration of the geomagnetic polarity timescale for the Late Cretaceous and Cenozoic. J. Geophys. Res. 100: 6093–6095.

62. Kukla G, An ZS, Melice JL, Gavin J, Xiao JL (1990) Magnetic susceptibility record of Chinese loess. Trans. R. Soc. Edinb. Earth Sci. 81: 263–288.

63. Puisségur JJ (1976) Mollusques continentaux quaternaires de Bourgogne. Significations stratigraphiques et climatiques. Rapports avec d'autres faunes boréales de France: Université de Dijon Mémoires Géologiques, 3: 241 p.

64. Shannon CE (1948) A mathematical theory of communication. Bell Syst. tech. J. 27: 379–423.

65. Rousseau DD (1992) Terrestrial mollusks as indicators of global aeolian dust fluxes during glacial stages. Boreas 21: 105–109.

66. Rousseau DD, Limondin N, Puissegur JJ (1993) Holocene environmental signals from mollusk assemblages in Burgundy (France). Quat. Res. 40: 237–253.

67. Rousseau DD, Kukla G (1994) Late Pleistocene climate record in the Eustis loess section, Nebraska based on land snail assemblages and magnetic susceptibily. Quat. Res. 42: 176–187.

68. Zhang CF, Wang Y, Deng T, Wang XM, Biasatti D, et al. (2009) C4 expansion in the central Inner Mongolia during the latest Miocene and early Pliocene. Earth Planet. Sci. Lett. 287: 311–319.

69. Rea DK, Snoeckx H, Joseph LH (1998) Late Cenozoic eolian deposition in the North Pacific: Asian drying, Tibetan uplift, and cooling of the northern hemisphere. Paleoceanography 13: 215–224.

70. Eronen JT, Kaakinen A, Liu LP, Passey BH, Tang H, et al. (2014) Here be Dragons: Mesowear and tooth enamel isotopes of the classic Chinese "Hipparion" faunas from Baode, Shanxi Province, China. Ann. Zool. Fennici 51: 227–244.

71. Zheng HB, Powell CM, Rea DK, Wang JL, Wang PX (2004) Late Miocene and mid-Pliocene enhancement of the East Asian monsoon as viewed from the land and sea. Glob. Planet. Change 41: 147–155.

72. Wang PX, Clemens S, Beaufort L, Braconnot P, Ganssen G, et al. (2005) Evolution and variability of the Asian monsoon system: state of the art and outstanding issues. Quat. Sci. Rev. 24: 595–629.

73. Wu YS (2001) Palynoflora at late Miocene-early Pliocene from Leijiahe at Lingtai, Gansu Province, China. Acta Botanica Sinica 43: 750–756 (in Chinese with English abstract).

74. Wang L, Lu HY, Wu NQ, Li J, Pei YP, et al. (2006) Palynological evidence for Late Miocene-Pliocene vegetation evolution recorded in the red clay sequence of the central Chinese Loess Plateau and implication for palaeoenvironmental change. Palaeogeogr. Palaeoclimatol. Palaeoecol. 241: 118–128.

75. An ZS, Liu TS, Lu YC, Porter SC, Kukla G, et al. (1990) The long-term paleomonsoon variation recorded by the loess-paleosol sequence in Central China. Quat. Int. 7: 91–96.

76. Ding ZL, Liu TS, Rutter NW, Yu ZW, Guo ZT, et al. (1995) Ice-volume forcing of East Asian winter monsoon variations in the past 800,000 years. Quat. Res. 44: 149–159.

77. Hovan SA, Rea DK, Pisias NG, Shackleton NJ (1989) A direct link between the China loess and marine $\delta^{18}O$ records: Aeolian flux to the North Pacific. Nature 340: 296–298.

78. Sun JM, Xu QH, Huang BC (2007) Late Cenozoic magnetochronology and paleoenvironmental changes in the northern foreland basin of the Tian Shan Mountains. J. Geophys. Res. 112: B04107, doi:10.1029/2006JB004653.

79. Sun JM, Zhang ZQ, Zhang LY (2009) New evidence on the age of the Taklimakan Desert. Geology 37: 159–162.

80. Sun DH, Zhang YB, Han F, Zhang Y, Yi ZY, et al. (2011) Magnetostratigraphy and palaeoenvironmental records for a Late Cenozoic sedimentary sequence from Lanzhou, Northeastern margin of the Tibetan Plateau. Glob. Planet. Change 76: 106–116.

81. Jansen E, Sjøholm J (1991) Reconstruction of glaciation over the past 6 Myr from ice-borne deposits in the Norwegian Sea. Nature 349: 600–603.

82. deMenocal PB (1993) Wireline logging on the North Pacific transect. JOIDES J. 19: 29.

83. Wolf-Welling TCW, Cremer M, O'Connell S, Winkler A, Thiede J (1996) Cenozoic Arctic gateway paleoclimate variability: indications from changes in coarse-fraction composition. Proc. Ocean Drill. Program, Sci. Results 151: 515–567.

84. Thiede J, Winkler A, Wolf-Weilling T, Eldholm O, Myhre AM, et al. (1998) Late Cenozoic history of the polar north Atlantic: results from ocean drilling. Quat. Sci. Rev. 17: 185–208.

85. St. John KEK, Krissek LA (2002) The late Miocene to Pleistocene ice-rafting history of southeast Greenland. Boreas 31: 28–35.

86. Hao QZ, Wang L, Oldfield F, Peng SZ, Qin L, et al. (2012) Delayed build-up of Arctic ice sheets during 400000-year minima in insolation variability. Nature 490: 393–396.

87. Kutzbach JE, Guetter PJ, Ruddiman WF, Prell WL (1989) Sensitivity of climate to late cenozoic uplift in southern Asia and American West: Numerical experiments. J. Geophys. Res. 94: 18393–18407.

88. Ruddiman WF, Kutzbach JE (1990) Late Cenozoic plateau uplift and climate change. Trans. R. Soc. Edinb. Earth Sci. 81: 301–314.

89. Liu XD, Yin ZY (2002) Sensitivity of East Asian monsoon climate to the uplift of the Tibetan Plateau. Palaeogeogr. Palaeoclimatol. Palaeoecol. 183: 223–245.

90. Liu XD, Kutzbach JE, Liu ZY, An ZS, Li L (2003) The Tibetan Plateau as amplifier of orbital-scale variability of the East Asian monsoon. Geophys. Res. Lett. 30: 16, 1839, doi:10.1029/2003GL017510.

91. Haywood AM, Dowsett HJ, Valdes PJ, Lunt DJ, Francis JE, et al. (2009) Introduction. Pliocene climate, processes and problems. Phil. Trans. R. Soc. A, 367: 3–17.

92. Rind D, Chandler M (1991) Increased ocean heat transports and warmer climate. J. Geophys. Res. 96: 7437–7461.

93. Raymo ME, Rau G.H (1992) Plio-Pleistocene atmospheric CO_2 levels inferred from POM $\delta^{13}C$ at DSDP Site 607. Eos Trans. AGU 73 (Suppl.), 95.

94. Ravelo AC, Andreasen DH (2000) Enhanced circulation during a warm period. Geophys. Res. Lett. 27: 1001–1004.

95. Cane MA, Molnar P (2001) Closing of the Indonesian seaway as a precursor to east African aridification around 3–4 million years ago. Nature 411: 157–162.

96. Dowsett HJ, Cronin TM, Poore PZ, Thompson RS, Whatley RC, et al. (1992) Micropaleontological evidence for increased meridional heat-transport in the North Atlantic Ocean during the Pliocene. Science 258: 1133–1135.

97. Philander GS, Fedorov AV (2003) Role of tropics in changing the response to Milankovich forcing some three million years ago. Paleoceanography 18: PA1045, doi:10.1029/2002PA000837.

98. Wara MW, Ravelo AC, Delaney ML (2005) Permanent El Nino-like conditions during the Pliocene warm period. Science 309: 758–761.

99. Fedorov AV, Dekens PS, McCarthy M, Ravelo AC, deMenocal PB, et al. (2006) The Pliocene paradox (mechanisms for a permanent El Nino). Science 312: 1485–1489.

100. Haywood AM, Valdes PJ (2004) Modelling Middle Pliocene warmth: contribution of atmosphere, oceans and cryosphere. Earth Planet. Sci. Lett. 218: 363–377.

101. Hall R (2002) Cenozoic geological and plate tectonic evolution of SE Asia and the SW Pacific: Computer-based reconstructions and animations. J. Asian Earth Sci. 20: 353–434.

102. Maier-Reimer E, Mikolajewicz U, Crowley T (1990) Ocean general circulation model sensitivity experiment with an open central American isthmus. Paleoceanography 5: 349–366.

103. Mikolajewicz U, Crowley TJ (1997) Response of a coupled ocean/energy balance model to restricted flow through the central American isthmus. Paleoceanography 12: 429–441.

104. Mikolajewicz U, Maier-Reimer E, Crowley TJ, Kim KY (1993) Effect of Drake and Panamanian gateways on the circulation of an ocean model. Paleoceanography 8: 409–426.

105. Chaisson WP, Ravelo AC (2000) Pliocene development of east-west hydrographic gradient in the equatorial Pacific. Paleoceanography 15: 497–505.

106. Li BH, Wang JL, Huang BQ, Li QY, Jian ZM, et al. (2004) South China Sea surface water evolution over the last 12 Myr: A south-north comparison from Ocean Drilling Program Sites 1143 and 1146. Paleoceanography 19: PA1009, doi:10.1029/2003PA000906.

107. Raymo ME, Ruddiman WF, Froelich PN (1988) Influence of late Cenozoic mountain building on ocean geochemical cycles. Geology 16: 649–653.

108. Raymo ME, Ruddiman WF (1992) Tectonic forcing of late Cenozoic climate. Nature 359: 117–122.

2

Latitudinal Environmental Niches and Riverine Barriers Shaped the Phylogeography of the Central Chilean Endemic *Dioscorea humilis* (Dioscoreaceae)

Juan Viruel[1], Pilar Catalán[1,2], José Gabriel Segarra-Moragues[3]*

1 Departamento de Agricultura y Economía Agraria, Escuela Politécnica Superior de Huesca, Universidad de Zaragoza, Huesca, Spain, **2** Department of Botany, Institute of Biology, Tomsk State University, Tomsk, Russia, **3** Departamento de Ecología, Centro de Investigaciones sobre Desertificación (CIDE), Consejo Superior de Investigaciones Científicas (CSIC), Moncada, Valencia, Spain

Abstract

The effects of Pleistocene glaciations and geographical barriers on the phylogeographic patterns of lowland plant species in Mediterranean-climate areas of Central Chile are poorly understood. We used *Dioscorea humilis* (Dioscoreaceae), a dioecious geophyte extending 530 km from the Valparaíso to the Bío-Bío Regions, as a case study to disentangle the spatio-temporal evolution of populations in conjunction with latitudinal environmental changes since the Last Inter-Glacial (LIG) to the present. We used nuclear microsatellite loci, chloroplast (cpDNA) sequences and environmental niche modelling (ENM) to construct current and past scenarios from bioclimatic and geographical variables and to infer the evolutionary history of the taxa. We found strong genetic differentiation at nuclear microsatellite loci between the two subspecies of *D. humilis*, probably predating the LIG. Bayesian analyses of population structure revealed strong genetic differentiation of the widespread *D. humilis* subsp. *humilis* into northern and southern population groups, separated by the Maipo river. ENM revealed that the ecological niche differentiation of both groups have been maintained up to present times although their respective geographical distributions apparently fluctuated in concert with the climatic oscillations of the Last Glacial Maximum (LGM) and the Holocene. Genetic data revealed signatures of eastern and western postglacial expansion of the northern populations from the central Chilean depression, whereas the southern ones experienced a rapid southward expansion after the LGM. This study describes the complex evolutionary histories of lowland Mediterranean Chilean plants mediated by the summed effects of spatial isolation caused by riverine geographical barriers and the climatic changes of the Quaternary.

Editor: Tzen-Yuh Chiang, National Cheng-Kung University, Taiwan

Funding: Financial support for this study was provided by a Fundación BBVA BIOCON 05-093/06 project grant to PC and JGSM. JV was supported by a Fundación BBVA Ph.D. grant. JGSM was supported by two consecutive Spanish Aragón Government "Araid" and Spanish Ministry of Science and Innovation "Ramón y Cajal" postdoctoral contracts. PC was partially funded by a Bioflora (http://bifi.es/bioflora/) research team grant co-funded by the Spanish Aragón Government and the European Social Fund. The funders had no role in study design, data collection and analysis, decision to publish, or preparation of the manuscript.

Competing Interests: The authors have declared that no competing interests exist.

* Email: jogasemo@gmail.com

Introduction

Historical, geographical and climatic events have a strong influence on the genetic diversity of species [1]. In South America, most of the biogeographical studies of plants have focused on the effects of Pleistocene glaciations and postglacial climatic fluctuations on Andean species, having identified several lowland refugia [2–4]. However, the phylogeography of lowland species inhabiting ice-free areas during glaciations remains scarcely documented. Population genetic diversity and structure of lowland taxa are not expected to have been severely impacted by the direct effect of glaciations because of the absence of ice sheets in the central Chilean depression [5–6] and the North-to-South arrangement of the Andes, which allowed latitudinal migration [7]. Additionally, the central Chilean depression and its surrounding coastal areas provided the most suitable and stable environments for the establishment of plant and animal populations during Quaternary

glaciations [2,4,8–9]. Unlike the high Andean regions, the areas currently occupied by lowland species likely allowed *in situ* survival during glaciations; however, global temperature cooling during the glaciations could have also contributed to narrowing their geographical ranges to warmer areas.

During the Last Glacial Maximum (LGM, 25000–15000 years ago), ice sheets extended from 56°S to 35°S along the Andes [5–6]. These extensive glaciations affected the central Chilean valleys of Maipo and Aconcagua [10]. Although Quaternary glaciers reached down to 1200–2800 m.a.s.l. [10–11], their occurrence was coupled with a decrease in temperature and an increase in precipitation rates at lower altitudes [12–13].

In addition to the West-to-East barriers imposed by the Coastal Cordillera and the Andean mountains, it has been proposed that large rivers (e.g. Aconcagua, Maipo) that completely cross Chile may contribute to within-species differentiation [14]. Water volume carried by those rivers fluctuated concomitantly with

Pleistocene glaciations, increasing considerably due to ice-melting from the Andes. Accordingly, their potential barrier effect to species migration was stronger during the glacial periods than during the interglacials [15]. The genetic structure of central Chilean lowland species during these glaciations may have been affected by an East-to-West contraction of their distribution ranges towards the central Chilean depression and by their dispersal ability to bypass the transversal river barriers during latitudinal migration.

The *Epipetrum* group of *Dioscorea* is a small evolutionary lineage of the Dioscoreaceae including two species, *D. humilis* Colla and *D. biloba* (Phil.) Caddick & Wilkin, with two subspecies in each [16] that probably originated in the late Miocene (Viruel *et al.*, unpublished data). The diversification of this small group followed the retreat of the marine transgressions of the middle Miocene (15-11 Ma) which covered central Chile, providing new lands available for plant colonization from the late Miocene onwards [17]. *Dioscorea humilis* is a dioecious, diploid ($2n = 14$), dwarf geophyte with a widespread distribution spanning five central Chilean regions (530 km), from its northernmost limit in Valparaíso to its southernmost limit in Bío-Bío [16,18] (Fig. 1, Table 1). Its current distribution range is included within the Mediterranean-type bioclimatic region of Chile [19], which is bounded northwards by the Atacama Desert and southwards by temperate forests [8]. This North-to-South range covers three different climatic environments (Fig. 1): semi-arid, sub-humid and humid Mediterranean climates [20]. *Dioscorea humilis* occurs in the lowland depression between the coastal mountain range and the Andes. It includes two subspecies, the widespread *D. humilis* subsp. *humilis* and the narrow parapatric Maule coastal endemic *D. humilis* subsp. *polyanthes* (F. Phil.) Viruel, Segarra-Moragues & Villar [16] (Fig. 1).

Dioscorea humilis has a sprawling habit with shoots creeping among rock crevices. Flowers are tiny and inconspicuous; those of males are produced in pauciflorate racemes, and those of females are generally solitary. The pollination mechanisms are unknown, but flower morphology suggests the implication of a small-sized insect. The wingless seeds are produced in capsules which are sustained by spirally curled peduncles that attach capsules close to the ground or inside rock crevices, suggesting extremely short-distance seed dispersal [16].

We used nuclear microsatellite markers and cpDNA sequences to document the current patterns of population genetic diversity and structure in *D. humilis*. Additionally, Environmental Niche Modelling (ENM) was estimated on the current range extension of the infraspecific genetic groups and projected to two past scenarios, the Last Glacial Maximum (LGM) and Last Inter-Glacial (LIG). Phylogeographical patterns obtained from molecular markers, together with the estimated past variation in range extension, were investigated to elucidate the effect of Pleistocene glaciations and geological and hydrological barriers in the evolutionary history of lowland central Chilean species with limited dispersal abilities like *D. humilis*.

Materials and Methods

Ethic Statement

Necessary permits for fieldwork and sampling were obtained from the Corporación Nacional Forestal (CONAF-Chile).

Plant Sampling, DNA Extraction and Microsatellite Amplification

Fresh leaves from a total 558 individuals from 17 populations of *D. humilis* were collected throughout its entire distribution range.

Fifteen populations (Dhh01-Dhh15) corresponded to *D. humilis* subsp. *humilis* and two populations (Dhp01-Dhp02) to *D. humilis* subsp. *polyanthes* (Table 1, Fig. 1). Eight populations of *D. humilis* subsp. *humilis* (Dhh01-Dhh08) were located North of the Maipo river basin, growing in semi-arid Mediterranean-type climate areas, whereas the other seven populations (Dhh09-Dhh15) were located South of it (Table 1, Fig. 1). Five of these (Dhh09-Dhh14) were growing in sub-humid Mediterranean-type climate areas, and the southernmost population (Dhh15), together with two populations of *D. humilis* subsp. *polyanthes* (Dhp01-Dhp02), were growing in humid Mediterranean-type climate areas [20]. DNA extraction followed the procedure described in [21]. Individuals were genotyped for eight unlinked microsatellite following [22].

Plastid DNA Amplification and Sequencing

Two plastid regions, *trn*T-L and *trn*L-F [23] were amplified and sequenced in up to six individuals per population following [21]. Sequences were deposited in Genbank under the accession numbers KF357945-KF357955. A combined matrix of individual sequences of both plastid regions totalling 54 sequences was used in subsequent analyses.

Microsatellite Analysis

Allele frequencies and genetic diversity indices were calculated in all populations using GENETIX 4.05 [24]. Deviations from Hardy-Weinberg equilibrium were tested in all populations using GENEPOP v. 4.0 [25]. Different taxonomic and geographical population groups were compared to reveal differences in average values of allelic richness ($A*$), observed heterozygosity (H_O), genetic diversity within populations (H_S), inbreeding coefficient (F_{IS}) and population differentiation (F_{ST}) using FSTAT v. 2.9.3.2 [26], and differences were tested for significance with 10,000-permutation tests. Population pairwise differentiation (F_{ST}) was calculated with ARLEQUIN 3.11 [27] and tested for significance using 1000 replicates. ARLEQUIN was also used to generate a matrix of pairwise linearized F_{ST} values (i.e. $F_{ST}/(1-F_{ST})$; [28]), which was correlated to a log-transformed matrix of geographical distances between populations to test for Isolation By Distance (IBD) through Mantel tests. Significance of correlation was tested with 1000 permutations with NTSYSpc 2.11 [29].

Pairwise D_A genetic distances [30] between populations were calculated with POPULATIONS 1.2.3 [31] and used to conduct a Principal Coordinates Analysis (PCO) and a Minimum Spanning Tree (MST) that was superimposed on the PCO plots using NTSYSpc 2.11.

Population genetic structure was investigated by means of Analysis of Molecular Variance (AMOVA) which was performed in ARLEQUIN 3.11 to partition in different population groups according to the taxonomical or geographical membership. The significance of the analyses was tested with 1000 replicates.

Bayesian clustering was also used to infer population genetic structure using STRUCTURE 2.1 [32]. Analyses were based on an admixture ancestry model with correlated allele frequencies, for a range of K genetic clusters from one to 19, with ten replicates for each K. The analyses were performed with a burn-in period and a run length of the Monte Carlo Markov Chain (MCMC) of 7×10^5 and 7×10^6 iterations, respectively. The most likely number of genetic clusters (K) was determined according to Evanno *et al.* [33].

Plastid DNA Data Analyses

Haplotype polymorphism was estimated within populations and within genetic and geographical groups through the analysis of the number of segregating sites (S), the number of haplotypes (h), the

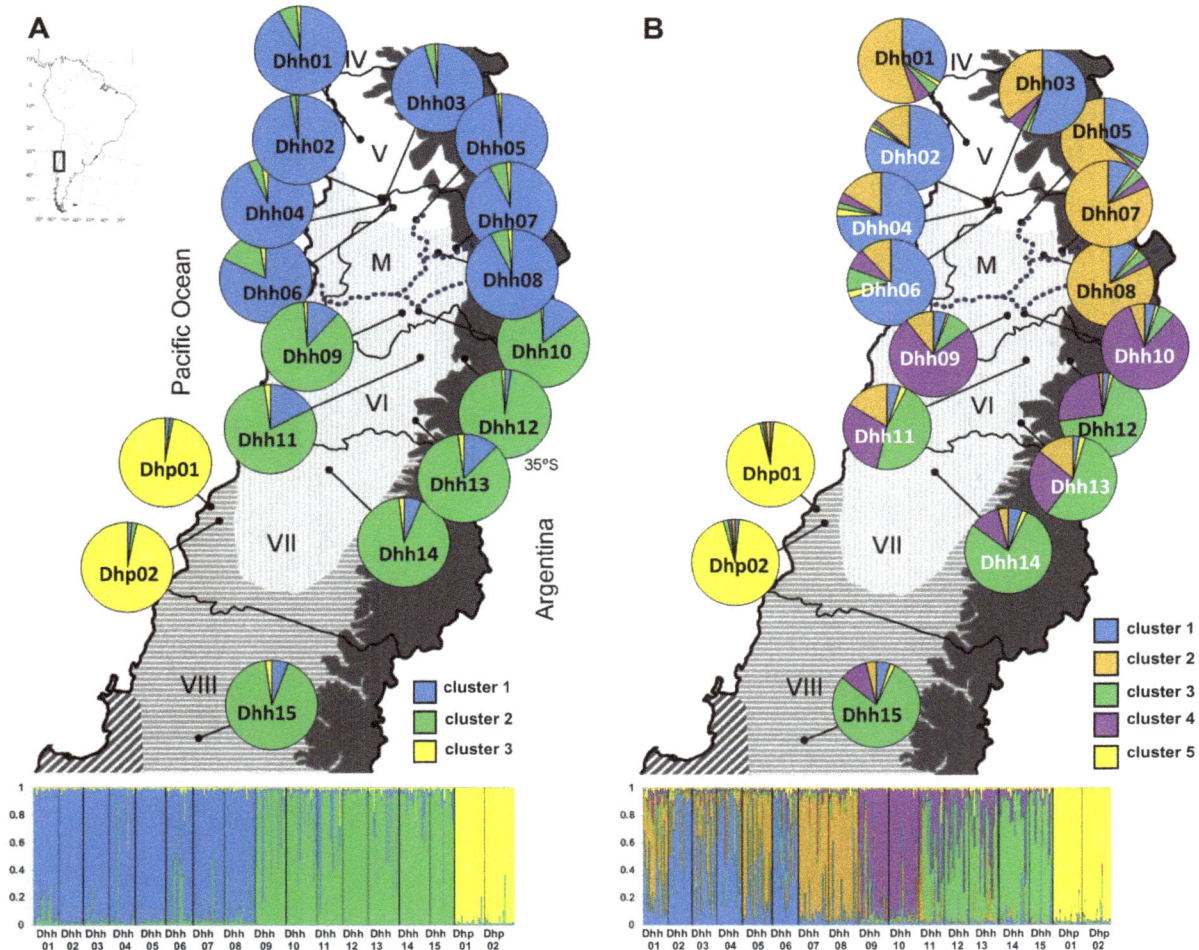

Figure 1. Geographical distribution of sampled populations of *Dioscorea humilis* **in Chile (Table 1) and Bayesian analyses of the genetic structure of 15 populations of** *D. humilis* **subsp.** *humilis* **and two populations of** *D. humilis* **subsp.** *polyanthes* **based on nuclear microsatellite data.** The mean proportion of membership of each predefined population to each of the A, three ($K = 3$), and B, five ($K = 5$), most likely inferred genetic clusters is shown. The dotted line indicates the location of the Maipo river. Chilean administrative regions: IV, Coquimbo, V, Valparaíso, M, Metropolitana, VI, Libertador General Bernardo O'Higgins, VII, Maule, and VIII, Bío-Bío. Geographical ranges of five climatic zones in central Chile from Castillo *et al.*, [20] are superimposed on the maps. From North to South: semiarid Mediterranean (white), sub-humid Mediterranean (vertical shading), humid Mediterranean (horizontal shading), hyper humid Mediterranean (diagonal shading) and eastern Andean Continental (solid grey). Map contour constructed from spatial data retrieved from http://www.diva-gis.org/gdata.

haplotype diversity index (Hd) and the average number of pairwise nucleotide differences between DNA sequences ($\theta\pi$) [34] with DnaSP5 [35]. Indels encompassing two to several nucleotides were reduced to single gaps and treated as a fifth nucleotide state for a statistical parsimony haplotype network analysis with TCS v. 1.21 software [36].

Environmental Niche Modelling Analyses

Environmental niche modelling (ENM) was conducted to evaluate the potential distribution of the geographical groups of *D. humilis* under current climatic conditions and under Last Glacial Maximum (LGM) and Last Interglacial (LIG) conditions. A set of 19 bioclimatic variables (Table S1 in Appendix S1) retrieved from WorldClim (www.worldclim.org) plus the altitude were used, and GIS layers with 30 sec resolution were clipped to the extent of central Chilean regions using DIVA-GIS [37]. Correlation among environmental variables was determined by Mantel tests using XLSTAT and tested for significance with 1000

random permutations (Table S1 in Appendix S1). Then we selected a reduced set of nine uncorrelated environmental variables with higher percent contribution (PC) and permutation importance (PI) based on jackknife pseudosampling on the ENM of *D. humilis* (Tables S1 and S2 in Appendix S1): altitude, bio3 (isothermality), bio4 (temperature seasonality), bio6 (minimum temperature of coldest month), bio7 (annual range temperature), bio9 (mean temperature of driest quarter), bio15 (precipitation seasonality), bio18 (precipitation of warmest quarter) and bio19 (precipitation of coldest quarter).

Additionally, we assessed pairwise correlations between all 20 environmental variables studied and pairwise D_A population genetic distances [30], and pairwise population linearized F_{ST} [28], among populations of *D. humilis*, and the correlation between the 20 environmental variables and latitude, using the Mantel test with 1000 random permutations.

The maximum entropy algorithm implemented in MAXENT v. 3.3.3k [38–39] was used to construct the models. Maxent is

Table 1. Population data and genetic diversity indices of 17 Chilean populations of *Dioscorea humilis* for eight nuclear microsatellite loci and overall estimates for taxonomic and geographical population groups.

Code	Locality	Latitude	Longitude	Altitude (m)	N^I	A^I	H_O^I	H_E^I	F_{IS}^I
Dioscorea humilis subsp. *Humilis* (Dhh01 to Dhh15)					488	3.969	0.468	0.425	-0.103***
Northern *D. humilis* subsp. *humilis* (Dhh01 to Dhh08)					257	3.800	0.466	0.410	-0.136***
Dhh01	Valparaíso: Catapilco, rincón de La Mestiza	32° 32' 33.4" S	71° 17' 37.7" W	90	30	4.125	0.471	0.405	-0.165**
Dhh02	Metropolitana: Granizo, Parque Nacional de La Campana, sector Granizo	32° 58' 54.9" S	71° 07' 58.2" W	450	28	4.000	0.424	0.386	-0.100**
Dhh03	Metropolitana: Granizo, Parque Nacional de La Campana, La Troya bridge.	32° 59' 05.3" S	71° 08' 17.8" W	420	30	4.125	0.425	0.371	-0.148**
Dhh04	Metropolitana: Parque Nacional de La Campana, sector Cajón grande	33° 00' 06.1" S	71° 07' 53.5" W	340	30	3.500	0.350	0.391	+0.107*
Dhh05	Metropolitana: Carretera a Til Til, Entrance to Parcelación de El tranque.	33° 08' 40.7" S	70° 52' 34.8" W	530	36	3.875	0.438	0.392	+0.036**
Dhh06	Metropolitana: Road from Til Til to Limache, detour to La Vega through el Camarico.	33° 03' 28.3" S	71° 02' 57.2" W	700	31	3.875	0.510	0.458	+0.019**
Dhh07	Metropolitana: Santiago de Chile, Cerro Manquehue, creek starting from Agua del Palo.	33° 21' 50.3" S	70° 34' 56.9" W	820	36	3.625	0.523	0.423	-0.241**
Dhh08	Metropolitana: Santiago de Chile, Cerro de Renca.	33° 23' 43.1" S	70° 42' 38.9" W	540	36	4.500	0.560	0.453	-0.135**
Southern *D. humilis* subsp. *humilis* (Dhh09 to Dhh15)					231	4.163	0.472	0.442	-0.068***
Dhh09	Metropolitana: Maipo, Cerro Cantillana, side of forest track from Rague to Pabellón.	33° 51' 10.8" S	70° 58' 44.5" W	630	36	4.625	0.460	0.467	+0.108**
Dhh10	Metropolitana : Maipo, Cerros de Aculeo.	33° 50' 04.4" S	70° 51' 14.0" W	380	36	4.714	0.549	0.521	-0.044ns

Table 1. Cont.

Code	Locality	Latitude	Longitude	Altitude (m)	N^1	A^1	H_O^1	H_E^1	F_{IS}^1
Dhh11	OHiggins: Road from Rancagua to Doñihue.	34° 11' 28.2" S	70° 50' 39.4" W	430	30	5.000	0.432	0.459	+0.078**
Dhh12	OHiggins: Road from Coya to Pangal.	34° 12' 03.2" S	70° 30' 58.4" W	850	29	3.250	0.453	0.357	−0.177***
Dhh13	OHiggins: Road from San Fernando to Tinguiririca, next to La Rufina forest track.	34° 40' 27.0" S	70° 52' 56.4" W	550	36	4.625	0.485	0.444	−0.073**
Dhh14	Maule: Road from Curicó to Sagrada Familia.	35° 03' 03.0" S	70° 31' 03.2" W	170	36	4.875	0.447	0.413	+0.046ns
Dhh15	Bio-Bío: Between Yumbel and Monteáguila	37° 05' 00.4" S	72° 28' 46.1" W	110	28	3.875	0.556	0.502	−0.064*
Dioscorea humilis subsp. *polyanthes* (**Dhp01-Dhp02**)					**70**	4.131	0.574	0.480	−0.195***
Dhp01	Maule: Constitución, just after the river Maule bridge to Putú.	35° 20' 05.6" S	72° 23' 17.2" W	20	35	4.250	0.598	0.489	−0.222**
Dhp02	Maule: de Constitución a San Javier	35° 26' 24.1" S	72° 20' 00.2" W	330	35	4.500	0.550	0.475	−0.162ns

[1]N, sample size; A, mean number of alleles per locus; H_O, H_E, observed and expected heterozygosity, respectively; F_{IS}, inbreeding coefficient. Significance: *, $P<0.05$, **, $P<0.01$, ***, $P<0.001$, ns, not significant.

optimal for ENM when using small sample sizes [40] and when environmental predictions are poorly influenced by the addition of irrelevant bioclimatic variables [41]. The *D. humilis* data fit these requirements since no significant increase in the area under the curve (AUC) values was observed when using all variables compared to those from the reduced set of variables (Table S3 in Appendix S1).

Occurrence data were split into training data (75%) to build the model and test data (25%) to test the accuracy of the model. Fifteen subsample replicates were performed in each run using the default options and 1000 iterations. Model accuracy was assessed with the AUC value of the receiver-operating characteristic curve (ROC) [38]. The contribution of each environmental variable to the ENM was evaluated through a Jackknife pseudosampling (see above). A tenth percentile threshold was applied for all models.

ENM were conducted for the two northern (Dhh01-Dhh08) and southern (Dhh09-Dhh15) population groups of *D. humilis* ssp. *humilis*. The low number of known populations of *D. humilis* subsp. *polyanthes* (Dhp01-02) precluded a confident ENM analysis of this taxon.

ENMs were projected to LGM (c. 21 ka BP), with 2.5 arc-minutes resolution [42], and to LIG (c. 120–140 ka BP), with 30 arc-seconds resolution [43] scenarios. Two palaeoclimatic layers simulated for two general atmospheric circulation models were used for LGM: the Community Climate System Model (CCSM, [44]) and the Model for Interdisciplinary Research on Climate (MIROC, [45]). Both CCSM and MIROC layers were combined following a conservative approach by including their overlapping predicted areas [46]. Current minimum predicted values were used to determine the past minimal predicted areas, assuming that the environmental requirements of *D. humilis* subsp. *humilis* have remained stable during at least since LIG.

A complementary ENM approach was done through a Principal Component Analysis (PCA), which was constructed with the raw data obtained from the 19 climatic variables and the altitude for each population of *D. humilis* using PAST 2.17c [47], Fig. S1 in Appendix S1).

Results

Microsatellite Genetic Diversity in *Dioscorea humilis*

All eight microsatellite loci were polymorphic and amplified a total of 79 alleles in the 17 studied populations of *D. humilis* (Table S4 in Appendix S2). The number of alleles per locus ranged from three (B633) to 24 (H442) with a mean of 9.88 ± 6.38 ($\pm$SD) alleles per locus. The mean number of alleles per locus and population ranged from 3.25 (Dhh12) to 5.00 (Dhh11, Table 1). Of the 79 microsatellite alleles scored, 34 (43.04%) were shared by both subspecies of *D. humilis*, while 36 (45.57%) and 9 (11.39%) were exclusive to *D. humilis* subsp. *humilis* and *D. humilis* subsp. *polyanthes*, respectively (Table S4 in Appendix S2).

Observed heterozygosities ranged from 0.350 (Dhh04) to 0.598 (Dhp01), and unbiased expected heterozygosities from 0.357 (Dhh12) to 0.521 (Dhh10) (Table 1). Five of the 17 populations showed HW deviations towards heterozygote deficiency; three, including one population of *D. humilis* subsp. *polyanthes*, showed non-significant departure from HW equilibrium, and the remaining eight populations of *D. humilis* subsp. *humilis* and one of *D. humilis* subsp. *polyanthes* showed a significant heterozygote excess (Table 1).

No significant differences were detected for the tested genetic diversity indices between *D. humilis* subsp. *humilis* and *D. humilis* subsp. *polyanthes*, except for observed heterozygosity (H_o). Surprisingly, the more restricted endemic *D. humilis* subsp.

polyanthes showed significantly higher ($p = 0.033$) average H_o (Table 1). Similarly, the comparison of northern and southern population groups of *D. humilis* subsp. *humilis* failed to find significant differences at any of the tested indices (Table 1).

Population Structure of *Dioscorea humilis*

Moderate but significant (different from zero; $p<0.05$) levels of population differentiation were observed among populations (results not shown). Higher average F_{ST} values were found between populations of both subspecies (average $F_{ST} = 0.295$) than among populations within subspecies (average Dhh $F_{ST} = 0.145$; Dhp $F_{ST} = 0.011$). Similarly, a higher average differentiation was observed between northern and southern populations groups of *D. humilis* subsp. *humilis* (average $F_{ST} = 0.198$), than among populations within northern populations (Dhh01-Dhh08 $F_{ST} = 0.069$) and southern populations (average Dhh09-Dhh15 $F_{ST} = 0.109$) of *D. humilis* subsp. *humilis* with differentiation between the groups not being statistically significant (Table 2).

Bayesian analysis of population structure showed a maximum $\Delta K = 1598.45$ value for $K = 3$ (Fig. S2 in Appendix S2). In this clustering, individuals of *D. humilis* subsp. *polyanthes* showed a high proportion of membership to cluster 3 and those of *D. humilis* subsp. *humilis* to cluster 1 (populations Dhh01-Dhh08) or to cluster 2 (populations Dhh09-Dhh15; Fig. 1a). Mean F_{ST} values corresponding to the divergence between clusters 1, 2 and 3 and the hypothetical ancestral population were 0.114, 0.201 and 0.266, respectively, indicating that populations showing a higher membership to cluster 1 were less diverged from the ancestral population. A further maximum $\Delta K = 127.90$ value was obtained for $K = 5$ (Fig. S2 in Appendix S2) which separated the populations of *D. humilis* subsp. *humilis* into two additional genetic clusters (clusters 1–4; Fig. 1b).

Non-hierarchical AMOVA attributed 19.03% of the total variation to among populations of *D. humilis s.l.*, and 15.00% of the total variation to among populations of *D. humilis* subsp. *humilis* (Table 2). In hierarchical AMOVA, the largest proportion of variation among groups (21.37%) was obtained for a taxonomical grouping of populations into subspecies. AMOVA based on a geographical grouping of populations attributed 12.15% of the variation to differences between northern and southern groups of *D. humilis* subsp. *humilis* and a lower proportion of variance (7.65%) to differences among populations within groups (Table 3). The grouping of *D. humilis* subsp. *humilis* populations into four genetic clusters did not increase the variance among groups (11.81%) but lowered the proportion of variance among populations within groups (5.34%).

PCO showed results consistent with STRUCTURE (Figs. 1 and 2). Populations of *D. humilis* subsp. *polyanthes* separated at a large distance from populations of *D. humilis* subsp. *humilis* (Fig. 2). Clustering of populations of this latter taxon was consistent with their geographical distribution (Fig. 2). PCO with superimposed MST analysis identified the closer relationship of *D. humilis* subsp. *polyanthes* to the southern populations of *D. humilis* subsp. *humilis* (Fig. 2).

A significant correlation between pairwise geographical distances and linearized F_{ST} values was found in both *D. humilis s.l.* ($r = 0.537$, $p = 0.001$) and *D. humilis* subsp. *humilis* ($r = 0.416$, $p = 0.004$) populations (Fig. 3a), thus showing significant isolation by distance (IBD). However, the pattern of IBD vanished when this analysis was separately conducted within both northern ($r = 0.296$, $p = 0.080$) and southern ($r = 0.005$, $p = 0.420$) geographical groups of *D. humilis* subsp. *humilis* (Fig. 3b).

Table 2. Analyses of molecular variance (AMOVA) of *Dioscorea humilis* populations based on microsatellite data.

Source of variation (groups)	Sum of squared deviations (SSD)	d.f.	Variance components	% of the total variance
1. *Dioscorea humilis* s.l.				
Among populations	443.049	16	0.39637	19.03
Within populations	1852.940	1099	1.68602	80.97
2. Taxonomic membership: *humilis* (Dhh01-Dhh15) *vs. polyanthes* (Dhp01. Dhp02)				
Among groups	150.915	1	0.53219	21.37
Among populations within groups	292.314	15	0.27222	10.93
Within populations	1852.940	1099	1.68602	67.70
3. *Dioscorea humilis* subsp. *humilis* s.l.				
Among populations	288.727	14	0.29176	15.00
Within populations	1589.255	961	1.65375	85.00
4. Geographical membership of *D. humilis* subsp. *humilis*: northern (Dhh01–08), *vs.* southern (Dhh09–15)				
Among groups	133.969	1	0.25056	12.15
Among populations within groups	154.758	13	0.15781	7.65
Within populations	1589.255	961	1.65375	80.20
5. Genetic membership (excluding *D. humilis* susbp. *polyanthes*): cluster Dhh01, Dhh05, Dhh07-Dhh08 *vs.* cluster Dhh02-Dhh04, Dhh06 *vs.* Dhh09-Dhh10 *vs.* Dhh11-Dhh15				
Among groups	194.783	3	0.23571	11.81
Among populations within groups	93.944	11	0.10659	5.34
Within populations	1589.255	961	1.65375	82.85

Plastid Haplotype Diversity in *Dioscorea humilis*

The combination of *trn*L–F and *trn*T–L plastid DNA regions produced eight haplotypes (Fig. 4; Table S5 in Appendix S2). Six and one haplotypes were restricted to *D. humilis* subsp. *humilis* and subsp. *polyanthes* respectively, and one haplotype was shared between both taxa. Haplotype IV was widespread in 12 populations of *D. humilis*, including one population of *D. humilis* subsp. *polyanthes*, and had the highest outgroup probability (0.771). Three haplotypes (I, II and III) were restricted to some western populations of northern *D. humilis* subsp. *humilis*, whereas haplotypes V, VI and VII were restricted to some southern populations of *D. humilis* subsp. *humilis* (Fig. 4a; Table S5 in Appendix S2).

Haplotype diversity was higher in *Dioscorea humilis* subsp. *humilis* ($S = 6$, $\theta\pi = 0.789$), than in *D. humilis* subsp. *polyanthes* ($S = 1$, $\theta\pi = 0.500$) (Table 3), as expected for the wider distribution range and population abundance of the former. The northern group of populations of *D. humilis* subsp. *humilis* (Dhh01-Dhh08) showed higher haplotype diversity ($S = 3$, $\theta\pi = 0.873$) compared to the southern group (Dhh09-Dhh15) of populations ($S = 3$, $\theta\pi = 0.534$) (Table 3).

TCS estimated a 95% maximum connection of 17 steps incorporating all eight haplotypes into the network and inferred three missing haplotypes (Fig. 4b). The haplotype network showed a star-like pattern with four of the six derived haplotypes directly connected to the most widespread one (Hap. IV; Fig. 4). Three haplotypes were connected to the central one at a larger number of mutations. Two of them were private to the northernmost population (Hap. I and II) whereas the other (Hap. V) was private to a population from the southern group (Fig. 4).

Environmental Niche Modelling

All rainfall-derived variables (bio12-bio19) and all but three temperature-derived variables (bio6, bio8 and bio11) were significantly correlated to genetic distances (Table S1 in Appendix S1). Also, latitude was highly correlated to all rainfall-derived variables (bio12-bio19), to two temperature-derived variables (bio2-bio3), and to both pairwise populations' D_A and F_{ST} genetic distances (Table S1 in Appendix S1).

According to response curves and Jackknife tests, the most informative variables for the ENM of *D. humilis* s.l. were altitude and three climatic variables derived from rainfall data (bio16, precipitation wettest quarter; bio18, precipitation warmest quarter; and bio19, precipitation coldest quarter). At the subspecies level, the variables with the largest contributions to the ENM of *D. humilis* subsp. *humilis* were altitude and the climatic variables bio8 (mean temperature of wettest quarter), bio9 (mean temperature of driest quarter) and bio15 (precipitation seasonality) (Table S2 in Appendix S1). Independent ENM for northern (Dhh01-Dhh08) and southern (Dhh09-Dhh15) genetic groups of *D. humilis* subsp. *humilis* revealed that the variables bio15 and bio18 were most informative for the northern group, whereas bio8, bio9 and bio15 were most informative for the southern group (Table S2 in Appendix S1). All projections (Figs. 5a–5c) showed excellent predictive success rates, with AUC values higher than 0.9 (Table S3 in Appendix S1). The PCA of environmental variables for the *D. humilis* populations (Fig. S1 in Appendix S1) separated northern from southern genetic groups of *D. humilis* subsp. *humilis*. The southernmost population, Dhh15, showed a distinct set of climatic conditions from the others, according to its separated position, and clustered together with the *D. humilis* subsp. *polyanthes* populations in the bidimensional PCA plot (Fig. S1 in Appendix S1).

Table 3. Plastid combined *trn*TL- *trn*LF haplotype diversity analysis of *D. humilis* populations and geographical/genetic groups.

Population/Group	N	S	h	Hd	θπ
D. humilis ssp humilis (Dhh01-15)	45	6	6	0.391	0.789 (0.000–2.404)
Northern range (Dhh01-08)	27	3	3	0.325	0.873 (0.000–2.610)
Dhh01-06	21	3	3	0.400	1.089 (0.000–3.124)
Dhh07-08	6	0	1	0.000	-
Dhh01	5	1	2	0.400	0.397 (0.000–1.800)
Dhh02	2	0	1	0.000	-
Dhh03	2	0	1	0.000	-
Dhh04	4	0	1	0.000	-
Dhh05	4	0	1	0.000	-
Dhh06	4	0	1	0.000	-
Dhh07	3	0	1	0.000	-
Dhh08	3	0	1	0.000	-
Southern range (Dhh09-15)	18	3	4	0.477	0.534 (0.000–1.863)
Dhh09-10	4	1	2	0.667	0.664 (0.000–2.500)
Dhh11-15	14	2	3	0.385	0.406 (0.000–1.560)
Dhh09	2	0	1	0.000	-
Dhh10	2	0	1	0.000	-
Dhh11	2	0	1	0.000	-
Dhh12	2	0	1	0.000	-
Dhh13	3	0	1	0.000	-
Dhh14	4	1	2	0.667	0.673 (0.000–2.500)
Dhh15	3	1	2	0.667	0.656 (0.000–2.667)
D. humilis ssp polyanthes (Dhp01-02)	9	1	2	0.500	0.506 (0.000–1.889)
Dhp01	3	0	1	0.000	-
Dhp02	6	0	1	0.000	-
Total	54	7	7	0.419	0.768 (0.000–2.379)

Population codes, sample size (*N*), and combined *trn*TL- *trn*LF haplotype frequency parameters: number of segregating sites (*S*), number of distinct haplotypes (*h*), and haplotype diversity (*Hd*) and molecular diversity (θπ) estimates (with 95% confidence intervals of θπ generated through 10,000 θ-based simulations under the coalescence model using the program DNAsp v.5 [35].

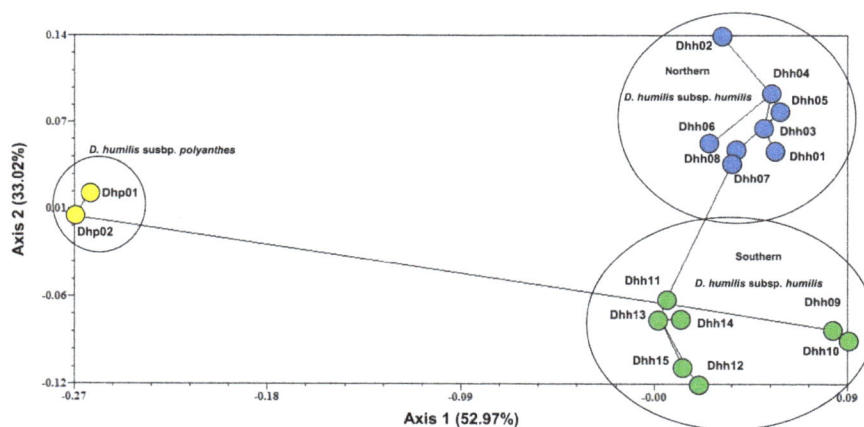

Figure 2. Principal Coordinates Analysis (PCO) showing the genetic relationships among populations of *Dioscorea humilis* **based on** D_A **genetic distance** [30]. Populations of *D. humilis* subsp. *polyanthes* (Dhp01, Dhp02), yellow circles; northern populations of *Dioscorea humilis* subsp. *humilis* (Dhh01 to Dhh08), blue circles; southern populations of *D. humilis* subsp. *humilis* (Dhh09 to Dhh15), green circles.

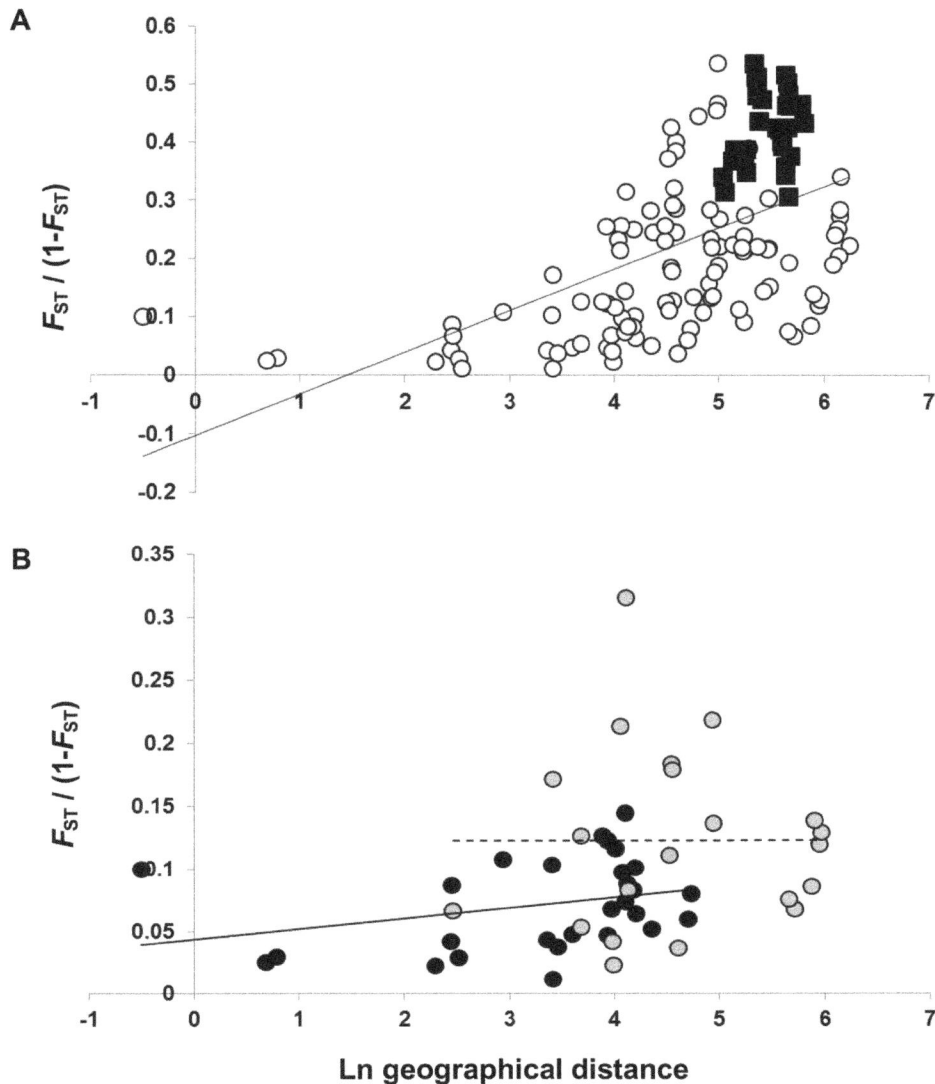

Figure 3. Isolation by distance analyses. Correlation between log-transformed pairwise geographical distances and linearized F_{ST} values [28] among populations of *D. humilis*. A. *D. humilis s.l.* where open circles represent pairwise comparisons among populations of *D. humilis*. subsp. *humilis* and black squares represent pairwise comparisons among populations of *D. humilis*. subsp. *humilis* and *D. humilis* subsp. *polyanthes*. Correlation between matrices was $r = 53.74\%$, $p = 0.001$ for *D. humilis s.l.* and $r = 41.61\%$, $p = 0.004$ for *D. humilis* subsp. *humilis* B. IBD analyses within geographical groups of *D. humilis* susbp. *humilis*, where black circles represent pairwise comparisons among populations of the northern group (Dhh01-Dhh08) and grey circles represent pairwise comparisons among populations of the southern group (Dhh09-Dhh15). Correlation between matrices was $r = 29.62\%$, $p = 0.080$ and $r = 0.47\%$, $p = 0.420$ for the northern and southern groups, respectively.

The ENM for current environmental conditions was mostly concordant with the current distribution of northern and southern genetic groups of *D. humilis* subsp. *humilis* (Fig. 5c), except for the southernmost population, Dhh15. Potential areas of contact were predicted on the eastern boundary of their distributions. Projections to the LIG (Fig. 5a) predicted a minimal extension area for the potential distribution of the southern group and a larger extension for the northern group. Projections to the LGM (Fig. 5b) predicted a substantial reduction in the southern group and, to a lesser extent, of the northern group. The present and the two historical models predicted areas with unsuitable environmental conditions between the northern and southern groups (Fig. 5), and

an increase in the potential areas of contact between the two groups since the LGM to the present.

Discussion

Genetic Diversity, Genetic Structure and Diversification of *Dioscorea humilis*

Genetic diversity and population structure in plant species is determined by various abiotic and biotic factors, some of which have triggered population differentiation [48] and speciation processes [49–50]. Biotic factors have been globally assigned to life-history (e.g. life form), and reproductive traits (e.g. reproductive systems, pollination and seed dispersal mechanisms [51]).

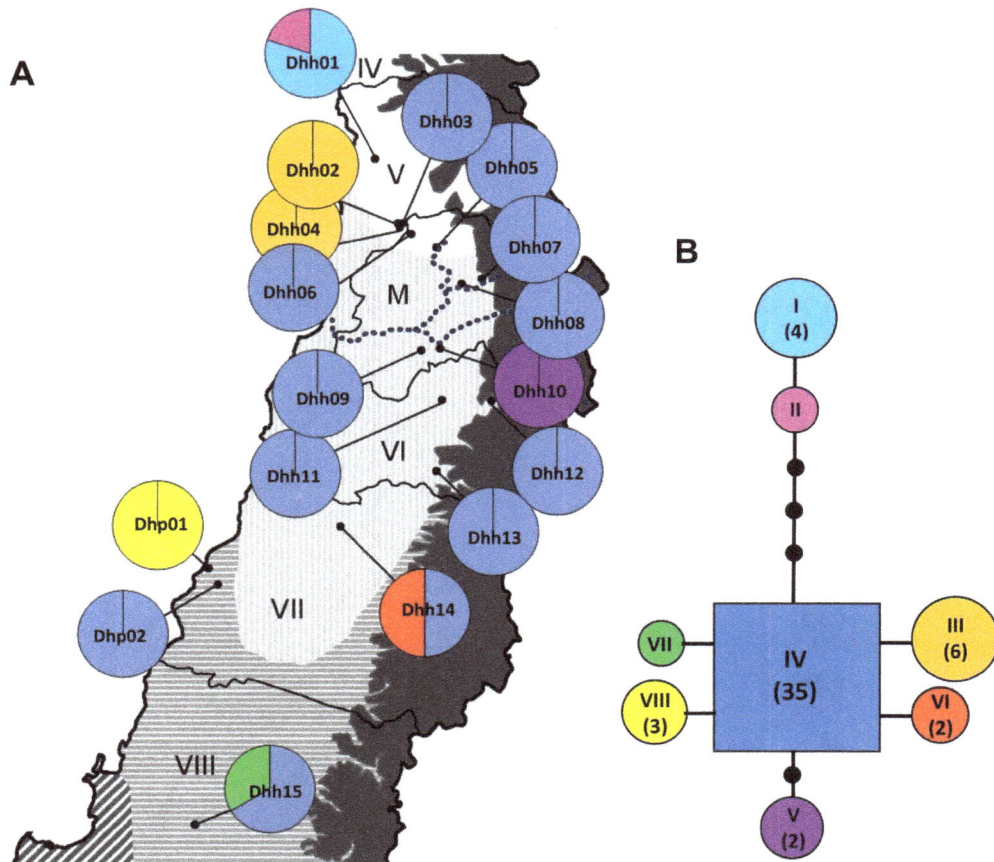

Figure 4. Plastid haplotype diversity in *Dioscorea humilis*. A. Geographical distribution of eight chloroplast haplotypes in 17 populations of *Dioscorea humilis*. Pie charts indicate relative frequencies of each haplotype in each population. The dotted line indicates the location of the Maipo river. Chilean administrative regions and climatic regions are indicated as in Fig. 1. B. Parsimony Network showing the relationships among eight haplotypes. Black dots indicate unsampled or extinct haplotypes. The size of the circles or squares is proportional to the number of sequences representing each haplotype, and is indicated in parentheses when higher than one. Map contour constructed from spatial data retrieved from http://www.diva-gis.org/gdata.

However, closely related taxa, such as the two subspecies of *D. humilis*, show similar biotic parameters. Abiotic factors include, as most relevant, climatic variables and barriers to dispersal. Our results suggest that the conjunction of these two later factors caused the intraspecific split within *D. humilis*.

Our analyses revealed moderate levels of allelic diversity and heterozygosity across populations of *D. humilis* ($A = 3.25–5.0$, $H_O = 0.350–0.598$, $H_E = 0.357–521$, Table 1), which were relatively lower than in the sister species *D. biloba* ($A = 5.14–7.29$, $H_O = 0.345–0.686$, $H_E = 0.458–0.706$, Table S6 in Appendix S3). However, genetic diversity parameters of *D. humilis* were in the range of other yam species with a different combination of life-history (climbers), reproductive (winged seeds) and distribution range (broad range, non-endemic), characteristics that, contrary to *D. humilis* should predispose them to higher levels of genetic diversity (Table S6 in Appendix S3). By contrast, *D. humilis* showed higher genetic diversity than those of species of the *Borderea* group of *Dioscorea* which are comparable in morphological and reproductive traits (Table S6 in Appendix S3), though the Pyrenean *Dioscorea* species differ from *D. humilis* in their even narrower distributions [52–53].

Widespread taxa tend to maintain higher levels of genetic diversity compared to geographically restricted congeners [54]. However, genetic diversity in *D. humilis* subsp. *polyanthes* was not

significantly lower than in *D. humilis* subsp. *humilis*, despite its more restricted geographical range (Table 1). This result could suggest equally or more efficient mechanisms buffering against genetic loss in *D. humilis* subsp. *polyanthes*.

Our study also revealed a strong geographical structure of nuclear microsatellite variation throughout the range of *D. humilis* (Fig. 1), with populations of *D. humilis* subsp. *polyanthes* clearly separate from those of *D. humilis* subsp. *humilis* that split into clusters 1 (northern populations Dhh01-Dhh08) and 2 (southern populations Dhh09-Dhh15; Fig. 1a). Clustering analyses (Fig. 2) and AMOVA (Table 3) also found a major differentiation between the two subspecies, in agreement with their morphological distinction [16]. However, plastid DNA haplotype sharing (haplotype IV) between the subspecies and the occurrence of a private plastid haplotype (VIII) in Dhp01, directly derived from the most common haplotype (IV), suggests recent diversification mediated by isolation by distance (Fig. 3a), with incomplete lineage sorting in *D. humilis* subsp. *polyanthes* (Fig. 4) or alternatively, introgression between subspecies.

Bayesian F_{ST} values supported an origin of the species in the northern region and a derived recent origin of *D. humilis* subsp. *polyanthes* from the southern group of *D. humilis* subsp. *humilis*. This was corroborated by the highest haplotypic diversity of the northern group (Table 3) and by the PCO-MST analysis, which

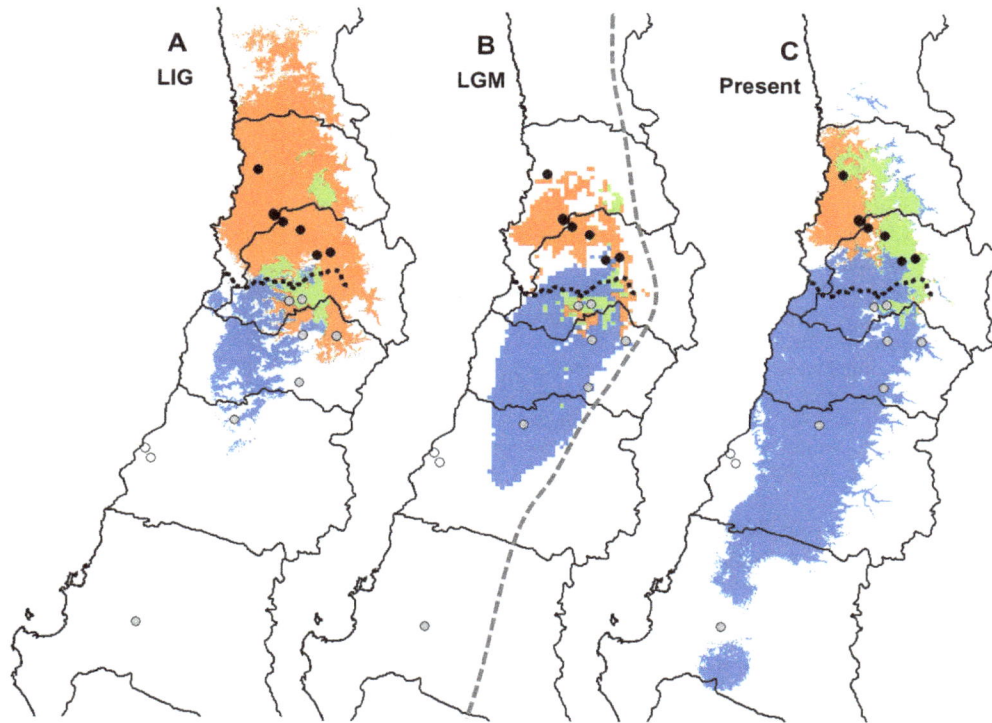

Figure 5. Environmental Niche Modelling (ENM) of *Dioscorea humilis* **estimated under Last Interglacial (LIG) (A), Last Glacial Maximum (LGM) (B) and current (C) climate conditions.** In orange, northern *D. humilis* subsp. *humilis* genetic group, in blue, southern genetic group and in bright green, overlapping areas among predicted distributions. A tenth percentile threshold was applied. Black circles: northern populations of *D. humilis* subsp. *humilis* (Dhh01 to Dhh08); grey circles: southern populations of *D. humilis* subsp. *humilis* (Dhh09 to Dhh15). The black dashed line indicates the location of the Maipo river. The grey dashed line in B indicates the approximate extent of the ice sheet during the LGM after [5]. Map contours constructed from spatial data retrieved from http://www.diva-gis.org/gdata.

indicated the closeness of *D. humilis* subsp. *polyanthes* to the southern *D. humilis* subsp. *humilis* populations (Fig. 2).

The intraspecific divergence of the southern *D. humilis* subsp. *polyanthes* from the southern group of *D. humilis* subsp. *humilis* could have been a consequence of local environmental adaptation. However, despite the lack of an ENM for the former taxon, the range of values for its 19 bioclimatic variables overlap with that of the latter group and are not significantly different from them (Table S2 in Appendix S1). Thus, the explanation for their divergence is other than a climatically driven speciation process; it may be rather the consequence of geographical isolation and incomplete plastid lineage sorting of *D. humilis* subsp. *polyanthes* from the central Chilean depression southern *D. humilis* subsp. *humilis* group during the last glacial and interglacial phases (Fig. 5).

Influence of Current and Past Latitudinal Climatic Heterogeneity on the Genetic Structure of *Dioscorea humilis* subsp. *humilis*

A noticeable finding was the strong geographical structure detected among *D. humilis* subsp. *humilis* populations, which separated into two North-to-South genetic groups (Fig. 1a). Such spatial patterns are usually driven by the effect of strong geographical or climatic barriers to dispersal, as proposed for *Hordeum chilense* Roem. & Schult., which is similarly distributed along a climatic gradient in Chile [20]. The distribution of the genetic groups of *D. humilis* subsp. *humilis* mostly paralleled those of the main Chilean latitudinal climatic zones (Fig. 1). Mountain

chains in the central Chilean depression show lower altitudes and may not significantly contribute to latitudinal isolation of populations. Our study showed that the genetic divergence of the two population groups occurred northwards and southwards of the Maipo river basin (Fig. 1). Indeed, river basins have been identified as efficient barriers to dispersal for seed plants, such as in Chinese populations of *Vitex negundo* L. (Verbenaceae) on opposite shores of the Yangtze river [55]. Specifically, the role of the Maipo river as a geographical barrier has been highlighted for other organisms with potentially higher dispersal capabilities than *D. humilis*, such as the snake *Philodryas chamissonis* Wiegmann [14]. The Maipo river acting as a geographical barrier to gene flow could contribute to explain the IBD pattern across the range of *D. humilis* subsp. *humilis* (Fig. 3a), and the abrupt difference in genetic structure between the two geographical groups (Fig. 1a). However, this IBD pattern does not apply for within-group pairwise population comparisons (Fig. 3b). The absence of IBD within the two geographical areas of *D. humilis* subsp. *humilis* contrasts with life-history and reproductive traits of the species which all point towards extremely short dispersal distances [16]. Therefore, the observed patterns are probably mirroring historical gene flow among populations within ranges preceding range expansions in the Holocene and a relatively rapid postglacial expansion by unknown vectors.

Geographical and historical variations of environmental variables have been demonstrated to greatly influence genetic divergence among populations [20,56]. Our ENM analyses indicate a strong latitudinal ecological differentiation throughout the current range of *D. humilis* subsp. *humilis* into two well defined environmental niches (Fig. 5c). Past projection of niche models indicate that this ecological niche differentiation likely originated earlier than LIG (Fig. 5a), and that ecological conditions have been maintained until present times. The current separation of the groups by the Maipo river basin [14] matches the ecogeographical division of the *D. humilis* range into northern semiarid and southern subhumid Mediterranean areas (Fig. 1, [20]). Ice-cover during the LGM, which reached to 35°S in the Andes [5−6]), together with its northwards influence that extended to approximately 33°S, could have strengthened the barrier effect of the Maipo river. Water volume of this river originating from the Andes was likely higher during LGM than in present times [14], which could account for the allopatric distribution of the two population groups, thereby contributing to the observed genetic differentiation between them (Fig. 5). Bayesian F_{ST} values supported the ancestry of the northern populations, suggesting a likely origin of the species in its northern range, in the overlapping area with its congener *D. biloba* [16], followed by further southwards expansion.

Nonetheless, the predicted extension of the potential distribution areas of the two population groups could have fluctuated both in latitude and longitude during glacial and interglacial episodes, as expected from changes in climatic parameters in those areas following periods of warming (LIG and present) and cooling (LGM, Fig. 5). The predicted distribution area of the northern population group showed a maximum extension during LIG (Fig. 5a), whereas the strong reduction during LGM (Fig. 5b) was maintained until present times (Fig. 5c). Contrastingly, the predicted distribution area of the southern group showed a progressive increase in extension from LIG (Fig. 5a) to present times (Fig. 5c). The potential overlap of distribution areas between the two groups reached its maximum extension during present times (Fig. 5c). However, it was restricted to the eastern range of the present distribution of the species, suggesting that lineage migration and admixture, as denoted by the occurrence of the common plastid haplotype IV, is likely to have occurred only along the eastern boundaries of both distribution areas (Fig. 5). Predicted environmental niche models of the northern and southern groups of *D. humilis* were also consistent with a contraction towards the central Chilean depression during the LGM (Fig. 5b), preceded by broader eastern and western distributions of the potential areas of the northern group during the LIG (Fig. 5a).

Bayesian analyses of nuclear microsatellite variation and of plastid haplotypes also revealed genetic signatures of postglacial population expansion within the northern and southern groups of *D. humilis* subsp. *humilis* (Figs. 1b, 4a). Concerning the northern group, western populations predominantly showed a microsatellite genetic membership to cluster 1, whereas eastern populations showed a predominant genetic membership to cluster 2 (Fig. 1a). This was paralleled to a lesser extent by the slower-evolving cpDNA data (Fig. 4a), where three northwestern populations showed three cpDNA haplotypes that were not represented in eastern populations (Fig. 4a). This would indicate a further isolation of the northwestern populations, which, unlike the eastern ones, did not admix with the southern ones.

Contrastingly, a North-to-South expansion was detected in the southern group of *D. humilis* subsp. *humilis*, supported by a gradual North-to-South decrease in microsatellite genetic membership to cluster 4, and an increase in membership to cluster 3 (Fig. 1b), agreeing with the predicted postglacial southwards expansion (Fig. 5c). Exclusive cpDNA haplotypes were scattered among populations in this range, and were all derived directly from the most common haplotype (Fig. 4b), suggesting recent divergence and dispersal [57−58].

Conclusions

Our study represents a significant contribution to the understanding of the phylogeography of lowland plants from the central Mediterranean area of Chile. Genetic and ENM analyses suggest that *D. humilis* subsp. *polyanthes* diverged from southern populations of *D. humilis* subsp. *humilis* due to local niche adaptation to coastal areas.

The study has also revealed a strong phylogeographical structure within *D. humilis* subsp. *humilis* and identified two highly differentiated genetic groups with distributions that match present latitudinal environmental heterogeneity in the area [20,59]. The genetic differentiation of these two groups could have been triggered by a coupled effect of adaptation to divergent ecological parameters of higher and lower aridity in the northern and southern geographical areas, respectively [60−62], enhanced by the permanent geographical barrier of the Maipo river basin between the two areas [14].

Supporting Information

Appendix S1   Environmental niche model analysis of *Dioscorea humilis*.

Appendix S2   Microsatellite allele frequencies, Bayesian estimation of genetic clusters and plastid haplotype frequencies in populations of *Dioscorea humilis*.

Appendix S3   Comparison of microsatellite genetic diversity in yam species.

Acknowledgments

We thank E. Pérez-Collazos for his help during fieldwork, D. López for her assistance in ENM analyses and Emily Lemonds for revising the English text. Environmental data was retrieved from The international modelling groups and the Laboratoire des Sciences du Climat et de l'Environnement (LSCE). The PMIP 2 Data Archive is supported by CEA, CNRS and the Programme National d'Etude de la Dynamique du Climat (PNEDC).

Author Contributions

Conceived and designed the experiments: PC JGSM. Performed the experiments: JV. Analyzed the data: JV JGSM. Contributed reagents/ materials/analysis tools: PC JGSM. Contributed to the writing of the manuscript: JV JGSM PC.

References

1. Schaal BA, Hayworth DA, Olsen KM, Rauscher JT, Smith WA (1998) Phylogeographic studies in plants: problems and prospects. Molecular Ecology, 7, 465–474.
2. Markgraf V, McGlone M, Hope G (1995) Neogene paleoenvironmental and paleoclimatic change in southern temperate ecosystems – a southern perspective. Trends in Ecology and Evolution, 10, 143–147.
3. Villagrán C (1991) Historia de los bosques templados del sur de Chile durante el tardiglacial y postglacial. Revista Chilena de Historia Natural, 64, 447–460.
4. Villagrán C (2001) Un modelo de la historia de la vegetación de la Cordillera de La Costa de Chile central-sur: la hipótesis glacial de Darwin. Revista Chilena de Historia Natural, 74, 783–803.
5. Clapperton C (1994) The Quaternary glaciation of Chile: a review. Revista Chilena de Historia Natural, 67, 369–383.

6. McCulloch RD, Bentley MJ, Purves RS, Hulton NRJ, Sugden DE, et al. (2000) Climatic inferences from glacial and palaeoecological evidence at the last glacial termination, southern South America. Journal of Quaternary Science, 15, 409–417.

7. Armesto J, Arroyo MTK, Hinojosa LF (2007) *The mediterranean Environment of Central Chile*. In: The Physical Geography of South America. Eds Veblen TT, Ypung KR, Orme AR, 184–199. Oxford University Press, New York.

8. Bull-Hereñu K, Martínez EA, Squeo FA (2005) Structure and genetic diversity in *Colliguaja odorifera* Mol. (Euphorbiaceae), a shrub subjected to Pleisto-Holocenic natural perturbations in a Mediterranean South American region. Journal of Biogeography, 32, 1129–1138.

9. Vásquez M, Torres-Pérez F, Lamborot M (2007) Genetic variation within and between four chromosomal races of *Liolaemus monticola* in Chile. Herpetological Journal, 17, 149–160.

10. Caviedes CN, Paskoff R (1975) Quaternary glaciations in the Andes of North-Central Chile. Journal of Glaciology, 14, 155–170.

11. Santana-Aguilar M (1973) La glaciation Quaternaire dans les Andes de Rancagua (Chili central). Bulletin of the Association of Geography France, 406/407, 473–483.

12. Graf K (1994) Discussion of palynological methods and paleoclimatical interpretations in northern Chile and the whole Andes. Revista Chilena de Historia Natural, 67, 405–415.

13. Lamy F, Hebbeln D, Wefer G (1999) High resolution marine record of climatic change in mid-latitude Chile during the last 28,000 years based on terrigenous sediment parameters. Quaternary Research, 51, 83–93.

14. Sallaberry-Pincheira N, Garin CF, González-Acuña D, Sallaberry MA, Vianna JA (2011) Genetic divergence of Chilean long-tailed snake (*Philodryas chamissonis*) across latitudes: conservation threats for different lineages. Diversity and Distributions, 17, 152–162.

15. Lamborot M, Eaton LC (1997) The Maipo River as a geographical barrier to *Liolaemus monticola* (Torpiduridae) in the mountain ranges of central Chile. Journal of Zoological Systematics and Evolutionary Research, 35, 105–111.

16. Viruel J, Segarra-Moragues JG, Pérez-Collazos E, Villar L, Catalán P (2010a) Systematic revision of the *Epipetrum* group of *Dioscorea* (Dioscoreaceae) endemic to Chile. Systematic Botany, 35, 40–63.

17. Donato M (2006) Historical biogeography of the family Tristiridae (Orthoptera: Acridomorpha) applying dispersal-vicariance analyses. Journal of Arid Environments, 66, 421–434.

18. Viruel J, Segarra-Moragues JG, Pérez-Collazos E, Villar L, Catalán P (2008) The diploid nature of the Chilean *Epipetrum* and a new base number in the Dioscoreaceae. New Zealand Journal of Botany, 46, 327–339.

19. Amigo J, Ramírez C (1998) A bioclimatic classification of Chile: woodland communities in the temperate zone. Plant Ecology, 136, 9–26.

20. Castillo A, Dorado G, Feuillet C, Sourdille P, Hernández P (2010) Genetic structure and ecogeographical adaptation in wild barley (*Hordeum chilense* Roemer and Schultes) as revealed by microsatellite markers. BMC Plant Biology, 10, 266–278.

21. Viruel J, Catalán P, Segarra-Moragues JG (2012) Disrupted phylogeographical SSR and cpDNA patterns indicate a vicariance rather than long-distance dispersal origin for the disjunct distribution of the Chilean endemic *Dioscorea biloba* (Dioscoreaceae) around the Atacama Desert. Journal of Biogeography, 39, 1073–1085.

22. Viruel J, Catalán P, Segarra-Moragues JG (2010b) New microsatellite loci in the dwarf yams *Dioscorea* gr. *Epipetrum* (Dioscoreaceae). American Journal of Botany, 97, e121-e12.

23. Taberlet P, Gielly L, Pautou G, Bouvet J (1991) Universal primers for amplification of three non-coding regions of chloroplast DNA. Plant Molecular Biology, 17, 1105–1109.

24. Belkhir K, Borsa P, Chikhi L, Raufaste N, Bonhomme F (2004) *GENETIX 4.05, logiciel sous Windows TM pour la génétique des populations*. Laboratoire Génome, Populations, Interactions, CNRS UMR 5000, Université de Montpellier II, Montpellier (France).

25. Rousset F (2008) GENEPOP'007: a complete re-implementation of the GENEPOP software for Windows and Linux. Molecular Ecology Resources, 8, 103–106.

26. Goudet J (2001) *FSTAT v. 2.9.3.2, a program to estimate and test gene diversities and fixation indices*. Available from: http://www2.unil.ch/popgen/softwares/fstat.htm.

27. Excoffier L, Laval G, Schneider S (2005) Arlequin ver. 3.0: An integrated software package for population genetics data analysis. Evolutionary Bioinformatics Online, 1, 47–50.

28. Slatkin M (1995) A measure of population subdivision based on microsatellite allele frequencies. Genetics, 139, 457–462.

29. Rohlf FJ (2002) *NtSYSpc, Numerical Taxonomy and Multivariate analysis System. Version 2.11a, User guide*. Exeter software, New York, 38 p.

30. Nei M, Tajima F, Tateno Y (1983) Accuracy of estimated phylogenetic trees from molecular data. Journal of Molecular Evolution, 19, 153–170.

31. Langella O (2000) *Populations (Logiciel de génétique des populations)*. CNRS, France.

32. Pritchard JK (2002) *Documentation for STRUCTURE software: version 2*. Available at: http://pritch.bsd.uchicago.edu/software/readme_2_1/readme.html.

33. Evanno G, Regnaut S, Goudet J (2005) Detecting the number of clusters of individuals using the software STRUCTURE: a simulation study. Molecular Ecology, 14, 2611–2620.

34. Tajima F (1983) Evolutionary relationship of DNA sequences in finite population. Genetics, 105, 437–460.

35. Librado P, Rozas J (2009) DnaSP v5: A software for comprehensive analysis of DNA polymorphism data. Bioinformatics, 25, 1451–1452.

36. Clement M, Posada D, Crandall KA (2000) TCS: a computer program to estimate gene genealogies. Molecular Ecology, 9, 1657–1659.

37. Hijmans RJ, Cameron SE, Parra JL, Jones PG, Jarvis A (2005) Very high resolution interpolated climate surfaces for global land areas. International Journal of Climatology, 25, 1965–1978.

38. Phillips SJ, Anderson RP, Schapire RE (2006) Maximum entropy modeling of species geographic distributions. Ecological Modelling, 190, 231–259.

39. Phillips SJ, Dudík M, Elith J, Graham CH, Lehmann A, et al. (2009) Sample selection bias and presence - only distribution models: implications for background and pseudoabsence data. Ecological Applications, 19, 181–197.

40. Petal RG, Raxworthy CJ, Nakamura M, Peterson AT (2007) Predicting species distributions from small numbers of occurrence records: a test case using cryptic geckos in Madagascar. Journal of Biogeography, 34, 102–117.

41. Pease KM, Freedman AH, Pollinger JP, McCormack JE, Buermann W, et al. (2009) Landscape genetics of California mule deer (*Odocoileus hemoinus*): the roles of ecological and historical factors in generating differentiation. Molecular Ecology, 18, 1848–1862.

42. Braconnot P, Otto-Bliesner B, Harrison S, Joussaume S, Peterschmitt J-Y, et al. (2007) Results of PMIP2 coupled simulations of the Mid-Holocene and Last Glacial Maximum - Part 1: experiments and large-scale features. Climate of the Past, 3, 261–277.

43. Otto-Bliesner B-L, Marshall SH, Overpeck JT, Miller GH, Hu A, et al. (2006) Simulating Arctic Climate Warmth and Icefield Retreat in the Last Interglaciation. Science, 24, 1751–1753.

44. Collins WD, Bitz CM, Blackmon ML, Bonan GB, Bretherton CS, et al. (2006) The community climate system model version 3 (CCSM3). Journal of Climate, 19, 2122–2143.

45. Hasumi H, Emori S (2004) *K-1 coupled GMC (MIROC) description* (ed. by H. Hasumi and S. Emori). K-1 Technical Report No. 1, September, 2004. Center for Climate System Research, University of Tokyo, Tokyo.

46. Waltari E, Hijmans RJ, Peterson AT, Nyari AS, Perkins SL, et al. (2007) Locating Pleistocene refugia: comparing phylogeographic and ecological niche model predictions. PLoS ONE, 2, e563.

47. Hammer Ø, Harper DAT, Ryan PD (2001) PAST: Paleontological statistics software package for education and data analysis. Palaeontologia Electronica, 4, 1–9.

48. Zattara EE, Premoli AC (2005) Genetic structuring in Andean landlocked populations of *Galaxias maculatus*: effects of biogeographic history. Journal of Biogeography, 32, 5–14.

49. Dillon MO, Tu T, Xie L, Quipuscoa-Silvestre V, Wen J (2009) Biogeographic diversification in *Nolana* (Solanaceae), a ubiquitous member of the Atacama and Peruvian Deserts along the western coast of South America. Journal of Systematics and Evolution, 47, 457–476.

50. Schmidt-Jabaily R, Sytsma KJ (2010) Phylogenetics of *Puya* (Bromeliaceae): placement, major lineages, and evolution of Chilean species. American Journal of Botany, 97, 337–356.

51. Hamrick JL, Godt MJW (1996) Effects of life history traits on genetic diversity in plant species. Philosophical Transactions of the Royal Society B: Biological Sciences, 351, 1291–1298.

52. Segarra-Moragues JG, Palop-Esteban M, González-Candelas F, Catalán P (2005) On the verge of extinction: genetics of the Critically Endangered Iberian plant species, *Borderea chouardii* (Dioscoreaceae) and implications for conservation management. Molecular Ecology, 14, 969–982.

53. Segarra-Moragues JG, Palop-Esteban M, González-Candelas F, Catalán P (2007) Nunatak survival *vs.* tabula rasa in the Central Pyrenees: a study on the endemic plant species *Borderea pyrenaica* (Dioscoreaceae). Journal of Biogeography, 34, 1893–1906.

54. Gitzendanner MA, Soltis PM (2000) Patterns of genetic variation in rare and widespread plant congeners. American Journal of Botany, 87, 783–792.

55. Zhang Z-Y, Zheng X-M, Ge S (2007) Population genetic structure of *Vitex negundo* (Verbenaceae) in Three-Gorge Area of the Yangtze River: The riverine barrier to seed dispersal in plants. Biochemical Systematics and Ecology, 35, 506–516.

56. Li Y, Fahima T, Korol AB, Peng J, Röder MS, et al. (2000) Microsatellite diversity correlated with ecological-edaphic and genetic factors in three microsites of wild emmer wheat in North Israel. Molecular Biology and Evolution, 17, 851–862.

57. Canestrelli D, Nascetti G (2008) Phylogeography of the pool frog *Rana* (*Pelophylax*) *lessonae* in the Italian peninsula and Sicily: multiple refugia, glacial expansions and nuclear–mitochondrial discordance. Journal of Biogeography, 35, 1923–1936.

58. Grill A, Amori G, Aloise G, Lisi I, Tosi G, et al. (2009) Molecular phylogeography of European *Sciurus vulgaris*: refuge within refugia? Molecular Ecology, 18, 2687–2699.

59. Pezoa LS (2003) *Recopilación y análisis de la variación de las temperaturas (período 1965–2001) y las precipitaciones (período 1931–2001) a partir de la información de estaciones meteorológicas de Chile entre los 33° y 53° de latitud*

sur. Tesis de grado Ingeniería Forestal, Facultad de Ciencias Forestales, Universidad Austral de Chile, Valdivia, Chile. 99 p.

60. Van Hulsen C (1967) Klimagliederung in Chile auf der Basis von Häufigkeits-verteilungen der Niederschlagssummen. Freiburger Geographische, 4, 1–109.

61. Veit H (1996) Southern Westerlies during the Holocene deduced from geomorphological and pedological studies in the Norte Chico, Northern Chile (27–33°S). Palaeogeography, Palaeoclimatology, Palaeoecology, 123, 107–119.

62. Bonilla CA, Vidal KL (2011) Rainfall erosivity in Central Chile. Journal of Hydrology, 410, 126–133.

3

First Record of Eocene Bony Fishes and Crocodyliforms from Canada's Western Arctic

Jaelyn J. Eberle[1]*, Michael D. Gottfried[2], J. Howard Hutchison[3], Christopher A. Brochu[4]

1 University of Colorado Museum of Natural History and Department of Geological Sciences, University of Colorado at Boulder, Boulder, Colorado, United States of America, 2 Department of Geological Sciences and Museum, Michigan State University, East Lansing, Michigan, United States of America, 3 University of California Museum of Paleontology, Berkeley, California, United States of America, 4 Department of Earth and Environmental Sciences, University of Iowa, Iowa City, Iowa, United States of America

Abstract

Background: Discovery of Eocene non-marine vertebrates, including crocodylians, turtles, bony fishes, and mammals in Canada's High Arctic was a critical paleontological contribution of the last century because it indicated that this region of the Arctic had been mild, temperate, and ice-free during the early – middle Eocene (~53–50 Ma), despite being well above the Arctic Circle. To date, these discoveries have been restricted to Canada's easternmost Arctic – Ellesmere and Axel Heiberg Islands (Nunavut). Although temporally correlative strata crop out over 1,000 km west, on Canada's westernmost Arctic Island – Banks Island, Northwest Territories – they have been interpreted as predominantly marine. We document the first Eocene bony fish and crocodyliform fossils from Banks Island.

Principal Findings: We describe fossils of bony fishes, including lepisosteid (*Atractosteus*), esocid (pike), and amiid, and a crocodyliform, from lower – middle Eocene strata of the Cyclic Member, Eureka Sound Formation within Aulavik National Park (~76°N. paleolat.). Palynology suggests the sediments are late early to middle Eocene in age, and likely spanned the Early Eocene Climatic Optimum (EECO).

Conclusions/Significance: These fossils extend the geographic range of Eocene Arctic lepisosteids, esocids, amiids, and crocodyliforms west by approximately 40° of longitude or ~1100 km. The low diversity bony fish fauna, at least at the family level, is essentially identical on Ellesmere and Banks Islands, suggesting a pan-High Arctic bony fish fauna of relatively basal groups around the margin of the Eocene Arctic Ocean. From a paleoclimatic perspective, presence of a crocodyliform, gar and amiid fishes on northern Banks provides further evidence that mild, year-round temperatures extended across the Canadian Arctic during early – middle Eocene time. Additionally, the Banks Island crocodyliform is consistent with the phylogenetic hypothesis of a Paleogene divergence time between the two extant alligatorid lineages *Alligator mississippiensis* and *A. sinensis*, and high-latitude dispersal across Beringia.

Editor: Peter Dodson, University of Pennsylvania, United States of America

Funding: This research was supported financially by the National Science Foundation (ARC 0804627 to JE; DEB 1257786 to CB). MDG thanks Michigan State University for travel support related to participating in the 2012 field component of this research. The funders had no role in study design, data collection and analysis, decision to publish, or preparation of the manuscript. Publication of this article was funded by the University of Colorado Boulder Libraries Open Access Fund.

Competing Interests: The authors have declared that no competing interests exist.

* E-mail: Jaelyn.Eberle@Colorado.edu

Introduction

Discovery of Eocene vertebrates, including alligators, turtles, fishes, and mammals, on Ellesmere and Axel Heiberg Islands in Canada's eastern High Arctic [1], [2], [3] (Figure 1) was a critical paleontological contribution of the last century, as it indicated that this region of the Arctic had been mild, temperate, and ice-free during the early – middle Eocene (~53–50 Ma), despite its location at ~76–77°N. paleolatitude [4]. Eocene vertebrate-bearing strata of the Eureka Sound Group crop out on islands across the Canadian Arctic; however, to date, discoveries of Eocene non-marine vertebrates are limited to Ellesmere and Axel Heiberg Islands. On Banks Island – Canada's westernmost Arctic Island – Eureka Sound strata are exposed extensively across northwestern parts of the island [5], but the paleoenvironment is interpreted as predominantly shallow marine, based upon abundant shark teeth [6], the trace fossil *Ophiomorpha*, marine microfossils, and the sedimentology [5]. Here, we describe the first Eocene non-marine vertebrates from northern Banks Island. Rare fossils of bony fishes, including the lepisosteid (gar) *Atractosteus*, an esocid (pike), and an amiid (bowfin), as well as a single vertebra of a crocodyliform were discovered in lower – middle Eocene strata of the Cyclic Member, Eureka Sound Formation near Eames River within Aulavik National Park (~76°N. paleolatitude). These fossils extend the known geographic ranges of Eocene Arctic lepisosteids, esocids, amiids, and crocodyliforms west by ~40° of longitude or ~1100 km [7]. Additionally, they provide a glimpse into the early – middle Eocene vertebrate fauna from Canada's western Arctic, hitherto known only from isolated sharks' teeth [6]. The shark fauna is being described elsewhere.

To date, the majority of paleoclimatic data for the Eocene Arctic has come from eastern Arctic localities [8] and a single

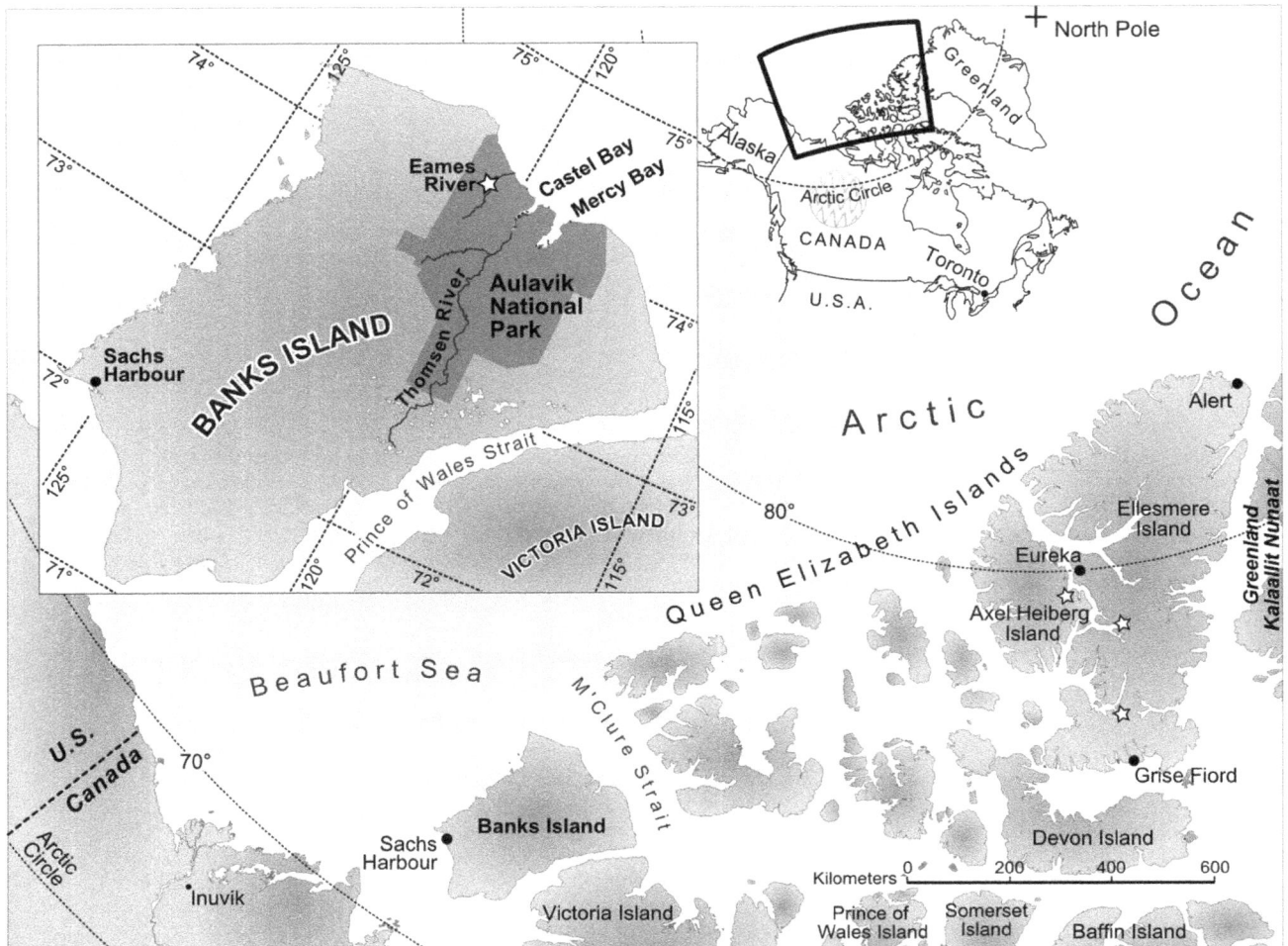

Figure 1. Map of Arctic Canada showing location of Eocene crocodyliform locality on northern Banks Island, NWT (inset). Stars on Ellesmere and Axel Heiberg Islands mark localities from which Eocene crocodylian and bony fish fossils were reported prior to this report [3]. Artwork by L. McConnaughey.

locality on Lomonosov Ridge in the central Arctic Ocean [9], [10]. By analogy with living crocodylians [11], the occurrence of a crocodyliform fossil on northern Banks implies that a regionally mild, temperate climate extended across the Canadian Arctic during early – middle Eocene time. Further paleoenvironmental, biogeographic, and phylogenetic implications are discussed below.

Geologic Setting and Age

The fossils were recovered alongside hundreds of shark teeth from CMN localities BKS04-16 and BKS04-19 in Eocene strata near Eames River, within the boundaries of Aulavik National Park on northern Banks Island, NWT, Canada (~N 74° 10'; W 120° 45–46'; Fig. 1). Because the locality is within the boundaries of a national park, we are not able to provide more precise coordinates in this paper. Qualified researchers should contact the Canadian Museum of Nature (CMN) in Ottawa, ON, Canada, to request the exact coordinates.

The Eocene bony fish and crocodyliform localities on northern Banks occur in strata initially mapped as the Cyclic Member of the Eureka Sound Formation [5], [12]. Subsequently, these strata were re-assigned to the Margaret Formation of the Eureka Sound

Group, and inferred to be correlative with Eocene, terrestrial vertebrate-bearing strata of the Margaret Formation on Ellesmere and Axel Heiberg Islands over 1,000 km away in Canada's eastern High Arctic [13], [14]. However, given the enormous distance from the type section of the Margaret Formation (at Strand Fiord on southern Axel Heiberg Island) [13], the variable lithology of the Eureka Sound Group across the Arctic, and deposition in multiple isolated basins separated by upwarps [13], the Eocene vertebrate-bearing sediments in Banks Basin on northern Banks Island are here left as the Cyclic Member. There are environmental differences that are consistent with taking this approach, namely that the Margaret Formation in the eastern Arctic is predominantly non-marine, producing palynomorphs and a terrestrial vertebrate fauna, while the Cyclic Member on northern Banks Island comprises coarsening-upward cycles of shale, silt, unconsolidated sand, paleosol, and lignitic coal that are interpreted as a deltaic sequence in a marginal marine setting [5]. At multiple localities, the Cyclic Member preserves abundant shark teeth, bivalves, and the trace fossil *Ophiomorpha*, interpreted as the burrow of a thalassinidean shrimp and generally indicative of shallow-water, moderately high energy, coastal marine environments [15].

Marine microfossils (foraminiferans and radiolarians) also are documented from the Cyclic Member [5].

The bony fish and crocodyliform fossils were recovered as float on unconsolidated sands in the Cyclic Member, a facies interpreted as distributary mouth bar deposits in the delta-front area [5]. Dry-screening of localities led to the recovery of additional smaller shark teeth, but did not turn up additional bony fish or croc material.

The Eocene age for the fossil localities near Eames River is based upon pollen samples initially analyzed by Hopkins [16,17], and reported by Miall [5]. Recent re-analysis of four pollen samples near Eames River (GSC samples C-26411, C-30610, C-30645, and C-30646) by Sweet [18] suggests that the localities are late early to middle Eocene in age and likely spanned the Early Eocene Climatic Optimum (EECO), based largely on overall species richness as well as abundance of *Caryapollenites* spp., *Ericipites, Intratriporopollenites* (Tilia), *Nyssapollenites* sp., and *Quercoidites* (oak) pollen. Presence of *Pistillipollenites* in three of the samples (absent from the coal of sample C-30646) suggests a probable minimum age of middle Eocene for the samples, while absence of *Aquilapollenites tumanganicus* Bolotnikova and closely allied species, infrequent occurrences of *Momipites* spp., and the richness of the angiosperm component of the assemblages precludes an earliest Eocene age for the samples [18].

Materials and Methods

The fossils were collected on northern Banks Island in 2004, 2010, and 2012, and permits to conduct paleontological field research in Aulavik National Park were provided by Parks Canada, Western Arctic Field Unit. All necessary permits were obtained for the described study, which complied with all relevant regulations. The fossils from Banks Island are curated at the Canadian Museum of Nature (CMN) in Ottawa, ON, and are on loan to the University of Colorado Museum of Natural History (UCM) for study. Identifications were made based upon comparison with specimens held in collections at the UCM, the University of California Museum of Paleontology (UCMP) in Berkeley, CA, and with published descriptions and images. Terminology used to describe the crocodyliform vertebra follows Romer [19], and for gar and amiid specimens follows Grande [20] and Grande and Bemis [21], respectively. Although additional bony fish material (isolated teeth and centra) was recovered from Banks Island, it could not be referred to familial level, and therefore we did not include these non-diagnostic specimens.

Systematic Paleontology

Actinopterygii Cope, 1887
Ginglymodi Cope, 1872
Lepisosteiformes Hay, 1929
Lepisosteidae Cuvier, 1825
Atractosteus Rafinesque, 1820

Referred Specimen. CMNFV 56070, lateral line scale, Fig. 2a, b.

Locality and Horizon. CMN Loc. BKS04-16, N 74° 10'; W 120° 45'; Eames River, Aulavik National Park, northern Banks Island, NWT; Cyclic Member, Eureka Sound Formation (early – middle Eocene).

Description. CMNFV 56070 is a complete lateral line scale and readily identifiable as a gar (Fig. 2a, b). The specimen measures 18×9 mm, and has the characteristic elongate rhombic (diamond) shape of a gar scale, with a thickened bony base and an outer surface mostly covered with ganoin that bears several dozen regularly-spaced shallow perforations. The edges of the ganoin

bear a relatively fine ornament of slightly serrated scalloping. The medial (inner) surface exhibits a well-formed canal, partially encased in the bony base of the scale, for carrying the mechanosensory lateral line through the scale.

The scale has a long, narrow, anteriorly-projecting anterodorsal process, which is a part of the scale that is overlapped by its neighbor, and therefore contributes to the rigidity of the interlocking scale cover in gars. The dorsal peg is, however, little more than a very low bump along the dorsal edge of the scale, just posterior to the base of the anterodrosal process. Grande [20] hypothesized that the absence of a prominent dorsal peg was a feature that may be diagnostic for gars in the genus *Atractosteus*, as opposed to its sister-genus *Lepisosteus* in which the scales typically have more prominent dorsal pegs. For this reason, as well as the close similarity in overall shape and proportions of the Banks Island scale to those of *Atractosteus* spp. illustrated by Grande [20], we assign the Banks Island scale to *Atractosteus*.

?Amiiformes Hay, 1929
?Amiidae Bonaparte, 1838

Referred Specimen. CMNFV 56069, vertebral centrum, Fig. 2c.

Locality and Horizon. CMN Loc. BKS04-19, N 74° 10'; W 120° 46'; Eames River, Aulavik National Park, northern Banks Island, NWT; Cyclic Member, Eureka Sound Formation (early – middle Eocene).

Description. CMNFV 56069 is a single nearly complete centrum, most likely from the mid-abdominal region and past the middle of the body, based on the relatively closely-spaced and ventrally directed parapophyses (Fig. 2c). The centrum is amphicoelous and slightly higher dorsoventrally (9.5–10 mm) than it is wide (8.5–9.0 mm), and varies from 4 to 5 mm in length, being slightly longer along its dorsal edge. The articular surfaces are rather shallowly concave, and the centrum overall has a simple, almost shark centrum-like appearance.

The bases of the neural arches are closely spaced and appear to be fused to the dorsal margin of the centrum. The larger, more prominent parapophyses (= the lateral components of the basiventral elements in the abdominal region, sensu Grande and Bemis [21]) are slightly more widely separated and project nearly straight ventrally, with only a hint of lateral divergence, which, as mentioned above, implies that this centrum is from a relatively posterior but still abdominal position in the vertebral column, and anterior to the caudal or ural centra.

While we cannot definitively rule out other possibilities, we note that this centrum has several features consistent with our determination that it is likely from an amiid. These include the overall proportions of the centrum (a relatively short length vs. diameter), the fact that it is perichordally ossified and amphicoelous with shallowly concave articular surfaces, the comparable shape and position of the parapophyses relative to those on posterior abdominal centra of amiids [21], the slightly pitted but otherwise relatively smooth external surface of the centrum, and the simple construction of the centrum without complexly elaborated features that are typical of centra from 'higher' actinopterygians. Finally, the centrum is bony, not calcified cartilage as it would be if it were from a shark, and amphicoelous, not opisthocoelous as it would be if it derived from a gar.

Teleostei sensu Patterson and Rosen 1977
Esocidae Cuvier, 1817

Referred Specimen. CMNFV 56071, 11 isolated teleost scales, one shown in Fig. 2d. CMN Loc. BKS04-19, N 74° 10'; W 120° 46'; Eames River, Aulavik National Park, northern Banks

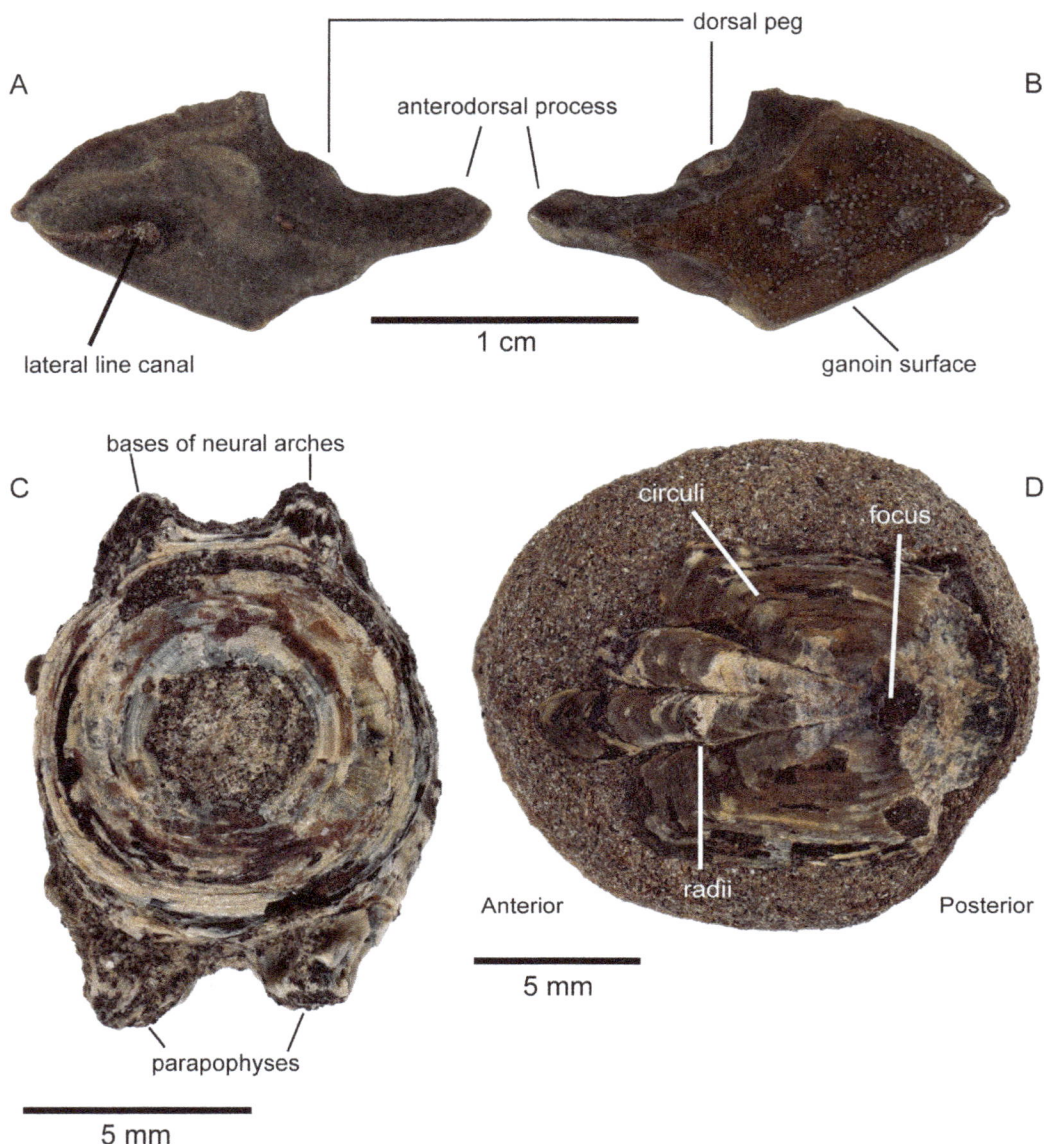

Figure 2. Fossils of Eocene bony fishes from northern Banks Island, NWT. CMNFV 56070, lateral line scale of *Atractosteus* from CMN Loc. BKS04-16, in medial (A) and lateral (B) views. (C) CMNFV 56069, vertebral centrum of ?Amiid. (D) CMNFV 56071, Esocid scale. C and D are from CMN Loc. BKS04-19.

Island, NWT; Cyclic Member, Eureka Sound Formation (early – middle Eocene).

Description. Eleven isolated teleost scales were recovered from 2012 fieldwork on northern Banks Island, all preserved in small siderite concretions. The clearest and best-preserved of these is figured here (Fig. 2d) and represents Esocidae (the family that includes pikes, pickerels, and muskellunges). Scale morphological terminology follows Patterson et al. [22]. The scale is cycloid and slightly longer craniocaudally than high dorsoventrally, measuring 17 mm by 14–15 mm. The lateral surface is exposed on the siderite concretion, and exhibits very fine concentric circuli that follow the contour of the outer margin of the scale. There is variation in the appearance of the circuli, suggesting that the more distinctly visible bands may be annuli that reflect seasonal differentiation in growth and growth checks. The anterior mid-region of the scale has a demarcated anterior field portion that

forms an elongate cone-shaped area set off by two distinct angled radii that reach the anterior margin of the scale; this field extends out from the center of formation (focus) of the scale, broadening towards the anterior border of the scale, and it is subdivided into two narrower regions within the anterior field by another, more medially positioned radii. The distinctive appearance of this scale very closely matches scales of extant esocids, including the Northern Pike *Esox lucius* [23].

Crocodyliformes Hay, 1930

Eusuchia Huxley, 1875

Referred Specimen. CMNFV 56059, incomplete vertebra, Fig. 3.

Locality and Horizon. CMN Loc. BKS04-19, N 74° 10′; W 120° 46′; Eames River, Aulavik National Park, northern Banks Island, NWT; Cyclic Member, Eureka Sound Formation (early – middle Eocene).

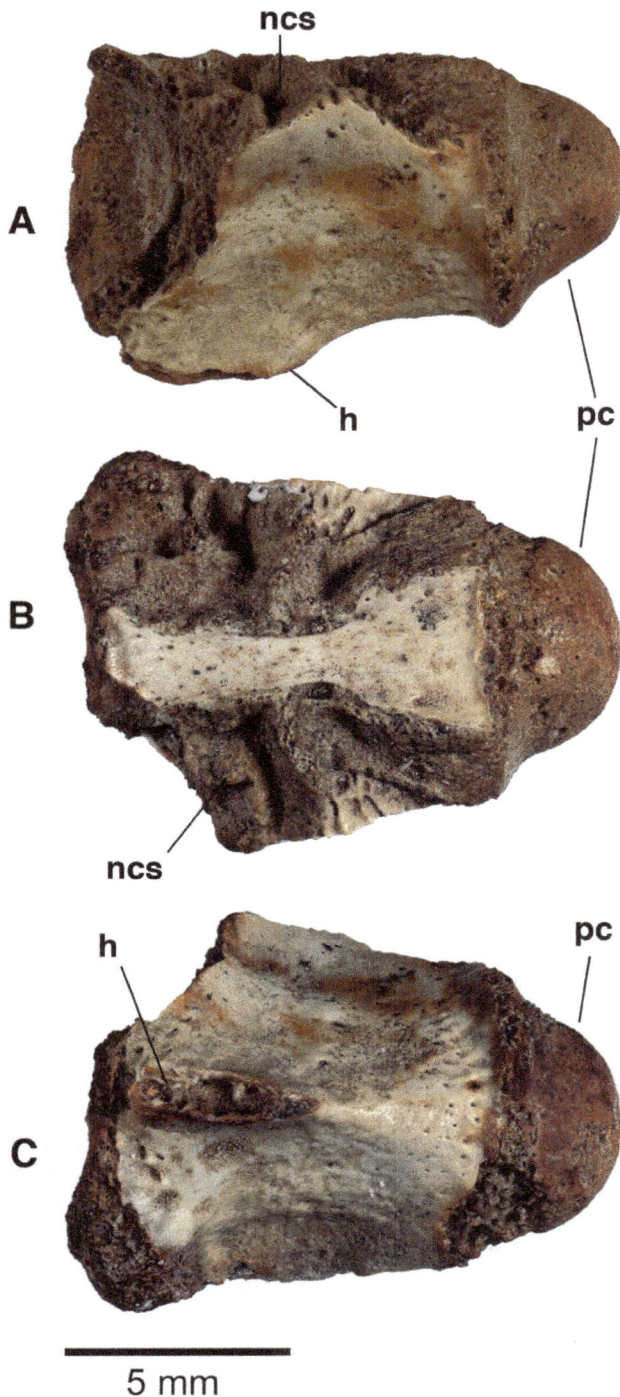

Figure 3. CMNFV 56059, vertebral centrum of an Eocene crocodyliform from CMN Loc. BKS04-19 on northern Banks Island, NWT. (A) Left lateral view; (B) dorsal view; (C) ventral view. h, hypapophysis; ncs, neurocentral sutural surface; pc, posterior cotyle. Scale bar equals 5 mm.

Description. CMNFV 56059 is a small procoelous vertebral centrum whose neural arch is no longer preserved (Length = 12.1 mm; Anterior width = 9.14 mm). The centrum bears a short hypapophysis on its anteroventral surface. The anterolateral surfaces are damaged on both sides, but the centrum

flares anteriorly in dorsal and ventral view (Fig. 3), which indicates that parapophyses were present on the centrum adjacent to the neurocentral suture.

Because of the shortness of the hypapophysis and inferred presence of a parapophysis on the centrum, we interpret CMNFV 56059 as coming from the posterior cervical part of the vertebral column. Although not preserved, the parapophysis was adjacent, or nearly so, to the neurocentral suture. The crocodyliform parapophysis is widely separated from the neural arch in the anteriormost cervical vertebrae except for the atlas, and it gradually adopts a more dorsal position on the centrum surface until, as one approaches the cervicodorsal transition, it straddles the neurocentral suture. More anteriorly located cervical vertebrae also have longer hypapophyses.

The size of CMNFV 56059 indicates a small animal, and the vertebra compares in size with anterior dorsals on a 2-ft long *A. mississippiensis* skeleton (UCM PTC-47). In extant crocodylians, closure of the neurocentral sutures in the vertebral column follows a caudal to cranial sequence during ontogeny, with the sutures of most caudal vertebrae closed at hatching while closure of the cervicals occurs near the end of ontogeny [24]. That the neurocentral suture surface is exposed on CMNFV 56059 (Fig. 3b) indicates that the neurocentral suture was not closed. The sutural surface, however, is very rugose. Together, these indicate an animal that was immature, but possibly approaching maturity.

Procoelous vertebrae occur in several crocodyliform lineages. In most, including some putative basal eusuchians, the anterior socket is shallow and the posterior cotyle is not very prominent [25], [26], [27], [28], [29], [30]. A deep socket and prominent hemispherical cotyle is most characteristic of Crocodylia and its closest relatives [31]. Because all known non-marine crocodyliforms from the Paleogene of North America are crocodylians, we expect more complete material from this locality to put CMNFV 56059 within Crocodylia.

Although posterior cervical and anterior crocodylian dorsal vertebrae are typically not diagnostic to family, it seems probable that CMNFV 56059 belonged to an alligatorid. Most Paleogene alligatorids were small animals between 2 and 3 m in length. Alligatorids are also more cold-tolerant than other crocodylians [32], [33] and more likely to occur at high latitudes. Further, alligatorid fossils referred to *Allognathosuchus* are relatively abundant at early Eocene localities on Ellesmere Island, known from dozens of teeth and osteoderms as well as jaws and an incomplete skull, whereas no other crocodyliform taxa are known from the Arctic [3], [8].

Discussion

Based largely on paleoclimate proxy data from the eastern and central Arctic, early – middle Eocene Arctic climate in this region has been characterized as having warm, wet summers and mild winters [9], [10], [34], [35] [36]. High-resolution carbon isotope analysis across tree rings in mummified wood from Muskox River on northern Banks Island (~50 km south of the Eames River locality) and Stenkul Fiord on southern Ellesmere Island allow the reconstruction of seasonal precipitation patterns in the Eocene Arctic [37]. Incorporation of intra-ring $\delta^{13}C$ values into a model based upon extant evergreen taxa [38] suggest that evergreen trees growing in the Eocene Arctic forests experienced three times more precipitation during summer than winter, a seasonal pattern analogous to today's temperate forests in eastern Asia [37].

The discovery at Eames River of a fossil from an immature crocodyliform, alongside rare turtle shell fragments (too small to be identified to family), indicates that mild temperatures extended across the Arctic during early – middle Eocene time. Specifically, based upon analogy with the geographic range and climatic preferences of living crocodylians [11], the Banks Island crocodyliform infers above-freezing, year-round temperatures. This is further reinforced by the presence of gars, which are associated with mild temperate to warm conditions, and are today restricted to freshwater environments in the southeastern USA, Central America, and Cuba [20].

Given their rarity among thousands of shark teeth, the crocodyliform and turtle fossils probably were washed into the coastal delta from fully freshwater upriver habitats. While this is also a plausible explanation for the paucity of lepisosteids, amiids, and esocids, it should be noted that gars, including the extant species *Atractosteus spatula*, are known to tolerate brackish coastal environments [20], and the extant Northern Pike *Esox lucius* enters brackish coastal wetland environments in the Baltic Sea, where it is anadromous [39]. Therefore, caution is required with bony fishes in this regard because salinity and overall environmental conditions of extant taxa may not always provide an accurate guide to the tolerances of their relatives in the fossil record.

Previous work from the Eocene Canadian Arctic, summarized in Eberle and Greenwood [8] and Estes and Hutchison [3], identified gars (cf. *Lepisosteus* sp.), amiids (*Cyclurus fragosa* and *Amia* cf. *A. pattersoni*), and esocids (cf. *Esox* sp.) on Ellesmere, which means that, at least to the family level, the low-diversity bony fish faunas of Ellesmere and Banks islands, separated by some 40 degrees of longitude and over 1,000 km, are essentially identical. This suggests a pan-High Arctic bony fish fauna of relatively basal groups that extended around the margin of the Eocene Arctic Ocean, inhabiting freshwater and probably low salinity marginal marine settings during the EECO. This family-level biogeographic hypothesis could be further refined if additional fish specimens are recovered that are identifiable to a finer taxonomic level, but at present our suggestion of a low-diversity, pan-High Arctic pattern in bony fishes is consistent with our current understanding of the Banks Island and comparable Arctic Eocene faunas. Along with the Banks and Ellesmere island occurrences, there is a relevant High Arctic fossil amiid record from the Svalbard Archipelago – *Pseudamiatus heintzi* (Lehman, 1951) [40], a partially articulated specimen collected at 78°N on the west coast of Spitsbergen. The specimen was recovered from a similar deltaic depositional environment as the Banks material, so it is uncertain whether the fish inhabited a marine, brackish, or freshwater environment. *Pseudamiatus* was first described as Eocene [40], but more recently the Firkanten Formation from which it derives has been re-interpreted as lower Paleocene (Danian) [41,21].

The survivors, in an Arctic context, of this Eocene High Arctic grouping are the esocids, today represented by the Northern Pike, *Esox lucius*, which is circumglobal in Holarctic freshwater environments as high as 74° N [42]. In contrast, the ranges of

gars and bowfins have retracted since the Eocene into their present-day lower-latitude and environmentally more mild distributions.

Arctic crocodylians could resolve several longstanding biogeographic questions, including the biogeographic origins of Asian alligatorids. There are two living species of *Alligator* – one in North America (the American alligator, *Alligator mississippiensis*) and another in China (the critically endangered Chinese alligator, *A. sinensis*). Because alligators are intolerant of salt water [43], a non-marine dispersal corridor, such as Beringia, probably explains the presence of an otherwise North American clade in eastern Asia [44], [45], [46]. Fossil evidence puts the minimum divergence time between the two lineages in the early Miocene [47], but high-latitude dispersal routes would not have been within crocodylian thermal preferences at that time [11]. Molecular data generally put the *mississippiensis-sinensis* split in the Paleogene [48], [49], [50], when climatic conditions were more favorable for high-latitude alligatorids, and presence of a crocodyliform (and probable alligatorid) in Canada's western Arctic is consistent with this hypothesis.

A longstanding biogeographic question regards the origin of Asian alligatorids. Phylogenetic analyses thus far have rejected a close relationship between Paleogene alligatorids and either living species of *Alligator* [44], [45], [46], [51], but if the *mississippiensis-sinensis* split occurred in the Paleogene and followed a Beringian route to Asia, we would predict the discovery of Arctic fossils within crown *Alligator*. Further field research in the region may uncover more diagnostic material that can resolve this biogeographic question.

Acknowledgments

The fossils were collected on expeditions led by J. Eberle to northern Banks Island in 2004, 2010, and 2012, and permits to conduct paleontological field research in Aulavik National Park were provided by Parks Canada, Western Arctic Field Unit. We thank Parks Canada personnel for the many ways that they supported the fieldwork. Logistical assistance and field gear was provided by the Polar Continental Shelf Program, a division of Natural Resources Canada, and the Aurora Research Institute. We thank A. Sweet (Geological Survey of Canada, Calgary, AB) for his re-analysis of the pollen samples initially collected and reported by A. Miall in the 1970s. K. Shepherd and M. Currie (Canadian Museum of Nature) curated the Banks Island fossils. K. Wolfson and M. Blanchard and provided imaging assistance, and L. McConnaughey created the Arctic map (Fig. 1). We sincerely appreciate the comments and constructive criticism of academic editor P. Dodson and reviewers M. Wilson and A. Fiorillo.

Author Contributions

Conceived and designed the experiments: JE MG JHH CB. Performed the experiments: JE MG JHH CB. Analyzed the data: JE MG JHH CB. Contributed reagents/materials/analysis tools: JE MG JHH CB. Wrote the paper: JE MG JHH CB.

References

1. Dawson MR, West RM, Raemakers P, Hutchison JH (1975) New evidence on the paleobiology of the Paleogene Eureka Sound Formation, Arctic Canada. Arctic 28: 110–116.
2. Dawson MR, McKenna MC, Beard KC, Hutchison JH (1993) An Early Eocene Plagiomenid Mammal from Ellesmere and Axel Heiberg Islands, Arctic Canada. Kaupia 3: 179–192.
3. Estes R, Hutchison JH (1980) Eocene lower vertebrates from Ellesmere Island, Canadian Arctic Archipelago. Palaeogeography, Palaeoclimatology, Palaeoecology 30: 325–347.
4. Irving E, Wynne PJ (1991) The paleolatitude of the Eocene fossil forests of Arctic Canada. In Christie, RL, McMillan, NJ, editors. Tertiary Fossil Forests of the Geodetic Hills, Axel Heiberg Island, Arctic Archipelago: Geological Survey of Canada Bulletin 403. 209–212.

5. Miall AD (1979) Mesozoic and Tertiary geology of Banks Island, Arctic Canada – The history of an unstable craton margin. Geological Survey of Canada Memoir 387. 235 p.
6. Dawson MR, West RM, Hickey LJ (1984) Paleontological evidence relating to the distribution and paleoenvironments of the Eureka Sound and Beaufort formations, northeastern Banks Island, Arctic Canada. Current Research, Part B, Geological Survey of Canada, Paper 84–1B: 359–361.
7. GEOMAR (2011) Ocean Drilling Stratigraphic Network Plate Tectonic Reconstruction Service: http://www.odsn.de/odsn/services/paleomap/paleomap.html (December 2011).
8. Eberle JJ, Greenwood DR (2012) Life at the top of the greenhouse Eocene world – A review of the Eocene flora and vertebrate fauna from Canada's High Arctic. GSA Bulletin 124: 3–23.

9. Sluijs A, Schouten S, Donders TH, Schoon PL, Röhl U, et al. (2009) Warm and wet conditions in the Arctic region during Eocene Thermal Maximum 2. Nature Geoscience 2: 1–4.

10. Moran K, Backman J, Brinkhuis H, Clemens SC, Cronin T, et al. (2006) The Cenozoic palaeoenvironment of the Arctic Ocean. Nature 441: 601–605.

11. Markwick PJ (1998) Fossil crocodilians as indicators of Late Cretaceous and Cenozoic climates: implications for using palaeontological data in reconstructing palaeoclimate. Palaeogeography, Palaeoclimatology, Palaeoecology 137: 205–271.

12. Thorsteinsson R, Tozer ET (1962) Banks, Victoria, and Stefansson Islands, Arctic Archipelago. Geological Survey of Canada Memoir 330. 83 p.

13. Miall AD (1986) The Eureka Sound Group (Upper Cretaceous – Oligocene), Canadian Arctic Islands. Bulletin of Canadian Petroleum Geology 34: 240–270.

14. Miall AD (1991) Late Cretaceous and Tertiary basin development and sedimentation, Arctic Islands. In Trettin HP, editor. Geology of the Innuitian Orogen and Arctic Platform of Canada and Greenland: Geological Survey of Canada, Geology of Canada, no. 3; (also Geological Society of America, The Geology of North America, v. E). 437–458.

15. Frey RW, Howard JD, Pryor WA (1978) Ophiomorpha: Its morphologic, taxonomic, and environmental significance. Palaeogeography, Palaeoclimatology, Palaeoecology 23: 199–229.

16. Hopkins WS (1974) Report on 36 Field Samples from Banks Island, District of Franklin, Northwest Territories, Submitted by A. Miall, 1973 (NTS 88C, F, 98D, E), Geological Survey of Canada, Paleontological Report KT-01-WSH-1974.

17. Hopkins WS (1975) Palynology Report on 44 Field Samples from Banks Island, Submitted by A.D. Miall,1974 (NTS 88B,C,F, 97H,98D,E); Geological Survey of Canada, Paleontological Report KT-10-WSH-1975.

18. Sweet AR (2012) Applied research report on 5 outcrop samples collected by Andrew Miall from northern Banks Island, NWT (NTS Map Sheets 098E/01, 08, 09): Geological Survey of Canada Paleontological report ARS-2012-01.

19. Romer AS (1956) Osteology of the Reptiles. Chicago: University of Chicago Press. 772 p.

20. Grande L (2010) An empirical synthetic pattern study of gars (Lepisosteiformes) and closely related species, based mostly on anatomy. American Society of Ichthyologists and Herpetologists Special Publication 6: 1–871.

21. Grande L, Bemis W (1998) A comprehensive phylogenetic study of amiid fishes (Amiidae) based on comparative skeletal anatomy. An empirical search for interconnected patterns of natural history. Society of Vertebrate Paleontology Memoir 4: 1–690.

22. Patterson RT, Wright C, Chang A, Taylor L, Lyons P, et al. (2002) Atlas of common squamatological (fish scale) material in coastal British Columbia and an assessment of the utility of various scale types in paleofisheries reconstruction. Palaeontologia Electronica 4: 1–88.

23. Gerdaux D, Dufour E (2012) Inferring occurrence of growth checks in pike (Esox lucius) scales by using sequential isotopic analysis of otoliths. Rapid Communications in Mass Spectrometry 26: 785–792.

24. Brochu CA (1996) Closure of neurocentral sutures during crocodilian ontogeny: Implications for maturity assessment in fossil archosaurs. Journal of Vertebrate Paleontology 16: 49–62.

25. Michard JG, Broin F, Brunet M, Hell J (1990) Le plus ancien crocodilien néosuchien spécialisé à charactéres "eusuchiens" du continent africain (Crétacé inférieur, Cameroun). Comptes Rendus de l'Academie des Sciences de Paris Ser. 2, 311: 365–370.

26. Brinkmann W (1992) Die Krokodilier-Fauna aus der Unter-Kreide (Ober-Barremium) von Una (Provinz Cuenca, Spanien). Berliner Geowissenschaftliche Abhandlungen E 5: 1–143.

27. Rogers JV (2003) Pachycheilosuchus trinquei, a new procoelous crocodyliform from the Lower Cretaceous (Albian) Glen Rose Formation of Texas. Journal of Vertebrate Paleontology 23: 128–145.

28. Salisbury SW, Frey E, Martill DM, Buchy MC (2003) A new crocodilian from the Lower Cretaceous Crato Formation of north-eastern Brazil. Palaeontographica Abt. A 270: 3–47.

29. Pol D, Turner AH, Norell M (2009) Morphology of the Late Cretaceous crocodylomorph Shamosuchus djadochtaensis and a discussion of neosuchian phylogeny as related to the origin of Eusuchia. Bulletin of the American Museum of Natural History 324: 1–103.

30. Buscalioni AD, Piras P, Vullo R, Signore M, Barbera C (2011) Early Eusuchia Crocodylomorpha from the vertebrate-rich Plattenkalk of Pietraroia (Lower Albian, southern Appenines, Italy). Zoological Journal of the Linnean Society 163: S199–S227.

31. Salisbury SW, Frey E (2001) A biomechanical transformation model for the evolution of semi-spheroidal articulations between adjoining vertebral bodies in crocodilians. In: Grigg GC, Seebacher F, Franklin CE, editors. Crocodilian Biology and Evolution. Chipping Norton: Surrey Beatty and Sons. 85–134.

32. Brisbin IL, Standora EA, Vargo MJ (1982) Body, temperatures and behavior of American alligators during cold winter weather. American Midland Naturalist 107: 209–218.

33. Thorbjarnarson J, Wang X (2010) The Chinese Alligator: Ecology, Behavior, Conservation, and Culture. Baltimore: Johns Hopkins University Press.

34. Eberle JJ, Fricke HC, Humphrey JD, Hackett L, Newbrey MG, et al. (2010) Seasonal variability in Arctic temperatures during early Eocene time: Earth and Planetary Science Letters 296: 481–486.

35. Greenwood DR, Basinger JF, Smith RY (2010) How wet was the Arctic Eocene rainforest? Estimates of precipitation from Paleogene Arctic macrofloras. Geology 38: 15–18. doi: 10.1130/G30218.1

36. Weijers JWH, Schouten S, Sluijs A, Brinkhuis H, Sinninghe Damste JH (2007) Warm Arctic continents during the Paleocene-Eocene thermal maximum. Earth and Planetary Science Letters 261: 230–238.

37. Schubert BA, Jahren AH, Eberle JJ, Sternberg LSL, Eberth DA (2012) A summertime rainy season in the Arctic forests of the Eocene. Geology 40: 523–526.

38. Schubert BA, Jahren AH (2011) Quantifying seasonal precipitation using high-resolution carbon isotope analyses in evergreen wood. Geochimica et Cosmochimica Acta 75: 7291–7303.

39. Nilsson J, Engstedt O, Larsson P (2013) Wetlands for northern pike (Esox lucius L.) recruitment in the Baltic Sea. Hydrobiologia 721: 145–154.

40. Lehman J-P (1951) Un novel amiid de l'Eocène du Spitzberg. Pseudamia heintzi. Arshefter Tromsø Museums 70(1947): 1–11.

41. Ohta YA, Hjelle A, Andresen A, Dallman WK, Salvigsen S (1992) Geological map of Svalbard 1: 100,000. Sheet B9G Isfjorden. Norsk Polarinstitutt. Temakart 16, Oslo.

42. Scott WB, Crossman EJ (1973) Freshwater fishes of Canada. Bulletin Fisheries Research Board of Canada 184: 1–96.

43. Taplin LE, Grigg GC (1989) Historical zoogeography of the eusuchian crocodilians: a physiological perspective. American Zoologist 29: 885–901.

44. Xu Q, Huang C (1984) Some problems in evolution and distribution of Alligator. Vertebrata PalAsiatica 22: 49–53.

45. Brochu CA (1999) Phylogeny, systematics, and historical biogeography of Alligatoroidea. Society of Vertebrate Paleontology Memoir 6: 9–100.

46. Snyder D (2007) Morphology and systematics of two Miocene alligators from Florida, with a discussion of Alligator biogeography. Journal of Paleontology 81: 917–928.

47. Brochu CA (2004) Alligatorine phylogeny and the status of Allognathosuchus Mook, 1921. Journal of Vertebrate Paleontology 24: 856–872.

48. Wu X, Wang Y, Zhou K, Zhu W, Nie J, et al. (2003) Complete mitochondrial DNA sequence of Chinese alligator, Alligator sinensis, and phylogeny of crocodiles. Chinese Science Bulletin 48: 2050–2054.

49. Roos JR., Aggarwal K, Janke A (2007) Extended mitogenomic phylogenetic analyses yield new insight into crocodylian evolution and their survival of the Cretaceous-Tertiary boundary. Molecular Phylogenetics and Evolution 45: 663–673.

50. Oaks JR (2011) A time-calibrated species tree of Crocodylia reveals a recent radiation of the true crocodiles. Evolution 65: 3285–3297.

51. Martin J, Lauprasert K (2010) A new primitive alligatorine from the Eocene of Thailand: relevance of Asiatic members to the radiation of the group. Zoological Journal of the Linnean Society 158: 608–628.

4

Wood Anatomy Reveals High Theoretical Hydraulic Conductivity and Low Resistance to Vessel Implosion in a Cretaceous Fossil Forest from Northern Mexico

Hugo I. Martínez-Cabrera[1]*, Emilio Estrada-Ruiz[2]

1 Estación Regional del Noroeste, Instituto de Geología, Universidad Nacional Autónoma de México, Hermosillo, México, 2 Laboratorio de Ecología, Departamento de Zoología, Escuela Nacional de Ciencias Biológicas – Instituto Politécnico Nacional, Ciudad de México, México

Abstract

The Olmos Formation (upper Campanian), with over 60 angiosperm leaf morphotypes, is Mexico's richest Cretaceous flora. Paleoclimate leaf physiognomy estimates indicate that the Olmos paleoforest grew under wet and warm conditions, similar to those present in modern tropical rainforests. Leaf surface area, tree size and climate reconstructions suggest that this was a highly productive system. Efficient carbon fixation requires hydraulic efficiency to meet the evaporative demands of the photosynthetic surface, but it comes at the expense of increased risk of drought-induced cavitation. Here we tested the hypothesis that the Olmos paleoforest had high hydraulic efficiency, but was prone to cavitation. We characterized the hydraulic properties of the Olmos paleoforest using theoretical conductivity (K_s), vessel composition (S) and vessel fraction (F), and measured drought resistance using vessel implosion resistance $(t/b)_h^2$ and the water potential at which there is 50% loss of hydraulic conductivity (P_{50}). We found that the Olmos paleoforest had high hydraulic efficiency, similar to that present in several extant tropical-wet or semi-deciduous forest communities. Remarkably, the fossil flora had the lowest $(t/b)_h^2$, which, together with low median P_{50} (-1.9 MPa), indicate that the Olmos paleoforest species were extremely vulnerable to drought-induced cavitation. Our findings support paleoclimate inferences from leaf physiognomy and paleoclimatic models suggesting it represented a highly productive wet tropical rainforest. Our results also indicate that the Olmos Formation plants had a large range of water conduction strategies, but more restricted variation in cavitation resistance. These straightforward methods for measuring hydraulic properties, used herein for the first time, can provide useful information on the ecological strategies of paleofloras and on temporal shifts in ecological function of fossil forests chronosequences.

Editor: Paul V. A. Fine, Berkeley, United States of America

Funding: This work was supported by funds from "Apoyos Complementarios para la Consolidación Institucional de Grupos de Investigación" CONACYT to HIMC. The funder had no role in study design, data collection and analysis, decision to publish, or preparation of the manuscript.

Competing Interests: The authors have declared that no competing interests exist.

* Email: hugomartinez2w@gmail.com.mx

Introduction

The expression of anatomical and morphological traits is subject to biophysical constraints imposed by environmental demands. Consequently, they can provide important information about ecological strategies of fossil assemblages [1,2] and paleoclimate (e.g. [3,4]). Because a large amount of water is needed to maintain plant growth, water is a major factor limiting plant distribution and trait expression. Water loss and carbon fixation are linked because during CO_2 uptake water is transpired to the atmosphere [5], and because diffusion coefficients for water are larger than for CO_2, efficient carbon fixation and plant growth requires a disproportionately high hydraulic supply to the leaves to meet evaporative demands during photosynthesis. This relationship between stem hydraulic capacity with tree growth rate [6,7] and leaf photosynthetic capacity is well established [8].

Across vegetation types and biomes, at low latitudes and altitudes, there is a positive relationship between water availability and xylem conduit size, such that in xeric environments vessel size is smaller on average than in tropical humid environments [9,10,11]. Plant water transport efficiency is directly related to the hydraulic conductivity of xylem, which is mostly determined by conduit size [12,13,14]. As warm wet environments are also more productive regions, hydraulic efficiency should be directly linked to carbon fixation, as has been empirically confirmed (e.g. [15]). In places with high water availability hydraulic capacity is increased by decreasing resistance to water flow in the xylem (e.g. increasing vessel diameter). In these warm wet environments, plants can maintain high transpiration rates and maximize carbon fixation and growth [16]. However, in dry conditions, wide, hydraulically efficient vessels are also more prone to experience mechanical failure [17,18,19] or drought induced cavitation [14,20,21].

Although drought induced cavitation occurs along the entire water availability continuum [22,23], plants from drier regions with smaller vessel diameters are in general more capable to cope with cavitation because it occurs at higher xylem tensions (lower water potential) than in wet adapted plants with larger vessel diameters. The consequence of vessel cavitation is the disruption of the water column, which reduces xylem hydraulic conductivity and the overall plant water supply to the photosynthetic surface [24]. Cavitation can thus be translated into a decrease in photosynthesis [15] and a drop in stomatal conductance [25], which further increase xylem tension and cavitation [24,26]. Cavitation resistance in extant plants is usually quantified using the water potential at which there is 50% loss of hydraulic conductivity (P_{50}). As P_{50} values are impossible to measure in fossil wood samples, we used a metric developed by Hacke et al. [18], the vessel resistance to implosion metric $(t/b)_h^2$, defined below) that can be used to approximate cavitation resistance. $(t/b)_h^2$ explains between 80% [18] and 95% [19] of P_{50} variation and is essentially a measure of vessel wall reinforcement. $(t/b)_h^2$ is entirely based on vessel anatomy and therefore can be used to determine drought tolerance thresholds (P_{50}) in fossil woods.

In this paper we used vessel anatomy to determine key functional traits related to drought resistance $((t/b)_h^2$ and $P_{50})$ and hydraulic capacity (potential conductivity K_s, vessel fraction F, and vessel size contribution metric S) of a fossil forest from the Olmos Formation (upper Campanian), Coahuila, Mexico. These two sets of functional traits provide information about the hydraulic functional strategies of the fossil forest. Because of their close link to the environment, they also provide hints as to the climate regime in which the fossil plants grew. With more than 80 different leaf morphotypes [27,28,29,30,31,32,33,34], mainly angiosperms (80%), the Olmos Formation is one of the richest Cretaceous paleoforests from Mexico and south-central Western Interior of North America (WINA). According to leaf physiognomy paleoclimate estimates, the Olmos paleoforest grew under a tropical climate (MAT 20–23°C) with high water availability (1.5 to 3 m) [28]. The Olmos paleoforest was probably more mesic-adapted than other floras from the southern part of North America during this period [28]. Based on these paleoclimate reconstructions and leaf physiognomy, we expected that this paleoforest would have had a very efficient hydraulic system to meet high evaporative demands, but we also expected that this hydraulic efficiency should be paired with a high risk of embolism formation. In other words, in the conduction efficiency-cavitation risk trade-off continuum, the Olmos paleoforest should represent a highly hydraulically efficient and embolism-prone ecological strategy only suitable under warm, wet environments. Moreover, large leaf area and tree heights, estimated to be up to 35 m in some species [28], suggest a highly productive environment and efficient carbon fixation, which would require an efficient hydraulic system.

Material and Methods

Geological setting

The outcrops of the Olmos Formation (upper Campanian) are found in the Sabinas Basin in the state of Coahuila in northern Mexico (Fig. 1) [35,36,37,38]. The Olmos Formation represents a fluvial-deltaic system with four main depositional sub-environments [36] that include: 1) swampy areas with restricted circulation, 2) floodplain environments and/or a lagoon system with open circulation, 3) fluvial environments, likely including braided rivers and 4) meandering rivers [36]. The angiosperm woods, along with numerous dinosaur bones, were collected in environments representing meandering rivers [36]. The woods were not found in growth position and despite the large size of some samples, we assumed that some transport occurred.

All necessary permits were obtained for the described study, which complied with all relevant regulations. These permits were issued by the Instituto Nacional de Antropología e Historia (INAH) and the material is housed at the National Collection of Paleontology (Universidad Nacional Autónoma de México).

Anatomical measurements

The anatomical characteristics of the 10 wood xylotypes thus far described for the Olmos Formation [27,39] were measured on transverse sections obtained using a standard thin-section technique. In 11 field trips we have collected nearly 100 samples, from these, we have recognized 10 dicot, 5 palms and 2 gymnosperm (Podocarpaceae and Taxodiaceae) xylotypes. Despite that most of the xylotypes were identified in the first two field visits, and no new xylotypes have been collected in recent visits, we do not discard the possibility of finding new morphospecies given the outstandingly high diversity of the leaf flora. The 10 wood xylotypes we studied here represent all the dicot fossil woods collected so far and therefore, despite their relatively low number, they are a good representation of the dicot flora. We measured more than one sample for six of the ten species/xylotypes (i.e., *Coahuiloxylon terrazasiae*, *Javelinoxylon* xylotype 2, *Javelinoxylon weberi*, *Metcalfeoxylon* xylotype 1, *Muzquizoxylon porrasii* and *Wheeleroxylon atascosense*). See Table 1 for species means and standard deviations, and Table S1 and Table S2 for values of each measured vessel and means per sample, respectively. Species means for each of the measured functional traits were calculated based on the mean of the each of the samples per xylotype.

To determine potential conductivity (K_s), vessel resistance to vessel implosion $(t/b)_h^2$, and S and F metrics, we calculated dimensions of 62 to 398 vessels per species (See Table S1 for the number of vessels measured per sample and species/xylotype). The number of vessels measured varied as a function of their density. To calculate vessel dimensions we first randomly selected radial sectors (cross sectional area limited by rays) and measured all the vessels contained in the sector. For species with a limited number of vessels (low vessel density) such as *Quecinium centenoae* and *Metcalfeoxylon* xylotype 1, we measured all the vessels that the preservation allowed. Vessel outlines were drawn using a graphics tablet (Intuous 3, Wacom, Kita Saitama-Gun, Saitama, Japan). Vessel area was calculated using XTools of ArcView (version 3.2, ESRI, Redlands, CA, USA). For details of the measuring protocol, see Martínez-Cabrera et al. [40]. We calculated vessel lumen diameter, including mean vessel diameter (d_{mean}) and hydraulic mean (d_h), using diameters of circles with the same area as the individual vessel lumens, following Kolb and Sperry [22]. d_h was calculated as the sum of the contribution of all conduit diameters $(\sum d^5)$ divided by the total number of vessels $(\sum d^4)$ [22,41]. d_{mean} was then used, together with other variables (see below), to calculate potential conductivity, while d_h was used for estimating the squared vessel-wall thickness-to-span ratio $(t/b)_h^2$, a proxy for the resistance to vessel implosion.

We calculated potential conductivity per stem cross sectional area (K_s) following Zanne et al. [13]: $K_s \propto F^{1.5} S^{0.5}$, where F is the vessel fraction and S is a vessel size contribution metric. Vessel fraction F is mean vessel area $\bar{A}$ times vessel density (N) ($F = \bar{A} * N$; mm^2·mm^{-2}), and S is the ratio between the same anatomical traits ($S = \bar{A}/N$; mm^4). $\bar{A}$ is the mean individual vessel cross sectional area and N is the vessel number per unit of sapwood area [13]. Vessel fraction F approximates the fraction of cross sectional area occupied by vessel space [13,42] and S measures the variation in vessel size composition. Higher values of

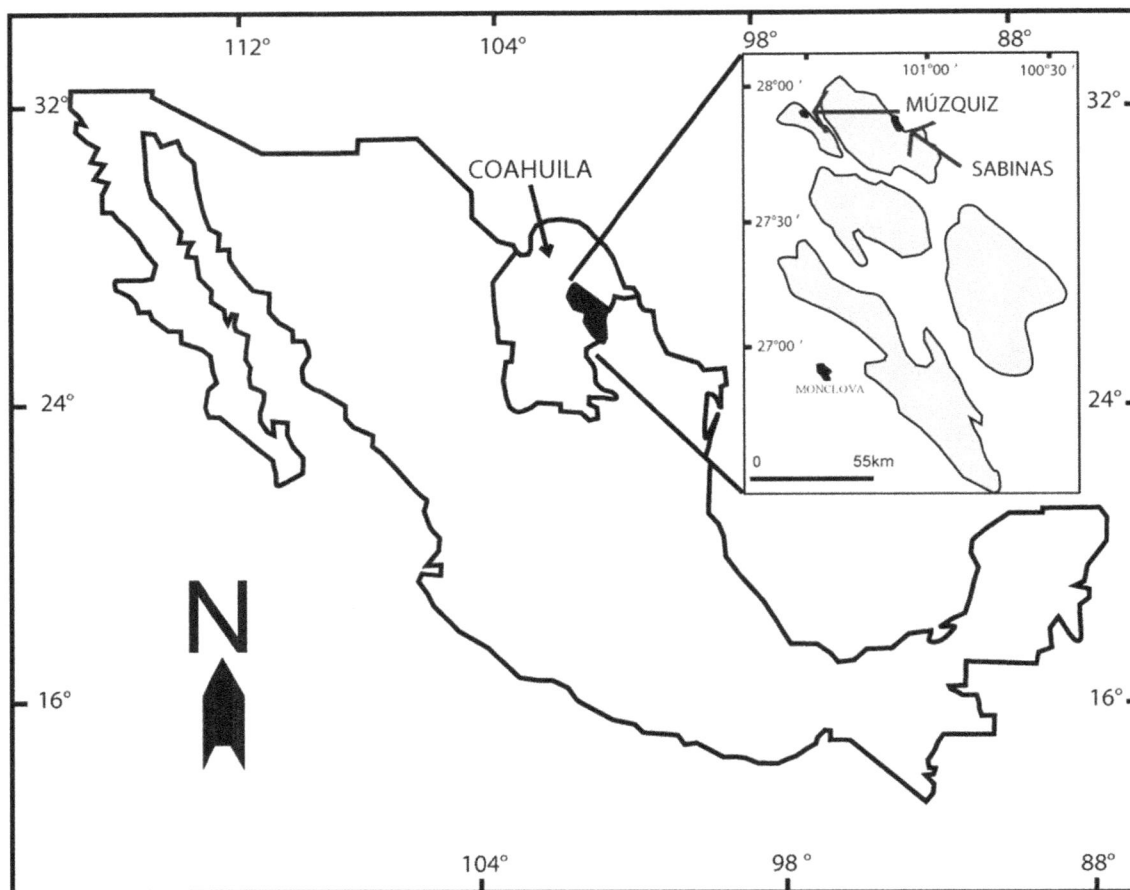

Figure 1. Location map of the Olmos Formation in the Sabinas Basin (in gray). Sampled sites in Sabinas and Múzquiz (Atascoso and Santa Elena Ranchs) are identified by arrows.

S indicate a greater contribution of wide vessels to water conduction in a given area [13].

The safety factor for vessel implosion [18] was calculated using the squared vessel-wall thickness-to-span ratio $(t/b)_h^2$, where the paired vessel wall thickness, t, is the thickness of the wall between two adjacent vessels and b is the diameter of the conduit closest to the hydraulic mean diameter (d_h). The $(t/b)_h^2$ was measured in 16 to 58 vessel pairs per sample, in which at least one of the vessel in a pair was within ± 5 μm of the hydraulic mean diameter (d_h). The number of vessels used for each sample and species/xylotype is presented in Table S2 and Table S3. b has a strong influence on $(t/b)_h^2$; therefore, we do not expect much variation between vessels. The relatively small sample size for some species is because of the insufficient number of vessel pairs within the desired vessel diameter. We also measured implosion resistance in species with solitary vessels that had conducting cells other than vessels (e.g. vascular and vasicentric tracheids) associated with them, as was the case for *Sabinoxylon pasac*, *Quercinium centenoae*, and *Metcalfeoxylon* xylotype 1. Since conduit $(t/b)_h^2$ captures between 80 and 95% of P_{50} variation, we used the relationship between these two variables to estimate cavitation resistance of the fossils $(P_{50} = -0.662 - 154.646\ (t/b)_h^2$; regression equation was kindly provided by Uwe Hacke, University of Alberta).

Comparison with extant communities

The values of the characteristics measured are a good proxy for the hydraulic function of the individual/species studied, in this sense they are by themselves measurements, not predictions of a value (as are the paleoclimate estimates using plant structures, which require a minimum number of species). We compared the calculated K_s, $(t/b)_h^2$, F and S values from the Olmos Formation woods with those from nine extant communities. All the species analyzed in these extant communities were dicots, and the anatomical information was taken from previous studies [1,40,43,44]. These extant communities included a tropical rain forest, a dry deciduous forest and a montane forest from Mexico [1,43] and several communities from sites with a wide range of precipitation in North and South America [40,44]. These last extant communities varied from hardwood forest to desert scrub. For this comparison, K_s, S and F were calculated for the entire set of extant communities. The lack of information necessary to calculate resistance to vessel implosion in the Mexican communities (tropical rain forest, Veracruz; montane forest, Estado de México; dry deciduous forest, Jalisco) restricted our comparisons for this metric to a smaller (6) set of extant localities. Because K_s, and S had non-homogeneous error variance term (Levene's test, P<0.0001 in both cases) with regard to vegetation type (i.e. variance of the error term is not constant throughout vegetation types) we used Kruskal-Wallis test to detect overall difference

Table 1. Means and standard deviations of the hydraulic and drought resistance traits for the Olmos Formation species/xylotypes.

Species	Affinities	K_s	F	S	$(t/b)_h^2$	P_{50}	N
Coahuiloxylon terrazasiae	Anacardiaceae/Burseraceae	4.28 (2.31)	0.22 (0.55)	0.00168 (0.0005)	0.0038 (0.0014)	−1.25 (0.22)	2
Javelinoxylon xylotype 1	Malvaceae	12.62	0.33	0.0045	0.0103	−2.26	1
Javelinoxylon xylotype 2	Malvaceae	1.63 (0.22)	0.16 (0.011)	0.00069 (0.00004)	0.0346 (0.0065)	−6.01 (1.005)	2
Javelinoxylon weberi	Malvaceae	6.97 (2.24)	0.36 (0.72)	0.0009 (0.00005)	0.0106 (0.0059)	−2.30 (0.091)	2
Metcalfeoxylon xylotype 1	Incertae sedis	5.8 (0.93)	0.15 (0.012)	0.0095 (0.0007)	0.0106 (0.0028)	−2.3 (0.043)	3
Muzquizoxylon porrasii	Cornaceae	0.92 (0.45)	0.17 (0.065)	0.00016 (0.00002)	0.064 (0.0005)	−10.58 (0.084)	2
Olmosoxylon upchurchii	Lauraceae	54.46	0.68	0.0094	0.00159	−0.909	1
Quercinium centenoae	Fagaceae	69.82	0.65	0.0175	0.00101	−0.82	1
Sabinoxylon pasac	Ericales	30.81	0.43	0.0116	0.00303	−1.13	1
Wheeleroxylon atascosense	Malvaceae	38.94 (3.94)	0.66 (0.033)	0.00524 (0.00026)	0.00607 (0.0004)	−1.6 (0.064)	2

K_s = potential conductivity; F = vessel fraction; S = vessel size to number ratio; $(t/b)_h^2$ = vessel implosion resistance; P_{50} = cavitation resistance; N = number of studied samples per xylotype.

among groups. Then, to detect pairwise differences between the Olmos paleoforest and the extant communities, we performed Mann-Whitney U test. The homoscedastic variables (F, implosion resistance and P_{50}) were compared using ANOVA. Similarly, we ran a Levene's test to assess the homoscedasticity of the error term of the functional traits at different MAP levels. The relationship of those traits with homogeneous variance and MAP (F and S) was then analyzed, prior a log transformation, using simple linear regression. For those traits with non-homogeneous variance (implosion resistance and K_s,) we ran weighed least squares regressions using variance at each MAP level as weights to fit the model.

Additionally, we used PCA to visualize 1) hydraulic properties (K_s, S and F) of the Olmos Formation woods with the full extant data set and 2) these same hydraulic properties plus implosion resistance in the smaller subset of communities. To determine whether vegetation type/communities were well discriminated by their hydraulic properties we carried out a between-group PCA and assessed the significance of the results using a Monte-Carlo permutation test on the between group inertia percentage [45,46]. Significance was calculated by comparing the between-group observed differences with the distribution of 999 permutations simulated. We used the R package ade4 [47,48] to perform this analysis.

Results

Potential hydraulic conductivity

There were significant differences of conduction capacities among all communities (Kruskal-Wallis chi squared = 106.3237, p<0.001). Although the Olmos paleoforest had the highest mean potential conductivity per stem cross sectional area ($K_s = 22.6$ g·mm^{-1}·Mpa^{-1}·s^{-1}), it was not significantly different from the hydraulic conductivity of the tropical rain forest (Mann-Whitney $Z = 265$, p = 0.93; $K_s = 17.81$ g·mm^{-1}·Mpa^{-1}·s^{-1}), the montane forest (Mann-Whitney $Z = 103$, p = 0.43; $K_s = 9.41$ g·mm^{-1}·Mpa^{-1}·s^{-1}) and the dry deciduous forest (Mann-Whitney $Z = 317$, p = 0.51; $K_s = 3.06$ g·mm^{-1}·Mpa^{-1}·s^{-1}) (Fig. 2a). The Olmos paleoforest had significantly higher K_s than the dry the North American hardwood forest (Mann-Whitney $Z = 14$, p = 0.005; $K_s = 2.71$ g·mm^{-1}·Mpa^{-1}·s^{-1}) and the remaining extant communities (Fig. 2a). The South American mesquite savanna had the lowest theoretical hydraulic capacity ($K_s = 0.28$ g·mm^{-1}·Mpa^{-1}·s^{-1}). Although the Olmos paleoforest had a high spread in K_s, ranging from 0.9 (g·mm^{-1}·Mpa^{-1}·s^{-1}) in *Muzquizoxylon* to 69.8 (g·mm^{-1}·Mpa^{-1}·s^{-1}) in *Quercinium* (Table 1), the tropical rain forest from Los Tuxtlas, Veracruz had a larger range in values.

Vessel composition and vessel fraction

Patterns of vessel composition S (not shown) were parallel to those for K_s. There was an overall difference in S values across the studied communities (Kruskal-Wallis chi squared = 124.5, p< 0.001). S values of the tropical rainforest and the Olmos paleoforest were not significantly different (Mann-Whitney $Z = 302$, p = 0.62; 0.006 mm^4 vs. 0.0054 mm^4). In these two communities, higher S indicates that fewer larger vessels have greater hydraulic contribution. The S metric was significantly lower in the dry-deciduous forest (Mann-Whitney $Z = 435$, p = 0.008; 0.0034 mm^4) and the in remaining of the extant communities. The montane forest (0.0014 mm^4), the juniper/mesquite savanna (0.00013 mm^4), North American hardwood forest (0.000119 mm^4), palm forest (0.000114 mm^4), desert (7.6e-05 mm^4), mesquite savanna (5.3e-05 mm^4) and sage scrub

$(4.9e-05 \text{ mm}^4)$ had S values several orders of magnitude lower than the Olmos paleoforest, indicating that numerous small vessels comprise the conducting area in the woods from these communities.

We detected an overall difference among vegetation types in F values (ANOVA, $F_{9,\ 198} = 4.5$, p<0.001). Vessel lumen fraction (F) was largest in the montane forest, where more than 50% $(0.5 \text{ mm}^2 \cdot \text{mm}^{-2})$ of the cross sectional area is occupied by vessels, followed by the dry-deciduous forest $(0.46 \text{ mm}^2 \cdot \text{mm}^{-2})$, hardwood forest $(0.44 \text{ mm}^2 \cdot \text{mm}^{-2})$, and tropical rainforest $(0.38 \text{ mm}^2 \cdot \text{mm}^{-2})$. The proportion of cross sectional area occupied by vessels in the Olmos paleoforest is around 38%, while drier communities had lower F values, ranging from 0.31 in the desert vegetation to 0.23 in palm forest, mesquite savanna and juniper mesquite savanna (Fig. 2b). The Olmos paleoforest was only significantly different from the montane forest ($F_{1,\ 34} = 5.88$, p = 0.02).

The safety factor for vessel implosion

There was a significant difference in resistance to implosion $(t/b)_h^2$ among the compared vegetation types (ANOVA, $F_{6,65} = 3.59$, p = 0.0038). Vessel implosion resistance was the lowest in the Olmos paleoforest (conduit $(t/b)_h^2 = 0.0145$) and ranged from 0.001 in *Quercinium* to 0.064 in *Muzquizoxylon* (Table 1). *Muzquizoxylon* drove mean paleocommunity conduit $(t/b)_h^2$ to higher values (Fig. 2c). As a result, the Olmos paleoforest was not significantly different ($F_{1,19} = 0.0027$, p = 0.95) from the North American hardwood forest and Argentinean palm forest ($F_{1,19} = 0.95$, p = 0.34) with a conduit $(t/b)_h^2 = 0.0139$ and 0.021, respectively. If we compare the medians, instead of the means, the vessel implosion thresholds of the fossil assemblage (0.0082) are lower to those of the hard wood (0.0145) and palm forests (0.0149). The desert ($F_{1,26} = 11.6$, p = 0.002) and the remaining vegetation types had significantly higher vessel implosion resistance than the Olmos paleoforest.

The estimated P_{50} values in the fossil assemblage, calculated with the regression equation describing its relationship with $(t/b)_h^2$, ranged from -0.082 MPa in *Quercinium* to -6 and -10.58 Mpa in *Javelinoxylon* xylotype 2 and *Muzquizoxylon*, respectively (Table 1). The mean paleoforest P_{50} (-2.9 Mpa, 95% CI = $-4.82, -1.01$) was driven to higher values by these two species. The median P_{50} of the fossil assemblage was -1.9 Mpa. We found a significant trade-off between K_s and resistance to vessel implosion ($R^2 = 0.14$, p<0.001; y = 0.014–0.008x), indicating that the very high hydraulic capacity of the Olmos fossils comes at the expense of high susceptibility to vessel implosion/cavitation (Fig. 3a). Low P_{50} values indicate a high risk of cavitation at relatively low water stress (Table 1).

In the extant species' data set, mean annual precipitation (MAP) was significantly related with resistance to vessel implosion ($R^2 = 0.32$, p<0.001; y = 3.8–1.72x, Fig. 3b). None of the vessel conduction metrics F ($R^2 = 0.02$, p = 0.27; y = 0.025+2.2e–5x), S ($R^2 = 0.003$, p = 0.67; y = 1.19–5.5e–5) or K_s ($R^2 = 0.01$, p = 0.78; y = 5.6–4.19x) was significantly associated with MAP, suggesting that resistance to implosion has a greater potential to infer precipitation than any other metric describing water conduction properties. Assuming that resistance to vessel implosion can provide at least a rough approximation of water availability and based on the lines in Figure 3b representing median and the first and third quartiles, we suggest that precipitation in the Olmos paleoforest was comparable to that of our wettest extant sites.

Between-PCA analysis

Species from the Olmos Formation occupied a different region of the functional space in the PCA analysis including hydraulic properties (K_s, S and F) and implosion resistance (Fig. 4). The three first PCA axes explained 99% of the variation. In this analysis, hydraulic capacity and implosion resistance are orthogonal (PC loadings for both PCA analysis are presented in Table S4). In the first PCA axis, K_s had the highest loading, while implosion resistance was highest in the second, and S and F in the third. For instance, in the PCA plot the Olmos paleoforest represents the hydraulically efficient highly prone to vessel implosion extreme. The PCA analysis also revealed that the Olmos paleoforest had greater variation along the hydraulic efficiency axis relative to the implosion resistance axis (the ellipse of the Olmos flora is elongated parallel to the efficient conduction axis). This pattern suggests that the Olmos paleoforest had a large variety of hydraulic strategies but a more constrained number of strategies in implosion/cavitation resistance. The drier communities showed the opposite pattern, with a large breadth in cavitation resistance variation (and also higher values) but highly constrained hydraulic capacity (with low absolute values) (see positive association between axes in Fig. 4). The Monte Carlo permutation test of the ratio of between-class and total inertia in the between PCA analysis supports the discrimination of these groups (ratio = 0.78, P<0.001; Fig. 4). In the second PCA analysis (only K_s, S and F; for all 9 extant communities and the Olmos forest), where the first two principal components explained 93% of the variation, significant differences in functional space among communities are also supported (ratio = 0.25, P<0.001; Fig. 5). In this analysis it is clear that the tropical rain forest exhibits a larger variation in hydraulic strategies than the rest of the communities, including the Olmos paleoforest. In addition, this second analysis reveals a positive relationship between the first two first axes in the tropical rain forest (high K_s is correlated with high S: large conduction capacity is reached with large vessels, see the ellipse orientation in Figure 5) while in drier communities the relationship between these two is negative (large K_s, although overall low in most of them, is reached by having many small vessels; low S). The Olmos paleoforest did not have any strong pattern in this regard.

Discussion

In this paper we set out to determine key properties of the hydraulic system of the Olmos Formation woods. Traits related to hydraulic capacity and vulnerability are important because, as biophysical constraints link them to environmental variation, they provide direct information on ecological strategies of fossil assemblages. For instance, given the warm, wet environment estimated for the Olmos Formation from leaves [28], we expected a highly efficient, highly prone to cavitation hydraulic system. Indeed, our analyses suggest that hydraulic capacity and vulnerability to cavitation at low water stress were high in the Olmos Formation fossil woods.

Hydraulic capacity of The Olmos Formation

The hydraulic capacity of the Olmos paleoforest was similar to several extant communities including wet tropical, dry deciduous, or montane forest. This suggests that the use of K_s in fossil woods provides information allowing characterization of conduction efficiency differences among communities, but that similar conduction efficiencies can be found under relatively different precipitation regimes. For instance, K_s, has enough power to detect differences in conduction capability between dry and wet floras and therefore may provide only rough estimates of water

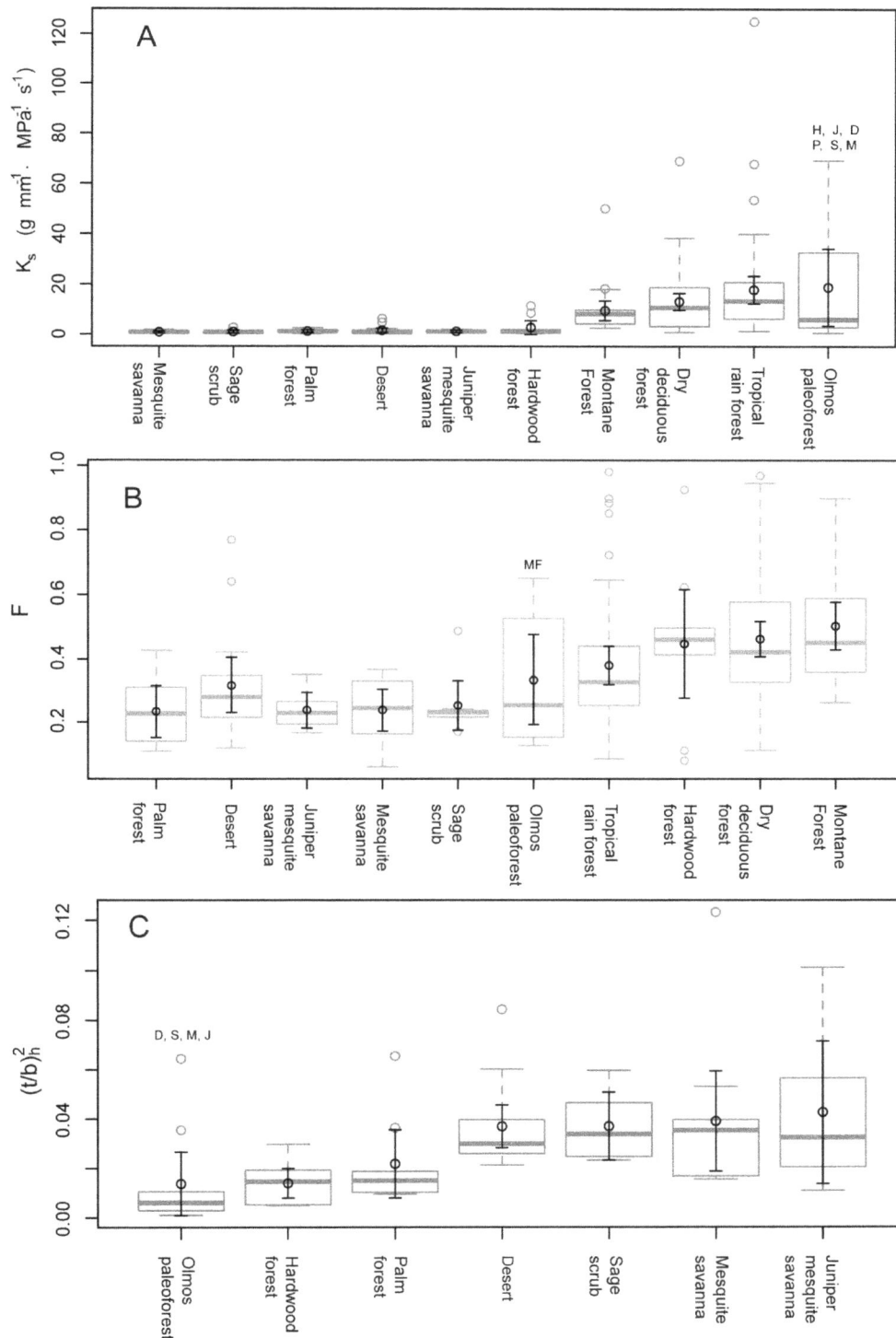

Figure 2. Comparison of a) potential conductivity, b) vessel fraction and c) implosion resistance between the Olmos Formation fossil woods and extant communities. Box plots in grey show the median and interquartile distance of each one of the variables. The black circle and error bars represent the mean and ±95% confidence intervals. Box plots showing implosion resistance only includes the drier extant communities and the Olmos formation flora. Sample sizes: mesquite savanna (M) = 11, sage scrub (S) = 8, palm forest (P) = 9, desert (D) = 17, juniper-mesquite savanna (J) = 7, hardwood forest (H) = 10, montane forest (MF) = 25, dry deciduous forest = 56, tropical rain forest = 54, Olmos flora = 10. The two different desert communities were pooled for this analysis. Significant difference between the Olmos pleoflora and extant vegetation types were determined using Mann-Whitney test for potential conductivity, and ANOVA for F and implosion resistance.

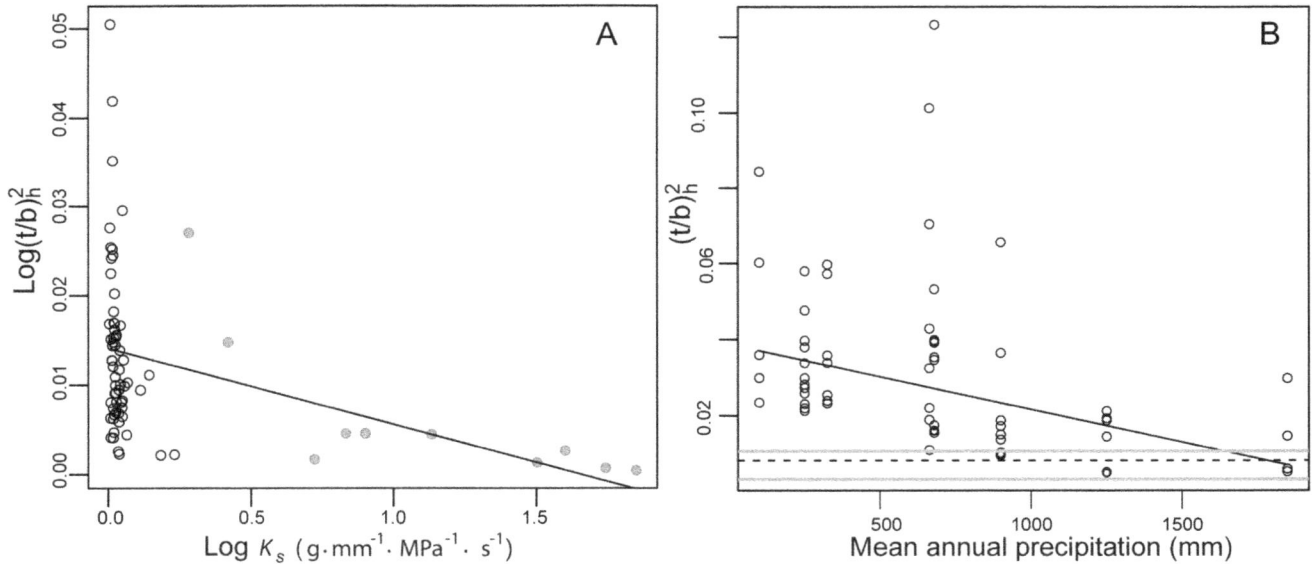

Figure 3. Implosion resistance as a function of a) potential conductivity, b) MAP. In figure 3a the Olmos formation xylotypes are in grey; black circles are the extant communities presented in Figure 2c. In figure 3b the horizontal lines show the median (dashed) and first and third quartiles (grey lines) of implosion resistance for the Olmos Formation, the circles are the species of all extant communities in Figure 2c. Regression line in 3a was fitted using simple linear regression on the log-transformed variables, while in 3b using weighed least square regression.

availability in paleoforests. This is further supported by the absence of a significant relationship between K_s and MAP in our analysis of extant communities. These two variables have been found to be independent in evergreen angiosperms [49]. There is, however, a significant inverse relationship (instead of positive as would be expected) between these two variables in deciduous angiosperms that has been interpreted as the adaptation to water limitation in plant with this ecological strategy [49]. The inverse relationship observed in that study is because most of the deciduous species analyzed [49] are winter deciduous, and thus, a product of the decrease in vessel size associated with higher thresholds to freezing induced cavitation of small vessels (e.g., [50,51]). It is also possible that K_s is maximized in seasonally dry habitats (e.g., tropical deciduous forest) because of the need for fast growth and carbon fixation during restricted periods of water availability [49,52]. In the context of our study, the absence of ring porous species and the faint growth rings in the Olmos Formation species indicate that the flora was not subject to sharp seasonal fluctuations of water transport and growth. Instead, high hydraulic efficiency of the woods analyzed, along with constancy of vessel diameter across growth rings, suggests environmental conditions allowing constant hydraulic capacity throughout the year.

As expected, variation in S and F yielded similar results. However, S and F describe different functional aspects of water conduction [13]. Low S indicates the presence of many small vessels, which are selective under conditions promoting freezing-induced embolism [13,16]. Thus the large S value of the Olmos woods, which was similar to the S value of the tropical rainforest, suggests the absence of freezing temperatures. F, on the other hand, is related to potential conductivity, but also describes the amount of cross sectional space occupied by lumen and, by extension, non-lumen (cell wall) area. This metric is bounded at higher values by mechanical support needs and at lower values by hydraulic requirements [13,42]. All else being equal, high F would indicate low construction cost because growth is achieved by greater gas fraction represented by vessel space [12,13,53]. The F

value of the Olmos paleoforest, which is not the highest in our study (around $0.38 \text{ mm}^2 \cdot \text{mm}^{-2}$ compared to up to 0.5 and $0.46 \text{ mm}^2 \cdot \text{mm}^{-2}$ in the montane and dry deciduous forest), suggests that construction cost was high relative to hydraulic efficiency. It is important to keep in mind, however, that the amount of variation of wood density (a proxy for construction cost) explained by vessel fraction or other vessel traits is low [13,19,40] and that density is mainly driven by fiber traits (e.g. [40]). High wood fraction is positively related with survival and slow growth rate in shade tolerant species, while increased gas fraction is related to high light requirements and adult stature in fast growing species [12]. In sum, our analysis indicate that the Olmos paleoforest had high hydraulic capacity (high K_s) carried out by few very efficient vessels (high S), but the amount of wood committed to water conduction was not very large (F).

Resistance to implosion and cavitation

Given the low estimated cavitation resistance (mean $P_{50} = -2.9$ MPa) and vessel wall reinforcement metric ($(t/b)_h^2 = 0.0145$) of the Olmos Formation plants, it seems that, on average, they were at high risk of cavitation even at high water potential (low water stress). This pattern is even more evident if the median P_{50} (-1.9 MPa) or the mean without outliers (-1.57 MPa) is considered. In a large study considering 230 observations belonging to 167 species from several vegetation types, Maherali et al. [49] showed that the median P_{50} for extant tropical rain forest was around -1 MPa (n = 41), and ranged from very close to 0 MPa to a little over -6 MPa. Therefore, despite most of species from tropical rain forests being very prone to experience cavitation at high water potential, this biome also has some species that can be quite resistant to drought-induced cavitation. Maherali et al. [49] also found that despite its adaptive significance, P_{50} exhibited large variation both within and across climates. In some species from the Olmos paleoforest such as *Quercinium* and *Olmosoxylon*, with $(t/b)_h^2$ values in the order of 0.001 and estimated P_{50} of -0.82 and -0.91 MPa, respectively, it is likely that mild water

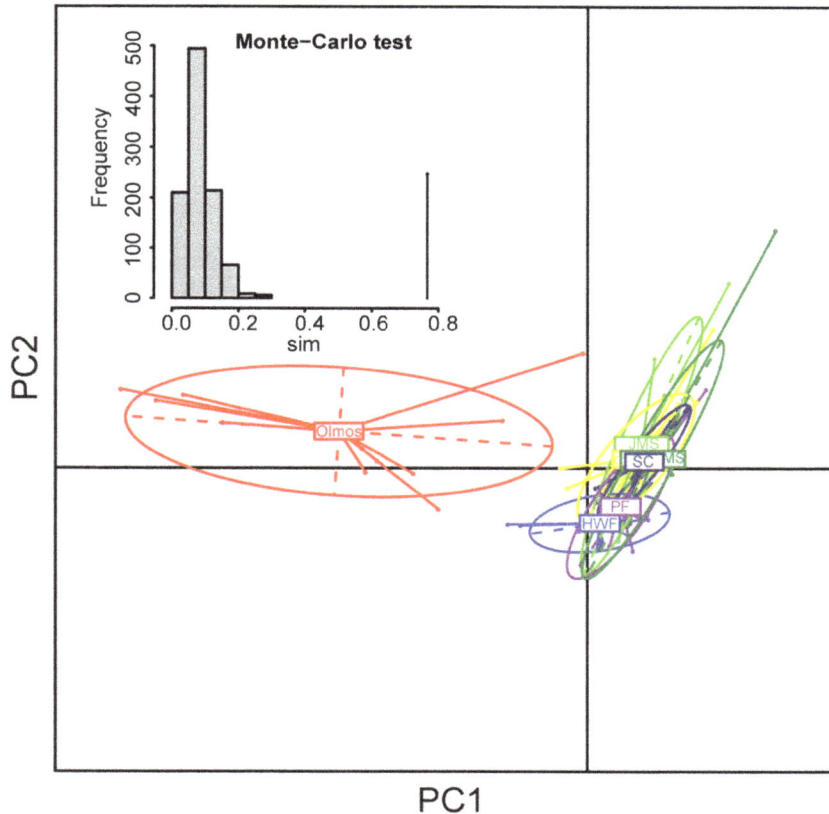

Figure 4. PCA plot portraying hydraulic strategies among communities. Between-Class principal component analysis showing ellipses and gravity centers for extant communities and the Olmos paleoforest. Wet tropical, dry deciduous and montane forest were not included in this analysis. The insert in the upper left corner is the histogram of 1000 simulated values for the between-groups PCA and the observed value (vertical line at sim = 0.78). Sim = ratio of between class and total inertia. HWF = hardwood forest, JMS = juniper-mesquite savanna, MS = mesquite savanna, PF = palm forest, SC = sage scrub.

stress would have driven embolism formation and/or vessel collapse. Median P_{50} for tropical dry forest in the study of Maherali et al. [49] was almost 1 MPa lower than our median (around -2.5 MPa, n = 19), but some species reached over -10 MPa. Of the Olmos Formation woods, only *Muzquizoxylon* reached such low values. It could be argued that the reason behind the extremely high $(t/b)_h^2$ and P_{50} values in *Muzquizoxylon* is deficiency in preservation, but our observations do not support this conclusion. As relatively high values of cavitation resistance are not absent from extant tropical communities, the high inferred cavitation resistance of *Muzquizoxylon* could also be attributed to intra-community variation.

We showed that there is a significant relationship between $(t/b)_h^2$ and MAP in extant communities and that the wall reinforcement parameter of the Olmos species were in general within the ranges of extant vegetations with higher precipitation. It has been found that P_{50} decreases (becomes more negative) with decreasing precipitation in extant evergreen angiosperms and conifers [49]. High resistance to embolism (more negative P_{50}) has therefore evolved as a strategy to cope with drought. It seems that the utility of $(t/b)_h^2$ and P_{50} in detecting water stress in fossil communities is higher than that for K_s, and the calculated values of these metrics for the Olmos paleoforest indicate high water availability.

Paleoclimate and vegetation

The high hydraulic capacity and low resistance to drought we calculated here reinforces evidence from foliar physiognomy indicating a wet warm climate for the Olmos paleoforest. Around 72% of the species are entire-margined, and 50% of the species with preserved leaf apices have drip tips [28]. Foliar physiognomy estimates a MAT of 20–23°C and a MAP of 1.5–3 m and growing season precipitation of 2 m. In addition, the occurrence of palms and the absence of growth rings in the dicot woods indicate cold month mean temperature >5°C [54]. Our results also agree with the current understanding of the climate regime of the southern WINA during the Late Cretaceous as having tropical temperatures (MAT 18–25°C) and above-freezing annual minimum temperatures [28,55,56]. Tropical rainforest is defined by a combination of climatic parameters and plant physiognomical features [57]. These include a MAP of over 1.8 m year^{-1}, high MAT (>18°C), a seasonal variation in temperature of less than 7°C and high percent of species with large, entire margined leaves and dip tips [57]. The climate calculated for the Olmos Formation and the physiognomic characteristics of the leaves suggest it was a tropical rainforest. Paleoclimate models show that the frost line during the Maastrichtian was well to the north [58], and based on these simulations [58], Lomax et al. [59] predicted high net primary productivity and leaf area index for the Olmos Formation region. Significantly, the leaf flora of the Olmos Formation also indicates greater precipitation levels than other North American and

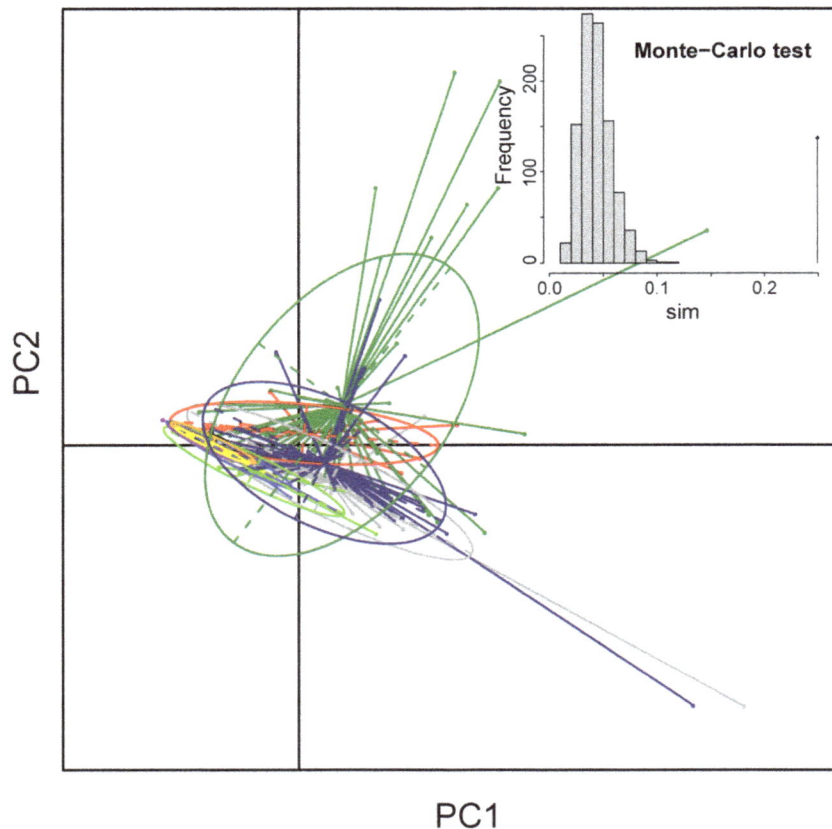

Figure 5. PCA plot portraying hydraulic strategies among all the studied communities. Between-Class principal component analysis shows the ellipses and gravity centers. This plot is based on only hydraulic traits (*F*, *S* and *K$_s$*). The insert in the upper right corner is the histogram of 1000 simulated values for the between-groups PCA and the observed value (vertical line at sim = 0.25). Sim = ratio of between class and total inertia. Red = Olmos paleoforest, dark green = tropical rain forest, light green = hardwood forest, light yellow = sage scrub, dark yellow = palm forest, light blue = desert, dark blue = dry deciduous forest, grey = montane forest, black = juniper mesquite savanna, purple = mesquite savanna.

Western Interior floras during the Late Cretaceous. Based on the small leaf size and low prevalence of drip tips in the paleofloras of the region, Wolfe and Upchurch [56] suggest a subhumid climate throughout the late Cretaceous. Other calculations, based on Climate Multivariate Leaf Analysis Program (CLAMP), support those estimates and propose that in the southern Western Interior precipitation reached up to 1.5 m but it could have been less than 1 m [60,61]. These estimates contrast with the more mesic conditions of the Olmos Formation flora, with up to 3 m of precipitation, and high prevalence of drip tips. In this sense, the Olmos paleoforest, is physiognomically closer to younger (Paleocene) southern Western Interior floras, where larger leaves, with up to 50% of them having drip tips, are present [56,28,36].

The high hydraulic capacity and low resistance to embolism agrees with the high water availability inferred by the leaf physiognomy [28]. Since the fossil woods we studied here were collected in the meandering rivers facies and some of them have pelecypod perforations, it possible that, given their very low cavitation resistance values, the assemblage could be riparian. However, at this point, this is uncertain since the fossil woods were not collected in growth position, despite that several of them reached over 40 m (*Metcalfeoxylon* and *Javelinoxylon* xylotype 1). Whether the woods represent a riparian environment or not, extreme humidity of the Olmos Formation is independently confirmed by the leaf flora as it was collected in the flood plain-lagoon lithofacies [28].

Globally, it has been suggested that an increase in xylem and leaf hydraulic capacity during the mid to late Cretaceous [62,63] likely influenced a significant amplification of angiosperm's forest biomass [64], and contributed to maximize carbon fixation [62] and expansion of tree size [64]. Indeed, our results indicate that the high conduction capacity of the woods from the Olmos paleoforest was necessary to sustain high leaf area [28] and high productivity predicted by paleoclimatic models [58,59] for the region. High hydraulic capacity in the Olmos Formation paleoforest supports the current understanding of late cretaceous angiosperm hydraulic function, which proposes an increased conduction efficiency compared to early cretaceous short statured trees/shrubs [62,63].

We suggest that the climate for the Olmos paleoforest, and likely other floras of the WINA, selected for an ecological strategy that maximized conductance and efficient carbon gain, and penalized high cavitation resistance because of its associated cost in hydraulic efficiency. As the probability of finding large pores in the pit membrane increases with vessel size [14,20,21], resistance to cavitation comes at expense of vessel size and conduction capacity. We suggest that the probability of drought in the Olmos paleoforest was low, otherwise the high vulnerability to cavitation of most species, together with a narrow range of ecological strategies along this functional aspect, would be an extremely risky strategy.

Supporting Information

Table S1 Vessel dimensions data for each of the samples/morphotypes.

Table S2 Sample means for each functional trait.

Table S3 Individual measurements of implosion resistance and estimated cavitation resistance.

Table S4 Trait loadings in the three first PC axes.

Acknowledgments

We thank Uwe Hacke for providing the regression equation to calculate P_{50} values, Jochen Schenk and Cynthia Jones for providing data for some of the extant communities, and Deborah Woodcock and Garland Upchurch Jr. for their comments on a previous draft. We thank Roberto Pujana, Paul Fine and an anonymous reviewer for their constructive criticism.

Author Contributions

Conceived and designed the experiments: HIMC EER. Performed the experiments: HIMC EER. Analyzed the data: HIMC EER. Contributed reagents/materials/analysis tools: HIMC EER. Wrote the paper: HIMC EER.

References

1. Martínez-Cabrera HI, Estrada-Ruiz E, Castañeda-Posadas C, Woodcock D (2012) Wood specific gravity estimation based on wood anatomical traits: Inference of key ecological characteristics in fossil assemblages. Review of Palaeobotany and Palynology 187: 1–10.
2. Royer DL, Sack L, Wilf P, Lusk CH, Jordan GJ, et al. (2007) Fossil leaf economics quantified: calibration, Eocene case study, and implications. Paleobiology 33: 574–589.
3. Wilf P, Wing SL, Greenwood DR, Greenwood CL (1998) Using fossil leaves as paleoprecipitation indicators. An Eocene example. Geology 26: 203–206.
4. Wolfe JA (1971) Tertiary climatic fluctuations and methods of analysis of Tertiary floras. Palaeogeography, Palaeoclimatology, Palaeoecology 9: 27–57.
5. Lambers H, Chapin FS, Pons TL (1998) Plant physiological ecology. New York: Springer-Verlag. 540 p.
6. Machado J-L, Tyree MT (1994) Patterns of hydraulic architecture and water relations of two tropical canopy trees with contrasting leaf phonologies: Ochroma pyramidale and Pseudobombax septenatum. Tree Physiology 14: 219–240.
7. Tyree MT, Sneidermann DA, Wilmot TR, Machado JL (1991) Water relations and hydraulic architecture of a tropical tree (Scheflera morototoni): data, models and a comparison to two temperate species (Acer saccharum and Thuja occidentalis). Plant Physiology 96: 1105–1113.
8. Santiago LS, Goldstein G, Meinzer FC, Fisher JB, Machado K, et al. (2004) Leaf photosynthetic traits scale with hydraulic conductivity and wood density in Panamanian forest canopy trees. Oecologia 140: 543–550.
9. Carlquist S (1975) Ecological strategies of xylem evolution. Berkeley: Univ. Calif. Press. 243 p.
10. Carlquist S (1988) Comparative wood anatomy. Berlin: Springer-Verlag. 436 p.
11. Wheeler EA, Baas P, Rodgers S (2007) Variations in dicot wood anatomy. A global analysis. IAWA Journal 28: 229–258.
12. Poorter L (2008) The relationships of wood-, gas-, and water fractions of tree stems to performance and life history variation in tropical trees. Annals of Botany 102: 367–375.
13. Zanne AE, Westoby M, Falster DS, Ackerly DD, Loarie SR, et al. (2010) Angiosperm wood structure: Global patterns in vessel anatomy and their relation to wood density and potential conductivity. American Journal of Botany 97: 207–215.
14. Zimmerman MH (1983) Xylem structure and ascent of the sap. Berlin: Springer-Verlag. 143 p.
15. Brodribb TJ, Feild TS (2000) Stem hydraulic supply is linked to leaf photosynthetic capacity: evidence from new Caledonian and Tasmanian rainforest. Plan Cell and Environment 23: 1381–1388.
16. Tyree MT (2003) Hydraulic limits on tree performance: transpiration, carbon gain and growth of trees. Trees 17: 95–100.
17. Hacke UG, Sperry JS (2001) Functional and ecological xylem anatomy. Perspectives in Plant Ecology, Evolution and Systematics 4: 97–115.
18. Hacke UG, Sperry JS, Pockman WT, Davis SD, Mcculloh KA (2001) Trends in wood density and structure are linked to prevention of xylem implosion by negative pressure. Oecologia 126: 457–461.
19. Jacobsen AL, Ewers FW, Pratt RB, Paddock WA, Davis SD (2005) Do xylem fibers affect vessel cavitation resistance? Plant Physiology 139: 546–556.
20. Jarbeau JA, Ewers FW, Davis SD (1995) The mechanism of water stress induced embolism in two species of chaparral shrubs. Plant Cell and Environment 18: 189–196.
21. Wheeler JK, Sperry JS, Hacke UG, Hoang N (2005) Inter-vessel pitting and cavitationin woody Rosaceae and other vesselled plants: a basis for a safety versus efficiency trade-off in xylem transport. Plant, Cell and Environment 28: 800–812.
22. Kolb KJ, Sperry JS (1999) Differences in drought adaptation between subspecies of sagebrush (Artemisia tridentata). Ecology 80: 2373–2384.
23. Sperry JS, Pockman WT (1993) Limitation of transpiration by hydraulic conductance and xylem cavitation in Betula occidentalis. Plant Cell and Environment 16: 279–288.
24. Meinzer FC, Clearwater MJM, Goldstein G (2001) Water transport in trees: current perspectives, new insights and some controversies. Environmental and Experimental Botany 45: 239–262.
25. Pratt RB, Ewers FW, Lawson MC, Jacobsen AL, Brediger M, et al. (2005) Mechanism for toleratin freeze-thaw stress of two evergreen chaparral species: Rhus ovata and Malosma laurina (Anacardiaceae). American Journal of Botany 92: 1102–1113.
26. Tyree MT, Sperry JS (1989) Vulnerability of xylem to cavitation and embolism. Annual Review of Plant Physiology and Plant Molecular Biology 40: 19–38.
27. Estrada-Ruiz E, Martínez-Cabrera HI, Cevallos-Ferriz SRS (2010) Fossil woods from the Olmos Formation (late Campanian-early Maastrichtian), Coahuila, Mexico. 97: 1179–1194.
28. Estrada-Ruiz E, Upchurch Jr GR, Cevallos-Ferriz SRS (2008) Flora and climate of the Olmos Formation (upper Campanian-lower Maastrichtian), Coahuila, Mexico: A preliminary report. Gulf Coast Association of Geological Societies Transactions 58: 273–283.
29. Estrada-Ruiz E, Upchurch Jr GR, Wolfe JA, Cevallos-Ferriz SRS (2011) Comparative morphology of fossil and extant leaves of Nelumbonaceae, including a new genus from the Late Cretaceous of Western North America. Systematic Botany 32: 337–351.
30. Serlin B, Delevoryas TH, Weber R (1980) A new conifer pollen cone from the Upper Cretaceous of Coahuila, Mexico. Review of Palaeobotany and Palynology 31: 241–248.
31. Weber R (1972) La vegetación maestrichtiana de la Formación Olmos de Coahuila, México. Boletín de la Sociedad Geológica Mexicana 33: 5–19.
32. Weber R (1973) Salvinia coahuilensis nov. sp. del Cretácico Superior de México. Ameghiniana 10: 173–190.
33. Weber R (1975) Aachenia knoblochii n. sp. an interesting conifer of the Upper Cretaceous Olmos Formation of Northeastern Mexico. Palaeontographica 152B: 76–83.
34. Weber R (1978) Some aspects of the Upper Cretaceous angiosperm flora of Coahuila, Mexico. Courier Forschungsinstitut Senckenberg 30: 38–46.
35. Eguiluz de Antuñano S (2001) Geologic evolution and gas resources of the Sabinas Basin in northeastern Mexico. In: Bartolini C, Buffler RT, Cantú-Chapa A, editors. The western Gulf of Mexico basin: Tectonics, sedimentary basins, and petroleum systems: American Association of Petroleum Geologists Memoir. pp. 241–270.
36. Estrada-Ruiz E (2009) Reconstrucción de los ambientes de depósito y paleoclima de la región de Sabinas-Saltillo, estado de Coahuila, con base en plantas fósiles del Cretácico Superior. Doctoral Thesis, Universidad Nacional Autónoma de México.
37. Flores Espinoza E (1989) Stratigraphy and sedimentology of the Upper Cretaceous terrigenous rocks and coal of the Sabinas-Monclova area, northern Mexico. Austin: University of Texas at Austin. 315 p.
38. Robeck RC, Pesquera VR, Ulloa AS (1956) Geología y depósitos de carbón de la región de Sabinas, Estado de Coahuila. XX Congreso Geológico Internacional, México. pp. 109.
39. Estrada-Ruiz E, Martínez-Cabrera HI, Cevallos-Ferriz SRS (2007) Fossil wood from the late Campanian-early Maastrichtian Olmos Formation, Coahuila, Mexico. Review of Palaeobotany and Palynology 145: 123–133.
40. Martínez-Cabrera HI, Jones CS, Espino S, Schenk HJ (2009) Wood anatomy and wood density in shrubs: Responses to varying aridity along transcontinental transects. American Journal of Botany 96: 1388–1398.
41. Davis SD, Sperry JS, Hacke UG (1999) The relationship between xylem conduit diameter and cavitation caused by freezing. American Journal of Botany 86: 1367–1372.
42. Preston KA, Cornwell WK, DeNoyer JL (2006) Wood density and vessel traits as distinct correlates of ecological strategy in 51 California coast range angiosperms. New Phytologist 170: 807–818.
43. Martínez-Cabrera HI, Cevallos-Ferriz SRS (2008) Palaeoecology of the Miocene El Cien Formation (Mexico) as determined from wood anatomical characters. Review of Palaeobotany and Palynology 150: 154–167.

44. Schenk HJ, Espino S, Goedhart CM, Nordenstahl M, Martínez-Cabrera HI, et al. (2008) Hydraulic integration and shrub growth form linked across continental aridity gradients. Proceedings of the National Academy of Sciences 105: 11248–11253.

45. Baty F, Facompré M, Wiegand J, Schwager J, Brutsche MH (2006) Analysis with respect to instrumental variables for the exploration of microarray data structures. BMC Bioinformatics 7: 422.

46. Dolédec S, Chessel D (1987) Rythmes saisonniers et composantes stationnelles en milieu aquatique I: Description d'un plan d'observations complet par projection de variables. Acta Oecologica, Oecologia Generalis 8: 403–426

47. Chessel D, Dufour AB, Thioulouse J (2004) The ade4 package-I: One-table methods. R News 4: 5–10.

48. Thioulouse J, Chessel D, Dolédec S, Olivier JM (1996) ADE-4: A multivariate analysis and graphical display software. Statistics and Computing 7: 75–83.

49. Maherali H, Pockman WT, Jackson RB (2004) Adaptive variation in the vulnerability of woody plants to xylem cavitation. Ecology 85: 2184–2199.

50. Sperry JS, Sullivan JEM (1992) Xylem embolism in response to freeze-thaw cycles and water stress in ring-porous, diffuse-porous, and conifer species. Plant Physiology 100: 605–613.

51. Wang J, Ives N, Lechowicz MJ (1992) The relation of foliar phenology to xylem embolism in trees. Functional Ecology 6: 469–475.

52. Reich PB, Ellsworth DS, Walters MB, Vose J, Gresham C, et al. (1999) Generality of leaf traits relationships: a test across six biomes. Ecology 80: 1955–1969.

53. Gartner BL, Moore JR, Gardiner BA (2004) Gas in stems: abundance and potential consequences for tree biomechanics. Tree Physiology 24: 1239–1250.

54. Greenwood MC, Wing SL (1995) Eocene continental climates and latitude temperature gradients. Geology 23: 1044–1048.

55. Spicer RA, Parrish JT (1986) Paleobotanical evidence for cool North Polar climates in middle Cretaceous (Albian-Cenomanian) time. Geology 14: 703–706.

56. Wolfe JA, Upchurch Jr GR (1987) North American nonmarine climates and vegetation during the Late Cretaceous. Palaeogeography, Palaeoclimatology, Palaeoecology 61: 33–77.

57. Jaramillo C, Cárdenas A (2013) Global warming and neotropical rainforests: A historical perspective. Annual Review of Earth and Planetary Sciences 41: 741–766.

58. Upchurch Jr GR, Otto-Bliesner BL, Scotese C (1998) Vegetation-atmosphere interactions and their role in global warming during the latest Cretaceous. Phil Trans R Soc Lond B 353: 97–112.

59. Lomax BH, Beerling DJ, Upchurch Jr GR, Otto-Bliesner BL (2000) Terrestrial ecosystem responses to global environmental change across the Cretaceous-Tertiary boundary. Geophysical Research Letters 27: 2149–2152.

60. Wolfe JA (1990) Paleobotanical evidence for a marked temperature increase following the Cretaceous-Tertiary boundary. Nature 343: 153–156.

61. Johnson KR, Reynolds ML, Wert KW, Thomasson JR (2003) Overvew of the Late Cretaceous, early Paleocene, and Early Eocene megafloras of the Denver Basin, Colorado. Rocky Mountain Geology 38: 101–120.

62. Feild TS, Brodribb TJ, Iglesias A, Chatelet DS, Baresch A, et al. (2011) Fossil evidence for Cretaceous escalation in angiosperm leaf vein evolution. Proceedings of the National Academy of Science 108: 8363–8366.

63. Feild TS, Wilson JP (2012) Evolutionary Voyage of Angiosperm Vessel Structure-Function and Its Significance for Early Angiosperm Success. International Journal of Plant Sciences 173: 596–609.

64. Upchurch Jr GR, Wolfe JA (1993) Cretaceous vegetation of the Western Interior and adjacent regions of North America. In: Kauffman EG, Caldwell WGE, editors. Cretaceous evolution of the Western Interior Basin. Geological Association of Canada Special Paper 39. pp. 243–281.

5

Prolonged Instability Prior to a Regime Shift

Trisha L. Spanbauer[1]*, Craig R. Allen[2], David G. Angeler[3], Tarsha Eason[4], Sherilyn C. Fritz[1], Ahjond S. Garmestani[4], Kirsty L. Nash[5], Jeffery R. Stone[6]

1 Department of Earth and Atmospheric Sciences and School of Biological Sciences, University of Nebraska–Lincoln, Lincoln, Nebraska, United States of America, 2 U.S. Geological Survey, Nebraska Cooperative Fish and Wildlife Research Unit, School of Natural Resources, University of Nebraska–Lincoln, Lincoln, Nebraska, United States of America, 3 Department of Aquatic Sciences and Assessment, Swedish University of Agricultural Sciences, Uppsala, Sweden, 4 Office of Research and Development, National Risk Management Research Laboratory, U.S. Environmental Protection Agency, Cincinnati, Ohio, United States of America, 5 Australian Research Council Centre of Excellence for Coral Reef Studies, James Cook University, Townsville, Queensland, Australia, 6 Department of Earth and Environmental Systems, Indiana State University, Terre Haute, Indiana, United States of America

Abstract

Regime shifts are generally defined as the point of 'abrupt' change in the state of a system. However, a seemingly abrupt transition can be the product of a system reorganization that has been ongoing much longer than is evident in statistical analysis of a single component of the system. Using both univariate and multivariate statistical methods, we tested a long-term high-resolution paleoecological dataset with a known change in species assemblage for a regime shift. Analysis of this dataset with Fisher Information and multivariate time series modeling showed that there was a~2000 year period of instability prior to the regime shift. This period of instability and the subsequent regime shift coincide with regional climate change, indicating that the system is undergoing extrinsic forcing. Paleoecological records offer a unique opportunity to test tools for the detection of thresholds and stable-states, and thus to examine the long-term stability of ecosystems over periods of multiple millennia.

Editor: John A. D. Aston, University of Cambridge, United Kingdom

Funding: The Nebraska Cooperative Fish and Wildlife Research Unit is jointly supported by a cooperative agreement between the U.S. Geological Survey, the Nebraska Game and Parks Commission, the University of Nebraska—Lincoln, the United States Fish and Wildlife Service, and the Wildlife Management Institute. This work was supported in part by the August T. Larsson Foundation of the Swedish University of Agricultural Sciences and the NSF's Integrative Graduate Education and Research Traineeship (IGERT) program (NSF #0903469) and the Sedimentary Geology & Paleobiology program (NSF #1251678). The funders had no role in study design, data collection and analysis, decision to publish, or preparation of the manuscript.

Competing Interests: The authors have declared that no competing interests exist.

* Email: tspanbauer@unl.edu

Introduction

Ecosystems can undergo regime shifts and reorganize into an alternative state when a critical threshold is exceeded [1–3]. Most quantitative regime shift research has focused on abrupt shifts that have occurred during a period of human observation; this has resulted in a better understanding of how fast variables (e.g. nutrient loading) erode resilience, but it hasn't addressed how slow variables (e.g. long-term changes in climate) can alter ecosystem state. Paleoecological records can provide insight on the frequency and duration of transitions between alternative states in systems that are affected by both fast and slow variables, at timescales not accessible in the observed record.

To test for regime shifts in the paleoecological record, we used a long-term high-resolution sedimentological record from Foy Lake (Montana, USA) that showed abrupt changes in diatom community structure at ~1.3 ka (thousands of years before present, with present defined as AD 1950). Foy Lake (48.1648°N, 1143589°W, 1005 m elevation) is a deep freshwater lake situated in the drought-sensitive Flathead River Basin in the Northern Rocky Mountains [4,5]. Diatom assemblages in this system are sensitive to changes in lake depth driven by changes in effective moisture [6] and represent one metric of ecological resilience. The percent abundances of 109 diatom species were collected from a lake sediment core that was sampled continuously at an interval of every ~5–20 years, yielding a ~7 kyr record of 800 time-steps.

To determine if regimes shifts could be anticipated in this paleoecological data set we (i) plotted several indicators proposed to be early-warning signals of approaching critical thresholds (increasing variance, skewed responses, kurtosis, and the autocorrelation at lag-1) [7] against time, (ii) collapsed the 109 species variables into the system's mean Fisher information (FI) [8], and (iii) used multivariate time series modeling based on canonical ordination [9]. Many of these statistical early-warning signals have been developed based on bifurcation theory, and they have successfully anticipated regime shifts in many [10–13], but not all [14] systems tested. Increasing variance, skewed responses, and kurtosis in time-series data may be indicative of flickering, the rapid alternating between two different states prior to a regime shift [15]. Along with autocorrelation at lag-1, increasing variance in time-series data can be caused by critical slowing down, where a system is slow to recover from minor disturbances as it approaches a critical transition [7]. These univariate metrics can be limited in their utility, because appreciable signals often occur at the onset of the regime shift, which is generally too late to implement effective management actions [16]. Hence, we sought methods (FI and multivariate time series modeling) that more effectively investigate the dynamics of complex multivariate systems. FI, an integrated index based on information theory, declines as it approaches a regime shift, indicating loss of order and increasing variability, and the regime shift is typically identified as a minimum FI value. Afterward, FI will often increase before settling into a new regime [8]. FI has been used to evaluate stability, regime shifts, and resilience in real complex systems, including ecosystems, climate data, urban systems, and nation states [8,17–25]. Multivariate time series modeling, which models the fluctuation of the frequencies of species or groups of species at distinct temporal scales [9], complements the FI approach. Multivariate time series modeling is sensitive to changes in the abundance and occurrence structure of species in the community. It is capable of identifying scale-specific temporal patterns (fluctuations at scales of decades, centuries, and millennia) in the data and therefore permits assessing how transitional and regime dynamics manifest across the modeled time scales. A key advantage of using these two methods with paleoecological data is that neither requires *a priori* knowledge of system structure or dynamics [8–9].

Results and Discussion

Of the indicators used, we found that univariate species-level indicators were weak predictors of regime shifts. Skewness, kurtosis, and critical slowing down showed minor changes in the frequency patterns of some variables. Several species showed increased variance prior to the abrupt change in species composition at ~1.3 ka. However, most of the species provided no warning signal; hence, conclusions about the dynamics of the overall system were unclear (Fig. 1). Since indicators must be computed for each variable (i.e., diatom species) individually, characterization of the overall system is difficult [8]. For example, the variance of two diatom species, *Cymbella cymbiformis* and *Amphora veneta*, showed very different patterns in variance (Fig. 2). The former would be a good candidate for anticipating the transition in community structure in Foy Lake, while variance in the latter species was random in relation to large scale community shifts. While some particular species might serve as a leading indicator of a regime shift in this system, it is impossible, *a priori*, to identify which species might be appropriate to monitor. In addition, an early-warning indicator species that is effective in Foy Lake may not be useful in other systems, because of differences in physical, chemical, or biological variables that affect community interactions. In summary, it was difficult to detect a community-level regime shift from any of the traditional indicators of early-warning signals, because of the multivariate nature of the study system and the univariate capacity of indicators.

Fisher information identified a substantial regime shift in the system prior to the abrupt community change. The mean FI results indicated that the system was in a steady state (regime one) from ~7.0 to ~4.5 ka. This was followed by a ~2 kyr period of instability, before it returned to a steady state (regime two) at ~1.3 ka (Fig. 3). The long period of instability was followed by an abrupt increase in mean FI at ~2 ka denoting a regime shift [23], which preceded the system regaining stability at ~1.3 ka, and, thus, returning to a steady state. Regimes one and two are considered stable states, because there is no overall directional trend in mean FI values during those periods [23]. During the ~2 kyr period of instability, the mean FI decreased steadily, indicating the system was losing dynamic order, and therefore resilience [8]; this slow period of change is a warning of the impending regime shift at ~2.0 ka.

Multivariate time series modeling revealed eight different temporal patterns in the diatom data set that were associated with eight significant canonical axes in the redundancy analysis (RDA) model. Each of these canonical axes reflects a modeled frequency pattern of individual species or groups of species in the diatom data set. The first three canonical axes capture 55% of the variance used to summarize the transitional dynamics and regime shifts (Fig. 4). The first axis explained the most important pattern in the data set (29% of adjusted variance explained); it separated regime two at ~1.7 ka from all prior time points (Fig. 4). Axes two and three, which explain 18% and 8% of the variability, respectively, separated the time series into three periods: the first regime from the beginning of the record to ~4.8 ka, the period of instability that lasts ~2 kyr, and a second regime that begins at ~1.7 ka. The frequency patterns in the three axes generated with RDA showed temporal patterns of change that are not exactly the same as those detected in the FI results, but that are complementary. The areas that differ most are the ages of both the onset of instability and of the regime shift; these differences likely occur because FI is a composite of all species, whereas the multivariate analysis partitions species into groups. RDA axis one is a long time interval that includes both regime one and the subsequent transition period between regimes one and two. The major axis break at the onset of regime two suggests that regime two is the stronger of the two stable-states in the system's history. This interpretation is supported by the higher mean FI and lower standard deviation in FI of the second regime (Fig. 4). This pattern was driven by a sudden shift in the relative abundance of diatoms, marked by the onset in numerical dominance of one species (*Cyclotella bodanica* var. *lemanica*) during the second regime (Fig. 1). The transitional period, delineated by mean FI, is not present in the first axis of the time series analysis (Fig. 4). However, it is evident in subsequent axes and reflects gradual changes in species composition and dominance patterns (Fig. 1, 4).

The regimes, transitional period, and regime shift detected by FI and time series modeling are consistent with ecological and regional climate patterns. Foy Lake was a moderately deep lake with a diverse planktic and benthic flora during regime one. Throughout the period of instability, the lake was much shallower and dominated by a benthic flora, and during the more recent regime two, Foy Lake was a deep lake dominated by *Cyclotella bodanica* var. *lemanica*, a planktic species [26]. It is possible that either intrinsic (e.g. nutrients) or extrinsic (e.g. climate change) drivers, or a combination of both are responsible for the abrupt ecological change [27]. However, synchronous change in multiple climate records from the region suggests that extrinsic drivers are likely the cause of the changes to the diatom community structure at Foy Lake. A pattern of recurrent multi-decadal drought in the Foy Lake region ended abruptly ~4.5 ka [26]; this is at the approximate time that regime one ends and the ~2 kyr period of instability begins. A shift in the dynamics of the climate system is also evident in multiple other mid-continental paleoclimatic records at ~4.2 ka [28]. At ~1.3 ka multiple regional lake records show a synchronous shift in diatom community structure [29], and regular patterns of reoccurring drought returned to the Foy Lake region [30]. Thus, the intervals of recurrent drought on multi-decadal scales coincide with the identified stable regimes in Foy Lake, whereas the onset of the period of instability occurs during a time of persistent severe drought in the mid-continent. There is a lag between the FI identified regime shift and the abrupt change in diatom community structure (from ~2 ka to 1.3 ka). This lag period is coincident with regional synchronous shifts in diatom communities at multiple lakes at ~2.2 ka, ~1.7 ka, and ~1.35 ka. This suggests that emerging from a period of instability may involve several smaller short-lived transitions in ecosystem state before long term stability is achieved.

Paleoenvironmental and paleoecological data provide a vital and fundamental perspective on the long-term functioning of

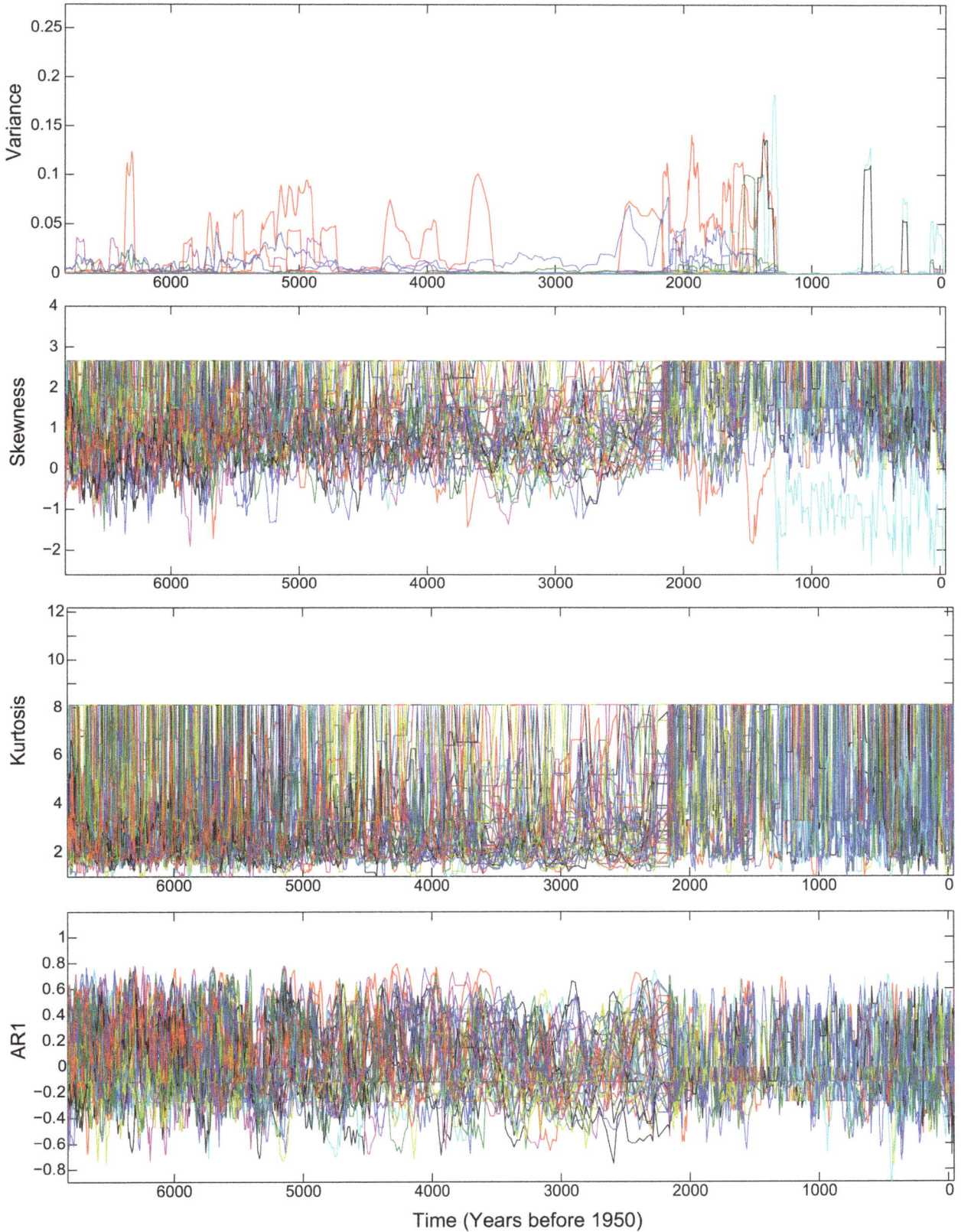

Figure 1. Early warning signals of regime shifts applied to 109 diatom species from Foy Lake. Several populations of species experienced increased variability in the Foy Lake record; this increased variability peaks prior to ~1.3 ka (**A**). Skewness (**B**), kurtosis (**C**), and critical slowing down (**D**) show no clear trends, although, slight frequency changes can be detected at approximately ~4.5 ka and ~2.0 ka.

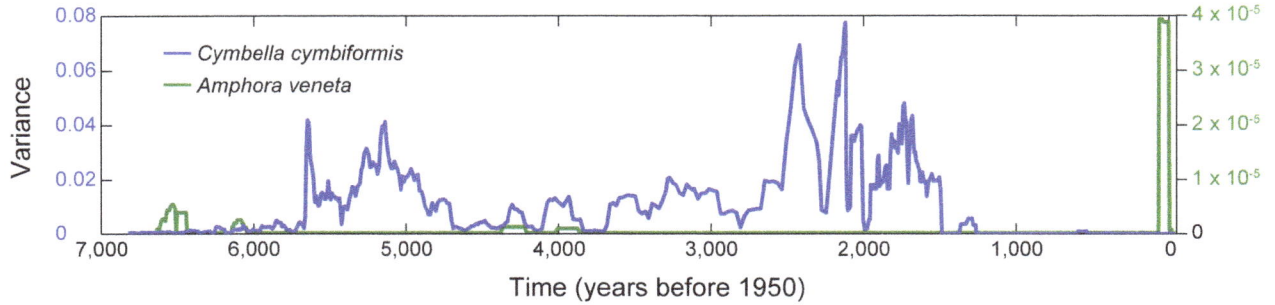

Figure 2. The variance of two diatom species. While *Cymbella cymbiformis* displayed a pattern of increasing variance prior to ~1.3 ka, *Amphora veneta* did not. Conflicting patterns make it difficult to use univariate statistics to characterize the behavior of a complex multivariate system.

complex ecological systems. Here we reveal that climate-driven regime shifts may be infrequent over time in systems not impacted by anthropogenic change, and that transitional periods leading to a regime shift can last a relatively long time (~2.0 kyr). Delayed responses and time lags have been found in other ecosystems [31–33], and these may provide a false sense that the ecosystems are stable, leading to their mismanagement [34]. It is likely that some ecosystems are currently in prolonged periods of instability, whereby they are losing resilience and are exposed to compounding

stresses driven by anthropogenic change. Moreover, when disturbance is large-scale and long-term, some early-warning signals may occur long before the system settles into an alternate stable regime, and the lag between signal and stability may be difficult to predict. Here we suggest effective tools (FI and multivariate time series modeling) to detect and understand changes in those ecosystems that are susceptible to periods of prolonged instability prior to regime shifts.

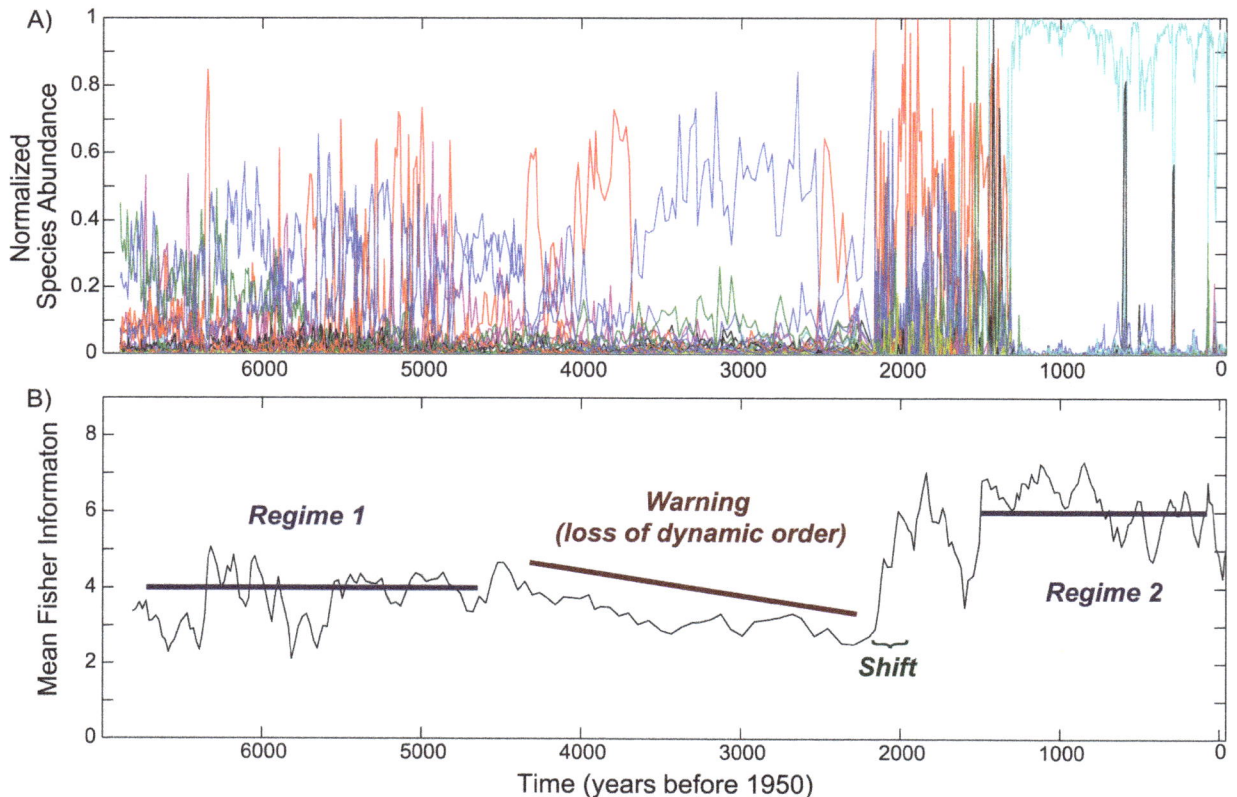

Figure 3. Normalized diatom species abundance for all species (A) and mean Fisher Information (B) for Foy Lake. Prior to ~4.5 ka the system had episodic fluctuations in species composition and mean FI, but the overall mean of the FI is unchanging; this suggests that this period was a stable regime characterized by high variability. At ~4.5 ka species evenness decreases, and the system begins a ~2 kyr gradual decrease in mean FI. Decreases in FI suggest the system is becoming unstable; as instability increases resilience decreases, warning of a possible regime shift. The system was in this unstable transitional period until ~2 ka, but it did not attain a new stable-state until ~1.3 ka.

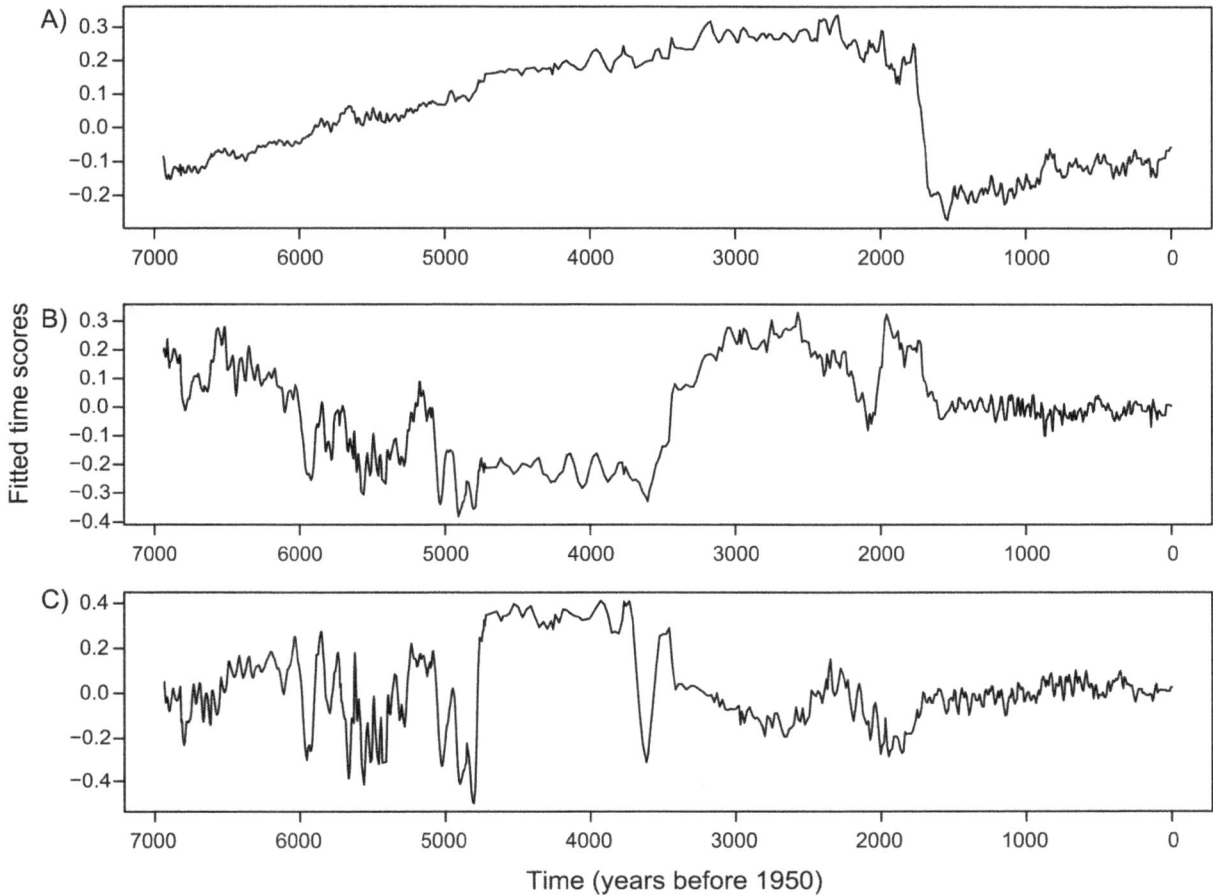

Figure 4. The first three significant axes of the multivariate time-series modeling. The proportion of variance explained by each axis is 29%, 18%, and 8% respectively. The amplitude of the frequency is low in axis one (**A**) with a major shift in score at ~1.7 ka, indicating a regime shift to an alternate state. This regime shift occurred when the lake changed from a shallow lake dominated by benthic taxa to a deep lake dominated by planktic taxa. Frequency pattern changes are present in axes two (**B**) and three (**C**) at ~4.8–5 ka and ~2–1.3 ka, at the beginning and the end of the period of instability.

Methods

Calculating Early Warning Signals

Rising variance, skewness, kurtosis, and critical slowing down are statistical measures that have been proposed and employed as indicators of impending regime shifts [11,16,35–37]. Most of the indicators (i.e., variance, skewness, and kurtosis) are straightforward and can be computed using readily available functions in standard statistical packages (e.g., the Matlab function for computing variance is var). Critical slowing down is estimated by using the lag-1 autocorrelation coefficient [11]. Hence, the autocorrelation function is used to calculate this indicator. For the sake of consistency, all statistical indicators were computed from the percent abundance of each diatom species given the same window size (10 time steps) over the 7 kyr record using Matlab (Release 2012a, Mathworks, Inc.).

Fisher Information

Fisher information (FI) can be used to evaluate the dynamic order of ecosystems, including regimes and regime shifts [23–24]. Unlike early warning signals, FI characterizes changes in complex system dynamics as a function of patterns in underlying variables (e.g., species abundances of diatoms) by collapsing their behavior

of into an index that can be tracked over time [23]. The form of Fisher information (*I*) used in this work was adapted by Fath et al. [18] and Mayer et al. [38].

$$I = \int \frac{1}{p(s)} \left[\frac{dp(s)}{ds} \right]^2 ds \qquad (1)$$

Here, p(s) is the probability of observing the system in a particular condition (state, s) of the system. This equation was adapted [17–18], such that FI could be computed analytically or estimated numerically [23]. The numerical approach of FI was applied in this work and calculated from the following expression (derived in detail by Karunanithi et al. [23]:

$$FI = 4 \sum_{i=1}^{n} [q_i - q_{i+1}]^2 \qquad (2)$$

where, the probability density p(s) is replaced by its amplitude ($q^2(s) \equiv p(s)$) in order to minimize calculation errors from very small p(s). From Equation 1, note that FI is proportional to the change in the probability of a system being in a particular state ($p(s)$) versus the change in state *ds*, i.e., $FI \propto dp/ds$ [18].

Calculating FI

Assessing the dynamic changes in system behavior requires gathering information on its condition (state) through time; hence, measurable variables (x_i) are selected such that a time varying system has a trajectory in a phase space defined by the n-dimensions of its system variables and time. Each point in the trajectory is defined by specific values for each of the n variables (i.e., a point at time i is defined as $[x_1(t_i), x_2(t_i) x_3(t_i)...x_n (t_i)]$). Since uncertainty is inherent in any measurement and system variables may fluctuate within a stable state, a state is defined as a region bounded by a level of uncertainty (or size of states for each dimension (i): $sost_i$), such that if $|x_i(t_i) - x_i (t_j)| \leq sost_i$ is true for all variables then the two points at times i and j are indistinguishable and are identified as being in the same state of the system. There are a number of methods for defining the sizes of states parameter, but the general idea is to assign a level of uncertainty for each variable based on either knowledge of the system (empirically or theoretically) or estimation [24]. Given this conceptual description of systems and states, the probability $p(s)$ of a system being in a particular state (s) can be estimated by counting the number of observational data points that meet the size of states criteria. Using this approach, it is possible to designate all possible states of the system over time.

The basic steps employed to compute FI for the Foy Lake system were as follows: (1) the diatom time series data (consisting of the relative abundances of all 109 species) were divided into a sequence of overlapping time windows with each window containing 10 time steps. Since the goal is to capture changing patterns, there is no particular window size that must be used to compute FI. The window size is set based on available data and from empirical studies, it is typically at least eight time steps [39]. (2) The level of uncertainty was estimated by searching for the window (i.e., 10 time steps) within the diatom time series with the least amount of variability. The standard deviation for each species was then calculated to establish the size of states criteria and bin points into states. (3) The binned points were then used to generate probability densities, $p(s)$, for each state. (4) Equation 2 was used to compute a unique FI for each window resulting in a sequence of FI values over time. The algorithm for computing FI was coded in Matlab (Release 2012a, Mathworks, Inc). Additional details of the FI derivation, calculation methodology, and computer code can be found in [23,39].

Interpreting FI

Assessing system behavior using FI is based on the fundamental idea that different regimes (set of system conditions) exhibit different degrees of dynamic order [23]. In practical terms, a regime fluctuates within a range of variation, such that the overall condition does not change from one observation to another. Hence, the resulting FI is non-zero and remains relatively stable through time. Steadily decreasing FI signifies loss of dynamic order and resilience of a regime and provides warning of an impending regime shift. A decrease in FI between two stable dynamic regimes denotes a regime shift [8,23]. This shift point is typically identified as a minimum FI value after which FI will often increase. While steadily rising FI is indicative of increasing dynamic order, it denotes a shift to a new regime, only if the increase is followed by a new stable regime (i.e., period in which $d\langle FI\rangle/dt \approx 0$). Note that there is no guarantee that the latter regime is more desirable than the former, i.e., while the condition of the system may be stable, the system could have organized into a less desirable regime (e.g., eutrophic lake). Hence, FI affords the ability to assess the stability of a system, not the quality of its condition [25]. Further evaluation of the underlying variables is required to determine whether the system state is desirable.

Multivariate time series modeling

To assess patterns and scales of diatom fluctuations, we constructed time series models based on redundancy analysis (RDA) [9], and used temporal variables extracted by PCNM (Principal Coordinates of Neighbor Matrices) analysis [40–41]. Briefly, the PCNM analysis converts the linear time vector that comprises the sampling frequency and length of the study period into a set of orthogonal temporal variables. In our study, the time vector consisted of 800 time steps during the 7 kyr study period. The PCNM analysis yielded 517 variables with sine-wave properties from the conversion of the linear time vector. Each PCNM variable corresponds to a specific temporal frequency in the diatom dynamics. That is, the first PCNM variable models the longest temporal frequency while the subsequent variables capture temporal variability from longer to increasingly shorter fluctuation frequencies in the community data over the study period. We constructed a parsimonious RDA model for diatom community dynamics by running a forward selection on the 517 PCNM variables.

The RDA retains significant PCNM variables, and these are linearly combined to extract temporal patterns from the Hellinger-transformed species matrices [42]; that is, the RDA identifies species with similar temporal patterns in the species × time matrix and uses their temporal patterns to calculate a modeled species group trend for these species based on linearly combined PCNMs. The significance of the temporal patterns of all modeled fluctuation patterns of species groups revealed by the RDA is tested by means of permutation tests. The RDA relates each modeled temporal fluctuation pattern with a significant canonical axis. The R software generates linear combination (lc) score plots, which visually present the modeled fit of temporal patterns of species groups that are associated with each canonical axis. Because the canonical axes are orthogonal (independent from each other), one can assess the number of temporal scales at which community dynamics unfold. All relevant steps in the time series analysis are carried out using the "quickPCNM" function in R 2.15.0 (R Development Core Team).

Supporting Information

Dataset S1 Percent abundances of diatom species from Foy Lake calculated relative to the total number of diatom valves counted in each sample. Time steps with no diatom data, due to poor preservation, were removed from the dataset. Time steps 301–312 were averaged for these analyses, because they were assigned the same age, as per the age model.

Acknowledgments

We thank two anonymous reviewers for comments that greatly enhanced this manuscript. Data reported in this paper are archived at the National Climatic Data Center, Data Contribution Series # 2008-070. Any use of trade, firm, or product names is for descriptive purposes only and does not imply endorsement by the U.S. Government. The views expressed in this paper are those of the authors and do not necessarily represent the views or policies of the U.S. Environmental Protection Agency.

Author Contributions

TLS led the writing of the manuscript with contribution from all of the authors. Conceived and designed the experiments: TLS CRA DGA TE. Performed the experiments: JRS. Analyzed the data: DGA TE KLN TLS CRA SCF ASG. Wrote the paper: TLS DGA TE.

References

1. Scheffer M, Carpenter SR, Foley JA, Folke C, Walker B (2001) Catastrophic shifts in ecosystems. Nature 413: 591–596.
2. Scheffer M, Carpenter SR (2003) Catastrophic regime shifts in ecosystems: linking theory to observation. Trends Ecol Evol 18: 648–656.
3. Folke C, Carpenter S, Walker B, Scheffer M, Elmqvist T, et al. (2004) Regime shifts, resilience, and biodiversity in ecosystem management. Annu Rev Ecol Evol Syst. 35: 557–581.
4. McCabe GJ, Palecki MA, Betancourt JL (2004) Pacific and Atlantic Ocean influences on multidecadal drought frequency in the United States. Proc Natl Acad Sci USA 101: 4136–4141.
5. Pederson GT, Fagre DB, Gray ST, Graumlich LJ (2004) Decadal-scale climate drivers for glacial dynamics in Glacier National Park, Montana, USA. Geophys Res Lett 31: doi:10.1029/2004GL019770.
6. Stone JR, Fritz SC (2004) Three-dimensional modeling of lacustrine diatom habitat areas: Improving paleolimnological interpretation of planktic:benthic ratios. Limnol Oceanogr 49: 1540–1548.
7. Scheffer M, Carpenter SR, Lenton TM, Bascompte J, Brock W, et al. (2012) Anticipating critical transitions. Science 338: 344–348.
8. Eason T, Garmestani A, Cabezas H (2014) Managing for resilience: early detection of catastrophic shifts in complex systems. Clean Techn Environl Policy, 16: 773–783.
9. Angeler DG, Viedma O, Moreno J (2009) Statistical performance and information content of time lag analysis and redundancy analysis in time series modeling. Ecology 90: 3245–3257.
10. Carpenter SR, Brock WA (2006) Rising variance: a leading indicator of ecological transition. Ecol Lett 9: 311–318.
11. Dakos V, Scheffer M, Van Nes E, Brovkin V, Petoukhov V, et al. (2008) Slowing down as an early warning signal for abrupt climate change. Proc Natl Acad Sci USA 105: 14308–14312.
12. Drake JM, Griffen BD (2010) Early warning signals of extinction in deteriorating environments. Nature 467: 456–459.
13. Dai L, Vorselen D, Korolev KS, Gore J (2012) Generic indicators for loss of resilience before a tipping point leading to population collapse. Science 336: 1175–1177.
14. Hastings A, Wysham DB (2010) Regime shifts in ecological systems can occur with no warning. Ecol Lett 13: 464–472.
15. Dakos V, Carpenter SR, Brock WA, Ellison AM, Guttal V, et al. (2012) Methods for Detecting Early Warnings of Critical Transitions in Time Series Illustrated Using Simulated Ecological Data. PLoS ONE 7: e41010. doi: 10.1371/journal.pone.0041010.
16. Biggs R, Carpenter SR, Brock WA (2009) Turning back from the brink: detecting an impending regime shift in time to avert it. Proc Natl Acad Sci USA 106: 826–831.
17. Mayer AL, Pawlowski CW, Cabezas H (2006) Fisher information and dynamic regime changes in ecological systems. Ecol Model 195: 72–82.
18. Fath BD, Cabezas H, Pawlowski CW (2003) Regime changes in ecological systems: an information theory approach. J Theor Biol 222: 517–530.
19. Fath BD, Cabezas H (2004) Exergy and Fisher information as ecological indices. Ecol Model 174: 25–35.
20. Cabezas H, Pawlowski CW, Mayer AL, Hoagland NT (2005) Sustainable systems theory: ecological and other aspects. J Clean Prod 13: 455–467.
21. Rico-Ramirez V, Reyes-Mendoza PA, Ortiz-Cruz JA (2010) Fisher Information on the performance of dynamic systems. Ind Eng Chem Res 49: 1812–1821.
22. Shastri Y, Diwekar U, Cabezas H, Williamson J (2008) Is sustainability achievable? Exploring the limits of sustainability with model systems. Environ Sci Technol 42: 6710–6716.
23. Karunanithi AT, Cabezas H, Frieden BR, Pawlowski CW (2008) Detection and Assessment of ecosystem regime shifts from Fisher information. Ecol Soc 13: 22 URL: http://www.ecologyandsociety.org/vol13/iss1/art22/.
24. Eason T, Cabezas H (2012) Evaluating the sustainability of a regional system using Fisher information, San Luis Basin, Colorado. J Environ Manage 94: 41–49.
25. Eason T, Garmestani AS (2012) Cross-scale dynamics of a regional urban system through time. Region et Developpement 36: 55–77.
26. Stone JR, Fritz SC (2006) Multidecadal drought and Holocene climate instability in the Rocky Mountains. Geology 34: 409–412.
27. Williams JW, Blois JL, Shuman BN (2011) Extrinsic and intrinsic forcing of abrupt ecological change: case studies from the late Quaternary. J Ecology 99: 664–667.
28. Booth RK, Jackson ST, Forman SL, Kutzbach JE, Bettis III EA, et al. (2005) A severe centennial-scale drought in mid-continental North America 4200 years ago and apparent global linkages. The Holocene 15: 321–328.
29. Bracht-Flyr B, Fritz SC (2012) Synchronous climatic change inferred from diatom records in four western Montana lakes in the U.S. Rocky Mountains. Quat Res 77: 456–467.
30. Stevens LR, Stone JR, Campbell J, Fritz SC (2006) A 2200-yr record of hydrologic variability from Foy Lake, Montana, USA, inferred from diatom and geochemical data. Quat Res 65: 264–274.
31. Frank KT, Petrie B, Fisher JAD, Leggett WC (2011) Transient dynamics of an altered large marine ecosystem. Nature 477: 86–89.
32. Krauss J, Bommarco R, Guardiola M, Heikkinen RK, Helm A, et al. (2010) Habitat fragmentation causes immediate and time delayed biodiversity loss at different trophic levels. Ecol Lett 13: 597–605.
33. Menéndez R, González Megías A, Hill JK, Braschler B, Willis SC, et al. (2006) Species richness changes lag behind climate change. Proc R Soc London Ser B 273: 1465–1470.
34. Hughes TP, Linares C, Dakos V, van de Leemput IA, van Nes EH (2013) Living dangerously on borrowed time during slow, unrecognized regime shifts. Trends Ecol Evol 28: 149–155.
35. Brock WA, Carpenter SR (2006) Variance as a leading indicator of regime shift in ecosystem services. Ecol Soc 11: 9 URL: http://www.ecologyandsociety.org/vol11/iss2/art9.
36. van Nes EH, Scheffer M (2007) Slow recovery from perturbations as a generic indicator of a nearby catastrophic shift. Am Nat 169: 738–747.
37. Scheffer M, Bascompte J, Brock W, Brovkin V, Carpenter S, et al. (2009) Early-warning signals for critical transitions. Nature 461: 53–59.
38. Mayer AL, Pawlowski CW, Fath BD, Cabezas H (2007) In: Frieden BR, Gatenby RA, editors. Exploratory Data Analysis Using Fisher Information. London: Springer-Verlag. pp. 217–244.
39. Cabezas H, Eason T (2010) Fisher information and order. In: Heberling MT, Hopton ME, editors. San Luis Basin Sustainability Metrics Project: A Methodology for Assessing Regional Sustainability. US EPA Report: EPA/600/R-10/182. pp. 163–222.
40. Borcard D, Legendre P (2002) All-scale spatial analysis of ecological data by means of principal coordinates of neighbour matrices. Ecol Model 153: 51–68.
41. Borcard D, Legendre P, Avois-Jacquet C, Tuomisto H (2004) Dissecting the spatial structure of ecological data at multiple scales. Ecology 85: 1826–1832.
42. Legendre P, Gallagher ED (2001) Ecologically meaningful transformations for ordination of species data. Oecologia 129: 271–280.

δ^{18}O in the Tropical Conifer *Agathis robusta* Records ENSO-Related Precipitation Variations

Bjorn M. M. Boysen[1], Michael N. Evans[2]*, Patrick J. Baker[3]

1 Department of Environmental and Primary Resources, State of Victoria, East Melbourne, Victoria, Australia, 2 Department of Geology and Earth System Science Interdisciplinary Center, University of Maryland, College Park, Maryland, United States of America, 3 Department of Forest and Ecosystem Science, Melbourne School of Land and Environment, University of Melbourne, Victoria, Australia

Abstract

Long-lived trees from tropical Australasia are a potential source of information about internal variability of the El Niño-Southern Oscillation (ENSO), because they occur in a region where precipitation variability is closely associated with ENSO activity. We measured tree-ring width and oxygen isotopic composition (δ^{18}O) of α-cellulose from *Agathis robusta* (Queensland Kauri) samples collected in the Atherton Tablelands, Queensland, Australia. Standard ring-width chronologies yielded low internal consistency due to the frequent presence of false ring-like anatomical features. However, in a detailed examination of the most recent 15 years of growth (1995–2010), we found significant correlation between δ^{18}O and local precipitation, the latter associated with ENSO activity. The results are consistent with process-based forward modeling of the oxygen isotopic composition of α-cellulose. The δ^{18}O record also enabled us to confirm the presence of a false growth ring in one of the three samples in the composite record, and to determine that it occurred as a consequence of anomalously low rainfall in the middle of the 2004/5 rainy season. The combination of incremental growth and isotopic measures may be a powerful approach to development of long-term (150+ year) ENSO reconstructions from the terrestrial tropics of Australasia.

Editor: Peter Wilf, Penn State University, United States of America

Funding: This research was supported through US National Science Foundation grants AGS0902794 and EAR0929983 to MNE and Australian Research Council grant DP0878744 to PJB. The funders had no role in study design, data collection and analysis, decision to publish, or preparation of the manuscript.

Competing Interests: The authors have declared that no competing interests exist.

* Email: mnevans@umd.edu

Introduction

The El Niño-Southern Oscillation (ENSO) is one of the leading sources of regional- and global-scale climate variability. A more complete understanding of longer-term, intrinsic variability in ENSO [1] is limited, however, by the absence of direct observations of surface climate before the second half of the 20th century; this is particularly true in the southwest Pacific [2–4] and across the West Pacific Warm Pool. The relationship between ENSO and regional climate variability requires a longer historical context, and paleoclimatic reconstructions provide a means of achieving this goal [3]. However, the most widely used high resolution tropical paleoclimatic archives – corals, speleothems, and tree rings – have significant limitations. Corals and speleothem archives are sparsely distributed and hence the paleodata acquired from them often have limited replication. Trees are a widely distributed terrestrial archive and may provide highly replicated paleodata, but in ENSO-affected tropical regions, may not reliably produce annually resolved tree rings [5,6].

Tropical and sub-tropical tree species and environments present a challenge for paleoclimatology because they often fail to generate the regular annual patterns of cambial activity and dormancy that produce anatomical features reliably identified as annual growth increments ("tree rings") [7,8]. Short-duration, transient climatic conditions may lead to opportunistic growth and dormancy cycles,

which may masquerade as annual growth increments (so-called "false rings") [9–13]. Furthermore, persistently poor growing conditions over long time periods may result in missing annual growth increments ("missing rings") [11,14]. The presence of false and missing growth increments undermines our ability to date materials accurately and precisely – a necessary precursor for paleoclimatic reconstructions from tropical trees. As a result, annually resolved tree rings may provide long, replicated records in the extratropics, but they are rarely reported in the tropical and sub-tropical regions that are directly influenced by ENSO dynamics [15].

Tropical environments are, however, often defined by pronounced and relatively regular variations in moisture, which can be indirectly observed in the oxygen isotopic composition of the α-cellulose component of tropical wood [16]. This is because the oxygen isotopic composition of α-cellulose primarily reflects the isotopic composition of soil moisture, modified by leaf-level evapotranspiration, isotopic exchange between leaf water and unmodified stem water, and biosynthetic fractionation [17,18]. In the tropics, the isotopic composition of soil moisture is in turn largely determined by the amount of precipitation, because removal of isotopically heavy condensate (at 25°C, the equilibrium fractionation factor for liquid relative to vapor is 1.0092) from a precipitating air mass leaves subsequent precipitation isotopically light [19–21]. The "tropical isotope dendroclimatology" hypothesis [16,22] predicts that sub-annual resolution sampling of tropical trees for isotopic composition can thereby permit detection of an annual cycle in precipitation amount and/or relative humidity, even in trees lacking well-defined annual ring structures [14,16,23–25]. For tropical trees with growth/dormancy cycles controlled by precipitation seasonality, and which can be therefore also be dated by tree-ring analysis [26–29], oxygen isotopic composition of α-cellulose may reflect interannual variations in precipitation amount and/or relative humidity, complementing information derived from analysis of growth increments, rates and anatomical features [30–34].

Northeastern Australia is of particular interest for dendrochronological studies of ENSO variability. Northern Queensland is in close proximity to the Western Pacific Warm Pool, over which ENSO causes changes in the position and strength of large-scale organized convective rainfall [35,36]. The trees in these areas are thus likely to record ENSO-related interannual variations in local rainfall and relative humidity [22]. If so, then high-quality tree-ring records from this region should have considerable potential as high-fidelity indirect observations of regional ENSO variability. Ogden et al (1981) [37] reported that the tropical forests on the slopes of the Atherton Tablelands may contain more than 100 tree species per hectare, and raised the possibility that these areas may harbor many species with untapped dendrochronological potential. A recent comprehensive survey examining the dendroclimatological potential of ~180 tropical tree species from northern Queensland identified several candidates, including *Agathis robusta* (Queensland Kauri), a long-lived conifer that reaches enormous sizes (up to 50 m tall and >3 m in diameter at breast height (1.3 m above the ground)), to be of potential value for climate analyses (P.J. Baker, unpublished data; [38]). Despite the well-established crossdating and interpretation of ring-width variations in New Zealand *Agathis australis* [39–44], early studies [45,46] on *A. robusta* in northern Queensland suggested that growth rings were not strictly annual, and that the major factor limiting growth was dry periods. In particular, occasional dry spells led to irregular patterns of cambial dormancy and thereby the formation of false rings. The presence of false rings in this and other tree species from northern Queensland has limited the progress of dendroclimatology in the region (but see [27,47]).

Here we combine classical dendrochronological techniques with oxygen isotope analyses at seasonal and annual resolution to establish a basis for paleoclimatological studies employing the tropical tree species, *A. robusta*. We assess and discuss the potential of the dual proxy approach for false ring detection and ENSO event reconstruction.

Materials and Methods

Study site and species

Wood sample collections were undertaken, and samples handled, processed and transported in accordance with permits and rules of the Parks and Wildlife Service of the Queensland Department of Environmental Protection, and the Animal and Plant Health Inspection Service of the US Department of Agriculture. No endangered or protected species were sampled in this study. We sampled live *A. robusta* at two locations within Dinden and Danbulla National Parks, situated on the Atherton Tablelands plateau of northern Queensland, Australia. *Agathis robusta* (Araucariaceae) is a large, fast-growing tree with a simple cylindrical growth form and branches that are mainly restricted to the crown. Such characteristics have made this species important for commercial logging, and also make it a prime candidate for dendrochronology. The sites were located at the interface of dry sclerophyll rainforest and wet tropical rainforest. Both National

Parks have a mean elevation of ~700 meters above sea level and are connected by continuous forest. The Atherton Tablelands experience a strong seasonal precipitation cycle (Figure 1), with 85% of annual precipitation occurring between November and April. Warmest temperatures are between October and April, and highest solar exposures are from September through November.

Dendrochronology

Within each National Park we identified areas in which *A. robusta* was relatively abundant. The sites (in Dinden National Park: 16° 58'45.96"S, 145° 36'11.40"E, and in Danbulla National Park: 17° 8'44.74"S, 145°35'14.49"E) were typically steep, well-drained, south-facing slopes composed of coarse granitic soils, and often contained groups of ~10–30 *A. robusta* each. We sampled 31 trees at Danbulla and 28 trees at Dinden during early October 2010, before the onset of the wet season. Three cores were taken from each tree at a height of ~1.2 m using a Haglöf 5.15 mm increment corer. Tree cores were mounted on wooden trays using water-soluble glue for protection and stability during processing. The core surfaces were then sanded using progressively finer grades of sandpaper to obtain a final high quality surface for anatomical analysis, and scanned using an 1800 dpi Microtek 1000XL digital scanner.

Because of the complex nature of anatomical features in the samples, careful visual microscopic examination of the materials was used to check age assignment features. The program COFECHA [48] was used to probabilistically identify dating errors and biases arising from missing and false rings. Following standard practice in dendrochronology, if dating uncertainties for individual series in the sample could not be rectified, those samples were excluded from the site average. Ring-width measurements were made using the image analysis software WinDendro (ver. 2009b, Regent Instruments). The mean interseries correlation (MIC; [49]) of each resulting site composite indicates the level of consistency between all possible pairwise tree core comparisons within a sample set and indicates the level of site-wide (potentially climatic) growth-increment responsiveness.

Oxygen isotopic analysis

Sample preparation. Isotopic sampling was performed at the Department of Geology, University of Maryland, U.S.A. on individual cores from three trees from Dinden National Park. The analysis included dated annual increments corresponding to the period 1995–2010, and leveraged the best dated samples as the basis for isotopic sampling in time. Wood samples for isotopic analysis were obtained from the tree cores using a rotary microtome. Slices of 20 μm thickness were taken in succession, and 20 slices were accumulated into one sample for isotopic analysis. This gave 4–8 samples per annual growth increment as determined by dendrochronological analysis described above. For the section of our cores that covered the period 2004–06, we doubled the sampling intensity (every 10th slice) because our dating process had identified the likelihood of a false ring at this time. The more detailed examination of this period enabled us to examine whether the isotopic composition of α-cellulose was consistent with the false-ring interpretation developed *a priori* from the anatomical analysis described previously.

Microtome slices were ground using a small metal rod to increase efficacy of the cellulose extraction chemistry. Cellulose extraction was carried out using the Brendel technique [50] modified for small sample extractions [16,25]. Samples of α-cellulose (mass: 350±20μg each) were encapsulated in silver foil and converted to carbon monoxide in a 99.999% helium environment over glassy carbon at 1080°C [51] in a Costech

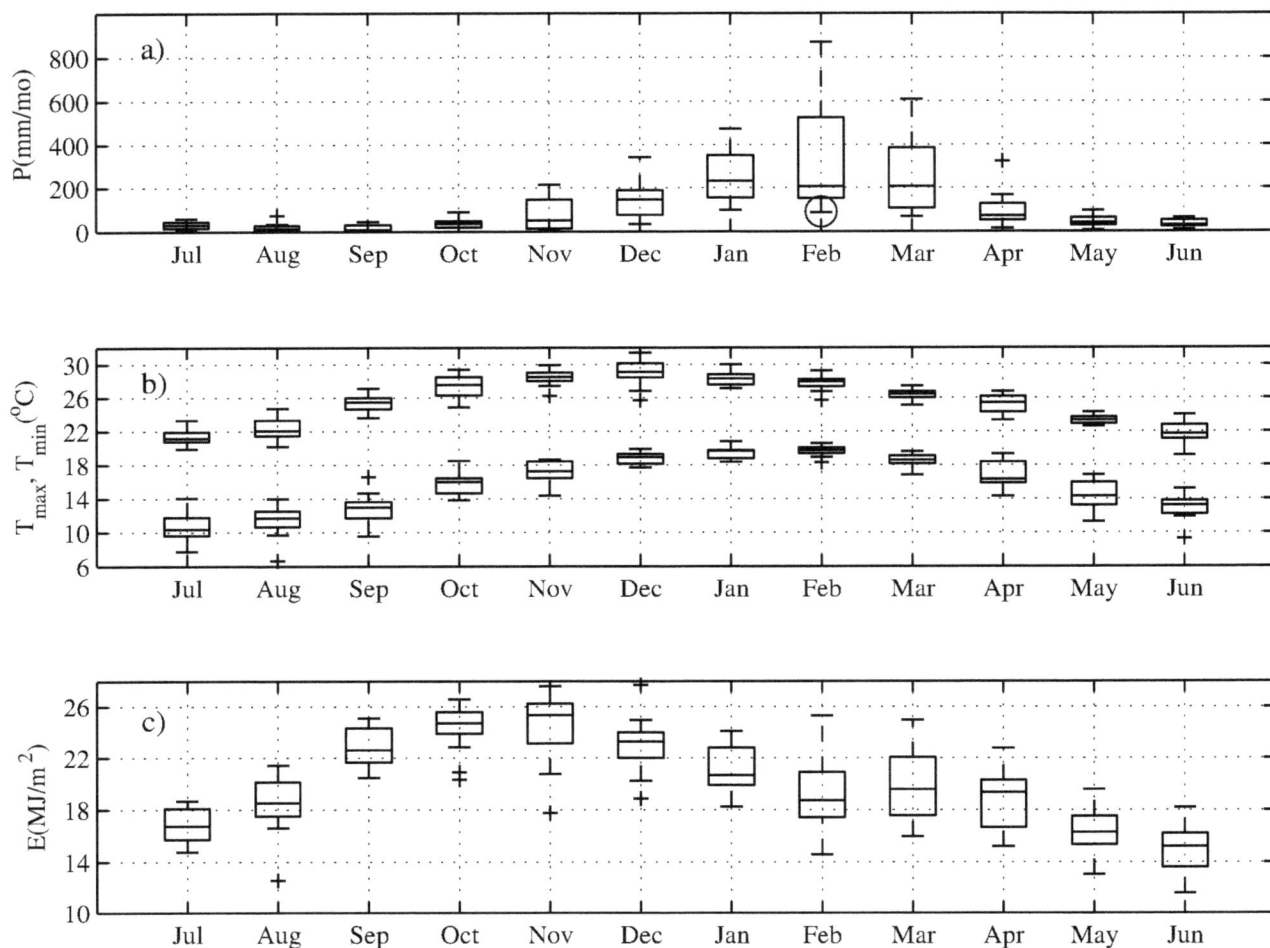

Figure 1. Climatological conditions near the study sites. Monthly averages of precipitation (mm/month; top) daily minimum and maximum temperature (°C; middle), and daily solar exposure (MJ/m²; bottom) averaged (as available) from Tinaroo Falls Dam (31075) and Kairi Research Station (31034) for 1996–2010 [61]. Boxes indicate 25th and 75th percentile range of data, with the line within indicating the median; whiskers correspond to approximately the 99% range of data assuming a normal distribution; outliers to this range are plotted as small crosses. The exceptionally dry month of February 2005 is indicated by the large open circle.

elemental analyzer, passed through a water/CO_2 trap and a molecular sieve 5A gas chromatographic column to separate the CO analyte from any N_2 in the sample stream, and then introduced into an Elementar Isoprime mass spectrometer via continuous flow interface. Reported isotopic values reflect within-sample drift correction via monitoring gas measurements, intersample drift correction by periodic measurement of two working standards, and correction for mean and variance bias by within-batch measurement of two α-cellulose working standards calibrated to the SMOW reference at known $\delta^{18}O$ values of 21.3‰ and 31.0‰, respectively [52,53]. Long-term precision of measurements, based on replicate analyses of cellulose working standards, are <0.3‰.

Isotopic age modeling and compositing. To develop a composite, annually resolved, calendar-dated oxygen isotope time series, we used the tropical isotope dendroclimatology hypothesis to assign January/February calendar age to isotopic minima within each dendrochronologically dated growth increment, assuming that each minimum in the isotope chronology represents the climatologically average wettest month of the November-April

rainy season in northern Queensland (Fig 1; [14,16,22,25]). For each data series, we then linearly interpolated each 4–8 point set of intra-seasonal $\delta^{18}O$ values to produce a uniform four $\delta^{18}O$ interpolates within each November-April growing season. This permitted us to composite data across the three sample series on a common intraseasonal time scale. The composite $\delta^{18}O$ time series is defined as the median of the interpolated isotopic data within each growing season and across the three replicate data series (Figure 2a).

Isotopic modeling

To provide a basis for interpretation of the composite $\delta^{18}O$ observations, we simulated the isotopic composition of Dinden α-cellulose $\delta^{18}O$ using the model of Evans (2007) [22]. Modeling inputs are monthly temperature, precipitation amount, and relative humidity, and 18 fixed parameters. Because specific or relative humidity data were not available from the closest Australian Bureau of Meteorology stations (Figure 1), we used the CRU TS3.10.01 [54] estimates of specific humidity and

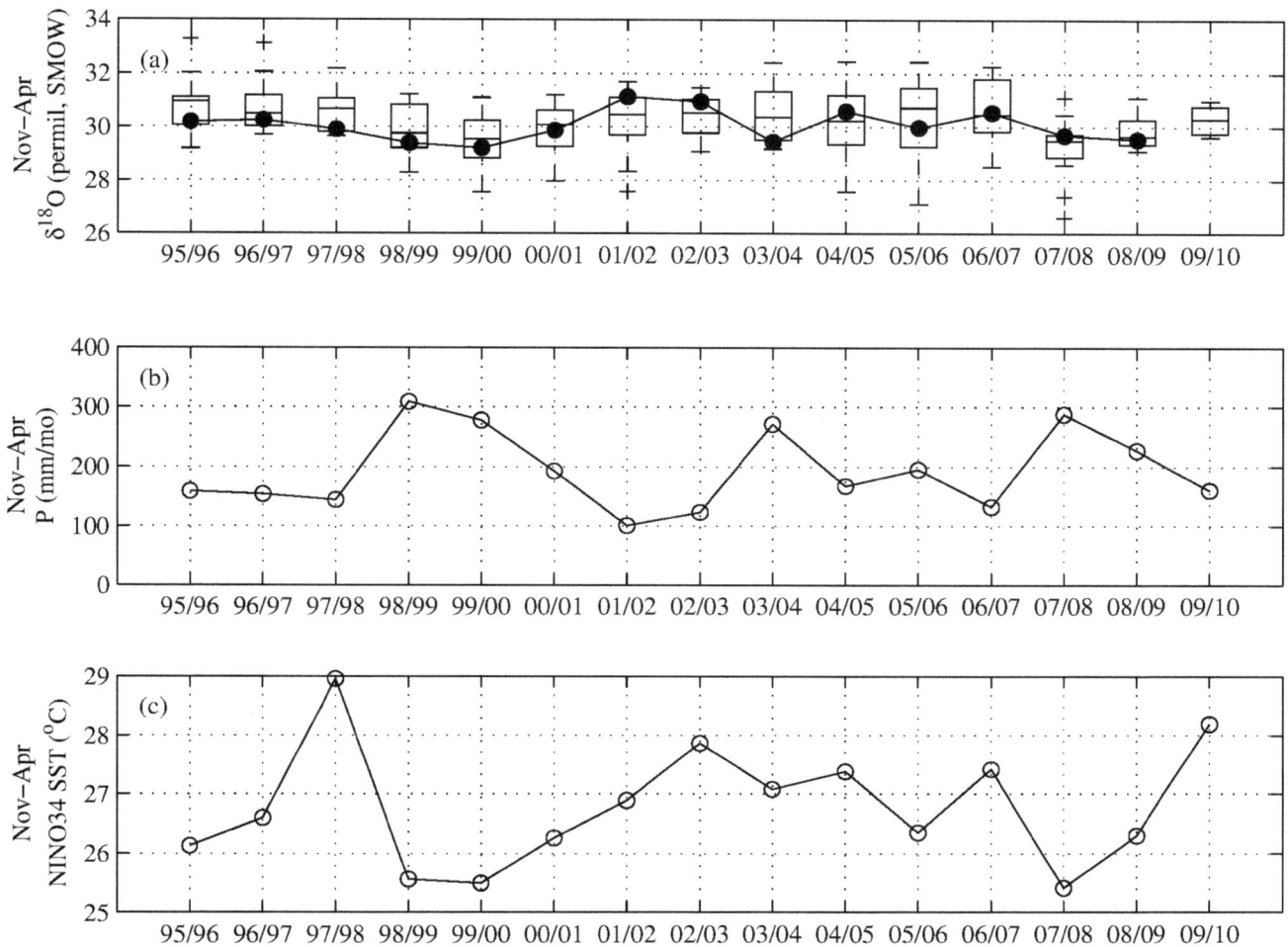

Figure 2. Observed composite δ^{18}O record from the Atherton Tablelands compared to local precipitation and NINO34 SST index. (a): Box-whisker plots (as described for Fig 1) show the distribution of values for each sample of 12 observations (4 interpolated values per season from each of three trees) in each growing season composite. Calendar age assignments (x-axis) are from the crossdating of the retained dendrochronological subset of tree cores from Dinden National Park (Table 2). Solid circles are process-modeled δ^{18}O estimates (see text for details.) (b) November–April average composite precipitation record from Tinaroo Falls and Kairi Research Station (Fig 1). (c) November–April average NINO34 SST.

temperature available for 1995–2010 for the 0.5 x 0.5 degree gridpoint closest to the study site, and calculated relative humidity following [55]. The δ^{18}O model is most sensitive to the specification of two parameters that determine the amount effect in precipitation [21,22]. Accessing all Australian data for oxygen isotopes in precipitation in the GNIP/WISER database [56], we found data for the δ^{18}O and amount of precipitation from one tropical station, Darwin (Station 9412000, 12.43°S, 130.87°E, 26 m). The data available from all years (1962–2002 inclusive) were used to construct the statistically significant linear regression (df = 177, r^2 = 0.22, F = 50.6, p < 0.001) of precipitation isotopic composition on precipitation amount as: δ^{18}O (‰, SMOW) = − 2.56(±0.66)–0.0068(±0.0019)*P (mm/mo) with ±2σ uncertainties on the slope and intercept estimates as indicated. We specified the fraction variance in the simulations associated with precipitation to 0.85 [22], and we set the simulation-mean Nov-April α-cellulose δ^{18}O to that of the composite time series. All other model parameters values were specified as in Tables 1 and 2 of [22], and parameter uncertainty for Monte-Carlo sampling was set to

±40% of parameter values. Because there are many parameters in the model, and the true values of these parameters are largely unknown for this application, we only assess the significance of the linear correlation between the median November–April averaged values from 1000 simulations of the 1995–2009 interval, and the corresponding observed composite medians from the cellulose δ^{18}O data series. Interpretation of the correlation of the simulated and observed medians is independent of specified simulated mean and variance, and relies only on the unspecified interannual coherence of the standardized simulation and observational data series. The isotopic modeling represents our best working hypothesis for explaining variance in the composite δ^{18}O data series; results are shown in Fig 2a.

Statistical Analysis

To assess whether isotopic observations, isotopic simulations and meteorological data are consistent with the hypothesized interpretation of the composite δ^{18}O timeseries, we performed correlation analysis between November–April averages for the

Table 1. ENSO state for the November-April season between 1995/1996 and 2009/2010.

Warm Phase	Neutral Phase	Cold Phase
1997/1998 (28.96)	2003/2004 (27.08)	2000/2001 (26.26)
2009/2010 (28.19)	2001/2002 (26.89)	1995/1996 (26.13)
2002/2003 (27.87)	1996/1997 (26.60)	1998/1999 (25.56)
2006/2007 (27.42)	2005/2006 (26.35)	1999/2000 (25.49)
2004/2005 (27.38)	2008/2009 (26.30)	2007/2008 (25.41)

Estimated using tercile analysis of the NINO34 SST Index for Nov-Apr 1995/6 through Nov-Apr 2009/2010. Average SST for each seasonal estimate (e.g. Nov 1997 - Apr 1998) is given in parentheses. The NINO34 SST index is an oceanographic indicator of large-scale ENSO activity [58], and is higher (lower) during ENSO warm (cold) phase conditions; warm (cold) phase years are expected to be associated with drier (wetter) conditions in northern Queensland [35,36].

common period November 1995–April 2010 (November 1995–April 2009 inclusive for correlation of observed and simulated $\delta^{18}O$ time series). The null hypothesis is that there is no significant correlation between variables, with significance at the $p = 0.05$ level evaluated as one-tailed for the correlation between observed and modeled $\delta^{18}O$ time series, and two-tailed otherwise.

To assess the response of Dinden precipitation amount and composite $\delta^{18}O$ to ENSO state, we also performed Type I analyses of variance (ANOVA) on November-April averages of monthly precipitation averaged from two nearby Australian Bureau of Meteorology observing stations (Figs 1a, 2b), and on observed composite November-April average cellulose $\delta^{18}O$ data (Fig 2a). The treatment groups were defined by the lower, middle and upper terciles of the November-April average NINO34 index, defined as sea surface temperature averaged over the region: 120°W–170°W, 5°N–5°S [57]), for Nov 1995–April 2010 inclusive (Table 1). The NINO34 SST index is an oceanographic indicator of large-scale ENSO activity [58], and is higher (lower) during ENSO warm (cold) phase conditions. The two-tailed null hypotheses for these tests were that there were no observed differences in treatment means among treatment groups, defined as cold, warm and neutral phase ENSO years within the period 1995/6–2009/10 (Table 1).

Results and Discussion

Dendrochronology of A. robusta

The development of a crossdated tree-ring chronology for *A. robusta* proved to be extremely difficult, because of the widespread presence of putatively false rings. Up to 70% of cores from some sites had to be excluded from the crossdating analysis because it was not possible to accurately assign dates to specific growth

increments. A summary of chronology statistics for *A. robusta* from Danbulla National Park (sites 1, 2 & 3) and Dinden National Park (Site 4), is given in Table 2. The difficulties in cross-dating *A. robusta* are indicated by the low mean interseries correlation (MIC) values for each site, despite the exclusion of cores with obvious dating problems. Without additional information, we could not determine whether these features corresponded to within or between growing season growth responses.

Interpretation of the composite $\delta^{18}O$ record

The composite, growing-season averaged $\delta^{18}O$ series is shown together with seasonally averaged station precipitation and NINO34 SST in Figure 2. The 15-year composite $\delta^{18}O$ chronology has an interannual standard deviation of 0.45‰ and within-year composite standard deviation of 0.18‰. Correlations and significances between NINO34 SST (Table 1), Atherton climatological data (Fig 1), observed composite median $\delta^{18}O$ series ($\delta^{18}O_{obs}$) and median simulated $\delta^{18}O$ ($\delta^{18}O_{sim}$) are given in Table 3. As expected from the observed influence of ENSO activity on northern Queensland climate, NINO34 SST is significantly correlated with Atherton precipitation ($p < 0.01$). Consistent with predictions of the tropical isotope dendroclimatology hypothesis, $\delta^{18}O_{obs}$ is negatively correlated with precipitation amount and positively correlated with $\delta^{18}O_{sim}$. The correlation of NINO34 with $\delta^{18}O_{sim}$, however, is not quite statistically significant ($p = 0.06$). These results are consistent with those of the Type (I) ANOVAs (Table 4), which show that the covariance of $\delta^{18}O_{obs}$ with ENSO state ($p = 0.12$) is less significant than the covariance of Atherton precipitation with ENSO state ($p = 0.04$). These two results together suggest that additional $\delta^{18}O_{obs}$ data, used either to improve growing season composite replication and/or to increase

Table 2. Chronology statistics for *A. robusta* at four sites in the Atherton Tablelands, northern Queensland, Australia.

Site	Location	Trees sampled (cores)	Crossdated trees (cores)	MIC
1	Danbulla NP	11 (36)	6 (14)	0.272
2	Danbulla NP	11 (29)	8 (15)	0.298
3	Danbulla NP	9 (28)	5 (12)	0.307
4	Dinden NP	28 (86)	8 (18)	0.276

Mean interseries correlation (MIC) is the average of all possible correlations between width series from individual cores within each site chronology, and is a measure of the quality of the ring width chronology at each site.

Table 3. Correlation between climatological data, composite median δ^{18}O and simulated median δ^{18}O.

Variable	P	T_{min}	T_{max}	S	$\delta^{18}O_{obs}$	$\delta^{18}O_{sim}$
NINO34	−0.67**	0.33	−0.47	0.48	0.57*	0.51
P		−0.37	0.11	−0.40	−0.74**	−0.88**
T_{max}			−0.27	0.69**	0.44	0.53
T_{max}				−0.32	−0.34	−0.11
S					0.19	0.50
$\delta^{18}O_{obs}$						0.59*

NINO34 is NINO34 SST; P, T_{max}, T_{min}, and S are precipitation, maximum and minimum temperature and solar exposure data compiled from two Atherton meteorological stations (Fig 1). Correlation significances are based on two-tailed t-tests with 13 degrees of freedom (12 degrees of freedom for correlations with $\delta^{18}18O_{sim}$). * indicates correlation is significant at $p<0.05$, ** indicates correlation is significant at $p<0.01$.

the number of degrees of freedom, should result in increased statistical confidence in our ability to detect ENSO state from the isotopic data.

Detection of false rings

Comparison of limited intraseasonal-resolution δ^{18}O measurements with wood anatomy suggests the potential for detecting false rings in dendrochronological samples. Figure 3 shows a small section of a tree core spanning two complete growing seasons (2004–05 and 2005–06) for which δ^{18}O was measured at a resolution of 8–12 samples per growing season. The 2005–06 growing season shows a simple annual growth ring and an annual cycle of δ^{18}O values consistent with the prediction of the tropical isotope dendroclimatology hypothesis, with minimum δ^{18}O in the middle of the growth increment, and maximum δ^{18}O at the beginning and end of the growing season (Figure 1).

In contrast, the 2004–05 ring presents a problem. In the image of the core (Fig 3, upper panel), there are two anatomically distinct bands of dark, high-density xylem cells in close proximity – one of which, if the isotopic age model hypothesis is correct – is a false ring. In the absence of the isotope data, there is no a priori reason to identify either one of these features from 2004–05 as a false ring. The earlier feature might be a false ring if the tree stopped growing near the end of the preceding growing season and then briefly reinitiated growth, due to, for example, an unseasonably late rainfall event. Alternatively, the later feature might be the false ring, if the tree started growing early in the growing season due to an unusually early rainfall event, then stopped due to restored dry season conditions, before the full onset of the wet season, and after which growth was reinitiated. The timing and amount of precipitation delivered during the 2003–2004 rainy season was not unusual. However, in February 2005, in the middle of the rainy season, precipitation in the study region was 89.1 mm, much lower than typical (Fig 1, open circle). Dry conditions at this time may have induced sufficient physiological stress in the trees to initiate the cessation of cambial activity and the formation of latewood-like cells. Many studies across a range of climates have shown that, as a consequence of dry conditions, smaller tracheids with thicker cell walls may be formed by the cambium (e.g. [59]). Tracheids that subsequently form when wetter conditions return are larger in diameter and have thinner

Table 4. Type (I) ANOVAs: Atherton precipitation and δ^{18}O of A. robusta, for the NINO34 sea surface temperature terciles defined in Table 1.

	SS	df	MS	F	p
Atherton precipitation					
Treatment	25048.04	2	12524.02	4.17	0.04
Error	36052.60	12	3004.38		
Total	61100.64	14			
δ^{18}O of A. robusta					
Treatment	0.90	2	0.45	2.51	0.12
Error	2.15	12	0.18		
Total	3.05	14			

Test of the null hypothesis that there is no significant difference between the mean precipitation (composite observed δ^{18}O) observed during above normal, normal, and below normal NINO34 SSTs, 1995/1996–2009/2010 (Table 1). SS = sum of squares; MS = mean square; df = degrees of freedom; F = F statistic; p = p-value.

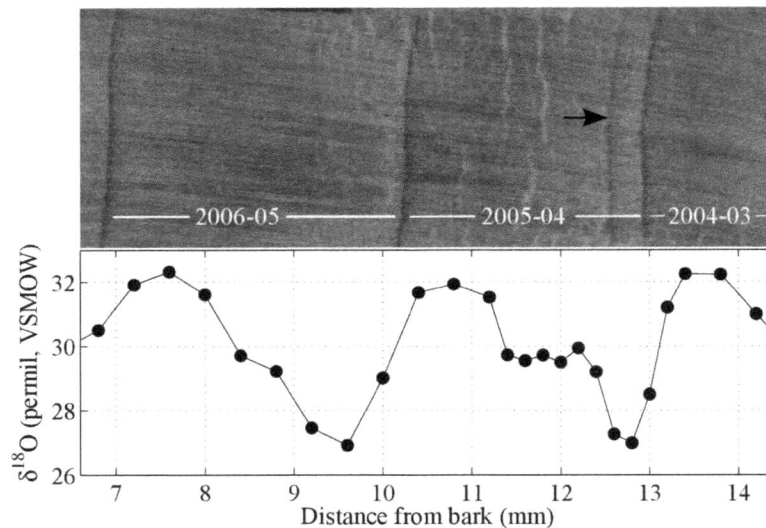

Figure 3. High resolution analysis of a section of *A. robusta.* Top: Section of scan of core A (BB11.17W) from Dinden National Park, including crossdated growth increments (white lines) and suspected false ring (black arrow). Bottom: $\delta^{18}O$ (‰, SMOW) measured vs. distance from tree bark, scale in millimeters.

walls, typical of earlywood [49]. The effect of this sequence of cambial responses to a transient, but relatively intense, climate anomaly is the formation of a false ring.

High resolution isotopic data (Fig 3, lower panel) support the interpretation of this feature as a false ring within the 2004/2005 growth increment. Relative to the $\delta^{18}O$ observations within the 2005/2006 growing season, the 2004/2005 $\delta^{18}O$ data appear to indicate an extended period within the growth increment and just following the putative false ring, with isotopic composition intermediate between values indicative of rainy season onset/termination and maximum (Fig 3). This result is consistent with the tropical isotope dendroclimatology hypothesis, which leads us to expect growth through the temporary precipitation hiatus of February 2005 (Fig 1a) within the otherwise normal 2004/2005 rainy season (Fig 2b), albeit with elevated cellulose $\delta^{18}O$ relative to values observed during rainy season maxima before and after (Fig 3). If the false ring indeed represents February 2005 conditions, the results indicate that about 80% of incremental growth occurred in the last third of the growing season, suggesting that the assumption of linear incremental growth rate within the growing season in this species and environment should be treated with caution [16].

Prospects

A key limitation in our analyses is the small number of replicates and events studied here. However, the results suggest that the combination of wood anatomical and isotopic analysis might be used to reconstruct tropical paleoclimates from locations and species for which classical methods of dendrochronology alone are unsuccessful. In addition, the superposition of ring width and oxygen isotopic composition measurements may permit the estimation of intraseasonal incremental growth rates in tree species and environments in which growth is episodic. Replication of false ring detection by this dual observational approach would provide additional confidence in our ability to use both increment growth and isotopic composition to more accurately reconstruct

wood increment chronology, and thereby subsequently develop improved paleoecological and paleoclimatological interpretations. Conversely, increased confidence in the chronology of tropical species may improve our ability to use the isotopic indicator of rainfall variation to accurately resolve the timing of interannual precipitation variations associated with ENSO activity in the terrestrial tropics. Century-scale oxygen isotope chronologies that crossdate to an acceptable level could be built using long-lived *A. robusta* trees. The results presented here lay the groundwork for that effort.

Conclusions

If the occurrence of false rings is frequent and undetectable, ring-width chronologies of *A. robusta* from northern Queensland may not be usable as a stand-alone climate indicator. However, the analysis of ring widths and oxygen isotopic composition in this study proved mutually complementary [60]. The small subset of samples that were successfully ring-dated provided the basis for oxygen isotopic analysis of extracted α-cellulose, which produced statistically significant correlations with local ENSO-induced variations in precipitation consistent with the tropical isotope dendroclimatology hypothesis. In turn, intraseasonal-resolution oxygen isotope analyses illustrated the basis for more confident detection of false rings in the ring-width data, potentially creating a pathway by which improvements in crossdating accuracy may be made.

We recommend that future studies of *A. robusta* or other species that present ambiguous ring formation use a combination of standard ring-width analysis and oxygen isotopic measurements. By this approach, data developed from these archives may be used to supplement the few existing ENSO reconstructions from the Australasian terrestrial tropics and other regions with climatic variations associated with ENSO activity. We expect that further work should eventually permit development of replicated 150–200 year records.

Acknowledgments

This work is distilled from the B.Sc. (Hons.) thesis of the first author at Monash University, Melbourne. We would like to thank Rohan Simkin for his help in the field. In Maryland, Nick Gava, Grant Jiang and Yongbo Peng assisted with preparation, analysis, calibration and interpretation of isotopic analyses. Dr Kathryn Allen helped with dendrochronological analysis and her advice regarding ring measurements was extremely valuable to this study. We would also like to thank the Atherton CSIRO and the University of Maryland for helping to facilitate this research.

Author Contributions

Conceived and designed the experiments: BMMB MNE PJB. Performed the experiments: BMMB. Analyzed the data: BMMB MNE. Contributed reagents/materials/analysis tools: PJB MNE. Wrote the paper: BMMB MNE PJB.

References

1. Collins M, An SI, Cai W, Ganachaud A, Guilyardi E, et al. (2010) The impact of global warming on the tropical Pacific Ocean and ENSO. Nat Geo 3: 391–397.
2. D'Arrigo RD, Cook ER, Wilson RJ, Allan R, Mann ME (2005) On the variability of ENSO over the past six centuries. Geophys Res Lett 32: L03711.
3. Gergis J, Braganza K, Fowler A, Mooney S, Risbey J (2006) Reconstructing El Niño Southern Oscillation (ENSO) from high-resolution palaeoarchives. J Quat Sci 21: 707–722.
4. Baker PJ, Palmer JG, D'Arrigo R (2008) The dendrochronology of *Callitris intratropica* in northern Australia: annual ring structure, chronology development and climate correlations. Aust J Botany 56: 311–320.
5. Dunbar RB, Cole JE, editors (1999) Annual Records of Tropical Systems: Recommendations for Research, 99–1. PAGES/CLIVAR.
6. Jones PD, Briffa KR, Osborn TJ, van Ommen TD, Vinther BM, et al. (2009) High-resolution palaeoclimatology of the last millennium: a review of current status and future prospects. The Holocene 19(1): 3–49.
7. Worbes M (2002) One hundred years of tree-ring research in the tropics – a brief history and an outlook to future challenges. Dendrchronologia 20: 217–231.
8. McCarroll D, Loader NJ (2004) Stable isotopes in tree rings. Quat Sci Rev 23: 771–801.
9. LaMarche VC, Holmes RL, Dunwiddie PW, Drew LG (1979) Chile, volume 2 of *Chronology Series V*. Tucson: University of Arizona.
10. Stahle DW, D'Arrigo RD, Krusic PJ, Cleaveland MK, Cook ER, et al. (1998) Experimental dendroclimatic reconstruction of the Southern Oscillation. AMS Bull 79: 2137–2152.
11. Pearson SG, Searson MJ (2002) High resolution data from Australian trees. Aust J Botany 50: 431–439.
12. Anchukaitis KJ, Evans MN, Lange T, Smith DR, Leavitt SW, et al. (2008) Consequences of a rapid cellulose extraction technique for oxygen isotope and radiocarbon analyses. Anal Chem 80(6): 2035–2041.
13. Pearson S, Hua Q, Allen K, Bowman DMJS (2011) Validating putatively cross-dated *Callitris* tree-ring chronologies using bomb-pulse radiocarbon analysis. Aust J Botany 59: 7–17.
14. Anchukaitis KJ, Evans MN (2010) Tropical cloud forest climate variability and the demise of the Monteverde Golden Toad. Proc Natl Acad Sci 107: 5036–5040.
15. Stahle DW, Mushove PT, Cleaveland MK, Roig F, Haynes G (1999) Management implications of annual growth rings in *Pterocarpus angolensis* from Zimbabwe. Forest Ecol Manag 124: 217–229.
16. Evans MN, Schrag DP (2004) A stable isotope-based approach to tropical dendroclimatology. Geochim et Cosmochim Acta 68(16): 3295–3305.
17. Roden JS, Lin G, Ehleringer JR (2000) A mechanistic model for interpretation of hydrogen and oxygen ratios in tree-ring cellulose. Geochimica et Cosmochimica Acta 64: 21–35.
18. Barbour MM, Roden JS, Farquhar GD, Ehleringer JR (2004) Expressing leaf water and cellulose oxygen isotope ratios as enrichment above source water reveals evidence of a Péclet effect. Oecologia 138: 426–435.
19. Dansgaard W (1964) Stable isotopes in precipitation. Tellus 16: 436–468.
20. Faure G (1986) Isotope Geology. Wiley, 2nd edition.
21. Gat JR (1996) Oxygen and hydrogen isotopes in the hydrologic cycle. Ann Rev Earth Planet Sci 24: 225–262.
22. Evans MN (2007) Toward forward modeling for paleoclimatic proxy signal calibration: a case study with oxygen isotopic composition of tropical woods. Geochemistry, Geophysics, Geosystems 8: Q07008.
23. Verheyden A, Helle G, Schleser GH, Dehairs F, Beeckman H, et al. (2004) Annual cyclicity in high-resolution stable carbon and oxygen isotope ratios in the wood of the mangrove tree *Rhizophora mucronata*. Plant, Cell and Environ 27: 1525–1536.
24. Poussart PF, Schrag DP (2005) Seasonally resolved stable isotope chronologies from northern Thailand deciduous trees. Earth Plan Sci Lett 235: 752–765.
25. Anchukaitis KJ, Evans MN, Wheelwright NT, Schrag DP (2008) Stable isotope chronology and climate signal calibration in neotropical montane cloud forest trees. J Geophys Res 113: G03030.
26. Baker PJ, Bunyavejchewin S, Oliver CD, Ashton PS (2005) Disturbance history and historical stand dynamics of a seasonal tropical forest in western Thailand. Ecol Monographs 75: 317–343.
27. Heinrich I, Weidner K, Helle G, Vos H, Banks JCG (2008) Hydroclimatic variation in Far North Queensland since 1860 inferred from tree rings. Paleog Paleocl Palaeoecol 270: 116–127.
28. Cook ER, Anchukaitis KJ, Buckley BM, D'Arrigo RD, Jacoby GC (2010) Asian monsoon failure and megadrought during the last millennium. Science 328: 486–489.
29. Buckley BM, Anchukaitis KJ, Penny D, Fletcher R, Cook ER, et al. (2010) Climate as a contributing factor in the demise of Angkor, Cambodia. Proc Nat Acad Sci 107: 6748–6752.
30. Vincent L, Pierre G, Michel S, Robert N, Masson-Delmotte V (2007) Tree-rings and the climate of New Caledonia (SW Pacific): Preliminary results from Araucariaceae. Palaeogeogr, Palaeocl, Palaeoecol 253: 477–489.
31. Ballantyne AP, Baker PA, Chambers JQ, Villalba R, Argollo J (2011) Regional differences in South American Monsoon precipitation inferred from the growth and isotopic composition of tropical trees. Earth Int 15: 1–35.
32. Brienen RJW, Helle G, Pons TL, Guyot JL, Gloor M (2012) Oxygen isotopes in tree rings are a good proxy for Amazon precipitation and El Niño-Southern Oscillation variability. Proc Nat Acad Sci 109(42): 16957–16962.
33. Drew DM, Allen K, Downes GD, Evans R, Battaglia M, et al. (2013) Wood properties in a long-lived conifer reveal strong climate signals where ring-width series do not. Tree Phys 33: 37–47.
34. Schollaen K, Heinrich I, Neuwirth B, Krusic PJ, D'Arrigo RD, et al. (2013) Multiple tree-ring chronologies (ring width, δ¹³C, and δ¹⁸O) reveal dry and rainy season signals of rainfall in Indonesia. Quat Sci Rev 73: 170–181.
35. Ropelewski CF, Halpert MS (1987) Global and regional scale precipitation patterns associated with the El Niño/Southern Oscillation. Mon Wea Rev 114: 2352–2362.
36. Ropelewski CF, Halpert MS (1989) Precipitation patterns associated with the high index phase of the Southern Oscillation. J Climate 2: 268–284.
37. Ogden J (1981) Dendrochronological studies and the determination of tree ages in the Australian tropics. J Biogeogr 8: 405–420.
38. Farjon A (2010) Handbook of the world's conifers. Boston: Brill.
39. Buckley B, Ogden J, Palmer J, Fowler A, Salinger J (2000) Dendroclimatic interpretation of tree-rings in *Agathis australis* (kauri): 1. Climate correlation functions and master chronology. J Roy Soc New Zealand 30(3): 283–275.
40. Fowler A, Palmer J, Salinger J, Ogden J (2000) Dendroclimatic interpretation of tree-rings in *Agathis australis* (kauri): 2. Evidence of a significant relationship with ENSO. J Roy Soc New Zealand 30(3): 277–292.
41. Boswijk G, Fowler A, Lorrey A, Palmer J, Ogden J (2006) Extension of the New Zealand kauri (*Agathis australis*) chronology to 1724 BC. The Holocene 16: 188–199.
42. Fowler A, Boswijk G, Gergis J, Lorrey AM (2008) ENSO history recorded in *Agathis australis* (kauri) tree rings. Part A: kauri's potential as an ENSO proxy. Int J Clim 28(1): 1–20.
43. Fowler A (2008) ENSO history recorded in *Agathis australis* (kauri) tree rings. Part B: 423 years of ENSO robustness. Int J Clim 28(1): 21–35.
44. Fowler AM, Boswijk G, Lorrey AM, Gergis J, Pirie M, et al. (2012) Multi-centennial tree-ring record of ENSO-related activity in New Zealand. Nat Clim Change 2(3): 172–176.
45. Ash J (1983) Tree rings in tropical *Callitris macleayana* F. Muell. Austr J Bot 31: 213–229.
46. Ash J (1983) Growth rings in *Agathis robusta* and *Araucaria cunninghamii* from tropical Australia. Clim Dyn 31: 269–275.
47. Heinrich I, Weidner K, Helle G, Vos H, Lindesay J, et al. (2009) Wood anatomical features in tree-rings as indicators of environmental change. Clim Dyn 33: 63–73.
48. Holmes R (1983) Computer assisted quality control in tree-ring dating and measurement. Tree-Ring Bulletin 44: 69–75.
49. Fritts HC (1976) Tree Rings and Climate. New York: Academic Press.
50. Brendel O, Iannetta PPM, Stewart D (2000) A rapid and simple method to isolate pure α-cellulose. Phytochemical Analysis 11: 7–10.
51. Werner RA, Kornexl BE, Roβmann A, Schmidt HL (1996) On-line determination of δ¹⁸O values of organic substances. Anal Chim Acta 319: 159–164.
52. Coplen TB, Brand WA, Gehre M, Groning M, Meijer HAJ, et al. (2006) New guidelines for δ¹³C measurements. Anal Chem 78: 2439–2441.
53. Evans MN, Tolwinski-Ward SE, Thompson DM, Anchukaitis KJ (2013) Applications of proxy system modeling in high resolution paleoclimatology. Quat Sci Rev 76: 16–28.
54. Harris I, Jones PD, Osborn TJ, Lister DH (2014) Updated high-resolution grids of monthly climatic observations – the CRU TS3.10 Dataset. Int J Clim 34: 623–642.

55. Bolton D (1980) The computation of equivalent potential temperature. Mon Wea Rev 108: 1046–1053.

56. Birks J (2012) Global Network for Isotopes in Precipitation: The GNIP/WISER Datbase. Technical report, International Atomic Energy Agency. Accessed via Internet: http://www-naweb.iaea.org/napc/ih/index.html.

57. CPC (2012). Monitoring & Data: Current Monthly Atmospheric and Sea Surface Temperature Index Values. Data accessed via internet: http://www.cpc.ncep.noaa.gov/data/indices/sstoi.indices.

58. Trenberth KE (1997) The definition of El Niño. Bull Amer Met Soc 78: 2771–2777.

59. Wimmer R (2006) Wood anatomical features in tree-rings as indicators of environmental change. Dendrchronologia 20: 21–36.

60. Heinrich I, Allen K (2013) Current issues and recent advances in Australian dendrochronology: Where to next? Geog Res 51: 180–191.

61. ABOM (2012). Australian Bureau of Meteorology Climate Information. Data accessed via internet: http://www.bom.gov.au/climate/data/.

7

Mixed Fortunes: Ancient Expansion and Recent Decline in Population Size of a Subtropical Montane Primate, the Arunachal Macaque *Macaca munzala*

Debapriyo Chakraborty[1,2]*, Anindya Sinha[1,2,3], Uma Ramakrishnan[2]

1 Nature Conservation Foundation, Gokulam Park, Mysore, India, 2 National Centre for Biological Sciences, GKVK Campus, Bangalore, India, 3 National Institute of Advanced Studies, Indian Institute of Science Campus, Bangalore, India

Abstract

Quaternary glacial oscillations are known to have caused population size fluctuations in many temperate species. Species from subtropical and tropical regions are, however, considerably less studied, despite representing most of the biodiversity hotspots in the world including many highly threatened by anthropogenic activities such as hunting. These regions, consequently, pose a significant knowledge gap in terms of how their fauna have typically responded to past climatic changes. We studied an endangered primate, the Arunachal macaque *Macaca munzala*, from the subtropical southern edge of the Tibetan plateau, a part of the Eastern Himalaya biodiversity hotspot, also known to be highly threatened due to rampant hunting. We employed a 534 bp-long mitochondrial DNA sequence and 22 autosomal microsatellite loci to investigate the factors that have potentially shaped the demographic history of the species. Analysing the genetic data with traditional statistical methods and advance Bayesian inferential approaches, we demonstrate a limited effect of past glacial fluctuations on the demographic history of the species before the last glacial maximum, approximately 20,000 years ago. This was, however, immediately followed by a significant population expansion possibly due to warmer climatic conditions, approximately 15,000 years ago. These changes may thus represent an apparent balance between that displayed by the relatively climatically stable tropics and those of the more severe, temperate environments of the past. This study also draws attention to the possibility that a cold-tolerant species like the Arunachal macaque, which could withstand historical climate fluctuations and grow once the climate became conducive, may actually be extremely vulnerable to anthropogenic exploitation, as is perhaps indicated by its Holocene ca. 30-fold population decline, approximately 3,500 years ago. Our study thus provides a quantitative appraisal of these demographically important events, emphasising the ability to potentially infer the occurrence of two separate historical events from contemporary genetic data.

Editor: Cédric Sueur, Institut Pluridisciplinaire Hubert Curien, France

Funding: The study was funded by a Department of Science and Technology, Government of India grant to AS, a Department of Biotechnology, Government of India grant to both UR and AS and an International Primatological Society research grant to DC. The funders had no role in study design, data collection and analysis, decision to publish, or preparation of the manuscript.

Competing Interests: The authors have declared that no competing interests exist.

* Email: debapriyoc@gmail.com

Introduction

The Pleistocene epoch, with its frequent climatic fluctuations, is known to have driven range expansions and population decline/ expansion of many species [1]. The climatic fluctuations consisted of episodes with relatively high and low global ice volumes. The lowest of ice volumes characterised the interglacial periods with relatively warmer climate [2]. The Last Glacial Maximum (LGM), which occurred near the end of the Pleistocene, about 20,000 years ago [3], was a particularly dramatic period of glacial advance and global cooling resulting in demographic bottlenecks for many animal populations. This period was eventually followed by a period of warm and humid climate that lasted through the Early Holocene period, with the subsequent expansion of many previously bottlenecked populations [4]. Importantly, the Holocene warming created opportunities for colonisation of new regions by modern humans as well, which further triggered profound alterations in world ecosystems and, in turn, the demography of several other species [5]. One of the ways we

can corroborate the past occurrence of these large-scale changes is by tracking changes in the population size of species over evolutionary time.

Investigations into past complex dynamic events have been facilitated, in the absence of more accurate ancient DNA, by recent advances in model-based hypothesis testing of current molecular data. Thus, it is now possible to employ an inferential framework to derive statistically robust inferences of the timing and magnitude of past changes in the effective population size (Ne) of species over clearly defined time periods [6,7]. While it is crucial to identify and separate ancient from more recent demographic events to obtain insights into the long-term population dynamics of species, this is often difficult to achieve in practice as similar genetic patterns can result from different demographic histories [8]. Nevertheless, the use of different types of genetic markers, such as mitochondrial DNA and microsatellites, combined with new kinds of statistical analytical methods, have achieved significant success in detecting genetic signatures of major demographic events, such as population bottlenecks, expansions and admixtures that may have occurred in different historical time periods [6].

Although a vast majority of studies have performed these tests and estimated demographic parameters across taxa, very few studies have been conducted on tropical or subtropical ecosystems (but see [9,10,11]). This gap in our knowledge is clearly problematic as the detrimental effects of climatic and other environmental changes on subtropical and tropical biodiversity across much of the world are only likely to increase in the near future [12].

In order to address this knowledge gap, we sought to investigate the past demographic history of a primate, the Arunachal macaque *Macaca munzala*, an endangered species recently reported from and believed to be endemic to the state of Arunachal Pradesh in northeastern India. It is not at all clear how the complex climatic history of the Indian subcontinent might have shaped the population history of this primate, particularly given its distribution in the subtropical mountainous habitat within the Eastern Himalayan biodiversity hotspot.

Pleistocene climate change and/or physical barriers have often been described as important forces driving the demographic history of many species from other parts of the subtropics [13,14,15]. A few studies that have specifically examined the demographic history of primate species, however, suggest instead a primary role for anthropogenic factors driving relatively recent demographic changes in primate populations [9,16,17,18]. The only reported exceptions to this trend are two macaque species from the northern hemisphere, the Barbary macaque *M. sylvanus* and the Japanese macaque *M. fuscata*. Their demographic history were shown to be shaped relatively more strongly by Pleistocene glaciation events than later anthropogenic influences [19,20]. Consequently, these macaque populations revealed signatures of genetic bottlenecks after the LGM, around 20,000 years ago [19,20]. We thus thought it important to explore the history of a macaque species in the Eastern Himalaya, which also marked the southernmost limit of the Pleistocene glaciers.

What also characterises this biodiversity-rich region is the decimation of its indigenous wildlife by the local human communities, a well-documented historical phenomenon. The state of Arunachal Pradesh is inhabited by a number of animistic tribes who continue to depend heavily on wildlife for their subsistence as well as for sport [21]. Our knowledge of the histories of these people is, however, fragmentary and incomplete. The peopling of this region seems to have occurred rather late in the history of the Indian subcontinent. According to variously recorded folklore and traditional knowledge, the ancestral Tani people migrated to central Arunachal Pradesh approximately 1,500 years ago from the north, through the Siang region. These people then settled in the Himalayan valleys and, due to the unique topography of the region as well as internecine warfare among them, were persistently isolated from one another. They were subsequently believed to have given rise to the many sub-tribes of today, such as the Adi and Apatani [22], which largely continue to follow their ancestral animistic culture. The Arunachal macaque is also known to be hunted across its distribution range, either in retaliation of its crop-raiding in the western districts of Tawang and West Kameng [23] or for food, sport and trade in the central districts of Upper Subansiri and West Siang [24].

In this study, therefore, we investigate the impact of two important factors, Pleistocene glaciation events and more recent anthropogenic activities, on the demographic history of the Arunachal macaque. More specifically, we ask whether the species was able to maintain its ancestral population size throughout its history on the southern subtropical fringes of the Tibetan Plateau.

If, however, there were significant changes in the population size of this large mammalian species, were these wrought more by Pleistocene climate change, which is known to have affected the population history of several other species in the region, or were humans the principal driving force? Answers to these questions are critical not only to predict the future trajectories of populations of primates and other mammalian species to impending climatic changes in the Eastern Himalayan biodiversity hotspot but also to develop conservation strategies for the increasingly endangered species of the region.

Methods

Ethics statement

This study was conducted in accordance with all relevant Indian laws, with due permits from the State Forest Department of Arunachal Pradesh. We collected only small pieces of dried skin samples from local communities without any kind of payment. As these were from old hunting trophies and no fresh hunting was reported from our study sites, we are confident that our sample collection did not encourage killing of the species in any way. Additionally, we have led a conservation programme for the species in this region over the last seven years.

Study area and population sampling

We obtained dried skin samples of Arunachal macaque individuals, killed and preserved as hunting trophies, from 14 villages across the districts of Tawang, Upper Subansiri and West Siang in the state of Arunachal Pradesh (Fig. S1). While Tawang forms the western edge of the state, both Upper Subansiri and West Siang are located in remote central Arunachal Pradesh. These samples were preserved in 95% ethanol at ambient temperature till they were transported to the laboratory. A single blood sample was also obtained from a captive individual in Typee, Tawang District. We considered the samples to belong to three tentative populations – Tawang (n = 5), Upper Subansiri (n = 12) and West Siang (n = 9) on the basis of their geographic origins (Table S1).

DNA extraction and PCR amplification

We extracted genomic DNA from skin and blood samples using the QIAGEN DNeasy Blood & Tissue Kit (Qiagen GMBH, Hamburg, Germany), following the manufacturer's protocols. We used extraction blanks as negative controls in downstream polymerase chain reaction (PCR) amplifications.

In order to sample a non-coding region of the Arunachal macaque mitochondrial genome, we amplified a 534bp-long D-loop (hyper-variable segment 1) region using the primer set from Li and Zhang [25]. We conducted standard 35-cycle PCR to amplify the target regions [26].

We amplified 22 fluorescently labelled microsatellite loci (DXS571, DXS6810, DXS8043, DXS6799, D20S171, D4S243, D12S372, D8S1466, D9S934, D7S794, D10S611, D8S1106, D15S823, D19S255, D2S146, D17S791, D18S869, D18S537, D6S2419, D16S403, D5S1457, D10S179, D11S2002), already established for other macaque species [27,28]. PCR amplification of 35 cycles was conducted for up to five loci simultaneously with combinations selected on the basis of fragment size, annealing temperature, and the fluorescent dye set DS - 33 components used (dy6FAM, VIC, PET or NED). The PCR products were resolved with an ABI 3130xl automated sequencer and analysed with GeneMapper software (version 4.0; Applied Biosystems, Foster City, USA). We could successfully sequence 24 of the individual samples collected – 5 for Tawang, 10 for Upper Subansiri and 9 for West Siang. The accession numbers for the mitochondrial DNA sequences have been given in Table S1. We are also open to sharing the microsatellite data for further analyses, if requested.

Population demographic history

Bayesian skyline plots. We used Bayesian skyline plots of mitochondrial DNA (mtDNA) to examine the past population dynamics of the study populations.

Bayesian skyline plots allow for the estimation of effective population size, Ne, through time without specifying population change models such as constant size or exponential growth. The age of 'the most recent common ancestor' or TMRCA of each clade [29] – Tawang: 0.12 mya (95% Highest Posterior Density or HPD $0.05 - 0.21$), Upper Subansiri: 0.46 mya (95% HPD $0.27 - 0.68$) and West Siang: 0.89 mya (95% HPD $0.53 - 1.3$) – were employed as priors for the three populations. Furthermore, we selected the HKY+Γ+I mutation model for the analysis with a mean HVS1 substitution rate of 0.1643 substitutions per nucleotide per million year (Myr) for the mean rate prior [29]. We used BEAST, version 1.7 [30] for constructing the Bayesian skyline plots. Two independent runs were conducted for 10^7 steps, with a sampling of the parameters once every 10^3 steps. Ten percent of the samples were discarded as burn-in. The estimated sample size (ESS) for all parameters was found to be above 200 at the end of each run, indicating good convergence of the Markov chain Monte Carlo (MCMC) simulations. The independent runs were combined using LogCombiner, version 1.5.4 [30], with resampling every 10^5 steps. After combining all the MCMC runs, the ESS for all parameters was observed to be more than 200.

The EWCL (Ewens, Watterson, Cornuet and Luikart) method. Demographic events such as recent population bottlenecks are known to leave distinct genetic signatures in the distributions of allele size, expected heterozygosity and in the genealogy of microsatellite loci [33,9]. Here, we used the allelic frequency spectrum, namely the number of alleles (n_A) and the expected heterozygosity (H_e), to determine the patterns of genetic diversity expected for a demographically stable population [32]. We performed simulations in Bottleneck, version 1.2.02 [33], for both the species separately, to obtain the distribution of H_e, conditional on n_A and on the sample size for each population and locus. These H_e values were then compared to those obtained from the real dataset. Three mutation models were used: the infinite-allele model (IAM), the stepwise-mutation model (SMM) and the two-phase model (TPM), with various amounts (70% to 95%) of single-step mutations [34]. Any departures from the null hypothesis were explained as departures from the model, including selection, population expansion or decline. Consistency across independent loci, however, was unlikely to be caused by selection but rather by demographic events. This approach allowed us to detect population size changes and confirm that the signal was consistent across mutation models. The demographic event, however, could not be dated.

The Approximate Bayesian Computation approach. We further inferred the demographic history of the Arunachal macaque, based on microsatellite data, using the approximate Bayesian computation (ABC) approach [35], implemented in the programme DIY-ABC, version 1.0.4.46b [36]. This approach allowed us to choose a demographic scenario among many that best fits the data and infer the posterior probability distributions for the parameters of interest under this preferred scenario. The different steps of the ABC parameter estimation procedure [37] are briefly described here.

We compared three demographic scenarios, graphically depicted in Fig. 1. Scenario 1 consisted of a null hypothesis that assumed a population whose effective size (N_I) remained stable over time. Scenario 2 assumed a population of size N_A that declined instantaneously to its current effective size (N_I), T_I generations ago. Conversely, Scenario 3 assumed a population of effective size N_B that increased instantaneously T_I generations ago to reach its current effective size (N_I). For Scenarios 2 and 3, we considered T_I to be between 20 and 4000 generations, during which time these events could have occurred.

For each of the three models, we simulated one million datasets based on a demographic history that described the model, using the programme DIY-ABC. Some or all parameters that defined each model (such as population sizes, timing of the demographic events or mutation rates) were considered as random variables for which some prior distributions were defined, as shown in Table 1. For each simulation, the parameter values were drawn from their prior distributions, defining a demographic history that was used to build a specific input file for the DIY-ABC programme. Coalescent-based simulations were run to generate a genetic diversity for each sample, with the same number of gene copies and loci as those originally observed. Summary statistics (S) were then computed for the simulated datasets for each of the observed dataset (S*). Following the method of [38], a Euclidean distance, δ, was calculated between the normalised S and S* for each simulated dataset.

The 22 microsatellite loci were assumed to follow a generalised stepwise mutation model [39] with two parameters: the mean mutation rate (μ_{mic}) and the mean parameter of the geometrical distribution assumed for the length in repeat numbers of mutation events (P) drawn from uniform prior distributions of 10^{-4} to 10^{-3} and 0.1 to 0.3, respectively. Each locus has a possible range of 40 contiguous allelic states, except Locus 12 (43 states) and was characterised by individual μ_{loc} and P_{loc} values, drawn from gamma distributions with respective means of μ_{mic} and P, and shape parameter 2 in both cases [40]. Using this setting, we allowed for large mutation rate variance across loci (range of 10^{-5} to 10^{-2}). We also considered mutations that insert a single nucleotide into or delete one from the microsatellite sequence. We used default values for all other mutation model settings. The details on model parameterisation and prior settings have been provided in Table 1.

The summary statistics of genetic diversity for the microsatellite loci, calculated with the programme DIY-ABC, included: the mean number of alleles per locus (A), mean expected heterozygosity (H_e), mean allele size variance (V), and mean GW Index across the loci [31] (Garza and Williamson 2001).

The posterior probability of each competing scenario was estimated using a polytomous logistic regression [41,36] on 1% of the simulated datasets closest to the observed dataset. We evaluated the ability of our ABC methodology to discriminate between scenarios by analysing simulated datasets with the same number of loci and individuals as in our real dataset. Following the method of Cornuet et al. [36], we estimated the Type I error probability as the proportion of instances where the selected scenario did not exhibit the highest posterior probability, as compared to the competing scenarios for 500 simulated datasets generated under the best-supported model. We similarly estimated the Type II error probability by simulating 100 datasets for each of two alternative scenarios and calculating the mean proportion of instances in which the best-supported model was incorrectly selected as the most probable model.

We estimated the posterior distributions of the demographic parameters under the best demographic model, using a local linear regression on the closest 1% of 10^6 simulated datasets, after the application of a logit transformation, the inverse of the logistic

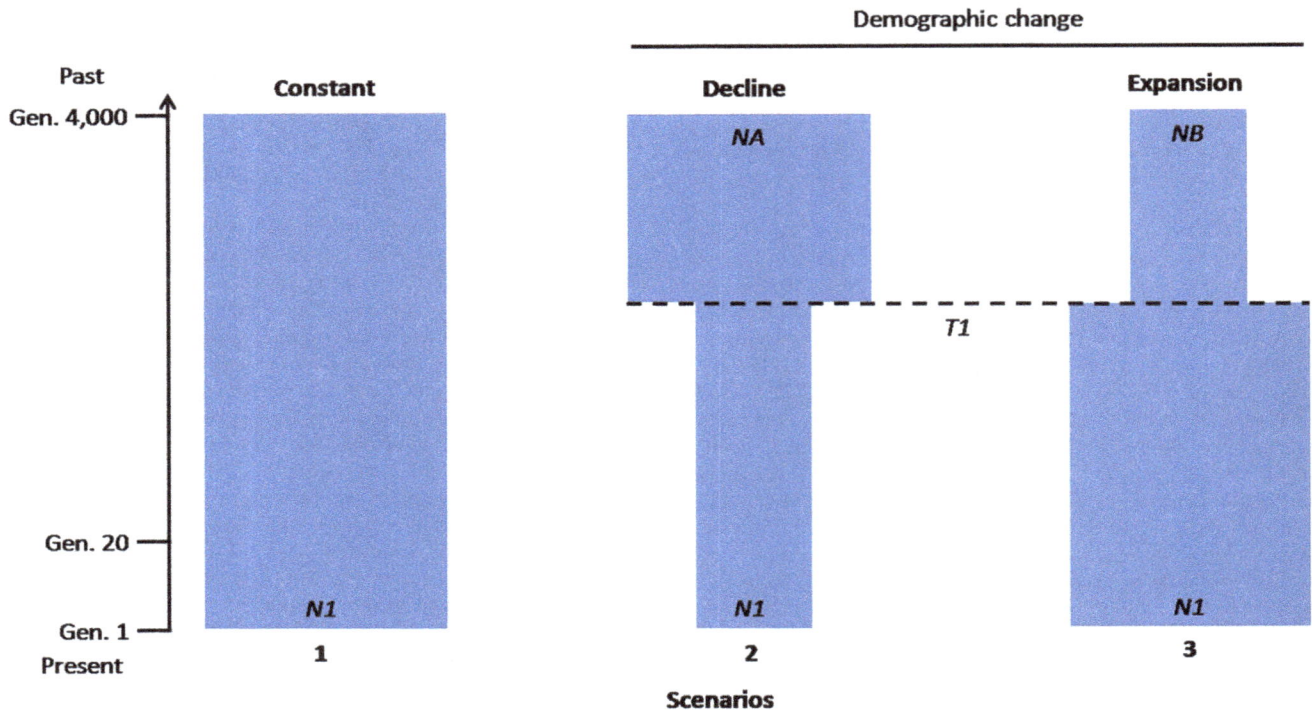

Figure 1. Possible alternative scenarios of the demographic history of the Upper Subansiri population. When tested using the ABC approach, Scenario 2 was best fit with the data (Table 2). The details of each scenario parameterisation have been given in the Methods. The time-scale is indicated by the arrow on the left. T_1 ranges between 20 and 4000 generations. Time has been measured backward in generations before the present. Gen: Generation.

Table 1. Model specifications and prior distributions for demographic parameters and locus-specific mutation model parameters.

Priors for the Demographic Parameters	
N_1	UN ~ [50, 5000]
T_1	UN ~ [20, 4000]
N_A	UN ~ [10000, 70000]
N_B	UN ~ [10, 5000]
Priors for the Mutation Model	
Autosomal microsatellites	
MEAN – μ_{mic}	UN~[1×10^{-4}, 1×10^{-3}]
GAM – μ_{mic}	GA~[1×10^{-5}, 1×10^{-2}, 2]
MEAN – P	UN~[0.10, 0.30]
GAM – P	GA~[0.01, 0.9, 2]
MEAN – SNI	LU~[1×10^{-8}, 1×10^{-4}]
GAM – SNI	GA~[1×10^{-9}, 1×10^{-3}, 2]

UN: Uniform distribution, with two parameters – minimum and maximum values; GA: Gamma distribution with three parameters – minimum and maximum values and shape parameter value; LU: Log-Uniform distribution with two parameters – minimum and maximum values. See Figure 2 for the demographic parameters of each model tested. The mutation model parameters for the microsatellite loci were the mutation rate (μ_{mic}), the parameter determining the shape of the gamma distribution of individual loci mutation rate (P), and the Single Insertion Nucleotide rate (SNI).

function, to the parameter values [35,41]. Finally, following the method of Gelman *et al.* [42], we evaluated whether, under the best model-posterior combination, we were able to reproduce the observed data using the model-checking procedure available in DIY-ABC [36]. Model-checking computations were processed by simulating 1000 pseudo-observed datasets under each studied model-posterior combination, with sets of parameter values drawn with replacement among the 1000 sets of the posterior sample. This generated a posterior cumulative distribution function for each summary statistic, allowing us to estimate the P values for the observed values of these summary statistics. In addition, a principal components analysis (PCA) was performed on the summary statistics. Principal components were computed from the 15,000 datasets simulated with parameter values drawn from the prior. The target (observed) dataset, as well as the 1,000 datasets simulated from the posterior distributions of parameters, was then added to each plane of the PCA.

Results

Bayesian skyline plots

Our mitochondrial DNA analysis (see also [29]) demonstrated that the three Arunachal macaque populations, Tawang, Upper Subansiri and West Siang, are geographically distinct. Macaque populations, being female philopatric, tend to exhibit such distinctive population genetic structure for maternally inherited mitochondrial DNA [19,20]. Given this population structure, we treated each population separately for demographic analyses. Conducting a skyline plot with data from all three populations combined would result in a putatively incorrect signature of population expansion since these population show significant structure [8]. Of these three populations, the Upper Subansiri had the largest number of sampled individuals and was, therefore, taken up for further analysis. It must be noted, however, that we could not rule out the existence of local migrants and their influence on our results, as we could not sample all the populations in the intervening time periods.

The Bayesian skyline plot for the Upper Subansiri population indicated a population expansion right after the Last Glacial Maximum (LGM) at around 15,000 years before present (Fig. 2). The population, however, appeared to have maintained a constant population size before this expansion. The plot also revealed a small reduction in effective population size in more recent times, right after the Middle Holocene (between 5,000 and 7,000 years before present; Fig. 2), an indication that the distribution may have stabilised around the mean, given that there is no detectable change in trends for the upper bound. Finally, it must be noted that such horizontal distribution can also be drawn within the credibility interval and, as a result, the mean effect may represent a simple trend rather than true statistical evidence of actual demographic shifts.

The EWCL (Ewens, Watterson, Cornuet and Luikart) method

This test, which attempted to detect the presence of a population bottleneck, indicated that, regardless of the mutation model assumed, the three Arunachal macaque populations, taken together, exhibited a significant signal of bottleneck for a majority of the microsatellite loci tested. It should, however, be remembered that this can be an artefact of the presence of structure in the tested populations. We, therefore, decided to treat the populations separately and found that only the Upper Subansiri site had an acceptable sample size for the test. For this population, the infinite-allele model (IAM) and two-phase model (TPM) showed statistical significance for heterozygosity excess, in both one-tailed and two-tailed tests (both $P \ll 0.05$), thus providing clear evidence for a genetic bottleneck in this population in the past. The application of a stepwise-mutation model (SMM), however, yielded non-significant results for both one-tailed ($P = 0.13$) and two-tailed ($P = 0.064$) tests.

The Approximate Bayesian Computation approach

We then tested three alternative scenarios of demographic change – demographic expansion, decline or constant size – using a model-based approximate Bayesian inferential framework. We first evaluated the relative posterior probability of each competing scenario using a polytomous logistic regression on 1% of the simulated datasets closest to the observed dataset. The resulting PCA unambiguously pointed to the scenario of population decline (Scenario 2), which assumed an ancient, large population size that declined at some point in the past to reach the present, much smaller population size (Fig. 3).

We next evaluated the power of the model choice procedure using the method implemented in DIY-ABC, following the recommendations of Robert *et al.* [43]. For this purpose, we first simulated 500 random datasets under the selected scenario (Scenario 2) and computed the proportion of cases in which this scenario did not display the highest posterior probability among all scenarios. This empirical estimate of the Type I error was only 16.6%. We then empirically estimated the Type II error rate by simulating 100 random datasets under each alternative scenario (Scenarios 1 and 3) and computing the proportion of cases in which Scenario 2 was incorrectly selected as the most likely scenario in these simulated datasets. The average Type II error rate was only 8%, indicating a statistical strength of 92%. Hence,

this simulation-based evaluation of the performance of the ABC model-choice procedure [43] clearly showed that, given the size and polymorphism of our dataset, the method had statistical properties of both power and robustness to distinguish between the alternative demographic scenarios that we investigated.

We estimated the marginal posterior probability density for each parameter of the decline scenario (Scenario 2) using 1% of the closest simulated datasets from the observed dataset (Table 2). Under this model, we estimated that a large population with an effective size of approximately 50,600 macaque individuals (95% Highest Probability Density or HPD 19,600 – 68,300) declined to a present effective population size of approximately 1,700 individuals (95% HPD 476 – 4,090), that is, approximately 30-fold. Assuming an average generation time of five years for macaques [44], we estimated that this decline occurred approximately 3,500 (95% HPD 520 – 13,800) years before present.

Finally, to assess the goodness-of-fit of the model (Scenario 2) to the data, we simulated 1,000 datasets under each of the three scenarios tested, drawing the values of their parameters into the marginal posterior distributions of these parameters. We thus identified which model was the most capable of reproducing the observed summary statistics computed from the real data, following the model-checking procedure described by Cornuet *et al.* [36]. Of the three demographic scenarios tested, the datasets simulated under Scenario 2 were most compatible with the observed summary statistics. This can be observed by comparing the values of summary statistics computed from the simulated datasets for each tested scenario against the real values. With the exception of Scenario 2, all competing scenarios generated large numbers of summary statistics that displayed highly significant outlying values (Fig. 4). This finding thus strongly supported the very high posterior probability values obtained for Scenario 2 of population decline, using the model-choice procedure.

Discussion

Classical bottleneck tests are often used by both evolutionary and conservation biologists to evaluate whether species have experienced historical demographic declines. In contrast, more recently developed Bayesian approaches offer the potential to draw more detailed inferences with respect to both bottleneck timing and severity [35,36] but have not yet been thoroughly evaluated. Consequently, we analysed a multi-marker dataset using both sets of approaches to elucidate the possible impacts of historical climate change and more recent anthropogenic effects on an "edge" primate species – a species that inhabits the undulating edges of a plateau instead of its flat centre [14] – the Arunachal macaque, in the northeastern Indian state of Arunachal Pradesh. This region, situated in the southern part of the Tibetan Plateau, also forms part of the Eastern Himalaya biodiversity hotspot.

Limited effect of Pleistocene glaciations before the Last Glacial Maximum (LGM)

Pleistocene climate change is known to have influenced the demographic history of many species across a wide variety of taxa. The landmass of Europe and a large part of northern America were under continental ice sheets repeatedly over the last three million years, forcing most species to either go extinct or shift their distribution range. Those, which shifted their range, also suffered a concomitant reduction in their population size. The effects of glaciation on species abundance and distribution, however, become more complex in eastern Eurasia where ice cover was never continuous [45,46].

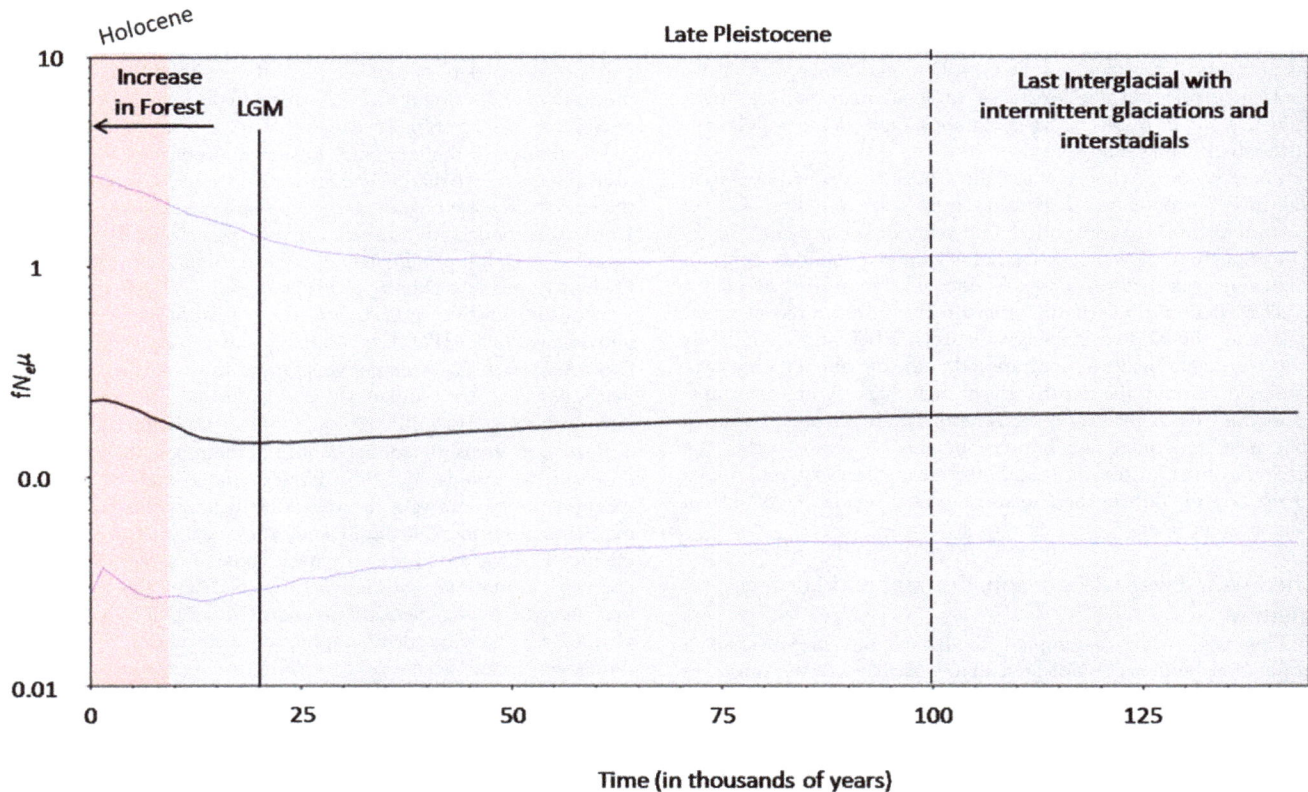

Figure 2. Bayesian skyline plot reconstruction of past population size trajectory for the Upper Subansiri population. The plot is the product of female effective population size (fN_e) and mutation rate (μ) through time, assuming a substitution rate of 0.1643 substitutions per nucleotide per million years [29]. The lower and upper 95% confidence interval for the Upper Subansiri TMRCA is also shown. LGM, Last Glacial Maximum, approximately 18 to 20 thousand years before present.

The southern part of the Tibetan Plateau is an interesting example in this regard. The southern edge of the plateau was characterised by complex orogenesis, as compared to that experienced by the flatter central regions. This area also marks the southern limit of the glaciers that occur on the plateau. During the Quaternary period, the Tibetan Plateau had undergone four to five glaciations but was less affected by the ice sheets than were other regions in Asia at that time [47]. The largest glacier on the plateau occurred in the middle of the Pleistocene (about 0.5 mya) and continued until 0.17 mya [47]. During that time, the ice cover may have been permanent in the higher altitudes and central regions of the plateau [48,49] while the southern and eastern regions experienced relatively less glaciation [50]. Both these sets of conditions created a fragmented ice cover where most of the high-elevation areas on the mountain ranges were ice-clad. Consequently, the prolonged glaciation periods during the late Pleistocene appear to have had a less severe effect on the species inhabiting these areas, as compared to those inhabiting Europe or North America.

It is now being suggested that the ice sheets possibly retreated more rapidly on the Tibetan Plateau than they did in Europe, as seems to be evident from the demographic history of species like the great tit *Parus major* from the plateau [46]. It was observed that contrary to the post-LGM expansion of European animal populations, the demographic histories of species on the Tibetan Plateau indicate expansions only before the LGM; they remained relatively stable or grew slowly subsequently through the LGM [46]. The edges of the plateau would have been less under the

influence of the glaciations, if at all, at least on the lower elevations. We find support for such a hypothesis in the Bayesian skyline plots, which suggest that the size of the Arunachal macaque populations appeared to remain constant during the prolonged pre-LGM climatic fluctuations. A recent comparative study on five avian species from the Tibetan Plateau [14] demonstrated that three species distributed on the central platform of the plateau experienced rapid population expansion after the retreat of the extensive glaciers during the pre-LGM (0.5–0.175 mya), results similar to that of Zhao *et al.* [46]. On the contrary, the population sizes of the other two species, the twite *Carduelis flavirostris* and the black redstart *Phoenicurus ochruros*, distributed on the edges of the plateau, remained at stable levels throughout the same pre-LGM period. It was concluded that the comparatively ice-free habitats on the edges of the plateau might have experienced milder climates during the glaciation period and this allowed the local species populations to persist in this stable niche [14]. Our data also allow us to arrive at similar conclusions for a mammalian species, the Arunachal macaque, in the Eastern Himalayas on the southern fringes of the Tibetan Plateau.

Effect of elevation and animal physiology

Altitude appears to have played an extremely important role in the demographic history of animal species during the past glacial periods with species at various elevations being affected differently. A very good example of this is in the Alps [51,52]. Many present-day species in the high elevations colonised their present range by expanding their range from the lower altitudes and latitudes

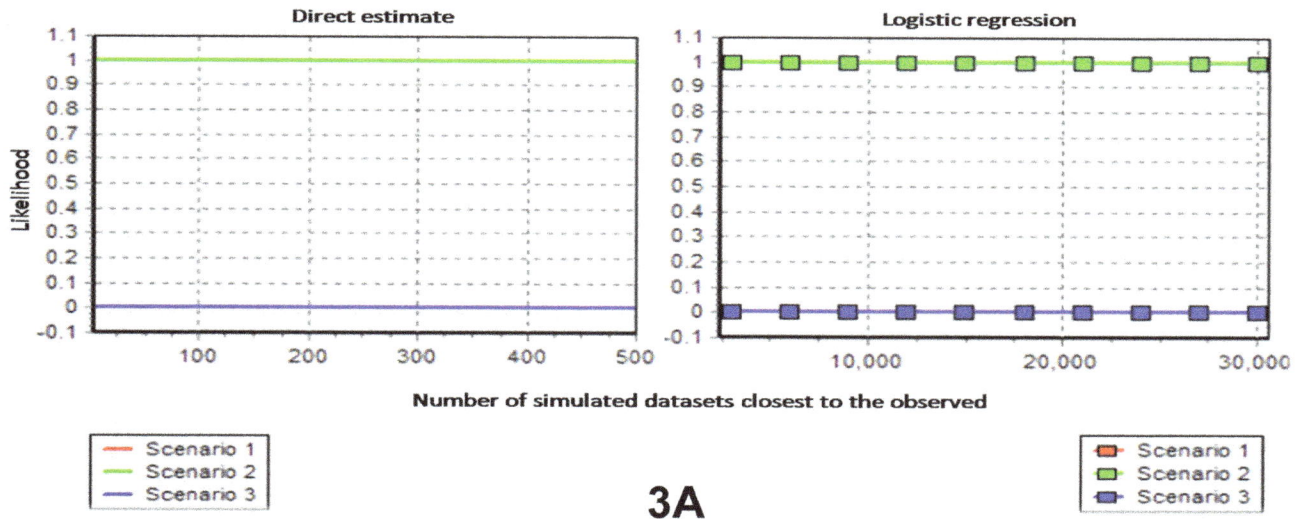

3A

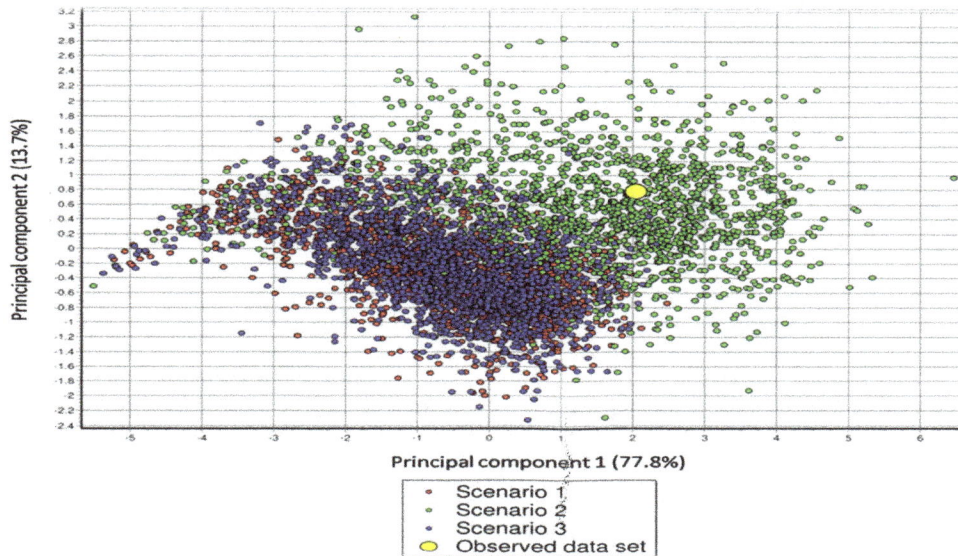

3B

Figure 3. Comparison of the three possible alternative demographic scenarios for the Upper Subansiri population. (A) Estimates of the posterior probability of each scenario and their comparison. Direct estimate: The number of times that a given scenario is found closest to the simulated datasets once the latter, produced under several scenarios, have been sorted by ascending distances to the observed dataset. Logistic regression: A polytomic weighted logistic regression was performed on the first closest datasets with the proportion contributed by each scenario as the dependent variable and the differences between the summary statistics of the observed and simulated datasets as the independent variables. The intercept of the regression, corresponding to an identity between simulated and observed summary statistics, was taken as the point estimate. In addition, 95% confidence intervals were computed [41]. In all these cases, Scenario 2 explained the observed data best. (B) Principal component analysis: Visual information on how data sets simulated under each scenario are close to the observed data set. Here too, Scenario 2 fits the observed data best [41].

during specific interglacial periods. In contrast, species with Arctic–Alpine distributions, which were not annihilated during peaks of glaciation, descended during these periods and may have spread more widely across the cold tundra and steppe plains of continental Europe [51,52]. The impact of Pleistocene glacial cycles is thus expected to have varied among species across geographical regions, in part, perhaps due to their biology of differential cold tolerance [15].

The first indication of such a historical change in the size of the study Arunachal macaque populations was detected through the mismatch distributions of the sampled individuals (data not shown) which, however, could be unreliable due to small sample size of

Table 2. Demographic parameters estimated under the best-supported demographic scenario (Scenario 2) of a recent population decline in the Upper Subansiri population of Arunachal macaques.

Parameter	Median	25% HPD	95% HPD
Current effective population size (N_1)	1720	476	4090
Population size before the decline (N_A)	50600	19600	68300
Magnitude of the population decline (N_1/N_A)	0.03	0.02	0.06
Time since the decline (T_1), years before present	3575	520	13800

Population sizes are given in effective number of diploid individuals. Time estimates were calibrated by assuming a generation time of 5 years [44].

the present study. Mismatch distribution analysis also has its limitations, as exemplified by many studies where contemporary samples of a bottlenecked population failed to recover a unimodal mismatch distribution [53]. Instead, coalescent theory, which incorporates information from genealogy may better able to evaluate the demographic histories of populations [54,55]. Bayesian skyline plots in our study also suggest expansion of the study macaque populations. These plots also propound changes in

population size only after the LGM, as has been established for several avian edge species on the Tibetan Plateau [14]. Moreover, the effective female population size of the Arunachal macaque seems to have increased approximately 15,000 years ago.

Although the climate on the Tibetan Plateau has been documented to have been colder during the pre-LGM extensive glacial period than during the LGM [56], mountain heights below the snowline were not apparently glaciated. The current snowlines

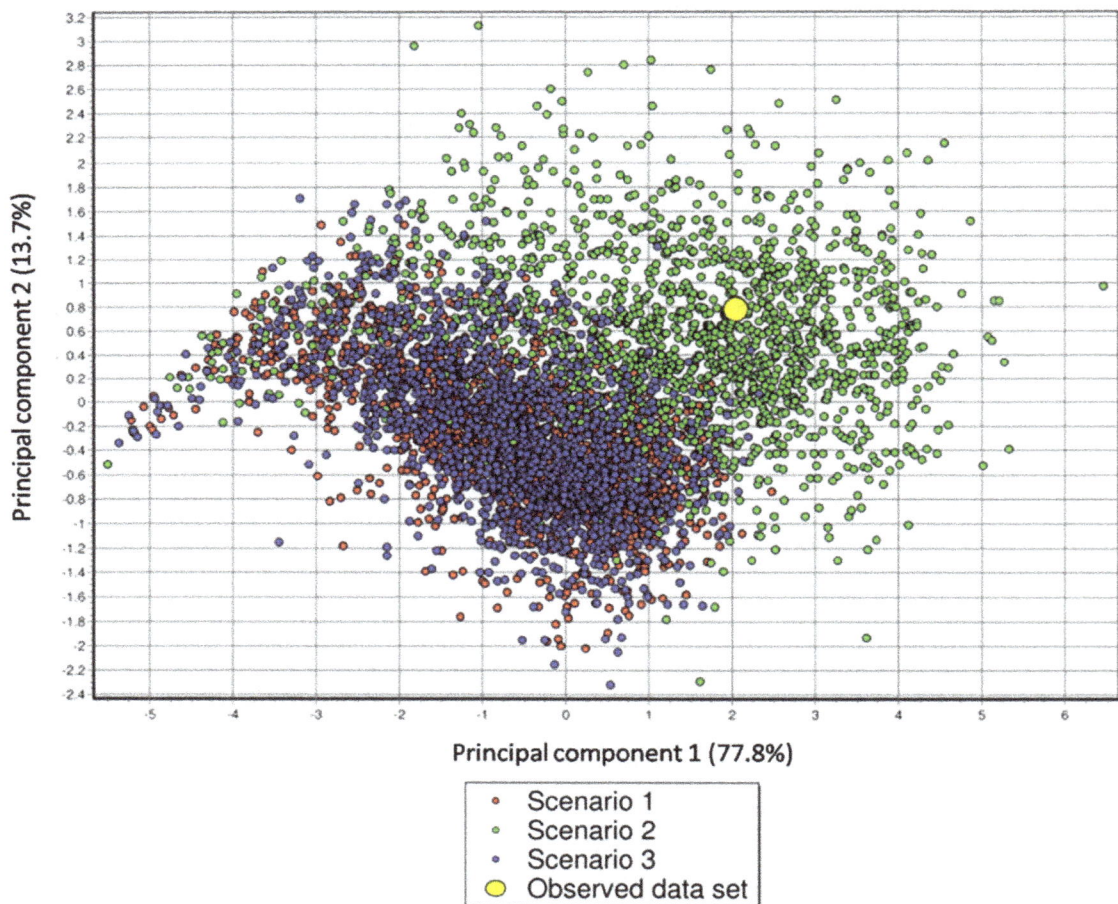

Figure 4. Model-checking to measure the discrepancy between the model parameter posterior combination and the real dataset for the three alternative demographic scenarios for the Upper Subansiri population of the Arunachal macaque. The summary statistics using datasets simulated with the prior distributions of the parameters, the observed data and the datasets from the posterior predictive distributions are represented on the plane of the first two principal components. The model (Scenario 2) fits the data well, as the cloud of datasets simulated from the prior (small green dots), datasets from the posterior predictive distribution (larger green dots) and the observed dataset (yellow circles) overlap completely.

on the eastern edge of the plateau extend from 4200 to 5200 m [57,58]. In Arunachal Pradesh, more specifically, the vegetation ends approximately at a 5,000-m elevation from where the present snowline starts [59]. During the LGM, however, the snowline descended to approximately 3,300 m on many mountain ranges on the Tibetan Plateau [56,57], thus contributing to cooler climates locally [47]. Arunachal macaques are known to presently occur at altitudes between 1800 – 3500 m [60] although they could also inhabit higher, unexplored, elevations in the region. It now seems possible that at least a part of the current habitat of the macaque was covered by ice during the LGM. We, therefore, speculate that this primate might have colonised even higher elevations after the LGM, resulting in the observed signature of population expansion.

The pattern that our results suggest have been conclusively established for another lower-elevation avian edge species from the Tibetan plateau, the black redstart *Phoenicurus ochruros* [14]. The relationship between glaciation, altitude and species demographic history is, however, far from straightforward. Interestingly, Lu *et al.* [15] found that the population size of the three low-elevation (1800 – 3200 m) stream salamander species from the Tibetan Plateau were significantly but slowly decreased from the beginning of the extensive glacial period long before the LGM, in striking contrast to what has been established for the Arunachal macaque and the black redstart. According to Lu *et al.* [15], the suitable habitats for the salamanders may have not been covered by ice during both the LGM and the extensive pre-LGM glacial period but the species may have suffered from climatic cooling because they are less cold-tolerant than their high-elevation counterparts. These ideas are consistent with the hypothesis that variation in ecological adaptations may affect geographical patterns of genetic variation in species populations [61,62]. The poikilothermic salamanders may be expected to have a much lower range of tolerance to changes in ambient temperatures than would homoeothermic animals such as birds and mammals. Thus, animal physiology is another important factor that may have differently affected the evolutionary history of species that otherwise occurred at similar elevations and latitudes in a particular geographical region.

Holocene population decline

The increase in effective female population size of the Arunachal macaque, as revealed by the Bayesian skyline plots, continues till approximately 5,000 years ago, after which it appears to have suffered a mild decline. Mitochondrial and nuclear autosomal microsatellite loci are known to be informative at different time scales, highlighting different episodes in the demographic history of a species [6]. The mtDNA diversity tends to reflect comparatively older demographic events in a species' history. Conversely, microsatellite data are more informative about the contemporary demography of a species. Such a difference in demographic signals, captured by each type of genetic marker, may arise from the differences in their respective mutation rates [36]. The relatively fast mutation rate of microsatellite loci enables them to capture recent and almost contemporaneous events but also increases homoplasy at these loci, which thereby reduces the signal of older demographic events [39]. These, more ancient, events can, however, still be detected using the slower evolving mtDNA sequences. That is why we employed our microsatellite data to validate the occurrence of a very recent population decline in the Arunachal macaque.

The classical heterozygosity excess test (the EWCL method) suggested a past population bottleneck in at least the Upper Subansiri population, as reflected in its microsatellite data

although the statistical significance of this result was highly dependent on the underlying mutational model. A bottleneck was inferred for this population for two particular models – the infinite-allele (IAM) and two-phase (TPM) models – but not for a third one, the stepwise-mutation (SMM) model, despite the IAM being unrealistic for most microsatellites [34]. Several other microsatellite-based studies of species, thought to have experienced severe, but temporary, reductions in population size, had either failed altogether to detect a bottleneck or, as with our study, yielded results that depended on the mutational model [63]. In at least some of these cases, a few specific markers may have disproportionately influenced the more conservative SMM model. However, in our study, we used 22 microsatellite loci, over twice the minimum number recommended [64]. To circumvent this problem, we applied the ABC method that clearly supported the population decline model (Scenario 2) over the constant size and population expansion models.

We could also estimate the extent and time of the decline from the posterior probabilities. Our simulations suggest the population decline started approximately 3,500 years ago, which was climatically a warm period and this is surprising as such climatic conditions are normally conducive for population growth. Several authors have postulated that such idiosyncratic mid-Holocene population declines may have been accelerated and enhanced by major expansions in ancient human civilizations [65] (ca. 1,500 BC), the rapid development of agriculture, and the resulting changes in landscapes in recent times. For example, it has been speculated that the expansion and migration of human populations into the erstwhile virgin mountains of the Sichuan region may have caused the decline of the giant panda population during the later part of the mid-Holocene [66].

In Arunachal Pradesh too, anthropogenic factors may have significantly influenced the demographic history of the Arunachal macaque over the last few decades or centuries. One of the largest indigenous groups of people that inhabit the districts of Upper Subansiri and West Siang, two areas from which our study samples were derived, and which hunt wildlife extensively for food and sport, is represented by the collective animist Adi people [67,68]. Recent ethnographic and population genetic studies reveal that the Adi, alternatively referred to as the Luoba Tibetan in Tibet, trace their ancestral migration and settlement history from southern Tibet into Arunachal Pradesh to different time periods during the 5th to 7th century AD, or to the last 1,300 to 1,500 years before present [67,68,69,70]. Although these estimates do not reject the possibility of a peopling of this region in even earlier times, it is entirely possible that the Arunachal macaque population declines may have actually begun at the time that central Arunachal Pradesh was being peopled by the Adi and other animistic tribes that continue to hunt rampantly even today. It must be reiterated, however, that there is currently no evidence to conclusively establish the definitive contribution of either climatic or anthropogenic factors to this decline in Arunachal macaque populations.

It should be noted that our investigation, which focuses on a rare, endangered, montane primate, has a smaller sample size than is typical for most population genetics studies. Although there appears to be little benefit in sampling more than 20 to 30 individuals per population for microsatellite-based studies that assess genetic diversity of populations [71,72], sample sizes that are even smaller, as is the case in the present study, may not precisely represent the populations and may yield large errors for descriptive statistical estimates such as allele frequencies and expected heterozygosity. It has, however, been suggested that where small sample sizes are unavoidable, as, for instance, in the case of rare and endangered species, the precision of the estimates can be significantly enhanced by increasing the number of genotyped microsatellite loci, which is precisely what we have done in our study [72]. Moreover, we have used approximate Bayesian computations to test whether the population size changes that we observe using other analytical methods are noteworthy or not. The approximate Bayesian computation framework manages any sample size, which may often be as low as 8 – 12 samples per population [73], by simulating the exactly observed sample sizes [74]. Small sample sizes would simply lead to wider credible intervals than would larger sample sizes [74]. We are aware of this limitation of our results but believe that the communication of our present study is of critical importance, in spite of its low sample size, for the conservation effort to save the species, particularly as the collection of more samples from the rare and heavily hunted macaque populations in Arunachal Pradesh is not currently feasible.

In conclusion, the Tibetan Plateau is an important region to study the population history of its native species for more than one reason. First of all, it is a highly variable region with its different regions – the flat central platform and the more variable mountainous edges based on altitude – differing significantly from one another in topography and climatic history. The edge regions of the plateau have had a tumultuous orogenic history and the species of these areas are, thus, also expected to show a more eventful evolutionary history. Our study of an edge primate species has, along with other studies from the region, hinted at a very complex image of how geographical features such as latitude and altitude, animal physiology, and climate change in the past may have together influenced its demographic history. Unlike in the tropics, where Pleistocene glacial fluctuations were less likely to have been extreme, this region seems to represent a balance between the tropics and the more severe, temperate arctic environments. This study also underscores the possibility that a cold-tolerant species like the Arunachal macaque, which could withstand historical climate change and grow once the climate became conducive, may actually be extremely vulnerable to anthropogenic exploitation, as is perhaps indicated by its more recent population decline. It is imperative that we understand the population dynamics of such a species at a much finer scale, which

could be possible with sampling more populations, particularly those that may connect the study populations. What is of greater concern, however, is the clear genetic signature of a serious decline in the populations of the species, possibly mediated by the extensive hunting that it continues to face across its distribution range. While these threats to the affected populations are currently being documented by on-ground field studies, it is entirely possible that unless immediate action is taken, genetic drift caused by hunting, epidemics or other natural calamities can rapidly eliminate the remaining genetic diversity of this less-known endemic primate from an extremely important biodiversity hotspot in the remote Himalayan mountains of northeastern India.

Supporting Information

Figure S1 Map of Arunachal Pradesh, northeastern India, with locations of the sampling sites. The triangles correspond to sampling sites in West Siang, circles to those in Upper Subansiri and rectangles to Tawang sampling sites. Inset: Location of the study site at the edge of Tibetan Plateau.

Table S1 Arunachal and bonnet macaque samples used in the study and their sites of origin.

Acknowledgments

Authors sincerely acknowledge positive criticisms from Cédric Sueur, Michael Huffman and three other unknown reviewers. Their contribution during the review process helped immensely in increasing the quality of the final manuscript. Debapriyo Chakraborty is grateful to many colleagues including Kakoli Mukhopadhyay, M D Madhusudan, Samrat Mondol and Shreejata Gupta for critical inputs that too improved the present work considerably. Authors also thank Shreejata Gupta for providing important help with presenting the figures and the organisation of references for the manuscript.

Author Contributions

Conceived and designed the experiments: DC AS UR. Performed the experiments: DC. Analyzed the data: DC. Contributed reagents/materials/analysis tools: UR. Wrote the paper: DC AS UR.

References

1. Davis MB, Shaw RG, Etterson JR (2005) Evolutionary responses to changing climate. Ecology 86: 1704–1714.
2. Kukla GJ (2000) The last interglacial. Science 287: 987–988.
3. Clark PU, Dyke AS, Shakun JD, Carlson AE, Clark J, et al. (2009) The last glacial maximum. Science 325: 710–714.
4. Williams JW, Shuman BN, Webb III T, Bartlein PJ, Leduc PL (2004) Late-Quaternary vegetation dynamics in North America: Scaling from taxa to biomes. Ecol Monogr 74: 309–334.
5. Barnosky AD (2008) Megafauna biomass tradeoff as a driver of Quaternary and future extinctions. Proc Natl Acad Sci USA 105: 11543–11548.
6. Fontaine MC, Snirc A, Frantzis A, Koutrakis E, Öztürk B, et al. (2012) History of expansion and anthropogenic collapse in a top marine predator of the Black Sea estimated from genetic data. Proc Natl Acad Sci USA 109: E2569–E2576.
7. Koblmüller S, Wayne RK, Leonard JA (2012) Impact of Quaternary climatic changes and interspecific competition on the demographic history of a highly mobile generalist carnivore, the coyote. Biol Lett 8: 644–647.
8. Chikhi L, Sousa VC, Luisi P, Goossens B, Beaumont MA (2010) The confounding effects of population structure, genetic diversity and the sampling scheme on the detection and quantification of population size changes. Genetics 186: 983–995.
9. Goossens B, Chikhi L, Ancrenaz M, Lackman-Ancrenaz I, Andau P, et al. (2006) Genetic signature of anthropogenic population collapse in orang-utans. PLoS Biol 4: e25.
10. Salmona J, Salamolard M, Fouillot D, Ghestemme T, Larose J, et al. (2012) Signature of a pre-human population decline in the critically endangered Reunion Island endemic forest bird Coracina newtoni. PLoS one 7: e43524.

11. Ting N, Astaras C, Hearn G, Honarvar S, Corush J, et al. (2012) Genetic signatures of a demographic collapse in a large-bodied forest dwelling primate (*Mandrillus leucophaeus*). Ecol Evol 2: 550–561.
12. Beaumont LJ, Pitman A, Perkins S, Zimmermann NE, Yoccoz NG, et al. (2011) Impacts of climate change on the world's most exceptional ecoregions. Proc Natl Acad Sci USA 108: 2306–2311.
13. Shapiro B, Drummond AJ, Rambaut A, Wilson MC, Matheus PE, et al. (2004) Rise and fall of the Beringian steppe bison. Science 306: 1561–1565.
14. Qu Y, Lei F, Zhang R, Lu X (2010) Comparative phylogeography of five avian species: implications for Pleistocene evolutionary history in the Qinghai-Tibetan plateau. Mol Ecol 19: 338–351.
15. Lu B, Zheng Y, Murphy RW, Zeng X (2012) Coalescence patterns of endemic Tibetan species of stream salamanders (Hynobiidae: *Batrachuperus*). Mol Ecol 21: 3308–3324.
16. Bergl RA, Bradley BJ, Nsubuga A, Vigilant L (2008) Effects of habitat fragmentation, population size and demographic history on genetic diversity: The Cross River gorilla in a comparative context. Am J Primatol 70: 848–859.
17. Chang ZF, Luo MF, Liu ZJ, Yang JY, Xiang ZF, et al. (2012) Human influence on the population decline and loss of genetic diversity in a small and isolated population of Sichuan snub-nosed monkeys (*Rhinopithecus roxellana*). Genetica 140: 105–114.
18. Storz JF, Beaumont MA, Alberts SC (2002) Genetic evidence for long-term population decline in a savannah-dwelling primate: Inferences from a hierarchical Bayesian model. Mol Biol Evol 19: 1981–1990.
19. Modolo L, Salzburger W, Martin RD (2005) Phylogeography of Barbary macaques (*Macaca sylvanus*) and the origin of the Gibraltar colony. Proc Natl Acad Sci USA 102: 7392–7397.

20. Kawamoto Y, Tomari K, Kawai S, Kawamoto S (2008) Genetics of the Shimokita macaque population suggest an ancient bottleneck. Primates 49: 32–40.

21. Aiyadurai A (2011) Wildlife hunting and conservation in Northeast India: a need for an interdisciplinary understanding. Int J Galliformes Conserv 2: 61–73.

22. Krithika S, Maji S, Vasulu TS (2008) A microsatellite guided insight into the genetic status of Adi, an isolated hunting-gathering tribe of Northeast India. PloS one 3: e2549.

23. Kumar RS, Gama N, Raghunath R, Sinha A, Mishra C (2008) In search of the *munzala*: distribution and conservation status of the newly-discovered Arunachal macaque *Macaca munzala*. Oryx 42: 360–366.

24. Kumar RS, Mishra C, Sinha A (2007) A preliminary survey for macaques in central Arunachal Pradesh, northeastern India. Mysore, India: Nature Conservation Foundation.

25. Li QQ, Zhang YP (2005) Phylogenetic relationships of the macaques (Cercopithecidae: *Macaca*), inferred from mitochondrial DNA sequences. Biochem Genet 43: 375–386.

26. Chakraborty D, Ramakrishnan U, Panor J, Mishra C, Sinha A (2007) Phylogenetic relationships and morphometric affinities of the Arunachal macaque *Macaca munzala*, a newly described primate from Arunachal Pradesh, northeastern India. Mol Phylogenet Evol 44: 838–849.

27. Rogers J, Bergstrom M, Garcia IV R, Kaplan J, Arya A, et al. (2005) A panel of 20 highly variable microsatellite polymorphisms in rhesus macaques (*Macaca mulatta*) selected for pedigree or population genetic analysis. Am J Primatol 67: 377–383.

28. Kanthaswamy S, Von Dollen A, Kurushima JD, Alminas O, Rogers J, et al. (2006) Microsatellite markers for standardized genetic management of captive colonies of rhesus macaques (*Macaca mulatta*). Am J Primatol 68: 73–95.

29. Chakraborty D (2013) Genes in space and time: Population genetic structure and demographic history of two primate species, the Arunachal macaque *Macaca munzala* and bonnet macaque *Macaca radiata*. Mysore, India: Nature Conservation Foundation - Manipal University.

30. Drummond AJ, Suchard MA, Xie D, Rambaut A (2012) Bayesian phylogenetics with BEAUti and the BEAST 1.7. Mol Biol Evol 29: 1969–1973.

31. Garza JC, Williamson EG (2001) Detection of reduction in population size using data from microsatellite loci. Mol Ecol 10: 305–318.

32. Cornuet JM, Luikart G (1996) Description and power analysis of two tests for detecting recent population bottlenecks from allele frequency data. Genetics 144: 2001–2014.

33. Piry S, Luikart G, Cornuet JM (1999) Computer note. BOTTLENECK: a computer program for detecting recent reductions in the effective size using allele frequency data. J Hered 90: 502–503.

34. Di Rienzo A, Peterson AC, Garza JC, Valdes AM, Slatkin M, et al. (1994) Mutational processes of simple-sequence repeat loci in human populations. Proc Natl Acad Sci 91: 3166–3170.

35. Beaumont MA, Zhang W, Balding DJ (2002) Approximate Bayesian computation in population genetics. Genetics 162: 2025–2035.

36. Cornuet JM, Ravigné V, Estoup A (2010) Inference on population history and model checking using DNA sequence and microsatellite data with the software DIYABC (v1. 0). BMC Bioinform 11: 401.

37. Beaumont MA (2010) Approximate Bayesian computation in evolution and ecology. Annu Rev Ecol Evol Syst 41: 379–406.

38. Storz JF, Beaumont MA (2002) Testing for genetic evidence of population expansion and contraction: an empirical analysis of microsatellite DNA variation using a hierarchical Bayesian model. Evol 56: 154–166.

39. Estoup A, Jarne P, Cornuet JM (2002) Homoplasy and mutation model at microsatellite loci and their consequences for population genetics analysis. Mol Ecol 11: 1591–1604.

40. Verdu P, Austerlitz F, Estoup A, Vitalis R, Georges M, et al. (2009) Origins and genetic diversity of pygmy hunter-gatherers from Western Central Africa. Curr Biol 19: 312–318.

41. Cornuet JM, Santos F, Beaumont MA, Robert CP, Marin JM, et al. (2008) Inferring population history with DIY ABC: A user-friendly approach to approximate Bayesian computation. Bioinform 24: 2713–2719.

42. Gelman A, Carlin JB, Stern HS, Rubin DB (1995) Bayesian Data Analysis. Boca Raton, Florida: Chapman & Hall/CRC.

43. Robert CP, Cornuet JM, Marin JM, Pillai NS (2011) Lack of confidence in approximate Bayesian computation model choice. Proc Natl Acad Sci USA 108: 15112–15117.

44. Harvey PH, Martin RD, Clutton-Brock TH (1987) Life histories in comparative perspective. In: Smuts BB, Cheney DL, Seyfarth RM, Wrangham RW, editors. Primate Societies. Chicago: University of Chicago Press. pp. 181–196.

45. Zhan X, Zheng Y, Wei F, Bruford MW, Jia C (2011) Molecular evidence for Pleistocene refugia at the eastern edge of the Tibetan Plateau. Mol Ecol 20: 3014–3026.

46. Zhao N, Dai C, Wang W, Zhang R, Qu Y, et al. (2012) Pleistocene climate changes shaped the divergence and demography of Asian populations of the great tit *Parus major*: Evidence from phylogeographic analysis and ecological niche models. J Avian Biol 43: 297–310.

47. Zheng B, Xu Q, Shen Y (2002) The relationship between climate change and Quaternary glacial cycles on the Qinghai–Tibetan Plateau: review and speculation. Quat Int 97: 93–101.

48. Shi Y, Zheng B, Li S (1990) Last glaciation and maximum glaciation in Qinghai-Xizang (Tibet) plateau. J Glaciol Geocryol 12: 1–15.

49. Shi YF (2002) A suggestion to improve the chronology of Quaternary glaciations in China. J Glaciol Geocryol 24: 687–692.

50. Zhang D, Fengquan L, Jianmin B (2000) Eco-environmental effects of the Qinghai-Tibet Plateau uplift during the Quaternary in China. Environ Geol 39: 1352–1358.

51. Hewitt GM (1996) Some genetic consequences of ice ages, and their role in divergence and speciation. Biol J Linn Soc 58: 247–276.

52. Hewitt GM (2004) Genetic consequences of climatic oscillations in the Quaternary. Philosl Trans R Soc Lond B 359: 183–195.

53. Hoffman JI, Grant SM, Forcada J, Phillips CD (2011) Bayesian inference of a historical bottleneck in a heavily exploited marine mammal. Mol Ecol 20: 3989–4008.

54. Felsenstein J (1992) Estimating effective population size from samples of sequences: inefficiency of pairwise and segregating sites as compared to phylogenetic estimates. Genet Res 59: 139–147.

55. Pybus OG, Rambaut A, Harvey PH (2000) An integrated framework for the inference of viral population history from reconstructed genealogies. Genetics 155: 1429–1437.

56. Shi Y, Zheng B, Li S, Ye B (1995) Studies on altitude and climatic environment in the middle and east part of Tibetan Plateau during Quarternery maximum glaciation. J Glaciol Geocryol 17: 97–112.

57. Shi Y, Zheng B, Yao T (1997) Glaciers and environments during the last glacial maximum (LGM) on the Tibetan Plateau. J Glaciol Geocryol 19: 97–113.

58. Liu T, Zhang X, Xiong S, Qin X (1999) Qinghai-Xizang Plateau glacial environment and global cooling. Quat Sci 5: 385–396.

59. Mishra C, Dutta A, Madhusudan MD (2004) The high altitude wildlife of western Arunachal Pradesh: A survey report. CERC technical report no. 8. Mysore, India: Nature Conservation Foundation

60. Sinha A, Datta A, Madhusudan MD, Mishra C (2005) *Macaca munzala*: A New Species from Western Arunachal Pradesh, Northeastern India. Int J Primatol 26: 977–989.

61. Gavrilets S (2003) Perspective: models of speciation: what have we learned in 40 years? Evol 57: 2197–2215.

62. Hewitt GM (2004) The structure of biodiversity–insights from molecular phylogeography. Front Zool 1: 1–16.

63. Spong G, Hellborg L (2002) A near-extinction event in lynx: do microsatellite data tell the tale? Conserv Ecol 6: 15.

64. Luikart G, Cornuet JM (1998) Empirical evaluation of a test for identifying recently bottlenecked populations from allele frequency data. Conserv Biol 12: 228–237.

65. Stavrianos LS (1998) A global history: From prehistory to 21st century. 7th ed. New Jersey: Prentice Hall Inc.

66. Zhang B, Li M, Zhang Z, Goossens B, Zhu L, et al. (2007) Genetic viability and population history of the giant panda, putting an end to the "evolutionary dead end"? Mol Biol Evol 24: 1801–1810.

67. Lego N (2006) History of the Adis of Arunachal Pradesh. Itanagar, Arunachal Pradesh, India: Jumbo Gumin Publishers and Distributors.

68. Tabi T (2006) The Adis. Pasighat, Arunachal Pradesh, India: Siang Literary Forum.

69. Krithika S, Maji S, Vasulu TS (2009) A microsatellite study to disentangle the ambiguity of linguistic, geographic, ethnic and genetic influences on tribes of India to get a better clarity of the antiquity and peopling of South Asia. Am J Phys Anthropol 139: 533–546.

70. Kang L, Li S, Gupta S, Zhang Y, Liu K, et al. (2010) Genetic structures of the Tibetans and the Deng people in the Himalayas viewed from autosomal STRs. J Hum Genet 55: 270–277.

71. Pruett CL, Winker K (2008) The effects of sample size on population genetic diversity estimates in song sparrows *Melospiza melodia*. J Avian Biol 39:252 – 256.

72. Hale ML, Burg TM, Steeves TE (2012) Sampling for microsatellite-based population genetic studies: 25 to 30 individuals per population is enough to accurately estimate allele frequencies. PloS one 7: e45170.

73. Fagundes NJ, Ray N, Beaumont M, Neuenschwander S, Salzano FM, et al. (2007) Statistical evaluation of alternative models of human evolution. Proc Nat Acad Sci 104: 17614–17619.

74. Beaumont MA, Nielsen R, Robert C, Hey J, Gaggiotti O, et al. (2010). In defence of model-based inference in phylogeography. Mol Ecol 19: 436–446.

8

What Story Does Geographic Separation of Insular Bats Tell? A Case Study on Sardinian Rhinolophids

Danilo Russo[1,2]*, **Mirko Di Febbraro**[3], **Hugo Rebelo**[2,4], **Mauro Mucedda**[5], **Luca Cistrone**[6], **Paolo Agnelli**[7], **Pier Paolo De Pasquale**[8], **Adriano Martinoli**[9], **Dino Scaravelli**[10], **Cristiano Spilinga**[11], **Luciano Bosso**[1]

1 Wildlife Research Unit, Dipartimento di Agraria, Università degli Studi di Napoli Federico II, Portici, Napoli, Italy, **2** School of Biological Sciences, University of Bristol, Bristol, United Kingdom, **3** EnvixLab, Dipartimento Bioscienze e Territorio, Università del Molise, Pesche, Italy, **4** CIBIO, Centro de Investigação em Biodiversidade e Recursos Genéticos da Universidade do Porto, University of Porto, Vairão, Portuga, **5** Centro per lo studio e la protezione dei pipistrelli in Sardegna, Sassari, Italy, **6** Forestry and Conservation, Cassino, Frosinone, Italy, **7** Museo di Storia Naturale dell'Università di Firenze, Sezione di Zoologia 'La Specola', Firenze, Italy, **8** Wildlife Consulting, Palo del Colle, Bari, Italy, **9** Unità di Analisi e Gestione delle Risorse Ambientali, Guido Tosi Research Group, Dipartimento di Scienze Teoriche e Applicate, Università degli Studi dell'Insubria, Varese, Italy, **10** Dipartimento di Scienze Mediche Veterinarie, Università degli Studi di Bologna, Ozzano dell'Emilia, Bologna, Italy, **11** Studio Naturalistico Hyla snc, Tuoro sul Trasimeno, Perugia, Italy

Abstract

Competition may lead to changes in a species' environmental niche in areas of sympatry and shifts in the niche of weaker competitors to occupy areas where stronger ones are rarer. Although mainland Mediterranean (*Rhinolophus euryale*) and Mehely's (*R. mehelyi*) horseshoe bats mitigate competition by habitat partitioning, this may not be true on resource-limited systems such as islands. We hypothesize that Sardinian *R. euryale* (SAR) have a distinct ecological niche suited to persist in the south of Sardinia where *R. mehelyi* is rarer. Assuming that SAR originated from other Italian populations (PES) – mostly allopatric with *R. mehelyi* – once on Sardinia the former may have undergone niche displacement driven by *R. mehelyi*. Alternatively, its niche could have been inherited from a Maghrebian source population. We: a) generated Maxent Species Distribution Models (SDM) for Sardinian populations; b) calibrated a model with PES occurrences and projected it to Sardinia to see whether PES niche would increase *R. euryale*'s sympatry with *R. mehelyi*; and c) tested for niche similarity between *R. mehelyi* and PES, PES and SAR, and *R. mehelyi* and SAR. Finally we predicted *R. euryale*'s range in Northern Africa both in the present and during the Last Glacial Maximum (LGM) by calibrating SDMs respectively with SAR and PES occurrences and projecting them to the Maghreb. *R. mehelyi* and PES showed niche similarity potentially leading to competition. According to PES' niche, *R. euryale* would show a larger sympatry with *R. mehelyi* on Sardinia than according to SAR niche. Such niches have null similarity. The current and LGM Maghrebian ranges of *R. euryale* were predicted to be wide according to SAR's niche, negligible according to PES' niche. SAR's niche allows *R. euryale* to persist where *R. mehelyi* is rarer and competition probably mild. Possible explanations may be competition-driven niche displacement or Maghrebian origin.

Editor: R. Mark Brigham, University of Regina, Canada

Funding: DR and LB were funded by Italian Ministry of the Environment and the Protection of Land and Sea, CIG nr. 464598541B (www.minambiente.it). HR was funded by the Programme Investigador FCT, IF/00497/2013 (www.fct.pt). The funders had no role in study design, data collection and analysis, decision to publish, or preparation of the manuscript.

* Email: danrusso@unina.it

Introduction

Species distribution patterns may potentially result from a range of causes, historical or current, involving abiotic factors as well as biotic interactions [1]. Identifying which factors determine species distribution among the several potential candidates may not be obvious.

A paradigm of ecology is that long-term coexistence is impossible for species sharing an identical ecological niche due to competitive and stochastic factors [2–4]. Opposite evolutionary pressures may act on sympatric species in the same guild. Ecomorphological convergence may take place as a result of

selective pressures associated with optimal exploitation of the same resources; on the other hand, if such resources are limiting, interspecific competition may occur, leading to niche segregation.

Several types of such mechanisms have been described, including spatial or temporal niche separation [5–7] and resource partitioning by morphological divergence [8–10]. Interspecific competition may lead to ecological character displacement: differences in morphological and behavioural traits between species are greater where the latter occur in sympatry, smaller or absent in allopatric conditions [2,11,12]. Character displacement is often referred to morphological divergence, whose relationship with resource utilization may sometimes be questionable [13,14]. However, other functionally important traits characterizing a species' ecological niche may undergo displacement, with crucial consequences for geographical distribution. Yet, the spatial dimension of competition – i.e., the large-scale alteration of species distribution due to biotic interactions – is a poorly explored issue. Interspecific competition might involve changes in the environmental niche of a species where the latter is sympatric to competitors, a process hereafter termed as "niche displacement" [14]. Assessing niche displacement constitutes a key approach to a better understanding of factors influencing species' geographical range and niche features, offering a major insight into present and future distributional dynamics [14]. Clearer patterns are expected where competition is especially harsh. This is the case with insular environments, where resources are often limiting [15]: thus, islands provide an ideal set to study these processes.

However, caution is needed when interpreting the current characteristics of the ecological niche: rather than resulting from forces acting in situ, they could have been shaped by historical processes occurred ex situ, i.e. in the population's geographical source, and then retained by their descendents in the newly established population (such as in a process of island colonization from the mainland). Although the rapid change of ecological traits have attracted the attention of scientists for centuries, there is increasing evidence that the tendency for many ecological traits to be retained over time, called niche conservatism [16] is an important, general phenomenon with major evolutionary and ecological consequences – among which, the stability of species assemblages [17].

Bats represent interesting models to test the effect of interactions between species that share habitats and ecomorphological traits due to adaptive convergence, phylogenetic relatedness or crypticism [18–20]. Although among bats several examples of niche segregation due to divergence in morphology, sensory ecology, foraging strategies or habitat partitioning are known [21–24], no evidence of competition-driven geographical displacement is available.

In this study we focus on two rhinolophid bat species, the Mediterranean (*Rhinolophus euryale*) and Mehely's (*Rhinolophus mehelyi*) horseshoe bats. These largely sympatric Mediterranean bats [25] may be regarded as sibling species as they derive from a close common ancestor and are morphologically very similar [26,27]. They are thought to have diverged only 3 My ago [26] and only in 1901 were they recognized as separate species by the German zoologist Paul Matschie. *R. euryale* is widely distributed from sea level to ca. 1,000 m a.s.l. in the south of the continent as well as north-west Africa, and the Near East [28]. Classified globally as near threatened, *R. euryale* populations are declining in most of the geographical range [28]. *R. mehelyi* is confined to the Mediterranean where it shows a patchy occurrence from north Africa and southern Europe through Asia Minor, Anatolia, to Transcaucasia, Iran and Afghanistan [29]. The species is classified as vulnerable on a global scale, and is reported to be extinct in north-east Spain, Mallorca [30], Croatia and Israel [31] and close to extinction in France [32] and Romania [33].

These species have been regarded as potential competitors when foraging in sympatry for marked similarities in morphology, echolocation and habitat selection [34–36] yet, provided environmental conditions are sufficiently heterogeneous, they may mitigate competition by fine-scale habitat partitioning [35,36].

The Italian distribution of these bats is puzzling. *R. mehelyi* is frequent and relatively abundant on Sardinia, whereas in the rest of Italy is almost absent – in fact on the brink of extinction, being restricted to two sites in Sicily where only small colonies occur, and one site on the mainland (Apulia, south-east Italy) where only in 2013 was a single individual observed after 40 years since the latest sighting [37]. *R. euryale* is widespread in most of the Italian peninsula and also occurs in Sicily. On Sardinia, although both species are present, *R. mehelyi* occurs in allopatry in most of the island while their sympatry is restricted to a small area. There, the two species show divergence in echolocation call frequency [38]. Specifically, Sardinian *R. euryale* shows lower frequencies than the peninsular conspecifics, a difference thought to represent an acoustic character displacement pattern driven by the dominant *R. mehelyi* probably to avoid interspecific frequency overlap and maintain separate communication frequency bandwidths [38].

Islands are ecological systems where spatial and trophic resources are often limited and may lead to increased competition [39,40]. In this study we used distributional data, maximum entropy models (Maxent) and Niche Analysis to test the main hypothesis that in an insular, food-limited environment (Sardinia), *R. euryale* may have at least partly accomplished geographical separation from its sibling species thanks to a distinct ecological niche which has allowed it to settle in an area where *R. mehelyi* is rare and competition probably negligible.

This hypothesis generates two predictions:

a) Significant overlap will occur between the environmental niches of Sardinian *R. mehelyi* and allopatric *R. euryale* populations from the mainland and Sicily (hereafter termed PES), setting the scene for interspecific competition;

b) although conspecifics, the niche of Sardinian *R. euryale* (hereafter termed SAR) will diverge from that of PES. This divergence will allow SAR to mitigate interspecific competition with *R. mehelyi*.

Assuming that SAR has originated from PES, once bats colonized Sardinia the original ecological niche may have undergone a niche displacement process driven by *R. mehelyi's* competition and generated the difference forecast by prediction b). However, the origin of *R. euryale's* population on Sardinia is unknown. Along with Europe, northern Africa represents an important geographical source for Sardinian bats [41–43]. Under a niche conservatism assumption [16], any niche difference spotted in SAR relative to PES might rather represent a legacy of an extra-European source population which colonized Sardinia and founded SAR. It may be hypothesized that once bats colonized the island, they retained their ecological niche by stabilizing selection [44] because it performed well in the southern region of Sardinia where competition with *R. mehelyi* was limited.

Accordingly, to search for clues on SAR's origin, we tested whether SAR's niche would perform better than PES' niche in the Maghrebian geographical set. This prediction would be consistent with a northern African origin of SAR. We tested this under different temporal scenarios: we trained distribution models with SAR and PES occurrences respectively and projected them to northern Africa in two snapshots – current time and Last Glacial Maximum (LGM, 21,000 years PB). We chose LGM because at that time geographical distances between islands and mainland were reduced by the emergence of land bridges favouring island colonization by bats, including that of Sardinia from northern Africa [42,45].

Materials and Methods

Study area

For this study we considered the entire Italian territory comprised ca. between latitudes 45° N–36° N and longitudes

Figure 1. Presence records for *Rhinolophus euryale* **(n = 65; black symbols) and** *R. mehelyi* **(n = 40; grey symbols) considered for this study.** The publicly available map layer was obtained from www.fao.org/geonetwork/srv/en/main.home and the image prepared with the Quantum Gis 2.2.0 Valmiera open source software.

6°E–18°E (corresponding to ca. 301.000 km², elevation range = 0–4810 m a.s.l.).

Presence species data

Presence records for *R. euryale* (n = 210) and *R. mehelyi* (n = 60) came from authors' personal databases (Figure 1). Most faunal records were taken by either direct observation or acoustic surveys – activities requiring no specific permission according to Italian laws and regulations. On Sardinia, when roosts were surveyed for

the first time the distinction between *R. euryale* and *R. mehelyi* was made by temporarily capturing bats under licence from the Italian Ministry of Environment (licence numbers: DPN/2D/2004/7489, DPN-2007-0003938, DPN-2010-0009609). Records were screened in ArcGis (version 9.2) for spatial autocorrelation using average nearest neighbour analyses and Moran's I measure of spatial autocorrelation to remove spatially correlated data points and guarantee independence. After this selection, 65 and 40

Table 1. Validation methods applied to Maxent Species Distribution Models for *Rhinolophus euryale* and *R. mehelyi*.

Model	AUC Training	SD	AUC Test	SD	AUC$_{diff}$	SD	TSS	SD
Current								
SAR	0.965	0.005	0.932	0.021	0.033	0.024	0.680	0.165
PES projected to Sardinia	0.997	0.000	0.993	0.003	0.004	0.003	0.804	0.170
R. mehelyi Sardinia	0.863	0.018	0.782	0.040	0.082	0.057	0.523	0.109
LGM								
SAR CCSM	0.944	0.004	0.926	0.014	0.018	0.022	0.768	0.065
SAR MIROC	0.987	0.006	0.917	0.016	0.070	0.005	0.754	0.045
PES CCSM	0.884	0.002	0.854	0.005	0.030	0.015	0.844	0.022
PES MIROC	0.885	0.003	0.877	0.002	0.008	0.022	0.799	0.056

SAR = Sardinian population of *R. euryale*; PES = Populations of *R. euryale* of Peninsular Italy and Sicily.

presence data respectively for *R. euryale* and *R. mehelyi* were used to generate SDMs.

Ecogeographical variables

To predict habitat suitability for the two species, we used a set of 21 Eco-Geographical Variables (EGVs). We included one topographical and 19 bioclimatic variables obtained from WorldClim database (www.worldclim.org/current) [46]. The latter variables are derived from the monthly temperature and rainfall values in order to generate more biologically meaningful factors [47]. Land cover was obtained from Global Land Cover 2000 (http://bioval.jrc.ec.europa.eu/products/glc2000/products.php). All variable formats were raster files (grid) with a resolution of 30 arc second (0.93×0.93 km $= 0.86$ km^2 at the equator) and 1,307,195 grid cells. In order to remove the highly correlated variables for the final distribution models, we calculated a correlation matrix using Pearson's technique and selected only the variables with r<0.5. We converted the eleven final EGVs used to model habitat suitability of both species in ASCII files.

Maximum entropy approach

We used Maxent – maximum entropy modelling of species geographic distributions [48] – to develop a geographic distribution model for *R. euryale* and *R. mehelyi*. Maxent is a machine learning method developed to detect habitat suitability of each grid cell as function of the interaction between EGVs and occurrence data [48]. This approach does not require absence data to model, an especially important feature for nocturnal, elusive animals such as bats. To build the models, we used Maxent ver. 3.3.3 k (http://www.cs.princeton.edu/~schapire/maxent), the presence record for *R. euryale* and *R. mehelyi* selected as described above, and the following EGVs: Altitude, Land cover, Mean Diurnal Range, Isothermality, Temperature Seasonality, Temperature Annual Range, Mean Temperature of Wettest Quarter, Mean Temperature of Driest Quarter, Precipitation Seasonality, Precipitation of Wettest Quarter and Precipitation of Coldest Quarter. Further details on EGV are given in Table S1. In the setting panel, we selected the following options: random seed; remove duplicate presence records; write plot data; regularisation multiplier (fixed at 1); 10,000 maximum number of background points; 1000 maximum iterations; and, finally, 20 replicate effects with cross-validate replicated run type. For the latter procedure, 80% of records were randomly extracted for training and 20% for testing the model and the procedure was repeated 20 times. The average final map obtained had a logistic output format with suitability values from 0 (unsuitable habitat) to 1 (suitable habitat). The 10th percentile (the value above which the model classifies correctly 90% of the training locations) was selected as the threshold value for defining the species' presence. This is a conservative value that is commonly used in species distribution modelling studies especially when considering datasets gathered over a long time by different observers and methods of collection. This threshold was used to reclassify our model into binary presence/absence maps [49].

We used Jacknife analysis to estimate the actual contribution that each variable provided to the geographic distribution models. During this process, Maxent generated three models: first, each EGV was excluded in turn and a model created with the remaining variables to check which of the latter was most informative. Second, a model was created using individually each EGV to detect which variable had the most information not featuring in the others. Third, a final model was generated based on all variables. Response curves derived from univariate models

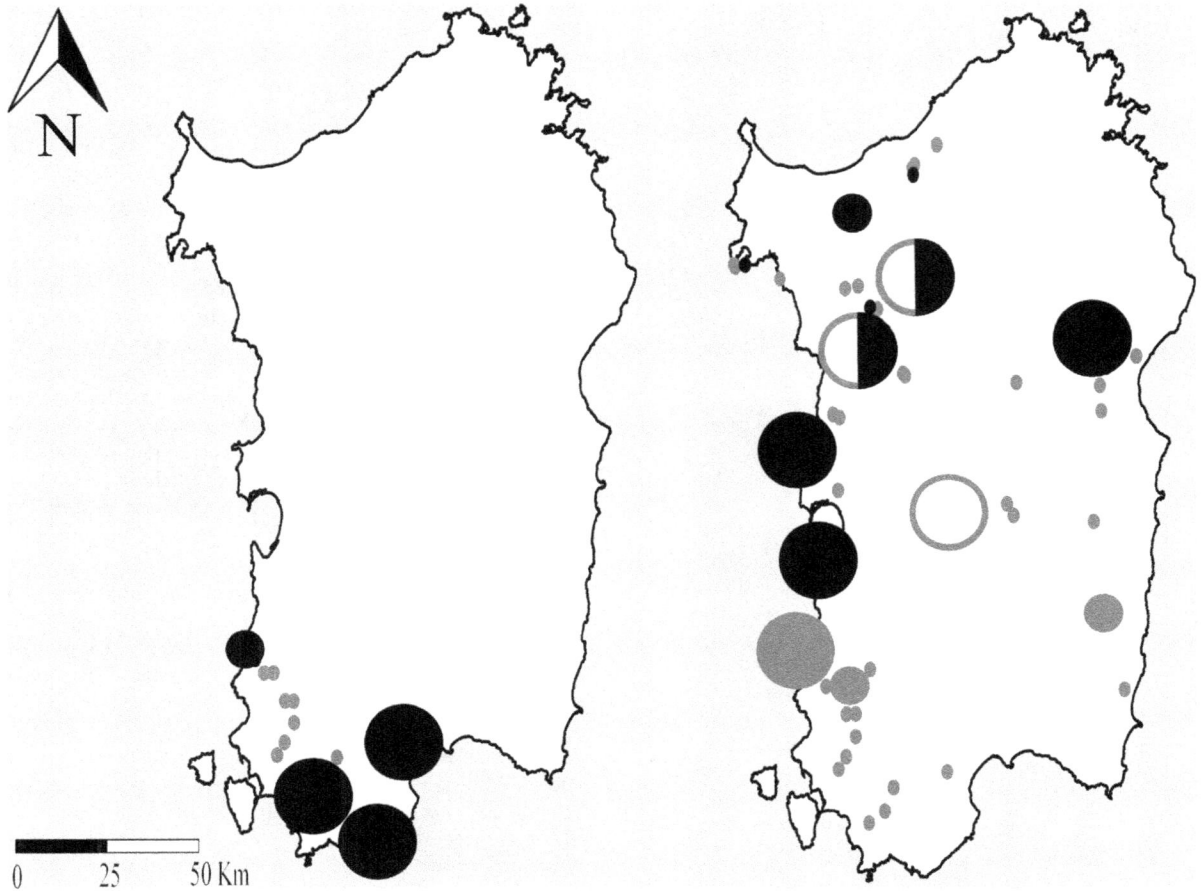

Figure 2. Distribution and colony size on Sardinia for *Rhinolophus euryale* **(left) and** *R. mehelyi* **(right).** Circle sizes are proportional to colony sizes. Black: nursery colonies; white: hibernacula; grey: other day-roost. Mixed-colour (white + black) symbols correspond to sites used by bats year round for both hibernation and reproduction. The publicly available map layer was obtained from www.fao.org/geonetwork/srv/en/main.home and the image prepared with the Quantum Gis 2.2.0 Valmiera open source software.

were plotted to know how each EGV influences the presence probability.

For *R. mehelyi* we generated a distribution model based on Sardinian occurrences only. For *R. euryale*, we generated two models: one calibrated with Sardinian occurrences only and projected to Sardinia (SAR), another based on occurrences from both the Italian peninsula and Sicily (PES) which was also projected to Sardinia. We also generated palaeo-distribution models based on bioclimatic variables only. These were trained with PES and SAR occurrences and projected to the Maghreb in the LGM (23,000–18,000 year BP). The two LGM models were based respectively on the Community Climate System Model, CCSM, and the Model for Interdisciplinary Research on Climate, MIROC [46,50]. Projecting SDMs to regions other than those on which models were calibrated, or to past or future times is a widespread approach to make inferences such as forecasting the spreading of alien organisms, providing palaeo-reconstructions or predicting distributional patterns in future epochs [51–53]. In order to project to a new area models calibrated elsewhere, whether in the current epoch or in the LGM, variables in the projection area must meet a condition of environmental similarity to the environmental data used for training the model. Therefore, we preliminarily ascertained that this condition was verified for both current and past projections, which were thus legitimate, by

inspecting Multivariate Environmental Similarity Surfaces [54] (data not shown in the results for brevity). All digital information had a resolution of 2.5 arc-minutes (4.6 km).

Model validation

We evaluated model performance with different methods: the receiver operated characteristics (ROC), analyzing the area under curve (AUC) [55]; the true skill statistic (TSS) [56]; and the minimum difference between training and test AUC data (AUC_{diff}) [57]. Such statistics were averaged across the 20 replicates run on the 80% (training) vs. 20% (testing) dataset split.

AUC established the discrimination ability of the models and may range from 0 (equalling random distribution) to 1 (perfect prediction). AUC values >0.75 correspond to high discrimination performances [58]. TSS compares the number of correct forecasts, minus those attributable to random guessing, to that of a hypothetical set of perfect forecasts. It considers both omission and commission errors, and success as a result of random guessing, and ranges from − 1 to +1, where +1 indicates perfect agreement and values of zero or less correspond to a performance no better than random [56]. By minimizing the difference between training and test AUC data, in fact we reduce the risk that models are over-parameterized in such a way as to be overly specific to the training data [57].

What Story Does Geographic Separation of Insular Bats Tell? A Case Study on Sardinian Rhinolophids

77

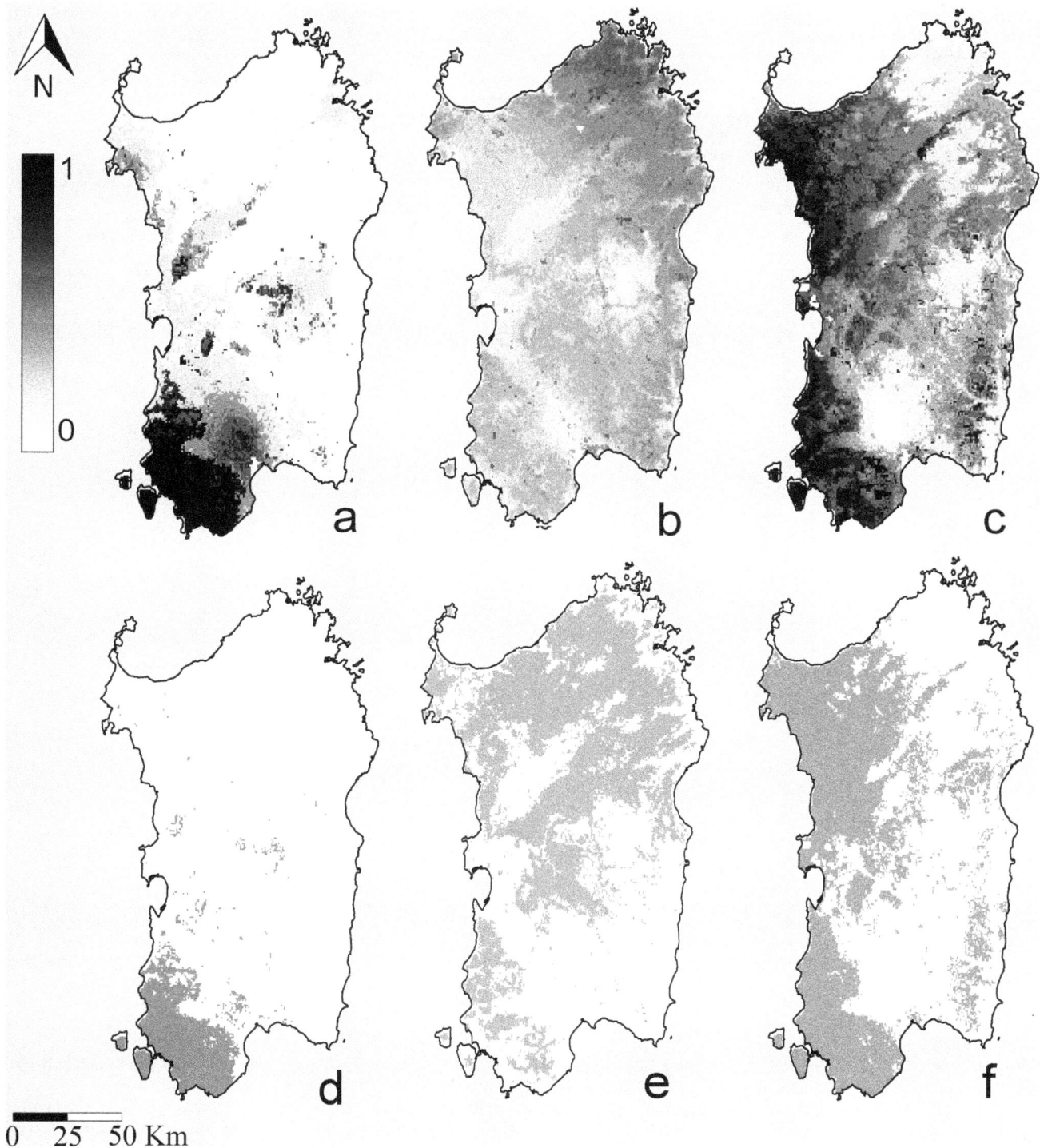

Figure 3. Maxent Species Distribution Models (SDM). a: SDM for *R. euryale* on Sardinia calibrated with Sardinian records only; b: SDM for *Rhinolophus euryale* on Sardinia calibrated with presence records from Italian populations except that of Sardinia and projected to the island; c: SDM for *R. mehelyi* on Sardinia calibrated with Sardinian records only; c: binary map for *R. euryale* on Sardinia calibrated with Sardinian records only; d: binary map for *Rhinolophus euryale* on Sardinia calibrated with presence records from Italian populations except that of Sardinia and projected to the island; e: binary map for *R. mehelyi* on Sardinia calibrated with Sardinian records only The publicly available map layer was obtained from www.fao.org/geonetwork/srv/en/main.home and the image prepared with the Quantum Gis 2.2.0 Valmiera and Maxent open source software packages.

Table 2. Schoener's D and Niche similarity significance levels relative to ordination methods and Species Distribution Models used to carry out niche comparison.

Comparison	Method	Schoener's D	Niche Similarity	
R. mehelyi vs. SAR			SAR → R. mehelyi	R. mehelyi → SAR
	Between group	0.473	0.039+	0.019+
	LDA	0.561	0.019+	0.019+
	Maxent1	0.594	0.019+	0.019+
	Maxent2	0.628	0.019+	0.019+
	MDS	0.205	0.019+	0.950
	PCA environmental	0.215	0.059	0.495
	PCA Occurrence	0.229	0.019+	0.099
	Within environmental	0.215	0.039+	0.455
	Within group	0.127	0.079	0.871
SAR vs. PES			PES → SAR	SAR → PES
	Between group	0.000	2.000	2.000
	LDA	0.000	2.000	2.000
	Maxent1	0.014	2.000	2.000
	Maxent2	0.014	0.792	0.019−
	MDS	0.000	2.000	1.584
	PCA environmental	0.000	2.000	2.000
	PCA Occurrence	0.000	0.673	0.495
	Within environmental	0.124	0.970	0.119
	Within group	0.124	0.891	0.733
R. mehelyi vs. PES			PES → R. mehelyi	R. mehelyi → PES
	Between group	0.409	0.554	0.019+
	LDA	0.060	0.831	0.198
	Maxent1	0.190	0.039+	0.415
	Maxent2	0.215	0.376	0.534
	MDS	0.103	0.594	0.019+
	PCA environmental	0.176	0.099	0.019+
	PCA Occurrence	0.170	0.178	0.019+
	Within environmental	0.216	0.396	0.019+
	Within group	0.243	0.039+	0.019+

LDA = Linear Discriminant Analysis; Maxent 1 and 2 = Maximum Entropy Algorithm analysis made using in turn one of the two Maxent outputs generated by either population (or species) as the comparison background against which the output of the remainder was contrasted; MDS = Multidimensional scaling; PCA environmental = Principal Component Analysis calibrated on the whole environmental space including the presence records where the species occur; PCA occurrence = Principal Component Analysis calibrated with EGV values associated with the occurrences of the species; Within environmental = Within-group calibrated on the whole environmental space. + = similarity; − = dissimilarity.

Niche analysis

We performed niche overlap analyses using the analytical framework proposed by [59] and recently adopted in different studies [60,61]. The procedure follows three steps: data pre-processing, calculation of the niche overlap measure and testing niche similarity. Further details are given in file S1.

To quantify niche overlap, we used the following ordination (for details see [59]) and SDMs methods: Principal Component Analysis calibrated with EGV values associated with the occurrences of the species (PCA-occ); Principal Component Analysis calibrated on the whole environmental space including the presence records where the species occur (PCA-env); Between-group and Within-group analyses (BETWEEN-occ and WITHIN-occ); Within-group calibrated on the whole environmental space (WITHIN-env); Linear Discriminant Analysis (LDA); Multidimensional scaling (MDS); and Maximum Entropy algorithm

(MAXENT). For the application of the latter, two tests were carried out (named Maxent 1 and 2) corresponding to the analysis made using in turn one of the two Maxent outputs generated by either population (or species) as the comparison background against which the output of the remainder was contrasted.

Results

Niche differences between R. mehelyi, SAR and PES

Maxent models showed high levels of predictive performance as can be seen from AUC, TSS and AUC_{diff} values (Table 1).

Distributional data showed that the two species are sympatric only in the southern portion of the island, to which R. euryale is confined (Figure 2). This distribution matches the prediction made by Maxent model for SAR (Figure 3). Besides, large colonies of R. mehelyi are found in most of the island (where only this bat, but not

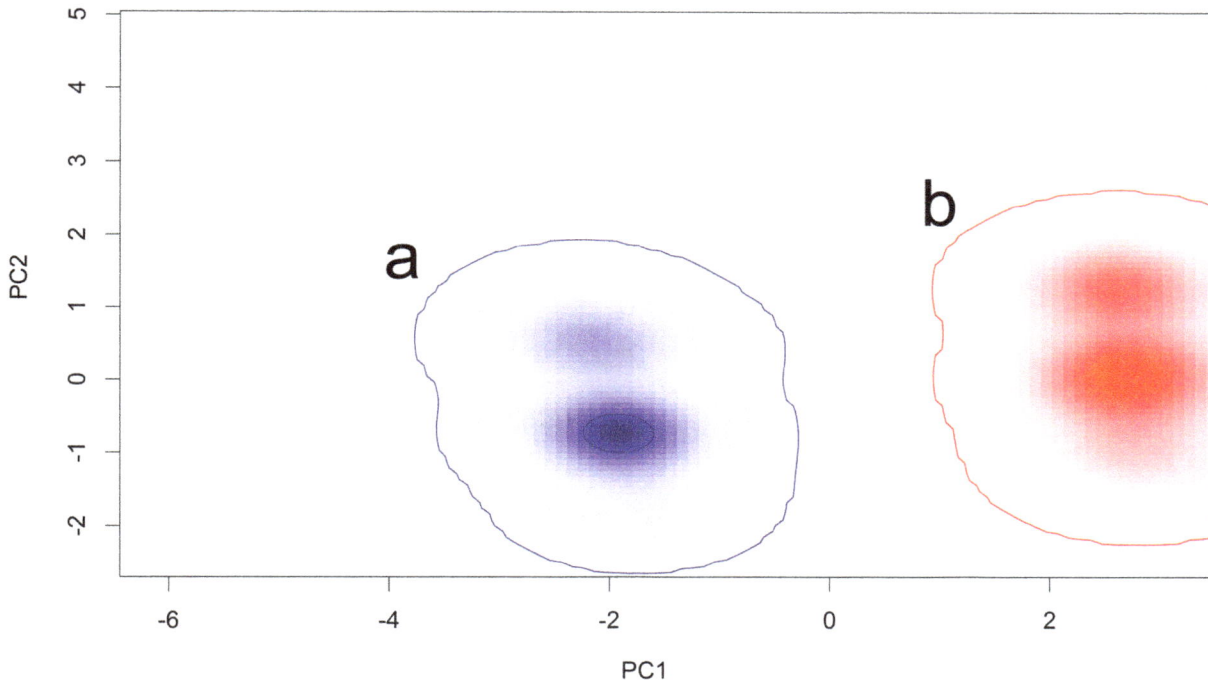

Figure 4. Graphical representation of the environmental niches for *Rhinolophus euryale*. a: Sardinian population; b: other Italian populations. In the example, niche were generated with Principal Component Analysis calibrated on the whole environmental space including the presence records where the species occur.

R. euryale occurs) except in the restricted area where *R. euryale* is present: there *R. mehelyi* only occurs with small numbers (Figure 2).

Maximum entropy models trained with PES presence data and projected to Sardinia show that according to PES ecological requirements, *R. euryale* would occupy a larger area and, compared to SAR, have a reduced probability of presence in the southern portion of the island (Figure 3). This distributional pattern would determine larger areas of sympatry with *R. mehelyi*, as also shown by Maxent's prediction for this species (Figure 3) hence increasing the likelihood of competition. From binary maps it can be derived that the predicted range overlap between *R. mehelyi* and PES is ca. 60%, while the former overlaps with SAR only by ca. 20% (Figure 3). The competition scenario is also supported by the fact that six out of nine niche analysis methods showed a significant similarity of *R. mehelyi* with PES (Table 2). Only two methods supported the similarity of PES with *R. mehelyi* (Table 2).

Niche comparison between SAR and PES showed no significant similarity (Table 2; Figure 4). The climatic variables that were most important to explain the potential distribution of SAR and PES were different. SAR is mainly localized in areas characterized by high isothermality values, mean temperature of wettest quarter of ca. 10–11°C, low standard deviation values of temperature seasonality and mean diurnal range between 8.5°C–10.5°C. SAR is also more likely to occur in areas of bare ground and mixed-leaved woodland at lower altitude. Such characteristics are found in SW Sardinia where SAR occurs. Suitability for PES decreases with increasing temperature seasonality. In the areas where the species' likelihood of occurrence is high (central and northern Sardinia), the mean diurnal temperature range is ca. 8°C and % precipitation seasonality is low.

Niche comparison carried out for *R. mehelyi* vs. SAR showed a substantial similarity, supported respectively by 7 (*R. mehelyi* vs. SAR) and 4 (SAR vs. *R. mehelyi*) methods in either direction (Table 2). As found for PES, *R. mehelyi* probability of occurrence on Sardinia decreased for increasing values of temperature seasonality (Figure S1) and was also associated to wooded habitats.

Overall, the analysis supports the existence of niche divergence between SAR and PES and shows that this results in a smaller overlap between the ranges of *R. mehelyi* and *R. euryale* on the island.

Niche difference as a legacy of biogeographic origin?

Palaeo-distribution models too showed excellent levels of predictive performance as can be seen from AUC, TSS and AUC_{diff} values (Table 1). Our reconstruction showed that both in the present time and under the LGM scenario SAR would largely occur in northern Africa whereas PES would be practically absent, in agreement with a Maghrebian origin of SAR (Figure 5).

Discussion

Niche differences between R. mehelyi, SAR and PES

We showed that *R. euryale* on Sardinia is confined to a small southern portion of the island where it occurs in sympatry with *R. mehelyi* although in that area the latter is far less numerous than in the north, where *R. euryale* is absent. The niches of PES and *R. mehelyi* are similar but PES and SAR niches are not. If SAR had shown a niche identical to that of PES, the geographical distribution of *R. euryale* and *R. mehelyi* on the island would be largely sympatric, potentially leading to stronger interspecific competition. We conclude that niche differences between *R. mehelyi* and SAR minimize sympatry and thus potential competition. This result is consistent with the hypothesis that Sardinian

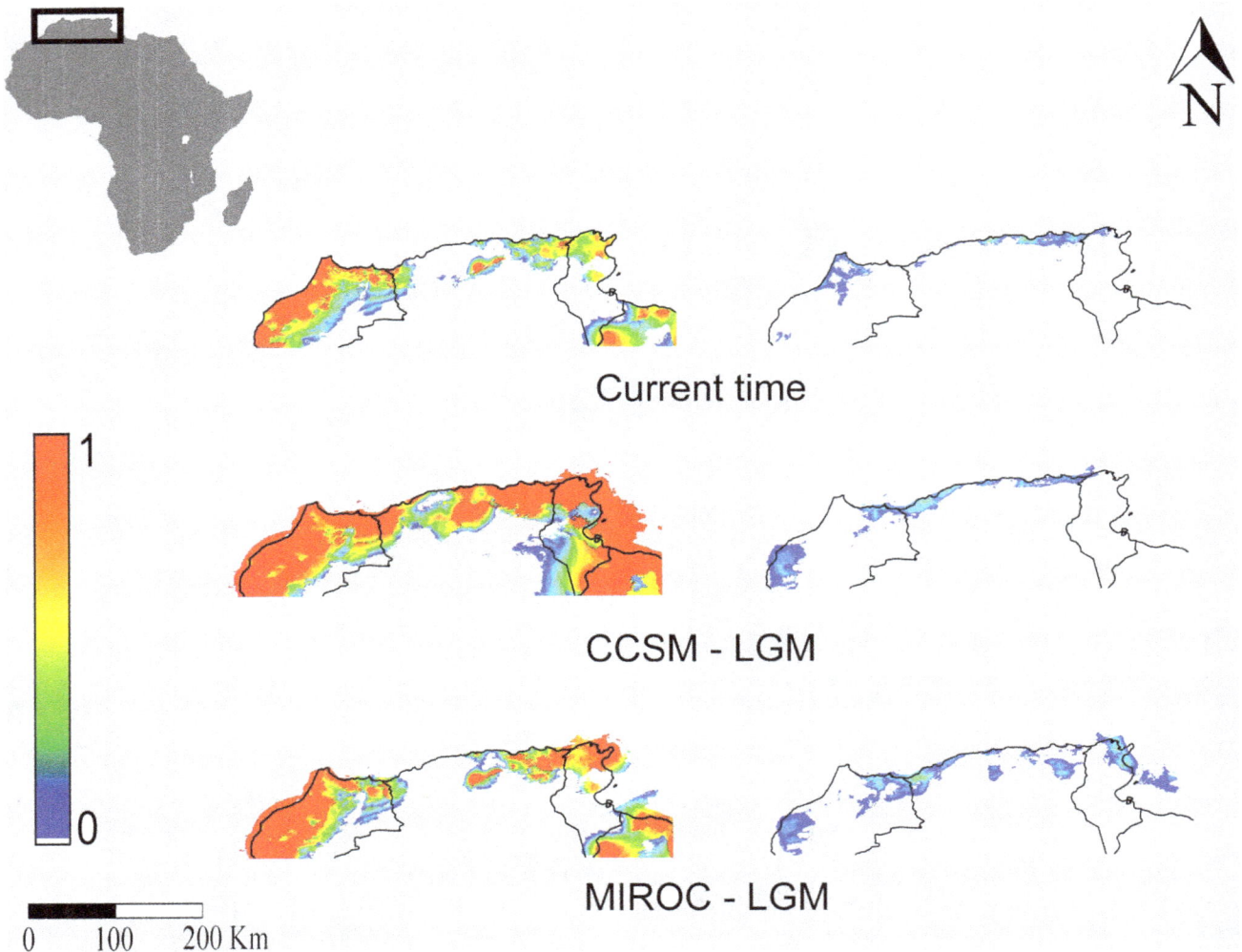

Figure 5. Maxent SDMs for *Rhinolophus euryale* **calibrated respectively on presence records of bats from Sardinia (left) and from the remaining Italian areas (right) and projected to northern Africa.** LGM = Last Glacial Maximum. CCSM = Community Climate System Model; MIROC = Model for Interdisciplinary Research on Climate. The publicly available map layer was obtained from www.fao.org/geonetwork/srv/en/main. home and the image prepared with the Quantum Gis 2.2.0 Valmiera and Maxent open source software packages.

R. euryale experienced a niche displacement process to mitigate competition with the numerically dominant *R. mehelyi*. Based on our models, the probabilities of occurrence of both *R. mehelyi* and SAR in the north-east sector of the island are small, whereas PES shows higher values. Why *R. euryale* has not occupied that region where competition would be low (at least based on PES ecological characteristics) appears less clear and perhaps explained by the biogeographic origin of Sardinian *R. euryale* (discussed below). Besides, based on the niche analysis results, the competition hypothesis would not be fully supported since despite the separation of SAR and PES niches, the former still partly overlaps with that of *R. mehelyi*. This may be explained by the fact that species distribution models were built from occurrence records but did not take colony size into account. Survey data instead showed that *R. mehelyi* occurs with large colonies in the areas of allopatry with SAR, but where the two species are sympatric, it is only present with smaller numbers. In other words, modelling based on presence records probably overestimated niche similarity by disregarding local population size: the difference between the

two niches of Sardinian *R. mehelyi* and *R. euryale* may thus be even larger than that estimated here.

Whatever the reason for its peculiarity, SAR niche must have allowed *R. euryale* to establish a viable population in an area where *R. mehelyi* appears to perform less well and thus be less competitive, as can be inferred from the smaller colony sizes of the latter in the southern area of sympatry. Noticeably, the two species are known to share roosting sites in their Mediterranean regions of sympatry [38], including Sardinia (this study). Bats often form mixed-species groups when they have common thermal preferences, and interspecific associations may result in mutual thermoregulatory (= energetic) benefits [62]. Based on such considerations, we rule out that the species we considered compete for roosting sites.

Food is much more likely to trigger competition. The diet of both species is well known, and is mostly made of moths both in sympatry and allopatry; the amounts of other prey only show small interspecific differences [35,63,64] In most areas of sympatry but not on Sardinia, echolocation call frequencies of *R. mehelyi* and *R. euryale* largely overlap each other [35,38,65] leading to the detection of similar prey [35,38]. Although Sardinian *R. euryale* have lower frequencies than local *R. mehelyi* [38], the difference is too small to account for niche partitioning. Since echolocation calls also convey individual information among conspecifics [66,67] this difference is best explained as a way to maintain separate communication bandwidth in the area of sympatry [38].

Foraging habitat use in these rhinolophids shows no interspecific differences in allopatric populations [35] but does differ in sympatry, where *R. mehelyi* performs better in less structurally complex habitats than in more closed vegetation [35,36] due to its lower flight manoeuvrability and agility [35,36,68]. Over the millennia, hundreds of human generations have shaped Sardinian landscapes and microclimates through deforestation, stock breeding and fires [69] so that much of the land is covered with Mediterranean scrubland and open forest, where *R. mehelyi* is probably more competitive than *R. euryale*. Reduced habitat heterogeneity such as that found on Sardinia as well as the limitedness of food, typical of insular systems [70] can be important factors increasing competition between the two rhinolophids.

Niche difference as a legacy of biogeographic origin?

In principle, our findings are in agreement with a niche displacement process: in fact, one possible scenario is that Sardinia was colonized by *R. euryale* from mainland Europe and that the newly established population shifted its ecological niche to counter competition pressures from heterospecific bats (*R. mehelyi*). However, questions arise on the geographical source of Sardinian *R. euryale* and its ecological consequences as an alternative explanation for SAR's niche distinctness. *R. euryale* might have colonized Sardinia from the Maghreb and the peculiar ecological niche of Sardinian bats could thus be a legacy of the African source population rather than the outcome of a niche displacement process. This niche may have been subjected to stabilizing selection [44] and conserved as it must have performed well to allow co-existence with insular *R. mehelyi*. By projecting respectively SAR and PES niches to the Maghreb both in the current time and in the LGM we found a striking difference in the probability of occurrence, much higher for SAR's projection. This result is in agreement with a possible SAR's Maghrebian origin.

Stretches of sea have been found to represent barriers to the movement of bats [71,72] although their permeability differs across species. The capacity of different species to overcome such barriers is not related to wing morphology and flight performances [73]. Colonization events would seem as difficult from mainland Italy as they would be from northern Africa. Sardinia lies ca. 200 km off the coasts of both regions, so both routes appear equally likely to explain the origin of Sardinian *R. euryale*. Colonization of Sardinia by bats was only possible across the sea since the end of the Messinian Event, ending 5.33 million years

ago [43,74]. Thus, if SAR had an African origin, its establishment would either date back to the Messinian Event or, if more recent, must have implied crossing the sea. The latter option is possible: there is evidence that after the Messinian Event Sardinia was subject to repeated bat colonization waves at different times (including recent ones) from Europe and northern Africa, as for the Maghrebian bat *Myotis punicus* [42], long-eared bats [41] and pipistrelles [43]. Glacial episodes that repeatedly occurred in the Pleistocene lowered sea levels and led to the emersion of land bridges [45], interrupting the isolation of Sardinia from the mainland during early and mid-Pleistocene, and favoured island colonization most probably via a stepping stone geographic system. This would be in agreement with the wide northern African distribution we obtained for *R. euryale* during the LGM by projecting SAR's niche to that region. Caution is needed when considering our LGM models for the Sahara area, where some overpredictions occurred. These were most likely due to the lack of solid information on the region's climate at that age [75] inevitably affecting the reliability of climatic variables.

Although these findings do not prove the biogeographical origin of SAR, we hope they will stimulate molecular studies investigating the phylogeography of *R. euryale* in the Mediterranean Basin for a final answer on the identity of SAR's geological source and a full reconstruction of this population's history.

Supporting Information

Figure S1 EGV response curves of Maxent SDMs for selected ecogeographic variables. a: Temperature seasonality for *R. mehelyi*; b: Temperature seasonality for PES; c: Isothermality for SAR.

Table S1 List of ecogeographical variables used for this study, their type and measurement unit.

File S1 Description of Niche Analysis.

Acknowledgments

We thank Ivy Di Salvo and Mara Calvini for providing information on rhinolophid occurrence respectively in Sicily and Liguria. We are grateful to two anonymous reviewers for their valuable comments on a first version of this manuscript.

Author Contributions

Conceived and designed the experiments: DR. Performed the experiments: LB MDF HR DR. Analyzed the data: LB MDF HR DR. Contributed reagents/materials/analysis tools: DR MM LC PA PPDP AM DS CS. Contributed to the writing of the manuscript: DR LB MDF HR.

References

1. Cox CB, Moore PD (2010) Biogeography: An Ecological and Evolutionary Approach, 8th edn. New York: John Wiley & Sons. 440 p.
2. Brown W, Wilson E (1956) Character displacement. Systematic Zoology 5: 49–64.
3. Hutchinson GE (1959) Homage to Santa Rosalia or why are there so many kinds of animals. The American Naturalist 93: 145–159.
4. Hardin G (1960) The competitive exclusion principle. Science 131: 1292–1297.
5. Kronfeld-Schor N, Dayan T (1999) The dietary basis for temporal partitioning: food habits of coexisting *Acomys* species. Oecologia 121: 123–8.
6. Albrecht M, Gotelli NJ (2001) Spatial and temporal niche partitioning in grassland ants. Oecologia 126: 134–141.
7. Navarro J, Votier SC, Aguzzi J, Chiesa JJ, Forero MG, et al. (2013) Ecological segregation in space, time and trophic niche of sympatric planktivorous petrels. PloS ONE 8: e62897.
8. Spencer LM (1995) Morphological correlates of dietary resource partitioning in the African Bovidae. Journal of Mammalogy 76: 448–471.
9. Castillo-Rivera M, Kobelkowsky A, Zamayoa V (1996) Food resource partitioning and trophic morphology of *Brevoortia gunteri* and *B. patronus*. Journal of Fish Biology 49: 1102–1111.
10. Albertson RC (2008) Morphological divergence predicts habitat partitioning in a Lake Malawi cichlid species complex. Copeia 2008: 689–698.
11. Grant P (1972) Convergent and divergent character displacement. Biological Journal of the Linnean Society of London 4: 39–68.
12. Goldberg E, Lande R (2006) Ecological and reproductive character displacement of an environmental gradient. Evolution 60: 1344–1357.
13. Dayan T, Simberloff D (2005) Ecological and community-wide character displacement: the next generation. Ecology Letters 8: 875–894.

14. Peers MJL, Thornton DH, Murray DL (2013) Evidence for large-scale effects of competition: niche displacement in Canada lynx and bobcat. Proceedings of the Royal Society B 280: 20132495.

15. Lomolino MV (2005) Body size evolution in insular vertebrates: generality of the island rule. Journal of Biogeography 32: 1683–1699.

16. Wiens JJ, Ackerly DD, Allen AP, Anacker BL, Buckley LB, et al. (2010) Niche conservatism as an emerging principle in ecology and conservation biology. Ecology Letters 13: 1310–1324.

17. Vamosi SM, Heard SB, Vamosi JC, Webb CO (2009) Emerging patterns in the comparative analysis of phylogenetic community structure. Molecular Ecology 18: 572–592.

18. Stadelmann B, Herrera LG, Arroyo-Cabrales J, Flores-Martínez JJ, May BP, et al. (2004) Molecular systematics of the fishing bat Myotis (Pizonyx) vivesi. Journal of Mammalogy 85: 133–139.

19. Stadelmann B, Lin LK, Kunz TH, Ruedi M (2007) Molecular phylogeny of New World Myotis (Chiroptera, Vespertilionidae) inferred from mitochondrial and nuclear DNA genes. Molecular Phylogenetics and Evolution 43: 32–48.

20. Jones G, Holderied MW (2007) Bat echolocation calls: adaptation and convergent evolution. Proceedings of the Royal Society of London B 274: 905–912.

21. Arlettaz R. (1999) Habitat selection as a major resource partitioning mechanism between the two sympatric sibling bat species Myotis myotis and Myotis blythii. Journal of Animal Ecology 68: 460–471.

22. Kingston T, Jones G, Zubaid A, Kunz TH (2000) Resource partitioning in rhinolophoid bats revisited. Oecologia 124: 332–342.

23. Siemers BM, Schnitzler HU (2004) Echolocation signals reflect niche differentiation in five sympatric congeneric bat species. Nature 429: 657–661.

24. Siemers BM, Swift SM (2006) Differences in sensory ecology contribute to resource partitioning in the bats Myotis bechsteinii and Myotis nattereri (Chiroptera: Vespertilionidae). Behavioral Ecology and Sociobiology 59: 373–380.

25. Mitchell-Jones AJ, Amori G, Bogdanowicz W, Krytufek B, Reijnder PJH, et al. (1999) The Atlas of European Mammals. T. and A.D. Poyser, London. 250 p.

26. Guillén A, Francis CM, Ricklefs RE (2003) Phylogeny and biogeography of the horseshoe bats. In: Csorba G, Ujhelyi P, Shropshire TN, editors. Horseshoe Bats of the World. pp. 7–24.

27. Zhou ZM, Guillent-Servent A, Lim BK, Eger JL, Wang YX, et al. (2009) A new species from southwestern China in the Afro-Paleartic lineage of the horseshoe bats (Rhinolophus). Journal of Mammalogy 90: 57–73.

28. Hutson AM, Spitzenberger F, Juste J, Aulagnier S, Alcaldé JT, et al. (2008a) Rhinolophus euryale. In: IUCN 2013. IUCN Red List of Threatened Species. Version 2013.1. Available: www.iucnredlist.org. Accessed 2013 November 14.

29. Hutson AM, Spitzenberger F, Juste J, Aulagnier S, Alcaldé JT, et al. (2008b) Rhinolophus mehelyi. In: IUCN 2013. IUCN Red List of Threatened Species. Version 2013.1. Available: www.iucnredlist.org. Accessed 2013 November 14.

30. Flaquer C, Puig X, Fàbregas E, Guixé D, Torre I, et al. (2010) Revisión y aportación de datos sobre quirópteros de Catalunya: Propuesta de Lista Roja. Galemys 22: 29–61.

31. Puechmaille SJ, Teeling EC (2014) Non-invasive genetics can help find rare species: a case study with Rhinolophus mehelyi and R. euryale (Rhinolophidae: Chiroptera) in Western Europe. Mammalia. Mammalia 78: 251–255.

32. Rombaut D, Haquart A (2002) Les Chiroptères de la Directive Habitats: le Rhinolophe de Mehely Rhinolophus mehelyi Matschie, 1901. Arvicola 14: 18–20.

33. Dragu A, Borissov I (2011) Low genetic variability of Rhinolophus mehelyi (Mehely's horseshoe bat) in Romania. Acta Theriologica 56: 383–387.

34. Russo D, Jones G, Migliozzi A (2002) Habitat selection by the Mediterranean horseshoe bat, Rhinolophus euryale (Chiroptera: Rhinolophidae) in a rural area of southern Italy and implications for conservation. Biological Conservation 107: 71–81.

35. Salsamendi E, Garin I, Arostegui I, Goiti U, Aihartza J (2012) What mechanism of niche segregation allows the coexistence of sympatric sibling rhinolophid bats? Frontiers in Zoology 9: 30.

36. Russo D, Almenar D, Aihartza J, Goiti U, Salsamendi E, et al. (2005) Habitat selection in sympatric Rhinolophus mehelyi and R. euryale (Chiroptera: Rhinolophidae). Journal of Zoology 266: 327–332.

37. Dondini G, Tomassini A, Inguscio S, Rossi E (2014) Rediscovery of Mehely's horseshoe bat (Rhinolophus mehelyi) in peninsular Italy. Hystrix 25: 59–60.

38. Russo D, Mucedda M, Bello M, Biscardi S, Pidinchedda E, et al. (2007) Divergent echolocation call frequencies in insular rhinolophids (Chiroptera): a case of character displacement? Journal of Biogeography 34: 2129–2138.

39. Krzanowski A (1967) The magnitude of islands and the size of bats (Chiroptera). Acta Zoologica Cracoviensia 12: 281–346.

40. McNab BK (2009) Physiological adaptation of bats and birds to island life. In: Fleming TH, Racey PA, editors. Island Bats. Evolution, Ecology and Conservation. pp. 153–175.

41. Kiefer A (2007) Phylogeny of Western-Palaearctic long-eared bats (Mammalia, Chiroptera, Plecotus): a molecular perspective. Ph.D. Thesis, Gutenberg Univeristy, Mainz.

42. Biollaz F, Bruyndonckx N, Beuneux G, Mucedda M, Goudet J, et al. (2010) Genetic isolation of insular populations of the Maghrebian bat, Myotis punicus, in the Mediterranean Basin. Journal of Biogeography 37: 1557–1569.

43. Veith M, Mucedda M, Kiefer A, Pidinchedda E (2011) On the presence of pipistrelle bats (Pipistrellus and Hypsugo; Chiroptera: Vespertilionidae) in Sardinia. Acta Chiropterologica 13: 89–99.

44. Russo D, Teixeira S, Cistrone L, Jesus J, Teixeira D, et al. (2009) Social calls are subject to stabilizing selection in insular bats. Journal of Biogeography 36: 2212–2221.

45. Rohling EJ, Fenton M, Jorissen FJ, Bertrand P, Ganssen G, et al. (1998) Magnitudes of sea-level low stands of the past 500,000 years. Nature 394: 162–165.

46. Hijmans RJ, Cameron SE, Parra JL, Jones PG, Jarvis A (2005) Very high resolution interpolated climate surfaces for global land areas. International Journal of Climatology 25: 1965–1978.

47. Busby JR (1991) BIOCLIM: A bioclimatic analysis and predictive system. In: Margules CR, Austin MP, editors. Nature Conservation: Cost Effective Biological Surveys and Data Analysis. pp. 64–68.

48. Phillips SJ, Anderson RP, Schapire R (2006) Maximum entropy modeling of species geographic distributions. Ecological Modelling 190: 231–259.

49. Bosso L, Rebelo H, Garonna AP, Russo D (2013) Modelling geographic distribution and detecting conservation gaps in Italy for the threatened beetle Rosalia alpina. Journal for Nature Conservation 21: 72–80.

50. Waltari E, Hijmans RJ, Peterson AT, Nyari AS, Perkins SL, et al. (2007) Locating Pleistocene refugia: comparing phylogeographic and ecological niche model predictions. PLoS ONE 2: e563.

51. Medley KA (2010) Niche shifts during the global invasion of the Asian tiger mosquito, Aedes albopictus Skuse (Culicidae), revealed by reciprocal distribution models. Global Ecology and Biogeography 19: 122–133.

52. Rödder D, Lawing AM, Flecks M, Ahmadzadeh F, Dambach J, et al. (2013) Evaluating the significance of paleophylogeographic species distribution models in reconstructing Quaternary range-shifts of Nearctic chelonians. PLoS ONE 8: e72855.

53. Keith DA, Elith J, Simpson CC (2014) Predicting distribution changes of a mire ecosystem under future climates. Diversity and Distributions 20: 440–454.

54. Elith J, Kearney M, Phillips SJ (2010) The art of modeling range-shifting species. Methods in Ecology and Evolution 1: 330–342.

55. Fielding AH, Bell JF (1997) A review of methods for the assessment of prediction errors in conservation presence/absence models. Environmental Conservation 24: 38–49.

56. Allouche O, Tsoar A, Kadmon R (2006) Assessing the accuracy of species distribution models: prevalence, kappa and the true skill statistic (TSS). Journal of Applied Ecology 43: 1223–1232.

57. Warren DL, Seifert SN (2011) Environmental niche modeling in Maxent: the importance of model complexity and the performance of model selection criteria. Ecological Application 21: 335–342.

58. Elith J, Graham CH, Anderson RP, Dudík M, Ferrier S, et al. (2006) Novel methods improve prediction of species' distributions from occurrence data. Ecography 29: 129–151.

59. Broennimann O, Fitzpatrick MC, Pearman PB, Petitpierre B, Pellissier L, et al. (2012) Measuring ecological niche overlap from occurrence and spatial environmental data. Global Ecology and Biogeography 21: 481–497.

60. Theodoridis S, Randin C, Broennimann O, Patsiou T, Conti E (2013) Divergent and narrower climatic niches characterize polyploid species of European primroses in Primula sect. Aleuritia. Journal of Biogeography 40: 1278–1289.

61. Di Febbraro M, Lurz PWW, Maiorano L, Girardello M, Bertolino S (2013) The use of climatic niches in screening procedures for introduced species to evaluate risk of spread: the case of the American Eastern grey squirrel. PLoS ONE 8: 1–10.

62. Bogdanowicz W (1983) Community structure and interspecific interactions in bats hibernating in Poznan. Acta Theriologica 28: 357–370.

63. Goiti U, Aihartza JR, Garin I (2004) Diet and prey selection in the Mediterranean horseshoe bat Rhinolophus euryale (Chiroptera, Rhinolophidae) during the pre-breeding season. Mammalia 68: 397–402.

64. Salsamendi E, Garin I, Almenar D, Goiti U, Napal M, et al. (2008) Diet and prey selection in Mehely's horseshoe bat Rhinolophus mehelyi (Chiroptera, Rhinolophidae) in the south-western Iberian Peninsula. Acta Chiropterologica 10: 279–286.

65. Russo D, Jones G, Mucedda M (2001) Influence of age, sex and body size on echolocation calls of Mediterranean (Rhinolophus euryale) and Mehely's (Rhinolophus mehelyi) horseshoe bats (Chiroptera: Rhinolophidae). Mammalia 65: 429–436.

66. Jones G, Siemers BM (2011) The communicative potential of bat echolocation pulses. Journal of Comparative Physiology A 197: 447–457.

67. Schuchmann M, Siemers BM (2010) Variability in echolocation call intensity in a community of horseshoe bats: a role for resource partitioning or communication? PLoS ONE 5: e12842.

68. Dietz C, Dietz I, Siemers BM (2006) Wing measurement variations in the five European horseshoe bat species (Chiroptera: Rhinolophidae). Journal of Mammalogy 87: 1241–1251.

69. Weiss S, Ferrand N (2007) Phylogeography of Southern European Refugia. Springer, Dordrecht. 377 p.

70. Krzanowski A (1967) The magnitude of islands and the size of bats (Chiroptera). Acta Zoologica Cracoviensia 12: 281–346.

71. Castella V, Ruedi M, Excoffier L, Ibanez C, Arlettaz R, et al. (2000) Is the Gibraltar Strait a barrier to gene flow for the bat Myotis myotis (Chiroptera: Vespertilionidae)? Molecular Ecology 9: 1761–1772.

72. Ruedi M, Walter S, Fischer MC, Scaravelli D, Excoffier L, et al. (2008) Italy as a major ice age refuge area for the bat *Myotis myotis* (Chiroptera: Vespertilionidae) in Europe. Molecular Ecology 17: 1801–1814.

73. García-Mudarra JL, Ibanez C, Juste J (2009) The Straits of Gibraltar: barrier or bridge to Ibero-Moroccan bat diversity? Biological Journal of the Linnean Society 96: 434–450.

74. Krijgsman W, Hilgen FJ, Raffi I, Sierro FJ, Wilson DS (1999) Chronology, causes and progression of the Messinian salinity crisis. Nature 400: 652–655.

75. Brito JC, Godinho R, Martínez-Freirería F, Pleguezuelos JM, Rebelo H, et al. (2014) Unravelling biodiversity, evolution and threats to conservation in the Sahara-Sahel. Biological Reviews 89: 215–231.

9

Amphibian Beta Diversity in the Brazilian Atlantic Forest: Contrasting the Roles of Historical Events and Contemporary Conditions at Different Spatial Scales

Fernando Rodrigues da Silva[1]*, **Mário Almeida-Neto**[2], **Mariana Victorino Nicolosi Arena**[1]

1 Departamento de Ciências Ambientais, Universidade Federal de São Carlos, Sorocaba, São Paulo, Brazil, **2** Departamento de Ecologia, Universidade Federal de Goiás, Goiânia, Goiás, Brazil

Abstract

Current patterns of biodiversity distribution result from a combination of historical and contemporary processes. Here, we compiled checklists of amphibian species to assess the roles of long-term climate stability (Quaternary oscillations), contemporary environmental gradients and geographical distance as determinants of change in amphibian taxonomic and phylogenetic composition in the Brazilian Atlantic Forest. We calculated beta diversity as both variation in species composition (CBD) and phylogenetic differentiation (PBD) among the assemblages. In both cases, overall beta diversity was partitioned into two basic components: species replacement and difference in species richness. Our results suggest that the CBD and PBD of amphibians are determined by spatial turnover. Geographical distance, current environmental gradients and long-term climatic conditions were complementary predictors of the variation in CBD and PBD of amphibian species. Furthermore, the turnover components between sites from different regions and between sites within the stable region were greater than between sites within the unstable region. On the other hand, the proportion of beta-diversity due to species richness difference for both CBD and PBD was higher between sites in the unstable region than between sites in the stable region. The high turnover components from CBD and PBD between sites in unstable *vs* stable regions suggest that these distinct regions have different biogeographic histories. Sites in the stable region shared distinct clades that might have led to greater diversity, whereas sites in the unstable region shared close relatives. Taken together, these results indicate that speciation, environmental filtering and limited dispersal are complementary drivers of beta-diversity of amphibian assemblages in the Brazilian Atlantic Forest.

Editor: Ricardo Bomfim Machado, University of Brasilia, Brazil

Funding: FRdS and MVNA were supported by Fundação de Amparo a Pesquisa no Estado de São Paulo ((http://www.fapesp.br/), Proc. 2013/50714-0 and Proc. 2011/22618-0, respectively). FRdS and MAN were supported by the Conselho Nacional de Desenvolvimento Científico e Tecnológico ((http://www.cnpq.br/), Proc. 483486/2011-6 and Proc. 306870/2012-6, respectively). The funders had no role in study design, data collection and analysis, decision to publish, or preparation of the manuscript.

Competing Interests: The authors have declared that no competing interests exist.

* Email: fernandors@ufscar.br

Introduction

Current patterns of biodiversity distribution result from a combination of historical and contemporary processes [1–2]. As a consequence, we could expect that the more current and past environmental conditions differ from each other, the less predictive models based on present-day variables will be. In order to understand the extent to which past conditions have affected current biodiversity distribution, several studies have contrasted the roles of historical processes and current environmental conditions in order to understand patterns of species distribution [3–6]. According to climate-diversity hypotheses, climatic variables that reflect present-day conditions are the key drivers of speciation, extinction and dispersal rates, thus also influencing current patterns of species distribution [7–8]. For example, distinct climatic conditions between habitats or regions might restrict the dispersal of individuals within the distribution limits of ancestral species - niche conservatism [7], or might promote the extinction

of some species due to differences in physiological tolerance - physiological tolerance hypothesis [9]. In contrast, the historical hypothesis postulates that the duration and extent of stable climatic conditions in Earth's history have allowed more opportunity for diversification due to high speciation and/or low extinction rates [4,6]. Furthermore, many studies have demonstrated that dispersal from the center of origin or from past refugia also play a key role in explaining current patterns of species distribution for many phylogenetic lineages [10–11]. According to Araújo *et al.* [6], narrow-ranging species of reptiles and amphibians occur preferentially in areas not covered by ice during the last glacial period in Europe, suggesting that low colonization ability limited the geographical distribution of these species during the interglacial period.

A practical challenge in comparing the effects of historical versus current conditions on the distribution of species diversity is to determine which components of diversity should reflect each process. Studies integrating compositional beta diversity (generally defined as variation in species composition among sites; hereafter CBD) and phylogenetic beta diversity (defined as the amount of shared phylogenetic history between two communities; hereafter PBD) have provided different insights into the ecological and evolutionary mechanisms that structure communities [12–15]. These frameworks allow us to evaluate how ecological processes, such as environmental filtering and dispersal, interact with historical processes (speciation and extinction), influencing extant patterns of distribution within and among different regions. For instance, high CBD together with low PBD indicates a high proportion of small-ranged species, reflecting speciation across regions [16]. In contrast, high CBD together with high PBD indicates low lineage dispersal across regions [16]. However, studies combining CBD and PBD to identify the lineages that are driving the patterns of turnover between regions are still scarce [5,12–15]. In order to test the relative importance of historical events and current environmental gradients, we investigated the spatial distribution of CBD and PBD on amphibians within the Atlantic Forest, a forest biome where humid forests retracted during the cooler, drier period of the Quaternary's climate oscillations without being covered by ice [17].

The Atlantic Forest is a highly threatened global biodiversity hotspot [18], with more than 500 known amphibian species, of which 88% are endemic [19]. The Pleistocene refuge hypothesis [20] suggests that during the cold dry conditions of the Last Glacial Maximum (LGM), approximately 21 000 yr BP, some areas (i.e. Pernambuco refuge, Bahia refuge and Southeastern refuge) in the Atlantic Forest experienced less variability in temperature and precipitation [17,21,22]. These areas not only served as a large climatic refugium for Neotropical species but also promoted local evolutionary differentiation and diversification, allowing for greater species turnover [22–25]. Based on this scenario, recent studies have modeled the spatial range of the Brazilian Atlantic Forest for different climatic scenarios (current and past) to examine whether the regions predicted to have remained stable across climatic fluctuations were consistent with current patterns of species endemism in coastal Brazil [21], as well as genetic diversity for certain populations [23–24]. However, these studies did not examine the current patterns of species diversity encompassing contemporary environmental conditions and different spatial extents among regions and local communities in this biome.

We compiled data on amphibian species composition to develop a thorough understanding of the processes driving the CBD and PBD patterns of amphibian distribution in the southern range of the Brazilian Atlantic Forest. First, we partitioned CBD and PBD into two distinct components: spatial turnover and nestedness. The partition of total beta diversity allowed us to infer the different processes (replacement or gain/loss of species) that structure communities [26–28]. Then, we assessed the relative roles of current environmental gradients, geographic isolation and long-term climatic conditions (Quaternary climatic oscillation) in shaping the present-day patterns of CBD and PBD. Specifically, we expected that the relative proportion of amphibian CBD and PBD that is due to species loss (extinction or limited dispersal) should be higher in areas where the effects of Quaternary climate changes were stronger (unstable regions), whereas the relative proportion of amphibian CBD and PBD that is due to spatial replacement (speciation) should be higher in areas that served as a large climatic refugium for Neotropical species (stable regions). Furthermore, based on physiological constraints and limited dispersal, two key characteristics of amphibians, we expected that variations in precipitation, temperature and/or altitude would be the environmental conditions that constrained the similarities in species composition.

Materials and Methods

Study area and climate data

We used checklists of local amphibian communities from 44 sites along the southern range of the Brazilian Atlantic Forest (see Appendix S1 in File S1 and Data S1). Because we obtained all data from literature surveys, no specific permissions were required. We limited our study to the southern range of the Brazilian Atlantic Forest because checklists of amphibians do not equally cover all the extent of the Atlantic Forest. The southeastern region is relatively over-represented while the northern and the extreme southern regions are under-represented. To reduce possible biases due to methodological differences in sampling procedures, we selected only sites whose checklists met the following criteria: (i) at least two out of four different survey methodologies (audio, active search, casual observations, and pitfall traps); (ii) samplings in all seasons for at least one year (Appendix S1 in File S1). Furthermore, anuran species not identified (e.g., *Leptodactylus* sp.) were excluded from the analysis. Although we recognize the difficulties posed by studies considering data from different inventories, we maintain that the above-cited criteria provide a more reliable checklist than geographic range maps [29].

We used the WorldClim data base [30] at a resolution of 2.5′ and DIVA-GIS 7.5 [31] to obtain the following climatic variables for each forest site: (1) annual mean temperature (ANNT); (2) maximum temperature of the warmest month (MAXT); (3) minimum temperature of the coldest month (MINT); (4) difference between MAXT and MINT (DIFT); (5) annual precipitation (PPT); (6) precipitation seasonality (coefficient of variation across months) (PPTS); (7) precipitation of wettest quarter (PPTW); (8) precipitation of driest quarter (PPTD); and (9) difference between PPTW and PPTD (DIFP). Furthermore, we used Google Earth to obtain the following topographical data: (10) maximum elevation (MAEL); (11) minimum elevation (MIEL); and (12) elevational range (difference between MAEL and MIEL: DIEL). These variables were used because they describe the average trends as well as variation in temperature, precipitation and elevational range which might represent physiological limits for amphibians [32–34].

Data analysis

Compositional Beta Diversity (CBD). We used two additive partitioning frameworks proposed by Baselga [26–27] and Carvalho *et al.* [35] to determine the CBD. Although these approaches are intended to measure species replacement and species richness differences, their methods can lead to radically different conclusions using the same dataset [see 36 for comparisons of these methods]. In short, both Baselga [26–27] and Carvalho *et al.*'s [35] approaches consist of decomposing the pairwise Jaccard dissimilarity index (β_{jac} and β_{cc} respectively) into two additive components. Specifically, Baselga's [27] approach consists of: i) the turnover component (β_{jtu}), which measures the proportion of unique species in two sites pooled together if both sites are equally rich, and ii) the nestedness-resultant component (β_{jne}), which measures how dissimilar the sites are due to a nested pattern. Carvalho *et al.*'s [35] approach consists of: i) species replacement (β_{-3}), which describes a species at one site that is substituted by a species at another site, and ii) richness disparities (β_{rich}), which reflect the absolute difference between the number of species that each site contains, irrespective of nestedness. In this study, Baselga and Carvalho *et al.*'s approaches produced qualitatively similar results. Therefore, we will present only the results from Baselga's approach (see Appendix S2 in File S1 for a discussion about the approaches).

Phylogenetic Beta Diversity (PBD). To assess the phylogenetic similarities among all species in our dataset, we constructed a cladogram based on the time-calibrated tree proposed by Pyron & Wiens [37] (Appendix S3 in File S1), which contains 2,871 species (40% of known extant species) from 432 genera (85% of the 500 currently recognized extant genera). We pruned the time-calibrated tree to include only the anuran species found on 44 sites used in this study. Only seven out of the 50 recorded genera (*Zachaenus*, *Megaelosia*, *Arcomover*, *Crossodactylodes*, *Myersiella*, *Stereocyclops* and *Euparkerella*; Appendix S4 in File S1) were not present on the time-calibrated tree of Pyron & Wiens [37]. Thus, the anuran species whose genera were present in the tree were inserted as within-genus polytomies while the species belonging to the genera not present in the tree were inserted based on phylogenetic relationships from other sources [38–39]. We acknowledge that polytomies under-sample branch length differences among species. However, polytomies are generally more sensitive to loss of resolution basally in the phylogeny and less sensitive to loss of resolution terminally [40], as represented in this study.

We quantified phylogenetic dissimilarities using the UniFrac index [41]. Values of UniFrac range from 0 (indicating that the two communities are composed of similar species) to 1 (indicating that the two communities are composed of distinct species). Like the beta diversity index, the UniFrac index has recently been decomposed into two components representing "true" phylogenetic turnover (UniFrac$_{Turn}$) and phylogenetic diversity gradients (UniFrac$_{PD}$) [28]. According to Leprieur *et al.* [28], UniFrac$_{Turn}$ measures the relative amount of gains and losses of unique lineages between communities that is not attributable to their differences in phylogenetic diversity. In contrast, UniFrac$_{PD}$ measures the amount of phylobetadiversity caused by differences in phylogenetic diversity between phylogenetically nested communities. Following Leprieur *et al.* [28], we tested whether pairs of assemblages were more or less phylogenetically dissimilar than expected by chance, using a null model in which species richness and the CBD between regions were fixed and only the identities of the species in the phylogeny were randomized 999 times. A standardized effect size (SES) was calculated for PBD and its components. SES values greater than 1.96 indicate a higher PBD than expected by CBD, while SES values below −1.96 indicate a lower PBD than expected by CBD [28]. If observed values of PBD do not differ from what would be expected by chance alone, then PBD is unlikely to be the result of historical processes.

Delimitation of stable and unstable regions. The major processes that control the distribution of species diversity may vary depending on the spatial or temporal extent of the study system [42] and the regional species pool [43]. To delimit the spatial extent of stable and unstable areas, we projected the annual precipitation distribution of the current time period onto two paleoclimate scenarios simulating the last glacial maximum period (LGM) 21,000 years ago: CCSM3 (Community Climate System Model, available at: http://www.worldclim.org/past) and MIROC (Model of Interdisciplinary Research on Climate, available at: http://www.worldclim.org/past) at a resolution of 2.5′ [30]. Then we produced a binary map by transforming 10% of grids with the highest values of annual precipitation into one and the remaining grids to zero. For last, we defined the stable climatic areas as intersections on maps for which the highest values of annual precipitation were inferred in all models (Fig. 1). The intersection map shows a coastal area of long-term climatic

stability (Fig. 1). Therefore, the spatially explicit predictions of climatic distribution in the southern range of the Atlantic Forest are consistent with the refugia areas described in other studies [22–24]. These data were used as predictor variables (unstable and stable regions) in the subsequent analyses.

Relative importance of geographical distance, current environmental gradients and long-term climatic conditions in explaining variations in CBD and PBD components. In order to reduce the data dimensionality and multicollinearity, we performed a principal component analysis (PCA) based on a correlation matrix of the data considering temperature variables (ANNT, MAXT, MINT and DIFT), precipitation variables (PPT, PPTS, PPTW, PPTD and DIFP) and elevation-related variables (MAEL, MIEL and DIEL). The first three axes of the principal component explained 85% of the variations in the climatic and topographic data (Appendix S5 in File S1). Therefore, for the subsequent analysis we calculated for each pair of sites the Euclidean distance based on all three PC axes - current environmental gradients [44].

The relative importance of current environmental gradients, geographical distance (Euclidean distance inferring the distance decay of similarity between sites [45]) and long-term climatic conditions in explaining variations in CBD and PBD components was examined using a variance partitioning technique where the total percentage of the variation of ordinary least-squares regressions is partitioned into unique and common contributions of the sets of predictors [46]. The total variation of CBD and PBD was divided into eight fractions: 1) variation explained purely by environmental gradients; 2) variation explained purely by geographical distance; 3) variation explained purely by long-term climatic conditions; 4) variation explained by environmental gradients and geographical distance together; 5) variation explained by environmental gradients and long-term climatic conditions together; 6) variation explained by geographical distance and long-term climatic conditions together; 7) variation explained by environmental gradients, geographical distance and long-term climatic conditions together; and 8) unexplained (residual). For these analyses, current environmental gradients and geographical distance were log-transformed.

All analyses were conducted in R 3.1.0 [47] using betapart [48], vegan [49] and ape [50] packages available at http://www.r-project.org/.

Results

Partitioning of CBD and PBD

A total of 238 amphibian species (Appendix S4 in File S1) were recorded in the 44 sites. The stable (14 sites) and unstable (30 sites) regions harbored 160 and 150 amphibian species, respectively, and only 71 amphibian species (29.8%) occurred in both regions. The partitioning of CBD and PBD revealed that the turnover component was the major reason for amphibian dissimilarity among sites (Fig. 2). Current environmental gradients and geographical distance together and these two variables with long-term climatic conditions explained on average about 25% and 21%, respectively, of the variation in turnover components for both CBD and PBD (Fig. 3). In fact, turnover components of CBD and PBD were highly correlated (Fig. 4), indicating that variation in PBD is explained by the turnover of CBD, or vice-versa. On the other hand, current environmental gradients, geographical distance and long-term climatic conditions explained a small proportion of the total variation of nestedness (β_{jne}–20%) and phylogenetic diversity (UniFrac$_{PD}$–2%) (Fig. 3).

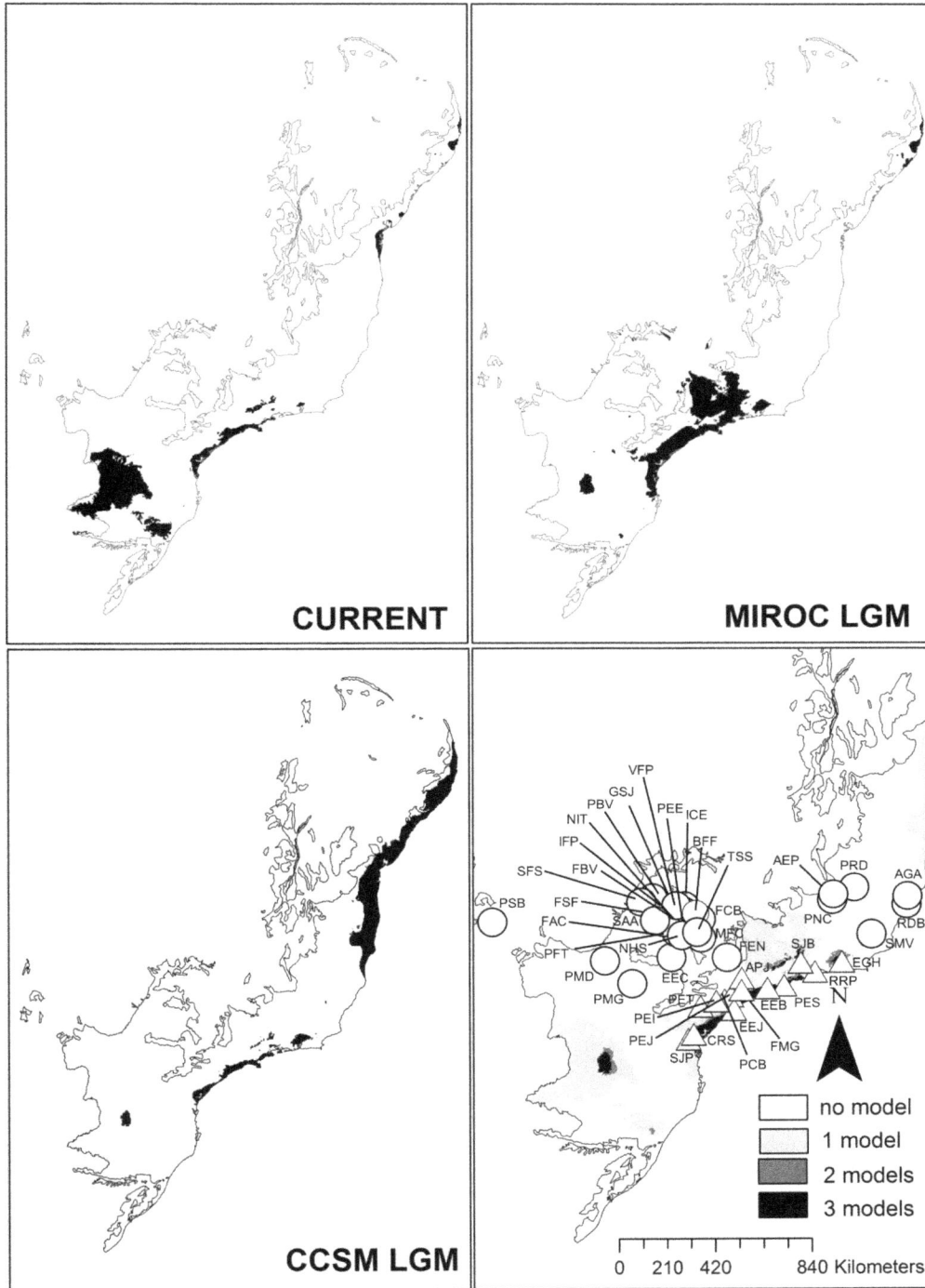

Figure 1. Distribution of the10% highest values of annual precipitation for the current time, the two last glacial maximum period (LGM), and the intersection of the three maps. Symbols (in the intersection map) indicate the 44 sites in the southern range of Brazilian Atlantic Forest. Dark shading indicates stable regions. White circles represent sites in an unstable region (Quaternary Climatic Oscillations). White triangles represent sites in a stable region (Quaternary Climatic Stability). Abbreviations of each site as in Appendix S1 in File S1.

We observed a positive relationship between the geographical distance and turnover components of CBD and PBD (Fig. 2). Values of β_{jtu} and UniFrac$_{Turn}$ between sites in different regions or between sites within stable region were on average 0.25 and 0.16 higher, respectively, than between sites within unstable region

(Fig. 2). Although the PBD pattern may be explained by species dissimilarities between sites, some UniFrac$_{Turn}$ values between sites in different regions were higher than expected by the null expectation (Fig. 5), indicating the influence of historic processes. On the other hand, a negative relationship was observed between

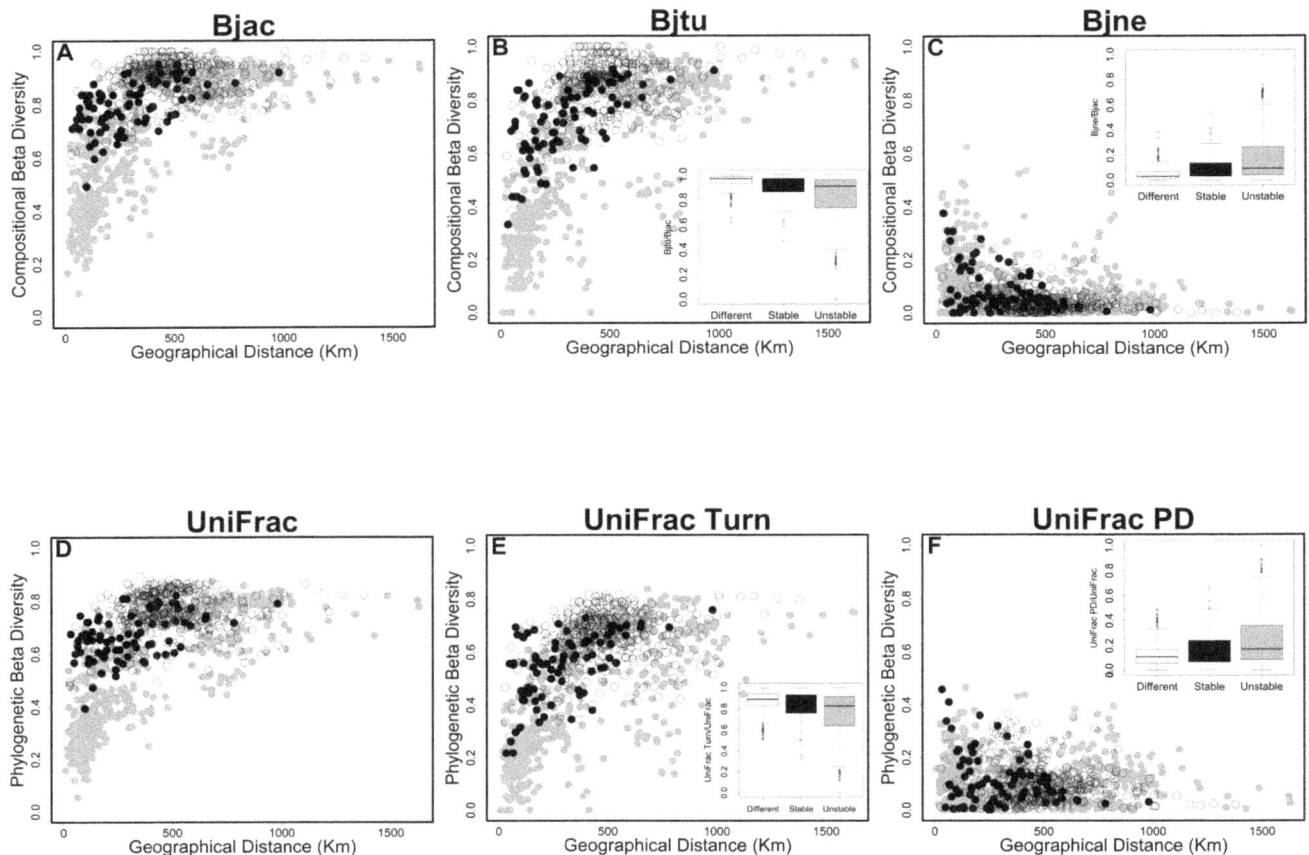

Figure 2. Relationships between geographical distance and Compositional Beta Diversity (CBD, A = β_{jac}, B = β_{jtu} and C = β_{jne}) and Phylogenetic Beta Diversity (PBD, D = UniFrac$_{Total}$, E = UniFrac$_{Turn}$ and F = UniFrac$_{PD}$) components. Subplots are showing the proportion explained by turnover (β_{jtu}/β_{jac} and UniFrac$_{Turn}$/UniFrac$_{Total}$) and nestedness (β_{jne}/β_{jac} and UniFrac$_{PD}$/UniFrac$_{Total}$) components. Symbol colors indicate the region where the sites occur in the southern range of Brazilian Atlantic Forest. Black circles indicate dissimilarity between sites within stable region (Quaternary Climatic Stability). Gray circles indicate dissimilarity between sites within unstable region (Quaternary Climatic Oscillations). White circles indicate dissimilarity between unstable and stable sites (different regions).

values of β_{jne} and UniFrac$_{PD}$ and geographical distance (Fig. 2). The average proportion of β_{jne} values between sites in different regions was 0.03, showing that CBD between sites in different regions are not subsets of each other. As expected, the proportion of nestedness components for both CBD and PBD were higher between sites in unstable region than between sites in stable region (Fig. 2). Furthermore, only some UniFrac$_{PD}$ values between sites in different regions and within stable region were higher than expected by the null expectation (Fig. 5), indicating that species composition in sites within unstable region are from the same lineages (i.e. same genera).

Discussion

Our results show that patterns of amphibian CBD and PBD in the southern range of Brazilian Atlantic Forest were largely determined by lineage turnover. Interestingly, the variance in turnover components from CBD and PBD is equally explained by current environmental gradients, geographical distance and long-term climatic conditions. However, the turnover components between sites from different regions and between sites within stable region were greater than between sites within unstable region. These results are consistent with previous studies that stressed that patterns of species distribution are influenced by different

processes depending on the regional species pool [42,51] and the spatial scale [52] considered.

Historical effects, environmental filtering and limited dispersal are recognized processes influencing the spatial distribution of turnover components of CBD and PBD among communities, and, consequently, determining which lineages reside in a particular region [16,42,53–54]. Even though stable and unstable regions are adjacent to each other, the high turnover components from CBD and PBD between sites from these distinct areas suggest that stable and unstable regions have different biogeographic histories. For example, eleven genera (*Arcovomer*, *Brachycephalus*, *Cycloramphus*, *Gastrotheca*, *Holoaden*, *Macrogenioglosus*, *Megaelosia*, *Myersiella*, *Paratelmatobius*, *Phrynomedusa* and *Scythrophrys*) were only recorded in sites within the stable region while nine genera (*Ameerega*, *Crossodactylodes*, *Dermatonotus*, *Eupemphix*, *Melanophryniscus*, *Phyllodytes*, *Pseudis Pseudopaludicola* and *Stereocyclops*) were only recorded in sites within the unstable region (Appendix S4 in File S1). According to Jansson [55], the higher is the long-term climatic stability in an area, the more new clades will persist without going extinct or reuniting with other clades. These results suggest that sites in stable regions may be areas of high diversification because of ameliorate conditions promoting high speciation and/or low extinction rates of anuran species with specialized reproductive modes (e.g. direct develop-

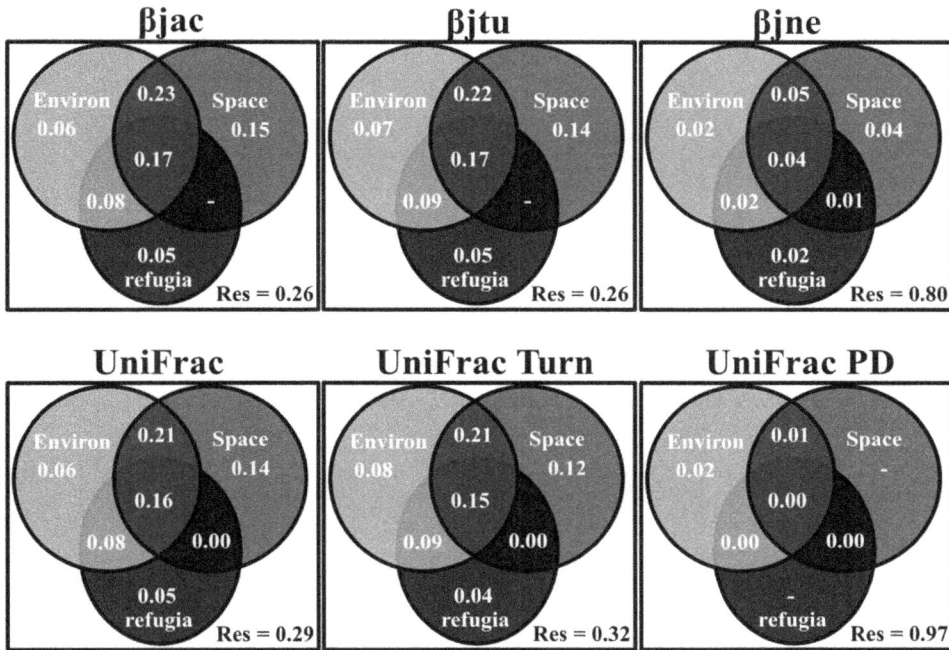

Figure 3. Partition of the variance of Compositional Beta Diversity (CBD, β_{jac}, β_{jtu} and β_{jne}) and Phylogenetic Beta Diversity (PBD, UniFrac$_{Total}$, UniFrac$_{Turn}$ and UniFrac$_{PD}$) components explained by geographical distance (Space), current environmental gradients (Environ) and long-term climatic conditions (refugia) for 44 sites in the southern range of Brazilian Atlantic Forest. Res = unexplained variance. "-" = variation explained <0.

ment of terrestrial eggs). Furthermore, current environmental gradients, temperature and precipitation in particular (see Appendix S4 in File S1), seem to act as filters to disperse species from stable (area of origin) to unstable regions. This result is similar to that found by da Silva *et al.* [34], who showed that moister sites in the Atlantic Forest harbored a greater phylogenetic diversity of amphibians than drier sites. Therefore, the interplay among historical effects, environmental filtering, and dispersal limitations might have contributed to the high values of turnover between sites in unstable and stable regions for both CBD and PBD. These results are in agreement with other studies showing that amphibian beta diversity is influenced by multiple factors [6,56–58]. For example, broad-scale amphibian richness is

strongly determined by historical constraints whereas regional patterns are determined by water and temperature [33].

At the regional scale, turnover components between sites within the stable region are larger than corresponding values between sites within the unstable region (Fig. 4). This result demonstrates that, although sites in stable region are relatively close, there are other factors driving CBD and PBD between them. Deviances from the expected distance-decay relationship usually happen if the distances are associated with marked geographical barriers to dispersal [15]. It is well-known that larger elevational ranges promote speciation through habitat specialization and isolation, thus increasing endemism and, consequently, discrepancies in species richness between sites within a region [59–64]. Sites in

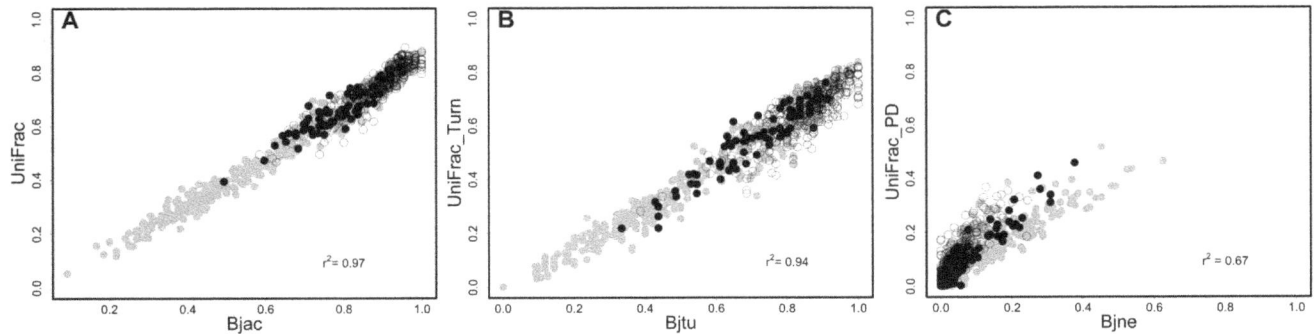

Figure 4. Relationship among Compositional Beta Diversity (CBD, A = β_{jac}, B = β_{jtu} and C = β_{jne}) and Phylogenetic Beta Diversity (PBD, A = UniFrac$_{Total}$, B = UniFrac$_{Turn}$ and C = UniFrac$_{PD}$) components. Symbol colors indicate the region where the sites occur in the Brazilian Atlantic Forest. Black circles indicate dissimilarity between sites within stable region (Quaternary Climatic Stability). Gray circles indicate dissimilarity between sites within unstable region (Quaternary Climatic Oscillations). White circles indicate dissimilarity between unstable and stable sites (different regions).

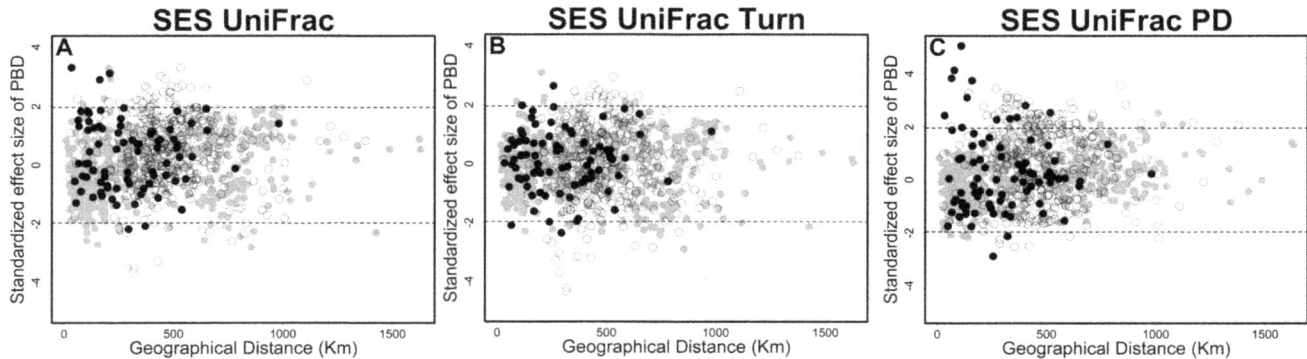

Figure 5. Standardized effect size (SES) values for phylogenetic beta diversity (PBD, A = UniFrac$_{Total}$, B = UniFrac$_{Turn}$ and C = UniFrac$_{PD}$) components among 44 sites in Brazilian Atlantic Forest. Values between dashed lines indicate that PBD components have no difference with respect to null expectation. Symbol colors indicate the region where the sites occur in the southern range of Brazilian Atlantic Forest. Black circles indicate SES between sites within stable region (Quaternary Climatic Stability). Gray circles indicates SES between sites within unstable region (Quaternary Climatic Oscillations). White circles indicate SES between unstable and stable sites (different regions).

stable regions are situated in the coastal Atlantic Forest that harbors the mountain complex of Serra do Mar and Serra da Mantiqueira, which may act as barriers to the dispersion of amphibians. Kozak & Wiens [65] showed that niche conservatism is the main factor promoting allopatric speciation and endemism in montane salamanders. According to these authors, allopatric sister taxa inhabiting similar climatic niches are not able to cross lowlands, which seemingly leads to geographic range fragmentation and speciation. Graham et al. [12] also found that the mountains are major dispersal barriers to hummingbird species, creating large differences in species composition despite the communities being phylogenetically similar. Therefore, these results suggest that topographic conditions and dispersal limitations can interact with evolutionary processes, influencing greater turnover components of CBD and PBD between sites within stable regions more than in unstable regions, usually associated to lowlands, within the Atlantic Forest.

As expected, the proportion of nestedness components from CBD was greater between sites in unstable rather than stable regions. This finding is similar to those reported by studies analyzing broad-scale effects of Quaternary climate conditions. For example, on a global scale, differences in fish faunas characterized by nestedness were greater in drainage basins that experienced larger amplitudes of Quaternary climate oscillations [66]. At the continental scale, beta diversity patterns at high latitudes of amphibians, birds and mammals from the New World were mostly determined by species richness differences in areas that were affected by glaciation until recently [57]. On the other hand, PBD between stable and unstable sites were not phylogenetically nested. Because nestedness can only occur between communities from the same overall species pool, the low values of PBD strengthened the idea that stable and unstable regions experience different biogeographic histories. Taken together, our results reinforce previous studies that highlight the importance of considering beta diversity components for CBD and PBD across multiple scales to infer how current environmental gradients and historical factors have influenced speciation, extinction and the dispersal of species throughout a region [12,14,16,28,58].

Conclusion

Our results show that the relative roles of geographical distance, current environmental gradients and long-term climatic conditions

in explaining the variations of amphibian CBD and PBD depend on the spatial scale under consideration. According to Belmaker & Jetz [52], the spatial scale at which environmental conditions constrain species richness will differ across clades with different home ranges and dispersal abilities. Therefore, this study demonstrates that historical events, current climatic conditions and geographical distances are complementary predictors of amphibian composition even for sites within the same biome. These results are in accordance with Buckley & Jetz [33], who analyzed the distribution of amphibians on a global scale and highlighted the importance of considering both the environment and historical events when attempting to understand gradients of amphibian distribution.

Furthermore, the variation in CBD and PBD among the amphibian assemblages indicates that stable and unstable regions have different biogeographic histories. We observed the highest values of CBD and PBD between sites of different regions (unstable vs stable), indicating low lineage dispersal across regions [16]. These results are in accordance with macroecological studies considering the Atlantic Forest. For example, Villalobos et al. [67] and Loyola et al. [68] showed that stable regions considered in our study harbored higher numbers of amphibian species with small range sizes (i.e. high endemism) and phylogenetic diversity than unstable regions. The same pattern was observed by Araújo et al. [6] to amphibian species with narrow ranges in Europe that occurred preferentially in areas that remained favorable during the last glacial period. We also observed the lowest values of CBD and PBD between sites within unstable regions, indicating homogenization of taxonomic and phylogenetic composition within a region that harbors amphibian species with broad range sizes. Evidence from phylogeographic studies suggest that populations within unstable regions were only recently colonized via dispersal of migrants from sites within stable regions in southern areas of Atlantic Forest [23–24]. Based on this scenario, we suggest that current patterns of amphibian species distribution in the southern range of Brazilian Atlantic Forest may be determined by high speciation rates in long-term climatic stability and limited dispersion to other regions due to the mountain complex of Serra do Mar and Serra da Mantiqueira, and precipitation gradients that act as barriers to some amphibian species with specific life history traits [34,69].

Supporting Information

Data S1 **Presence and absence of anuran species in the 44 sites of Brazilian Atlantic Forest used in the analysis.**

File S1 Appendix S1, Description of 44 sites of Brazilian Atlantic Forest used in the analysis. Appendix S2, Partition of the variance of Compositional Beta Diversity components based on Carvalho's *et al.* (2012, β_{cc}, β_3 and β_{rich}). Appendix S3, Cladogram of anuran demonstrating the phylogenetic relationships of our data-set based on the phylogenetic hypotheses proposed by Pyron & Wiens (2011). Appendix S4, Taxonomic classification and occurrence of amphibian species in stable and unstable sites from Brazilian Atlantic Forest. Appendix S5, Analysis summary of the principal component analysis (PCA) for 44 sites in the Brazilian Atlantic Forest.

Acknowledgments

We thank T.G. Souza and D.B. Provete for providing helpful suggestions on the first version of the manuscript.

Author Contributions

Conceived and designed the experiments: FRdS MAN. Performed the experiments: FRdS MVNA. Analyzed the data: FRdS. Contributed reagents/materials/analysis tools: FRdS. Wrote the paper: FRdS MAN MVNA.

References

1. Ricklefs RE (1987) Community diversity: relative roles of local and regional processes. Science 235: 167–171.
2. Pyron RA, Wiens JJ (2013) Large-scale phylogenetic analyses reveal the causes of high tropical amphibian diversity. Proc. R. Soc. B 280: http://dx.doi.org/10.1098/rspb.2013.1622
3. Oberdorff T, Hugueny B, Guégan J-F (1997) Is there an influence of historical events on contemporary fish species richness in rivers? Comparisons between Western Europe and North America. J. Biogeogr. 24: 461–467.
4. Svenning J-C, Skov F (2005) The relative roles of environment and history as controls of tree species composition and richness in Europe. J. Biogeogr. 32: 1019–1033.
5. Graham CH, Moritz C, Williams SE (2006) Habitat history improves prediction of biodiversity in rainforest fauna. Proc. Natl. Acad. Sci. USA 103: 632–636.
6. Araújo MB, Nogués-Bravo D, Diniz-Filho JAF, Haywood AM, Valdes PJ, et al. (2008) Quaternary climate changes explain diversity among reptiles and amphibians. Ecography 31: 8–15.
7. Wiens JJ, Donoghue MJ (2004) Historical biogeography, ecology and species richness. Trends Ecol. Evol. 19: 639–644.
8. Hua X, Wiens JJ (2013) How does climate influence speciation? Amer. Nat. 182: 1–12.
9. Currie DJ, Mittelbach GG, Cornell HV, Field R, Guégan J-F, et al. (2004) Predictions and tests of climate-based hypotheses of broad-scale variation in taxonomic richness. Ecol. Lett. 7: 1121–1134.
10. Araújo MB, Pearson RG (2005) Equilibrium of species' distributions with climate. Ecography 25: 693–695.
11. Kozak KH, Wiens JJ (2012) Phylogeny, ecology, and the origins of climate–richness relationships. Ecology 93: 167–18.
12. Graham CH, Parra JL, Rahbek C, McGuire J (2009) Phylogenetic structure in tropical hummingbird communities. Proc. Natl. Acad. Sci. USA 106: 19673–19678.
13. Gómez JP, Bravo GA, Brumfield RT, Tello JG, Cadena CD (2010) A phylogenetic approach to disentangling the role of competition and habitat filtering in community assembly of Neotropical forest birds. J. Anim. Ecol. 79: 1181–1192.
14. Fine VA, Kembel SW (2011) Phylogenetic community structure and phylogenetic turnover across space and edaphic gradients in western Amazonian tree communities. Ecography 34: 552–565.
15. Peixoto FP, Braga PHP, Cianciaruso MV, Diniz-Filho JA, Brito D (2013) Global patterns of phylogenetic beta diversity components in bats. J. Biogeogr. doi:10.1111/jbi.12241
16. Graham CH, Fine PVA (2008) Phylogenetic beta diversity: linking ecological and evolutionary processes across space and time. Ecol. Lett. 11: 1265–1277.
17. Martins FM (2011) Historical biogeography of the Brazilian Atlantic forest and the Carnaval Moritz model of Pleistocene refugia: what do phylogeographical studies tell us? Biol. J. Linn. Soc. 104: 499–509.
18. Mittermeier RA, Myers N, Mittermeier CG, Robles G.P. (2005) Hotspots: Earth's biologically richest and most endangered terrestrial ecoregions. Monterrey, Mexico: Cemex, Conservation International and Agrupación Sierra Madre.
19. Haddad CFB et al. (2013) Guide to the amphibians of the Atlantic Forest: diversity and biologia. – Anolisbooks.
20. Haffer J. (1969) Speciation in Amazonian forest birds. Science 165: 131–137.
21. Carnaval AC, Moritz C (2008) Historical climate modelling predicts patterns of current biodiversity in the Brazilian Atlantic forest. J. Biogeogr. 35: 1187–1201.
22. Porto TJ, Carnaval AC, Rocha PLB (2013) Evaluating forest refugial models using species distribution models, model filling and inclusion: a case study with 14 Brazilian species. – Divers. Distrib. 19: 330–340.
23. Carnaval AC, Hickerson MJ, Haddad CFB, Rodrigues MT, Moritz C (2009) Stability predicts genetic diversity in the Brazilian Atlantic Forest hotspot. Science 323: 785–789.
24. Thomé MTC, Zamudio KR, Giovanelli JGR, Haddad CFB, Baldissera Jr FA, et al. (2010) Phylogeography of endemic toads and post-Pliocene persistence of the Brazilian Atlantic Forest. Mol. Phylogenet. Evol. 55: 1018–1031
25. Rull V (2011) Neotropical biodiversity: timing and potential drivers. Trends Ecol. Evol. 10: 508–513.
26. Baselga A (2010) Partitioning the turnover and nestedness components of beta diversity. Global Ecol. Biogeogr. 19: 134–143.
27. Baselga A (2012) The relationship between species replacement, dissimilarity derived from nestedness, and nestedness. Global Ecol. Biogeogr. 21: 1223–1232.
28. Leprieur F, Albouy C, De Bortoli J, Cowman PF, Bellwood DR, et al. (2012) Quantifying phylogenetic beta diversity: distinguishing between 'true' turnover of lineages and phylogenetic diversity gradients. *PLoS ONE* 7: e42760. doi:10.1371/journal.pone.0042760
29. Ficetola GF, Rondinini C, Bonardi A, Katariya V, Padoa-Schioppa E, et al. (2013) An evaluation of the robustness of global amphibian range maps. J. Biogeogr. 41: 211–221.
30. Hijmans RJ, Cameron SE, Parra JL, Jones PG, Jarvis A (2005) Very high resolution interpolated climate surfaces for global land areas. Int. J. Climatol. 25: 1965–1978.
31. Hijmans RJ, Guarino L, Mathur P (2012) DIVA-GIS, version 7.5. A geographic information system for the analysis of biodiversity data. Manual. Available: http://www.diva-gis.org
32. Navas CA, Antoniazzi MM, Jared C (2004) A preliminary assessment of anuran physiological and morphological adaptation to the Caatinga, a Brazilian semi-arid environment. Int. Congr. Ser. 1275: 298–305.
33. Buckley LB, Jetz W (2007) Environmental and historical constraints on global patterns of amphibian richness. Proc. R. Soc. B 274: 1167–1173.
34. da Silva FR, Almeida-Neto M, Prado VHM, Haddad CFB, Rossa-Feres DC (2012) Humidity levels drive reproductive modes and phylogenetic diversity of amphibians in the Brazilian Atlantic Forest. J. Biogeogr. 39: 1720–1732.
35. Carvalho JC, Cardoso P, Gomes P (2012) Determining the relative roles of species replacement and species richness differences in generating beta diversity patterns. Global Ecol. Biogeogr. 21: 760–771.
36. Carvalho JC, Cardoso P, Borges PAV, Schmera D, Podani J (2013) Measuring fractions of beta diversity and their relationships to nestedness: a theoretical and empirical comparison of novel approaches. Oikos 122: 825–834.
37. Pyron RA, Wiens JJ (2011) A large-scale phylogeny of Amphibia including over 2800 species, and a revised classification of extant frogs, salamanders, and caecilians. Mol. Phylogenet. Evol. 61: 543–583.
38. de Sá RO, Streicher JW, Sekonyela R, Forlani MC, Loader SP, et al. (2012) Molecular phylogeny of microhylid frogs (Anura: Microhylidae) with emphasis on relationships among New World genera. Evolution. Biol. 12: 1–21.
39. Fouquet A, Blotto BL, Maronna MM, Verdade VK, Juncá FA, et al. (2013) Unexpected phylogenetic positions of the genera Rupirana and Crossodacty-lodes reveal new insights into the biogeography and reproductive evolution of leptodactylid frogs. Mol. Phylogenet. Evol. 67: 445–457.
40. Swenson NG (2009) Phylogenetic resolution and quantifying the phylogenetic diversity and dispersion of communities. PLoS ONE 4(2): e4390. doi:10.1371/journal.pone.0004390
41. Lozupone C, Knight R (2005) UniFrac: a new phylogenetic method for comparing microbial communities. Appl. Environ. Microbiol. 71: 8228–8235.
42. Whittaker RJ, Willis KJ, Field R (2001) Scale and species richness: towards a general, hierarchical theory of species diversity. J. Biogeogr. 28: 453–470.
43. Lessard J-P, Borregaard MK, Fordyce JA, Rahbek C, Weiser MD, et al. (2012) Strong influence of regional species pools on continent-wide structuring of local communities. Proc. R. Soc. B 279: 266–274.
44. Qian H, Swenson NG. Zhang J (2013) Phylogenetic beta diversity of angiosperms in North America. J. Biogeogr. 10: 1152–1161.
45. Nekola JC, White PS (1999) The distance decay of similarity in biogeography and ecology. J. Biogeogr. 26: 867–878.
46. Borcard D, Legendre P, Drapeau P (1992) Partialling out the spatial component of ecological variation. Ecology 73: 1045–1055.

47. R Development Core Team (2014) R: A language and environment for statistical computing, reference index version 3.1.0. R Foundation for Statistical Computing, Vienna, Austria. Available via DIALOG. Available: http://www.Rproject.org

48. Baselga A, Orme D, Villeger S, De Bortoli J, Leprieur F (2013). betapart: Partitioning beta diversity into turnover and nestedness components. R package version 1.3. Available: http://CRAN.R-project.org/package=betapart

49. Oksanen J. et al. (2013). vegan: Community Ecology Package. R package version 2.0-10. Available: http://CRAN.R-project.org/package=vegan

50. Paradis E, Claude J, Strimmer K (2004) APE: analyses of phylogenetics and evolution in R language. Bioinformatics 20: 289–290.

51. Cavender-Bares J, Kozak KH, Fine PVA, Kembel SW (2009) The merging of community ecology and phylogenetic biology. Ecol. Lett. 12: 693–715.

52. Belmaker J, Jetz W (2011) Cross-scale variation in species richness–environment associations. Global Ecol. Biogeogr. 20: 463–474.

53. Emerson BC, Gillespie RG (2008) Phylogenetic analysis of community assembly and structure over space and time. Trends Ecol. Evol. 23: 619–30.

54. Vamosi SM, Heard SB, Vamosi JC, Webb CO (2009) Emerging patterns in the comparative analysis of phylogenetic community structure. Mol. Ecol. 18: 572–92.

55. Jansson R (2003) Global patterns in endemism explained by past climatic change. Proc. R. Soc. B 270: 583–590.

56. Buckley LB, Jetz W (2008) Linking global turnover of species and environments. Proc. Natl. Acad. Sci. USA 105: 17836–17841.

57. Dobrovolski R, Melo AS, Cassemiro FAS, Diniz-Filho JAF (2012) Climatic history and dispersal ability explain the relative importance of turnover and nestedness components of beta diversity. Global Ecol. Biogeogr. 21: 191–197.

58. Baselga A, Gómez-Rodríguez C, Lobo JM (2012) Historical legacies in world amphibian diversity revealed by the turnover and nestedness components of beta diversity. PLoS ONE, 7, e32341. doi:10.1371/journal.pone.0032341

59. Janzen DH (1967) Why mountain passes are higher in the tropics. Amer. Nat. 101: 233–249.

60. Lynch JD (1986) Origins of the high Andean herpetofauna. In: Vuilleumier F, Monasterio M, editors. High altitude tropical biogeography. Oxford: Oxford University Press. pp. 478–499.

61. Jetz W, Rahbek C, Colwell RK (2004) Rarity, richness and the signature of history in centers or endemism. Ecol. Lett. 7: 1180–1191.

62. Kozak KH, Wiens JJ (2007) Climatic zonation drives latitudinal variation in speciation mechanisms. Proc. R. Soc. B 274: 2995–3003.

63. Smith SA, Nieto M.O.A., Reeder TW, Wiens JJ (2007) A phylogenetic perspective on elevational species richness patterns in Middle American treefrogs: why so few species in lowland tropical forests. Evolution 61: 1188–1207.

64. Graham CH et al. (2014) The origin and maintenance of montane diversity: integrating evolutionary and ecological processes. Ecography doi: 10.1111/ecog.00578

65. Kozak KH, Wiens JJ (2006) Does niche conservatism promote speciation? A case study in North American salamanders. Evolution 60: 2604–2621.

66. Leprieur F, Tedesco PA, Hugueny B, Beauchard O, Dürr HH, et al. (2011) Partitioning global patterns of freshwater fish beta diversity reveals contrasting signatures of past climate changes. Ecol. Lett. 14: 325–334.

67. Villalobos F, Dobrovolski R, Provete DB, Gouveia SF (2013)Is rich and rare the common share? Describing biodiversity patterns to inform conservation practices for South American anurans. PLoS ONE 8(2): e56073. doi:10.1371/journal.pone.0056073

68. Loyola RD, Lemes P, Brum FT, Provete DB, Duarte DS (2014) Clade-specific consequences of climate change to amphibians in Atlantic Forest protected areas. Ecography 37: 65–72.

69. Loyola RD, Becker CG, Kubota U, Haddad CFB, Fonseca CR, et al. (2008) Hung out to dry: choice of priority ecoregions for conserving threatened Neotropical anurans depends on life-history traits. PLoS ONE 3(5): e2120. doi:10.1371/journal.pone.0002120.

10

Region-Specific Sensitivity of Anemophilous Pollen Deposition to Temperature and Precipitation

Timme H. Donders[1]*, **Kimberley Hagemans[1,2]**, **Stefan C. Dekker[2]**, **Letty A. de Weger[3]**, **Pim de Klerk[4]**, **Friederike Wagner-Cremer[1]**

1 Palaeoecology, Department of Physical Geography, Faculty of Geosciences, Utrecht University, Laboratory of Palaeobotany and Palynology, Utrecht, The Netherlands, **2** Department of Environmental Sciences, Copernicus Institute, Faculty of Geosciences, Utrecht University, Utrecht, The Netherlands, **3** Department of Pulmonology, Leiden University Medical Centre, Leiden, The Netherlands, **4** Botany section, Staatliches Museum für Naturkunde Karlsruhe, Karlsruhe, Germany

Abstract

Understanding relations between climate and pollen production is important for several societal and ecological challenges, importantly pollen forecasting for pollinosis treatment, forensic studies, global change biology, and high-resolution palaeoecological studies of past vegetation and climate fluctuations. For these purposes, we investigate the role of climate variables on annual-scale variations in pollen influx, test the regional consistency of observed patterns, and evaluate the potential to reconstruct high-frequency signals from sediment archives. A 43-year pollen-trap record from the Netherlands is used to investigate relations between annual pollen influx, climate variables (monthly and seasonal temperature and precipitation values), and the North Atlantic Oscillation climate index. Spearman rank correlation analysis shows that specifically in *Alnus*, *Betula*, *Corylus*, *Fraxinus*, *Quercus* and *Plantago* both temperature in the year prior to (T_{-1}), as well as in the growing season (T), are highly significant factors (T_{April} r_s between 0.30 [P<0.05[and 0.58 [P<0.0001]; T_{Juli-1} rs between 0.32 [P<0.05[and 0.56 [P<0.0001]) in the annual pollen influx of wind-pollinated plants. Total annual pollen prediction models based on multiple climate variables yield R^2 between 0.38 and 0.62 (P<0.0001). The effect of precipitation is minimal. A second trapping station in the SE Netherlands, shows consistent trends and annual variability, suggesting the climate factors are regionally relevant. Summer temperature is thought to influence the formation of reproductive structures, while temperature during the flowering season influences pollen release. This study provides a first predictive model for seasonal pollen forecasting, and also aides forensic studies. Furthermore, variations in pollen accumulation rates from a sub-fossil peat deposit are comparable with the pollen trap data. This suggests that high frequency variability pollen records from natural archives reflect annual past climate variability, and can be used in palaeoecological and -climatological studies to bridge between population- and species-scale responses to climate forcing.

Editor: Robert Guralnick, University of Colorado, United States of America

Funding: This work is supported by Utrecht University and Leiden University Medical Centre. The funders had no role in study design, data collection and analysis, decision to publish, or preparation of the manuscript.

Competing Interests: The authors have declared that no competing interests exist.

* Email: t.h.donders@uu.nl

Introduction

Pollen production by wind pollinated (anemophilous) plants is characterized by inter-annual variation [1]. This variation is not random but related to biological processes (e.g. mast cycles, [2] and changes in vegetation dynamics such as tree line fluctuations [3]. Part of the variation is caused by climatic conditions, particularly the character of the seasonal cycle [3,4] but also lower-frequency climate variability [5]. Identification of the strength, sign and type of effect of climatic variables on the annual pollen deposition is significant in multiple fields of research, ranging from palaeoecology and palaeoclimate [6], global change biology [7], plant ecology [8], to allergology [9], and forensics [10].

Firstly, if annual pollen production is influenced by climate variables, then high resolution records of fossil pollen have the potential for quantitative reconstruction of past climate conditions on annual to decadal time scales [6,11–13]. Traditionally, palynological studies focus on population-scale successional changes to study variations in past vegetation cover and associated climate change [14]. A higher resolution is needed for understanding past variability of climate systems that vary on annual to decadal timescales such as the North Atlantic Oscillation (NAO) for Europe and North America [15], and climatic oscillations in other parts of the world. Beside vegetation succession, annual-scale studies of pollen production and deposition rates allow us to assess the influence of climatic factors on such short timescales, and identify how such signals are preserved in natural archives that are the source for vegetation and derived climate reconstructions.

Secondly, in allergology accurate and timely prediction of seasonal pollen production is important for sufferers of pollinosis (hay fever) [9]. Tree pollen allergens affect health in up to 15% of the population [2].While much effort is directed at observed and future changes on the timing of pollen release e.g. [16,17], or masting effects [2], it is vital to also establish the exact relation between total annual pollen production and climate variables to improve predictions of seasonal pollen concentrations [18]. Global climate changes will likely affect the intensity of the pollen season e.g. [7,19] whereby region-specific studies of climate parameters relevant for pollen production and deposition will lead to better long-term pollinosis scenarios. This involves quantification of both the annual-scale variability and decadal-scale climatic trends and pollen production to improve the long-term pollen predictions and region-specific seasonal forecast of pollinosis.

Thirdly, forensic studies benefit from well-documented relations between pollen deposition and climate parameters to reduce uncertainty between concentration and composition of pollen deposition and pollen assemblages collected from crime scenes [10]. As for pollen-based climate reconstructions, forensic paly-nological studies need to take into account both annual climatic conditions, as well as differential preservation (taphonomy) of pollen in soils compared to the atmospheric composition as factors that influence sample comparison.

In this study we aim to improve our understanding of the region-specific response of annual pollen deposition to climate variables for common anemophilous pollen types in NW Europe. We assess the regional variability between pollen trap records, the strength and type of climate-pollen correlations, the robustness of the correlations for the purpose of seasonal forecasts and climate reconstructions, and the preservation of the annual variability in natural peat archives in comparison to the trap data.

While timing of pollen release is commonly known to be climate dependent [17,20–22], the annual pollen production is also affected in several ways [3–5,23,24]. In most trees, reproductive structures are formed during the summer in the year prior to flowering [1,4]. Warmer conditions at that time can positively influence pollen production and thereby influence the intensity of the pollen season in the following year [4–6,25]. During the flowering season, weather conditions influence pollen release in tree species that typically flower in late winter and early spring. For example frost may damage the flower buds before pollen release and can therefore reduce the total annual pollen production [1]. These relations differ regionally, as pollen production can be limited by summer temperatures and temperature sum (boreal Finland and Denmark [4–6]), winter temperature (Jura Mountains [26]), or humidity (Australia [24], Central Europe and Caucasus [25]). Depending on the species, correlations can have opposite signs [25], and reveal autocyclic biological variations in time [2,23].

To achieve our aims and improve our understanding on climate-pollen relations between different regions and in different types of archives, multiple records of atmospheric and sediment-derived pollen deposition are needed [14]. Also, information on local land use changes and vegetation composition [26], as well as reliable precipitation and temperature data are required. A rare combination of such time series is available in The Netherlands, which presents a unique opportunity to investigate the impact and consistency of climate on annual pollen deposition with exact time control. The correlations and patterns between the different records allow better explanation and independent testing of the annual variability, and thereby help to strengthen long-term allergy forecasts, aid development and understanding of annual-resolved palaeoclimate records, and increase reliability of forensic studies.

Material and Methods

At Leiden University Medical Centre (LUMC, 52°23'N; 4°29E) in W Netherlands (Fig. 1, Table S1 in File S1) atmospheric pollen concentrations have been recorded daily since AD 1969 [17,27], resulting in one of the longest and most continuous pollen trapping stations worldwide. Since AD 1975 a second locality in The Netherlands, the Elkerliek Hospital in Helmond (SE Netherlands,

51°29'N; 5°38E, Fig. 1, Table S2 in File S1), produced daily pollen counts, although counts were not year round until AD 2008. Thus far studies on the pollen-climate relations in the Netherlands have only focused on the influence of climate on the long-term trends and timing of pollen release [17,27]. The records present a unique opportunity to investigate the impact of seasonal climate variables on annual pollen production as well as assess regional and plant-specific variability in the Netherlands. For example, the degree of consistency between records determines how regionally applicable a pollen forecast is and whether the climatic interpretation of high-resolution palaeovegetation records are robust, and ecosystem dependent. We test effects of seasonal variations in temperature and precipitation on the total annual pollen deposition of anemophilous plants in pollen traps from the Netherlands, and assess regional variations by comparison of both trapping stations and a natural sediment archive.

Comparison with a climate-pollen correlation study from southern Denmark [4] provides a supra-regional view on the most important climate drivers on pollen deposition in NW Europe. Extending the Danish study, we hypothesize that temperature of the growing season in the previous year regionally is the most important factor for annual pollen deposition (used here as the combined result of pollen production and transport processes), and that this factor will be consistent between the different pollen trap records in NW Europe. We expect the natural archive to contain elevated levels of local site-specific vegetation relative to the trapping stations, but the trends and annual-scale patterns to be comparable. Following recommendations by Joosten and De Klerk [28] we indicate pollen taxa in small capitals (e.g., BETULA) to distinguish them from taxonomic plant species.

We use simple linear correlation and multiple linear regression to determine, and statistically model, the influence of monthly and seasonal temperature and precipitation on total annual pollen production. Winter temperature and precipitation in the Netherlands (Fig. 2) are influenced by the NAO, defined by the difference of normalized sea level pressure between Lisbon, Portugal and Reykjavik, Iceland [15]. Therefore, the effect of the winter NAO (December through March) on pollen production is also analysed, as well as testing for cyclic patterns and autocorrelations in the data. Further, a well-dated near-annually resolved pollen record (±1953–1992) from a peat deposit in the vicinity of Helmond [29] provides insight on the reflection of annual-scale variations in pollen production and species-specific offsets in natural archives compared to the trap records. If similar climate-forced variations in annual pollen production can be identified for both the pollen traps and the natural sediment archives, the forcing is regionally relevant and preserves well in natural archives implying that similar high-frequency climate variations can be reconstructed for pre-instrumental periods.

Source data description

The LUMC trap is located in an urban area with non-natural vegetation in nearby parks, gardens and along streets. In these parks, *Alnus* and *Populus* are the most dominant trees [30]. The Elkerliek area is mostly surrounded by parks and gardens. Pollen was collected with a Hirst-type [31] volumetric, continuous pollen trap [27]. Initially a daily operated type (Castella) was used, while in 1976 (LUMC) and 1980 (Elkerliek) a weekly (Burkard) trap was installed on the roofs ~200 m from the original location [30]. Daily pollen concentrations are expressed as $n\ m^{-3}\ day^{-1}$ in the air, and are corrected for changes in counting and trapping method. Daily pollen counts were summed to determine total annual pollen influx ($n\ m^{-3}\ year^{-1}$). The pollen concentration in

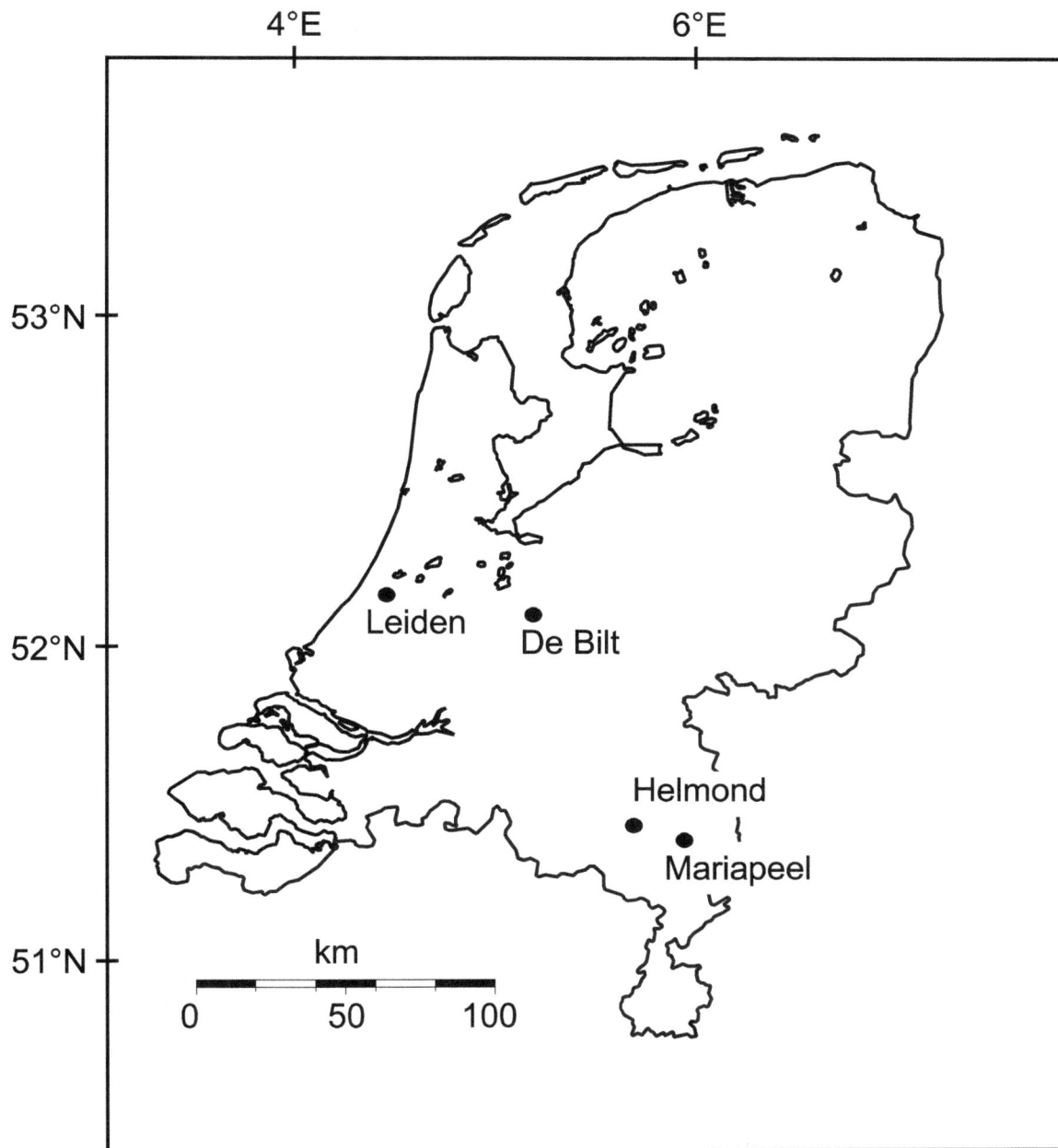

Figure 1. Map of the Netherlands showing the location of the LUMC (Leiden) and Elkerliek (Helmond) pollen traps, De Bilt meteorological station, and the peat deposit from Mariapeel Natural reserve.

the air was not measured whole year round at LUMC until 1977 (Table 1), and in Elkerliek until 2008. At LUMC, this caused missing or incomplete data for CORYLUS counts between 1969 and 1974, and in 1976, and for ALNUS counts in 1969, 1973, and 1976 since these species flowered before pollen counts started in those years. For calculation of correlations the missing values on annual pollen influx were replaced by the series mean. Plots of pollen influx for individual taxa against time (Fig. S1) were evaluated for data consistency and missing values, and if needed, corrected based on the original count sheets.

A summary pollen diagram from the peat deposit from Mariapeel Nature Reserve nearby Helmond (51° 25' 03.04" N,

5.55' 31.94" E, Fig. 1, Table S3 in File S1) was reported earlier [29], and includes age information and site description. The profile represents secondary peat growth AD ±1953 to 1992 in a small pit after the site was mined in the early 20th century and shows near-annual variability of local and regional vegetation. Age assessment was based on linear interpolation between three local well-documented land-use changes (AD 1961, 1973 and 1980), resulting in a time-scale with an estimated accuracy of ±1 year [29]. Pollen accumulation rates were calculated based on an added spike with a known amount of *Lycopodium clavatum* spores and a fixed sample volume. Initially, counts of 20 pollen types were selected based on anemophily and presence in all three data

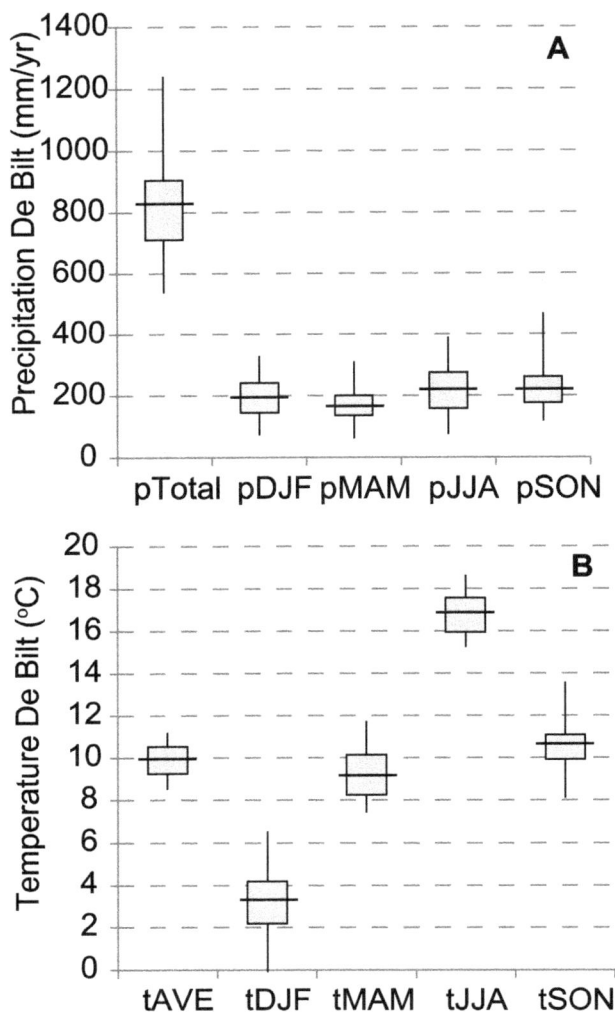

Figure 2. Annual and seasonal variability of the meteorological data from De Bilt. Values are expressed as the upper and lower quartiles, median, and minimum and maximum values of (a) cumulative precipitation and (b) mean temperature for AD 1969–2012.

Table 1. Availability of LUMC (day of the year, January 1st = day 1).

LUMC pollen data	
Year AD	**Measurement period (days)**
1969	99–259
1970	75–259
1971	75–259
1972	69–245
1973	71–167
1974	60–167
1975	61–260
1976	63–238
>1977	Year round

counts was performed to assess the internal relations of the pollen assemblages using CANOCO software version 4.5 [33]. A two-tailed Spearman rank correlation was performed to assess which pollen types (n = 20), as well as the PCA 1st axis values representative of the pollen assemblage, correlate significantly with climate. All variables, monthly and seasonal temperature (T_{month} and T_{season}), precipitation (P_{month} and P_{season}), and the NAO index, were tested for both the flowering year and the prior year (subscript $_{variable-1}$). Due to the large number of comparisons, care should be taken for correlations that are only significant at the 95% level as they are likely randomly occurring [4]. Common pollen types with significant correlation to multiple climate variables (see results, Tables 2, 3) were analysed further with a linear regression analysis with time as a predictor, and a multiple regression analysis with climatological variables as predictors.

An initial model was created for each selected pollen type (n = 6), including monthly temperature values of the year before flowering (i.e. lag −1), and monthly temperature during the current year until one month after the end of the flowering season (following ref. [34]) (Table 3). Precipitation variables were only included when the Spearman rank correlation showed a significant correlation to pollen influx. To prevent statistical over fitting, a stepwise backward multiple regression analysis was performed based on the initial models to reduce the number of redundant parameters. While stability of weather patterns can cause several consecutive months to correlate with pollen production, only a single or few months are relevant for pollen production [25]. Prediction skill of the resulting model was tested by random division of the dataset in a calibration and an independent test set. Predicted values were plotted against observed values, whereby R^2 and associated P-values <0.05 were considered to be statistically significant models. Cyclicity patterns and autocorrelations were evaluated through multitaper spectral analysis method incorporated in PAST software version 2.17 [35], which produces an F-value statistic for significance testing.

Results

General trends

The total pollen influx of all pollen types (n = 20) at LUMC ranges from 12.5×10^3 to 45.7×10^3 m^{-3} year^{-1} with an average pollen influx of 23.5×10^3 m^{-3} year^{-1} (Fig. 3a). Long-term influx of ALNUS, CARPINUS, CORYLUS, QUERCUS, JUGLANS and FRAXINUS

sources, eight of which are associated with hay fever. LUMC data were used to test correlations with climatic parameters as it is the most complete and consistent, the Elkerliek data are used to assess the regional consistency, and Mariapeel data for comparison between trapping stations and a natural sediment archive. Temperature and precipitation data were obtained from the Royal Netherlands Meteorological Institute (KNMI) at the automatic meteorological station in De Bilt, a central location in the Netherlands (52°06'N; 5°11'E) representative of mean conditions and at intermediate distance between both trapping sites (Figs. 1, 2). The NAO-index data were obtained from the Climate Analysis Section of the National Center for Atmospheric Research, Boulder, USA [32].

Statistical analysis

Pollen counts were square-root transformed as Q–Q plots and non-parametric Kolmogorov-Smirnov testing showed a non-normal distribution for most pollen types. A principal component analysis (PCA) of the LUMC square-root transformed pollen

Table 2. Selected pollen types of tree and herb species and their flowering season.

Family	Pollen type	Flowering season
Betulaceae	Alnus*	January, February, March
	Betula*	April, May, June
	Corylus*	January, February, March
Fagaceae	Quercus	May, June
Oleaceae	Fraxinus	April, May
Plantaginaceae	Plantago*	May, June, July, August, September

Pollen types associated with pollinosis are indicated by *.

pollen increases from about AD 1990, while TILIA is more dominant before AD 1990 (see Fig. 3b and Fig. S1). The herb pollen types POACEAE, ARTEMISIA, CHENOPODIACEAE, PLANTAGO and URTICACEAE show a phase of maximal influx between approximately AD 1990 and 2000. For 20 pollen types tested, only 6 show no significant correlation to any climate parameter (Tables S4–8 in File S2). Correlations are clearly stronger and more frequent for the temperature variables. In contrast to precipitation, within-season variability is much smaller than between-season variability for temperature (Fig. 2), but the within-season effectively controls the length of the seasons. Six common pollen types with multiple significant correlation coefficients and consistent records at both pollen trap stations, ALNUS, BETULA, CORYLUS, FRAXINUS, QUERCUS and PLANTAGO (Table 2), were selected for further regression analysis (Tables 4–6, Fig. 4). Year-to-year deposition of the six pollen types at LUMC and Elkerliek vary consistently with highly significant correlations

Table 3. Initial model parameters for multiple regression analysis using backward selection.

Pollen	Predictors
ALNUS	January – December Temperature$_{-1}$
	January – April Temperature
BETULA	January – December Temperature$_{-1}$
	January – June Temperature
	April Precipitation$_{-1}$
	June Precipitation$_{-1}$
	April Precipitation
CORYLUS	January – December Temperature$_{-1}$
	January – April Temperature
	Precipitation March$_{-1}$
FRAXINUS	January – December Temperature$_{-1}$
	January – June Temperature
QUERCUS	January – December Temperature$_{-1}$
	January – June Temperature
	Precipitation April
PLANTAGO	January – December Temperature$_{-1}$
	January – September Temperature
PCA 1st axis	January – December Temperature$_{-1}$
	January – September Temperature

Temperature$_{-1}$ stands for temperature in the year before flowering.

of the square root-transformed data (Fig. 4). A linear regression of the total influx shows significant long-term increase with time for all six selected pollen types except CORYLUS for the LUMC site, while only FRAXINUS increases significantly at Elkerliek (Table 4).

Correlation to climate

Positive significant relations were found between the NAO index and the annual pollen influx of QUERCUS and PLANTAGO, as well as the first axis of the PCA analysis, which represents mean composition of the pollen assemblage. Positive significant influence of mean growing season temperature, of the year prior to the flowering season ($T_{summer-1}$ and /or $T_{spring-1}$) is evident in all six pollen types (Table 5), where ALNUS, FRAXINUS and PLANTAGO influx show particularly strong relations with $T_{spring-1}$, while $T_{summer-1}$ is important for BETULA and, to a lesser degree, CORYLUS and QUERCUS. During the year of flowering, T_{spring} is dominant except in BETULA. Based on the monthly correlations, April and March are the most important months in T_{spring} and $T_{spring-1}$, while T_{jul-1} is the dominant summer month. The only herbaceous type, PLANTAGO is also significantly influenced by T_{may} and T_{may-1}. Correlations show much less influence of precipitation on total pollen influx values (Table 6). Only BETULA influx is positively influenced by $P_{spring-1}$ (mainly P_{apr-1}), while in both BETULA and CORYLUS P_{spring} (for BETULA: P_{apr} and CORYLUS: P_{mar}) has a negative influence, although not highly significant and possibly a result of random correlations.

Predictive regression model

The effect of a combination of meteorological variables on pollen influx, assessed through multiple regression analysis with backward selection, is shown in Table 7 for the LUMC data. Calibration and prediction skill of the resulting optimal model of pollen influx are shown in Fig. 5. The optimal regression models show a good performance based on their reported R^2. In BETULA and FRAXINUS some redundant parameters were discarded after the data splitting step, creating a model with slightly lower performance but less parameters, thereby reducing the complexity and statistical overfitting. Growing season temperature of the previous year again shows to be the most important factor, especially T_{mar-1}. In addition, winter temperatures prior to flowering show to be significant as well (T_{nov-1}, T_{dec-1}, T_{jan}, T_{feb}) and mostly have a negative loading (Table 7). Depending on the genus, T_{mar} has a negative (BETULA) or positive (ALNUS, PLANTAGO) loading in the model. Plotted against time (Fig. 5), the trends and phase relations between the observed and predicted influx values correspond well, especially for BETULA and PLANTAGO, although the variability in the predicted values is usually slightly lower than the observations. The multitaper spectral analysis showed signif-

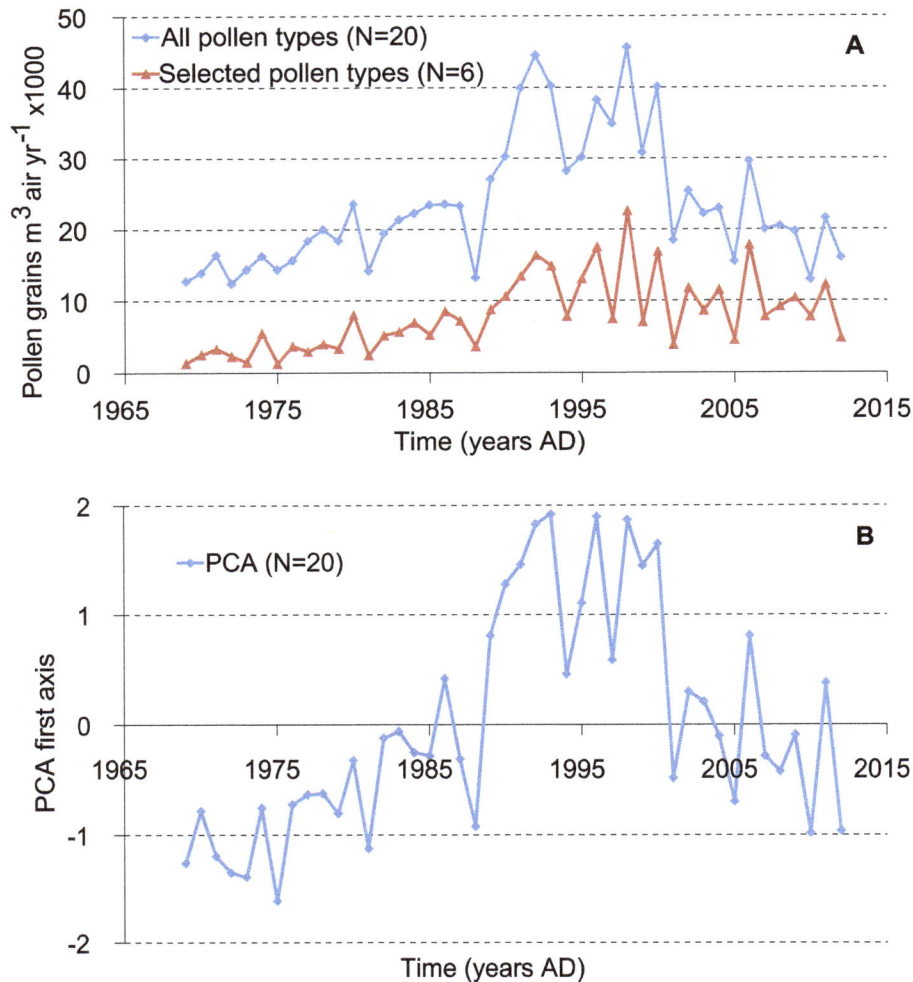

Figure 3. Total annual pollen influx values 1969–2012. (a) Total LUMC pollen influx for all pollen types (n = 20) and for selected pollen types (n = 6), and (b) sample values of the 1st PCA axis based on square-root transformed pollen count data.

icant cyclic patterns for BETULA (Fig. 6b) at a frequency of 0.19 (5.3 year) and 0.45 (2.2 year). Also CORYLUS (3.3 and 2 year, Fig. 6c) and PLANTAGO (3.2 and 2.6 year, Fig. 6f) show significant variability at short periodicities.

Discussion

General observations

The primary aim of this study was to investigate the possible effects of temperature and precipitation variables on the annual pollen production of wind-pollinated plants in the Netherlands. Earlier evaluations showed no significant long-term increase for BETULA and QUERCUS [30], however, ten years of extra data has revised this conclusion (Table 4). The long-term trend observed in Fig. 3b can in part be attributed to construction work at the LUMC site from 1985 until the mid-1990s, which provided favourable conditions for early successional herb species on cleared land. Increased planting in recent years of non-native *Alnus spaethi*, which is well-adapted to urban environments and has an early flowering season, probably adds to the long-term increase as well. Explanation of the increased pollen influx observed in most taxa (see Fig. S1) in terms of long-term

temperature change is tempting, considering the sensitivity to annual-scale temperature variations shown in Table 5, but other factors such as long-term increase of CO_2 [7,36], potentially increased production due to long-term atmospheric nutrient deposition [37,38], as well as non-documented changes in composition and management of the urban vegetation cannot be discarded.

The results of this study showed that the annual pollen influx of ALNUS, BETULA, CORYLUS, FRAXINUS and QUERCUS, was positively influenced by summer temperature in the year before flowering, which coincides with the production of flowering buds for most tree species [1,4] and confirms our hypothesis. The reported relation between annual influx of ALNUS, BETULA, FRAXINUS, QUERCUS and PLANTAGO and temperature during the flowering season is likely more related to the final stages of pollen ripening and deposition speed. In that light, it is surprising that precipitation has little effect on most taxa (Table 6), in contrast to the conclusions for central Europe [25]. Only for BETULA the effect of additional precipitation is beneficial for the following year, while spring rain during pollen release reduces atmospheric pollen content [4], although it is unclear why this is only for BETULA and CORYLUS in our data. The Netherlands rarely has significant water

Table 4. Linear regression of total annual of pollen influx with time (year) as predictor.

Pollen	LUMC, Leiden				Elkerliek, Helmond			
	M	S.E.	r²	β	M	S.E.	r²	β
Alnus	2926	354	.278	.527***	2241	234	.037	.191
Betula	2643	287	.144	.379*	7880	963	.010	.100
Corylus	147	16	.072	.268	233	22	.010	.102
Fraxinus	841	110	.387	.622***	653	96	.456	.675**
Quercus	1469	183	.225	.474***	3558	361	.091	.302
Plantago	180	11	.201	.448**	106	10	.095	−.308

Significance levels are indicated: * P<0.05; ** P<0.01; ***P<0.001. S.E= standard error, M= mean, β= slope.

shortage (Fig. 2), and water stress is therefore an unlikely limiting factor for pollen production for most taxa. Comparison of the correlation coefficients in Tables 5 and 6 with results from Southern Denmark [4] show that particularly pollen deposition of Betula, Fraxinus and Quercus are driven by largely the same climatic variables, while Corylus agrees in sensitivity to T_{apr} and T_{apr-1}, but not $T_{jan/feb}$ (Denmark) and T_{jul-1} (LUMC). Contrary to the LUMC data, Alnus in Denmark shows surprisingly little correlation with climatic variables. Clearly, these results highlight the need for region-specific correlation models for pollen-climate relationships, but do identify climate factors of regional relevance. Specifically for the medical treatment of pollinosis, the linear models (Table 7 and Fig. 5) are a first step in the development of a seasonal predictive model for pollen forecasting. A subsequent step should involve combining the total season pollen load with knowledge on the timing of pollen release, which is climate dependent [17,39]. For forensic palynology, the most important aspect of the study is the fact that the pollentrap record shows variability that is consistent across large distances (Fig.4), which corroborates a study from the Jura Mountains in which pollen composition from several high-resolution peat records were compared [26]. Annual variations in pollen deposition and composition are largely climate driven and, hence, it is meaningful to compare samples from a crime scene to a pollen trap record to estimate ages or relative timing (season).

Alnus. In *Alnus*, catkins are initiated in early July and meteorological conditions before this period can be expected to influence their formation [40]. In this study, the positive influence of T_{mar-1} suggests that warmer March temperatures stimulate catkin formation in *Alnus*. Catkin formation is followed by a dormancy period during autumn and winter, which protects them from frost damage, until temperature increases at the start of pollen season [40,41]. For the United Kingdom, it has been suggested that bud dormancy starts in August and lasts until February [40]. However, the results of this study indicated that T_{aug-1} affects Alnus production, suggesting that catkin formation continues in August and bud dormancy starts in the period after August in the Netherlands. A study on the Iberian Peninsula suggested that the chilling period for *Alnus* starts in November and continues until January [41].

In the Netherlands, anthesis (opening of flower buds) and pollination takes place from January until March [34]. The relation of T_{mar} with Alnus influx found here suggests that warmer temperatures during the flowering season stimulate pollen release or lengthens the flowering season. In the Netherlands, it has been demonstrated that T_{jan} affects the start of the Alnus pollen season [39], although the Danish results showed no clear relation with temperature [4]. Despite a reported biannual cyclicity for Alnus [23,41], our study (Fig. 6a) and results from Denmark and Spain showed no significant cyclic patterns [4,41], although some autocorrelation is present with a 2 year time lag (Fig. S2), pointing to a limited role of autocyclic processes in *Alnus* flowering intensity.

Betula. In the Netherlands *Betula* species occur: *Betula pendula* and *B. pubescens*. *B. pubescens* is closely related to *B. papyrifera* [18] and the phenology of the latter has been thoroughly investigated [42–44]. The phenology of *B. papyrifera* is also characteristic to *B. pendula* and *B. pubescens* [18] and can thus be used as a general description of flower formation of *Betula* trees in the Netherlands. The initiation of male catkins starts just before leaves develop in the year before flowering. They become visible in June and July and develop until August after which dormancy begins [18,43,44]. In the Netherlands, leaf unfolding generally begins in April [45]. It can thus be expected that the

Table 5. Correlation coefficients between annual total pollen influx per pollen type, temperature and NAO index as shown by a Spearman Rank Correlation.

			Spearman rank correlation (r_s), 2-tailed						
Climate variable			**ALNUS**	**BETULA**	**FRAXINUS**	**CORYLUS**	**QUERCUS**	**PLANTAGO**	**PCA 1st axis**
Monthly temperature	Previous year	T_{jan-1}	.240	.134	.129	-.024	.128	.248	.229
		T_{feb-1}	.195	-.006	.324*	.158	.046	.318*	.167
		T_{mar-1}	.530***	.197	.512***	.309	.360*	.487***	.459**
		T_{apr-1}	.360*	-.014	.513***	.048	.173	.360*	.145
		T_{may-1}	.170	.308*	.193	.224	.144	.444**	.297
		T_{jun-1}	.124	.317*	.037	-.037	.110	.019	.090
		T_{jul-1}	.318*	.564***	.371*	.562***	.440**	.192	.442**
		T_{aug-1}	.363*	.237	.290	.348*	.135	.048	.260
		T_{sept-1}	.054	.072	.064	.079	.124	.175	.062
		T_{oct-1}	.016	.093	-.010	.152	-.046	-.002	.031
		T_{nov-1}	.138	.200	.081	-.032	.005	-.036	.035
		T_{dec-1}	-.044	.109	.030	-.062	.070	.318*	.098
	Flowering year	T_{jan}	-.040	.078	.130	-.090	.201	.209	.084
		T_{feb}	.117	.320*	.167	.143	.494***	.278	.245
		T_{mar}	.329*	-.022	.385**	.277	.147	.500***	.330*
		T_{apr}	.507***	.463**	.393**	.325*	.576***	.303*	.430**
		T_{may}	.146	.093	.224	.150	.510***	.440**	.314*
		T_{jun}		.171	.241		.272	.244	.206
		T_{jul}						.318*	.143
		T_{aug}						.179	.152
		T_{sept}						.131	.150
Seasonal temperature	Previous year	$T_{winter-1}$	.225	.136	.227	.083	.169	.293	.260
		$T_{spring-1}$	.506***	.254	.597***	.295	.290	.592***	.422**
		$T_{summer-1}$	.365*	.534**	.306*	.399*	.362*	.113	.372*
		$T_{autumn-1}$	.146	.175	.135	.150	.031	.096	.080
	Flowering year	T_{winter}	-.004	.219	.153	-.027	.328*	.355*	.176
		T_{spring}	.534***	.200	.452**	.350*	.572***	.573***	.476***
		T_{summer}		.114	.314*		.139	.349*	.223
NAO	Previous year	NAO_{-1}	.178	.085	.054	-.117	-.041	.199	.168
		NAO_{DJFM-1}	.272	.242	.202	-.029	.184	.448**	.376*
	Flowering year	NAO	.197	.251	.296	.279	.373*	.466**	.401**
		NAO_{DJFM}	.100	.160	.114	.019	.209	.248	.290

Significance levels are indicated: * $P<0.05$; ** $P<0.01$; *** $P<0.001$.

Table 6. Correlation coefficients between annual total pollen influx per pollen type and precipitation as shown by a Spearman Rank Correlation.

Climate variable			Spearman rank correlation (r_s), 2-tailed						
			ALNUS	BETULA	FRAXINUS	CORYLUS	QUERCUS	PLANTAGO	PCA 1st axis
Monthly Precipitation	Previous year	P_{jan-1}	.161	-.051	.030	-.153	-.124	-.043	.017
		P_{feb-1}	.163	.095	.203	.190	.044	.144	.074
		P_{mar-1}	-.037	.238	-.074	-.042	.241	.110	.147
		P_{apr-1}	-.246	.407**	-.042	.288	.186	.044	.110
		P_{may-1}	-.089	.138	.029	-.172	.114	-.048	-.083
		P_{jun-1}	-.054	-.326*	.130	-.041	-.119	.173	-.098
		P_{jul-1}	-.038	-.283	.095	-.205	-.074	.030	-.208
		P_{aug-1}	.002	.064	-.021	.226	.181	.079	.016
		P_{sept-1}	.211	.090	.215	-.037	.218	.065	.185
		P_{oct-1}	.006	-.045	-.075	-.241	.056	.024	.070
		P_{nov-1}	-.198	-.090	-.102	.111	.048	-.048	-.064
		P_{dec-1}	.019	.196	.253	-.138	-.012	.176	.114
	Flowering year	P_{jan}	-.241	.066	.006	-.216	.058	-.053	-.068
		P_{feb}	-.117	.186	.057	.053	.144	.029	.021
		P_{mar}	-.085	-.052	-.139	-.340*	.195	.183	-.005
		P_{apr}	-.080	-.308*	-.148	-.143	-.333*	.175	-.072
		P_{may}		-.154	-.138		-.129	-.030	-.160
		P_{jun}		.037	-.120		.123	-.008	.104
		P_{jul}						-.136	-.114
		P_{aug}						.030	-.029
		P_{sept}						.229	.168
Seasonal Precipitation	Previous year	$P_{winter-1}$	.189	-.014	.083	-.076	-.033	.049	.061
		$P_{spring-1}$	-.204	.375*	-.054	-.004	.312*	.060	.086
		$P_{summer-1}$	-.025	-.208	.094	.034	.083	.208	-.091
		$P_{autumn-1}$	.066	-.014	.032	-.116	.195	.037	.130
	Flowering year	P_{winter}	-.188	.194	.143	-.166	.068	.069	.012
		P_{spring}	-.142	-.303*	-.259	-.334*	-.126	.162	-.139
		P_{summer}		.068	.052		.271	.003	.001

Significance levels are indicated: * P<0.05; ** P<0.01; ***P<0.001.

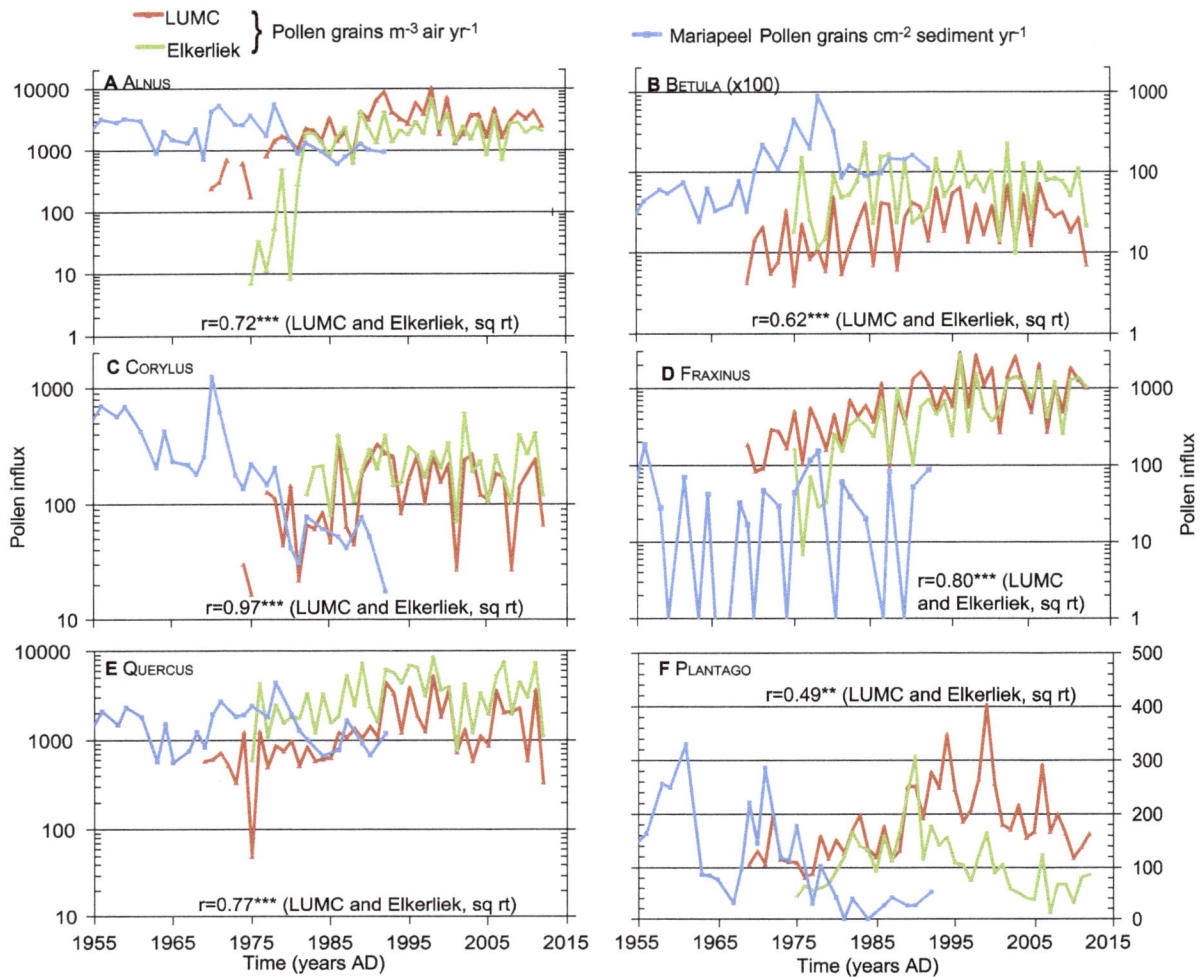

Figure 4. Annual pollen influx values 1969–2012 for LUMC, Elkerliek (n m^{-3} air yr^{-1}) and Mariapeel (n cm^{-2} sediment yr^{-1}). Values are shown for (a) ALNUS, (b) BETULA, (c) CORYLUS, (d) FRAXINUS, (e) QUERCUS, and (f) PLANTAGO. Note that a–e are on a logarithmic y-axis.

formation of male catkins begins in April and that the complete catkin development likely continues until August.

The here reported relation of BETULA influx with T_{mar-1}, T_{may-1} and T_{jul-1} coincides with the main period of catkin formation. Studies in Denmark show correlations to T_{may-1}, T_{jun-1}, T_{jul-1} [4,48], and in northern Finland to T_{jun-1} [6], suggesting an earlier start of catkin formation more to the south. The sensitivity of BETULA to precipitation ($P_{apr/jun-1}$) is identical to that in Denmark [4], and similar to central Europe [25], and might be related to its preference for relatively wet habitats, particularly in *B. pubescens*. The 5-year cyclicity found in BETULA only, possibly relates to precipitation as well (Fig. 5b). Here, we hypothesize that *Betula* invests in reproduction strategies (i.e. pollen production) during dry conditions, while in wetter conditions these trees invest in vegetative growth.

Including the temperature values of the October - March winter dormancy period in *Betula* [18] improves the predictive model (Table 7). It can be hypothesized that warmer temperatures in November may disturb this winter dormancy but as the sign of the correlation differs between November and December this needs independent confirmation. The negative relation with T_{mar} is remarkable as *Betula* flowering season is between April and June [34]. A possible explanation might be that although warmer

temperatures in early spring can advance the *Betula* pollen season [17], the increased chance of frost damage results in lower annual pollen influx.

Beside the strong 5-year cyclicity, a 2.4 year cycle in BETULA influx confirms earlier finds of bi-annual variability [18] and a three-year cycle [21]. As high pollen production likely results in a high energy-intensive fruit production this might inhibit the development of reproductive structures for the flowering season in the next year [30]. Ranta et al. [2] indeed conclude that "masting of birch species is regulated by weather factors together with the system of resource allocation among years". High inflorescence numbers might result in smaller and fewer leaves, lowering overall photosynthetic capacity of the tree including development of new flower buds. As a consequence, few inflorescences in the following year will relocate more energy for the development of leaves, and in turn stimulates the development of flower buds [18]. Pollen production is indeed correlated to catkin formation, and year-to year changes have previously been shown to be similar across large distances (up to 500 km) [49].

As temperature and precipitation influence photosynthetic capacity, climatic conditions are likely to exert control over this autocyclic pattern, which might explain variations between 2- and 3-year cycle lengths.

Table 7. Statistical parameter values of the regression model with backward selection method and the optimal regression model after data splitting.

Pollen type	Regression model after backward selection			Optimal regression model after data splitting analysis		
	R^2	Variable	β	R^2	Variable	β
ALNUS	.538***	$T_{March-1}$	.549***	.538***	$T_{March-1}$	.549***
		$T_{August-1}$	.303**		$T_{August-1}$	.303**
		$T_{January}$	−.265*		$T_{January}$	−.265*
		T_{March}	.230		T_{March}	.230
BETULA	.677***	$T_{January-1}$	−.258*	.565***	$T_{March-1}$	.286*
		$T_{March-1}$	.432***		T_{May-1}	.333**
		T_{May-1}	.411***		T_{July-1}	.486***
		T_{July-1}	.410**		$T_{August-1}$	.195
		$T_{August-1}$	.263*		$T_{November-1}$	−.248
		$T_{October-1}$	.209		$T_{December-1}$	.261*
		$T_{November-1}$	−.327*		T_{March}	−.358**
		$T_{December-1}$	.298*		P_{June-1}	−.357**
		$T_{February}$	.225			
		T_{March}	−.513***			
		P_{June-1}	−.363**			
CORYLUS	.616***	$T_{January-1}$	−.377**	.616***	$T_{January-1}$	−.377**
		$T_{March-1}$	.330**		$T_{March-1}$	.330**
		T_{May-1}	.276*		T_{May-1}	.276*
		T_{July-1}	.437***		T_{July-1}	.437***
		$T_{August-1}$	.219		$T_{August-1}$	.219
		$T_{October-1}$	.261*		$T_{October-1}$	.261*
		$T_{November-1}$	−.364**		$T_{November-1}$	−.364**
		P_{march}	−.224*		P_{march}	−.244*
FRAXINUS	.629***	$T_{January-1}$	−.298*	.386***	$T_{February-1}$	.325*
		$T_{February-1}$	.438**		$T_{April-1}$	.376**
		$T_{March-1}$	.392**		T_{July-1}	.298*
		$T_{April-1}$	.414***			
		T_{July-1}	.475***			
		$T_{November-1}$	−.232			
		$T_{January}$	−.340**			
QUERCUS	.703***	$T_{March-1}$	.296**	.523***	T_{July-1}	.493***
		T_{July-1}	.498***		T_{May}	.544***
		$T_{November-1}$	−.219*			
		$T_{January}$	−.242*			
		T_{April}	.242*			
		T_{May}	.514***			
PLANTAGO	.579***	$T_{March-1}$	.254*	.463***	$T_{March-1}$	.277*
		T_{May-1}	.211		T_{May-1}	.196
		$T_{November-1}$	−.212		T_{March}	.298*
		T_{March}	.230		T_{July}	.276*
		T_{May}	.264*			
		T_{July}	.289*			

Significance levels are indicated: * $P<0.05$; ** $P<0.01$; *** $P<0.001$.

CORYLUS. As in BETULA, CORYLUS pollen influx broadly depends on growing season temperature of the year before and shows an earlier start compared to Denmark [4]. *Corylus* flowers early in the year, from January until March [34], and unlike observations in Denmark [4], CORYLUS influx has no (negative) relation with respect to winter temperatures, which confirms their

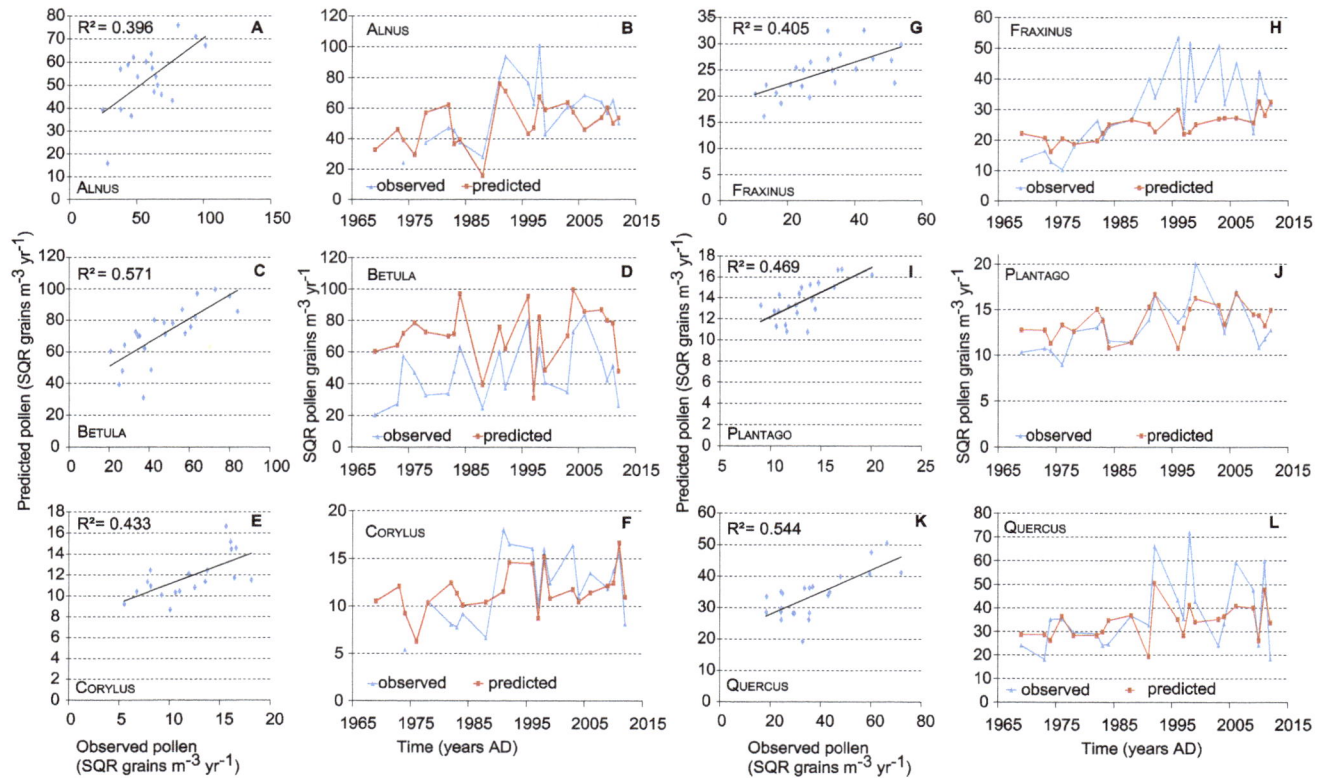

Figure 5. Observed versus climate predicted LUMC pollen influx based on the split data sets. Values are shown for (a,b) Alnus, (c,d) Betula, (e,f) Corylus, (g,h) Fraxinus, (i,j) Quercus, and (k,l) Plantago.

cold adaptation and relative insensitivity to temperature in that period in this region. The negative effect of P_{mar} on annual pollen influx is probably the consequence of rain washing the pollen out of the atmosphere during the pollination period. The 2.3 year cyclicity in the signal is of the same character as in Betula (Fig. 6c), and has also been observed in the United Kingdom [22]. Although no specific studies confirm this, we infer a similar mechanism as that described for *Betula* as both are Betulaceae.

Fraxinus. The factors T_{feb-1}, T_{apr-1} and T_{jul-1} best predict the annual Fraxinus influx suggesting a particularly long period of flower bud formation and sensitivity. Highly similar results from Denmark confirm this [4], while a study from Galicia (northwest Spain) involves also precipitation as an important factor for *Fraxinus* flowering [8]. The predictive model does not explain the majority of the variability (39%), suggesting edaphic factors and other internal biological processes play a significant role.

Quercus. In the temperate climate zone, *Quercus* forms flower buds during the summer of the previous year and enters a dormancy period during autumn and winter [46]. The sensitivity to T_{jul-1} at LUMC and T_{aug-1} in Denmark [4] suggest a short critical period for bud formation. In cork-oak (*Q. suber*), reports of steep temperature drop during initial stages of microsporogenesis resulted in catkin mortality and much lower pollen quantities [16].

Plantago. In perennial herbaceous plants, warm springs and summers in previous years are important for plant growth and flower development, as shown by the influence of T_{mar-1} and T_{may-1} in our analysis, whereas summer droughts may kill plants and thereby affect the pollen production [23]. Warmth during early spring and summer stimulates plant development and flower differentiation [23] and continued pollen generation typical of herbs [47], which can explain

the positive impact of T_{mar} and T_{Jul} on the annual influx of Plantago. Earlier studies of *P. lanceolata* from Poland did not reveal climate/pollen relationships, likely limited by the length of the record (10 years) [25]. The results of this study also show a positive correlation between the NAO index and the total annual Plantago influx. The multi-taper analysis of the annual Plantago influx reveals a particularly strong 3.2- year cycle (Fig. 6f), which, in combination with a positive correlation with the NAO index, points to a multi-annual climatic influence.

Palaeoecological interpretation

Our aim to compare trends in pollen trap data from LUMC and Elkerliek with those in the palaeoecological record are complicated by human management of the Mariapeel area and, despite the accurate depth-age model, slight age uncertainties that preclude a year-to-year correlation between the trapping and sub-fossil data. The two pollen trap sites agree surprisingly well in terms of phase relation and absolute numbers of pollen (Fig. 3). The Elkerliek site has generally higher amounts of Betula and Quercus, which is in agreement with the greater amount of tree cover, sandy soils and vicinity to natural reserves of that site. LUMC has a greater amount of Plantago, in agreement with a more urban setting. The offset in Alnus of the early part of the record in Elkerliek is related to incomplete counts in the first years. The consistency and significant correlations between both trap data (Fig. 4) shows that the stations produce data that are representative of a broad region. This suggests that the correlation with climate quantified at LUMC is a regionally relevant ecological factor that is responsible for the majority of the observed annual-scale pollen influx variability. This observation

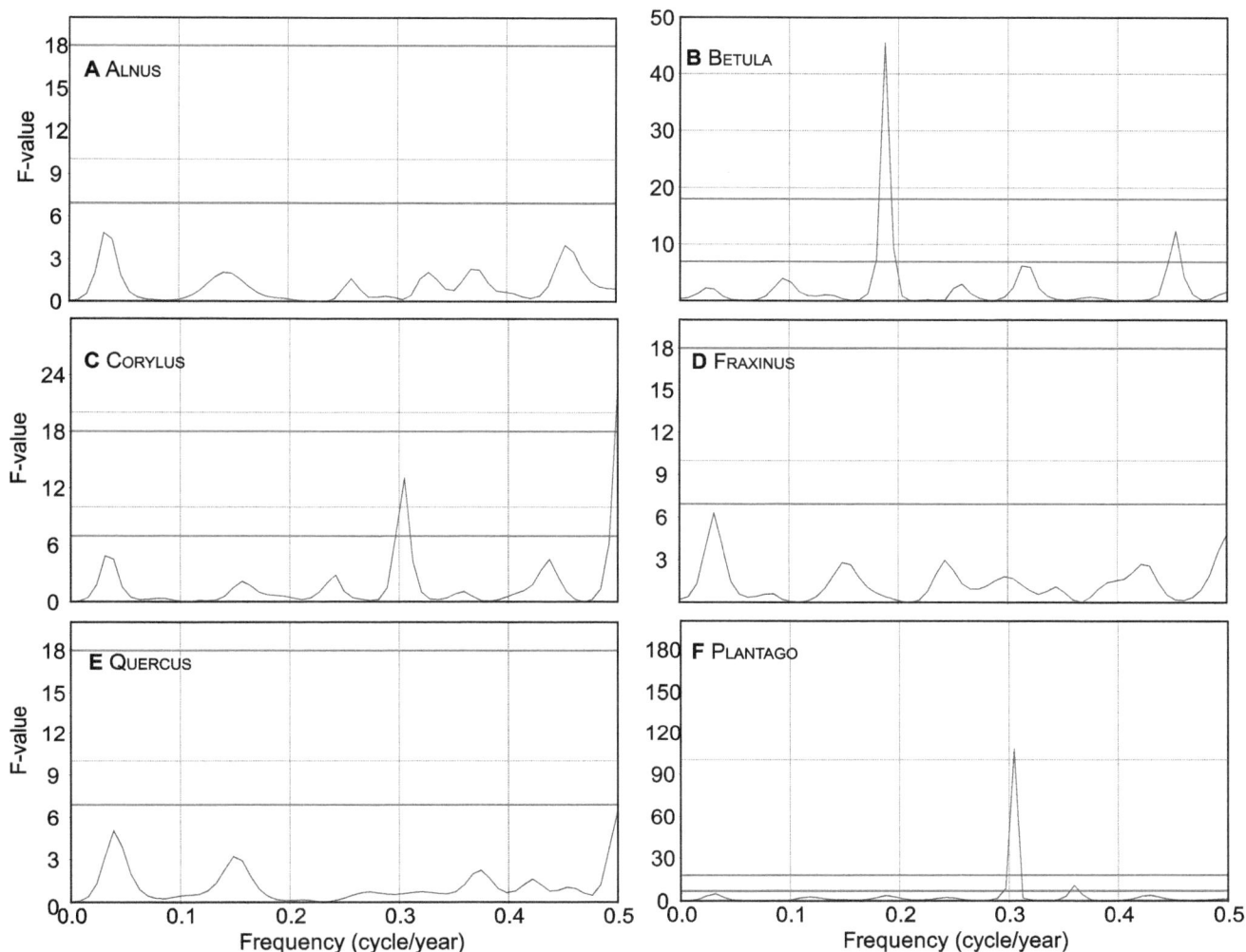

Figure 6. Spectral analysis output with 95% and 99% significance levels (upper and lower red line, respectively) of the 1969-2012 annual pollen influx time series in the LUMC trap. (a) Alnus, (b) Betula, (c) Corylus, (d) Fraxinus, (e) Quercus, and (f) Plantago.

also aides comparison of pollen assemblages with the trapping stations in forensic studies (e.g. for the purpose of dating a crime scene).

The Mariapeel site presently is a semi-open *Betula* forest with secondary *Sphagnum* growth in pits formerly mined for peat. The pollen accumulation rates from the sub-fossil peat deposit show a large decrease in all taxa around AD 1980 when the area was declared nature reserve and much of the local standing *Betula* tree vegetation has been cut and water levels were raised to stimulate peat regrowth. Visual inspection of the pre-1980 data (Fig. 3) from Mariapeel show variations of the same order (frequency and amplitude) as the variability present in the pollen trap records. Pollen from the more regionally occurring trees, such as *Quercus* and *Corylus*, show comparable trends and short-term changes as the trap data, but the local changes and short overlap period between unimpacted local vegetation and the traps preclude detailed correlation. The comparison does demonstrate that natural (peat) archives record and preserve high-order variations, with similar amplitude and frequencies as in the pollen trap data. Given the preserved high frequency variations and the identified climatic drivers (Tables 5 and 6) of annual pollen deposition, natural archives can provide significant insights in past climatic

variations at near annual scales (see e.g. [13]). Hence, our analysis provides a regional interpretation framework for climatic interpretations of high-frequency changes in pollen records in NW Europe.

Conclusion

This study showed that climate is an important factor in annual pollen production in the Netherlands and that annual pollen influx shows highly similar variability across a broad geographical area, which is driven by largely the same variables. Summer temperatures in the year before flowering, as well as temperature during the flowering season, are the primary climate variables that determine the annual pollen influx of wind-pollinated plants, while the effect of precipitation is minimal, except for Betula. Summer temperature influences the formation of reproductive structures, while temperature during the flowering season is thought to influence pollen release. The importance of long observational records is evident as shorter series often contain too much scatter to determine pollen-climate relations. Our results provide a first developmental step toward a region-specific predictive model for seasonal pollen forecasting for hay fever patients and forensic studies, and on short timescales (years to decades) predicts the likely impact of changing temperatures on annual pollen production due to global change.

The similar-scale high frequency variations observed in the peat record compared to the pollen traps suggest that, although influenced by local edaphic factors, natural archives can provide a proxy for quantitative reconstruction at high (e.g. annual) temporal resolution. However, the observed relation between climate and pollen production found for the pollen data from Leiden cannot be directly tested on the samples of from the Mariapeel, due to local landscape management changes and small differences in age.

Supporting Information

Figure S1   Total annual pollen accumulation rates of all recorded taxa in the pollen traps from LUMC, Leiden, and climate variables from De Bilt, The Netherlands from 1969 to 2012.

Figure S2   Autocorrelation of LUMC pollen accumulation rates for 0 to 20 years lags. Curved lines represent 95% significance level.

File S1   This file contains Table S1–Table S3. Table S1. Pollen count data for LUMC pollen traps. Table S2. Pollen count data for Elkerliek pollen traps. Table S3. The peat core from

Mariapeel Natural Reserve, The Netherlands. For locations see main text and Fig. 1.

File S2   Spearman rank correlation coefficients between annual pollen influx and climate variables of the pollen types not shown in Tables 5 and 6 of the main text (Tables S4–S8). Table S4. North Atlantic Oscillation Index (annual and winter). Table S5. Temperature in the flowering year. Table S6. Temperature in the year before flowering. Table S7. Precipitation in the flowering year. Table S8. Precipitation in the year before flowering.

Acknowledgments

The pollen data from Helmond were kindly made available by dr. A. O. de Graaf of the Elkerliek Hospital in Helmond. We thank two anonymous reviewers for the constructive comments that improved the manuscript, and thank Arnold van Vliet (Wageningen University) for his advice on the topic.

Author Contributions

Conceived and designed the experiments: THD FWC SD. Performed the experiments: KH THD FWC SD. Analyzed the data: THD KH SD. Contributed reagents/materials/analysis tools: LdW PdK. Wrote the paper: THD KH FWC SD LdW PdK.

References

1. Rogers CA (1993) Application of aeropalynological principles in palaeoecology. Rev Palaeobot Palynol 79: 133–140. http://dx.doi.org/10.1016/0034-6667(93)90043-T.
2. Ranta H, Oksanen A, Hokkanen T, Bondestam K, Heino S (2005) Masting by *Betula*-species; applying the resource budget model to north European data sets. Int J Biometeorol 49: 146–151. 10.1007/s00484-004-0228-0. Available: http://dx.doi.org/10.1007/s00484-004-0228-0.
3. Hicks S (2001) The use of annual arboreal pollen deposition values for delimiting tree-lines in the landscape and exploring models of pollen dispersal. Review of Palaeobotany and Palynology 117: 1–29.
4. Nielsen A, Möller P, Giesecke T, Stavngaard B, Fontana S, et al. (2010) The effect of climate conditions on inter-annual flowering variability monitored by pollen traps below the canopy in Draved forest, Denmark. Vegetation History and Archaeobotany 19: 309–323. 10.1007/s00334-010-0253-3. Available: http://dx.doi.org/10.1007/s00334-010-0253-3.
5. Barnekow L, Loader NJ, Hicks S, Froyd CA, Goslar T (2007) Strong correlation between summer temperature and pollen accumulation rates for *Pinus sylvestris, Picea abies* and *Betula* spp. in a high-resolution record from northern Sweden. Journal of Quaternary Science 22: 653–658. 10.1002/jqs.1096.
6. Autio J, Hicks S (2004) Annual variations in pollen deposition and meteorological conditions on the fell Aakenustunturi in northern Finland: Potential for using fossil pollen as a climate proxy. Grana 43: 31–47.
7. Ziska LH, Caulfield FA (2000) Rising CO_2 and pollen production of common ragweed (*Ambrosia artemisiifolia* L.), a known allergy-inducing species: Implications for public health. Functional Plant Biol 27: 893-898. Available: http://www.publish.csiro.au/paper/PP00032.
8. Jato V, Rodríguez-Rajo J, Dacosta N, Aira M (2004) Heat and chill requirements of *Fraxinus* flowering in Galicia (NW Spain). Grana 43: 217–223.
9. Shea KM, Truckner RT, Weber RW, Peden DB (2008) Climate change and allergic disease. J Allergy Clin Immunol 122: 443–453. Available: http://linkinghub.elsevier.com/retrieve/pii/S0091674908011810?showall = true
10. Walsh KAJ, Horrocks M (2008) Palynology: Its position in the field of forensic science. J Forensic Sci 53: 1053–1060. 10.1111/j.1556–4029.2008.00802.x.
11. Kuoppamaa M, Huusko A, Hicks S (2009) *Pinus* and *Betula* pollen accumulation rates from the northern boreal forest as a record of interannual variation in July temperature. Journal of Quaternary Science 24: 513–521. 10.1002/jqs.1276.
12. Donders TH, Punyasena SW, de Boer HJ, Wagner-Cremer F (2013) ENSO signature in botanical proxy time series extends terrestrial el niño record into the (sub)tropics. Geophys Res Lett 40: - 2013GL058038. 10.1002/2013GL058038.
13. Finsinger W, Schoning K, Hicks S, Lücke A, Goslar T, et al. (2013) Climate change during the past 1000 years: A high-temporal-resolution multiproxy record from a mire in northern Finland. Journal of Quaternary Science 28: 152–164. 10.1002/jqs.2598.

14. Brewer S, Guiot J, Barboni D (2007) Pollen methods and studies, use of pollen as climate proxies. In: Elias SA, editor.Encyclopedia of Quaternary Science.Ox-Oxford: Elsevier. pp. 2497–2508. Available: http://dx.doi.org/10.1016/B0-44-452747-8/00177-0.
15. Hurrell JW, van Loon H (1997) Decadal variations in climate associated with the North Atlantic Oscillation. Climatic change 36: 301–326.
16. García-Mozo H, Hidalgo PJ, Galán C, Maria TG, Domínguez E (2001) Catkin frost damage in mediterranean cork-oak (*Quercus suber* L.). Isr J Plant Sci 49: 42–47. 10.1560/JR25-EB31-9JTG-WTD6. Available: http://www.tandfonline.com/doi/abs/10.1560/JR25-EB31-9JTG-WTD6.
17. Van Vliet AJH, Overeem A, De Groot RS, Jacobs AFG, Spieksma FTM (2002) The influence of temperature and climate change on the timing of pollen release in the Netherlands. Int J Climatol 22: 1757–1767. 10.1002/joc.820.
18. Dahl-Strandhede ÅS (1996) Predicting the intensity of the birch pollen season. Aerobiologia 12: 97–106. 10.1007/BF02248133. Available: http://dx.doi.org/10.1007/BF02248133
19. Beggs PJ (2004) Impacts of climate change on aeroallergens: Past and future. Clinical & Experimental Allergy 34: 1507–1513. 10.1111/j.1365-2222.2004.02061.x.
20. Spieksma FT, Emberlin JC, Hjelmroos M, Jäger S, Leuschner RM (1995) Atmospheric birch (*Betula*) pollen in europe: Trends and fluctuations in annual quantities and the starting dates of the seasons. Grana 34: 51–57. 10.1080/00173139509429033. Available: http://dx.doi.org/10.1080/00173139509429033.
21. Detandt M, Nolard N (2000) The fluctuations of the allergenic pollen content of the air in Brussels (1982 to 1997). Aerobiologia 16: 55–61. 10.1023/A:1007619724282. Available: http://dx.doi.org/10.1023/A%3A1007619724282.
22. Emberlin J, Smith M, Close R, Adams-Groom B (2007) Changes in the pollen seasons of the early flowering trees *Alnus* spp. and *Corylus* spp. in Worcester, United Kingdom, 1996–2005. Int J Biometeorol 51: 181–191. 10.1007/s00484-006-0059-2. Available: http://dx.doi.org/10.1007/s00484-006-0059-2.
23. Andersen ST (1980) Influence of climatic variation on pollen season severity in wind-pollinated trees and herbs. Grana 19: 47–52. 10.1080/00173138009424986. Available: http://www.tandfonline.com/doi/abs/10.1080/00173138009424986.
24. Green D, Singh G, Polach H, Moss D, Banks J, et al. (1988) A fine-resolution palaeoecology and palaeoclimatology from south-eastern Australia. Journal of Ecology 76: 790–806.
25. Knaap WO, Leeuwen JN, Svitavská-Svobodová H, Pidek I, Kvavadze E, et al. (2010) Annual pollen traps reveal the complexity of climatic control on pollen productivity in Europe and the Caucasus. Vegetation History and Archaeobotany 19: 285–307. 10.1007/s00334-010-0250-6. Available: http://dx.doi.org/10.1007/s00334-010-0250-6.

26. Sjögren P, van Leeuwen JFN, van der Knaap WO, van der Borg K (2006) The effect of climate variability on pollen productivity, AD 1975–2000, recorded in a *Sphagnum* peat hummock. The Holocene 16: 277–286. 10.1191/0959683606hl924rp.

27. Spieksma FM, Nikkels AH (1998) Airborne grass pollen in Leiden, the Netherlands: Annual variations and trends in quantities and season starts over 26 years. Aerobiologia 14: 347–358. 10.1007/BF02694304. Available: http://dx.doi.org/10.1007/BF02694304.

28. Joosten H, de Klerk P (2002) What's in a name?: Some thoughts on pollen classification, identification, and nomenclature in quaternary palynology. Rev Palaeobot Palynol 122: 29–45. Available: http://dx.doi.org/10.1016/S0034-6667(02)00090-8.

29. Wagner F, Below R, Klerk PD, Dilcher DL, Joosten H, et al. (1996) A natural experiment on plant acclimation: Lifetime stomatal frequency response of an individual tree to annual atmospheric CO_2 increase. Proceedings of the National Academy of Sciences 93: 11705–11708.

30. Spieksma FT, Corden JM, Detandt M, Millington WM, Nikkels H, et al. (2003) Quantitative trends in annual totals of five common airborne pollen types (*Betula, Quercus, Poaceae, Urtica*, and *Artemisia*), at five pollen-monitoring stations in western Europe. Aerobiologia 19: 171–184. 10.1023/B:AERO.0000006528.37447.15. Available: http://dx.doi.org/10.1023/B%3AAERO.0000006528.37447.15.

31. Hirst JM (1952) An automatic volumetric spore trap. Annals Applied Biology 39: 257–265.

32. Anonymous (2012) North Atlantic Oscillation index data provided by the Climate Analysis Section, NCAR, Boulder, USA. Available: http://climatedataguide.ucar.edu/guidance/hurrell-north-atlantic-oscillation-nao-index-station-based.

33. Ter Braak CJF, Smilauer P (2002) CANOCO reference manual and CanoDraw for windows user's guide: Software for canonical community ordination (version 4.5). Ithaca, New York: Microcomputer Power.

34. van der Meijden R (2005) Heukels' flora van Nederland. Groningen, the Netherlands.: Wolters-Noordhoff.

35. Hammer O, Harper D, Ryan P (2001) PAST: Paleontological statistics software package for education and data analysis. Paleontologia Electronica 1: 1–9.

36. Ziello C, Sparks TH, Estrella N, Belmonte J, Bergmann KC, et al. (2012) Changes to airborne pollen counts across Europe. PLoS ONE 7: e34076. Available: http://dx.doi.org/10.1371%2Fjournal.pone.0034076.

37. Lau T, Stephenson AG (1993) Effects of soil nitrogen on pollen production, pollen grain size, and pollen performance in *Cucurbita pepo* (Cucurbitaceae). Am J Bot 80: 763–768. Available: http://www.jstor.org/stable/2445596.

38. Bobbink R, Hicks K, Galloway J, Spranger T, Alkemade R, et al. (2010) Global assessment of nitrogen deposition effects on terrestrial plant diversity: A synthesis. Ecol Appl 20: 30–59. 10.1890/08-1140.1. Available: http://dx.doi.org/10.1890/08-1140.1

39. Spieksma FT, Frenguelli G, Nikkels AH, Mincigrucci G, Smithuis LOMJ, et al. (1989) Comparative study of airborne pollen concentrations in central Italy and the Netherlands (1982–1985). Grana 28: 25–36. 10.1080/00173138909431009. Available: http://dx.doi.org/10.1080/00173138909431009.

40. McVean DN (1955) Ecology of *Alnus glutinosa* (L.) gaertn.: I. fruit formation. J Ecol 43: 46–60. Available: http://www.jstor.org/stable/2257118.

41. Rodríguez-Rajo F, Dopazo A, Jato V (2004) Environmental factors affecting the start of the pollen season and concentrations of airborne *Alnus* pollen in two localities of Galicia (NW Spain). Annals of Agricultural and Environmental Medicine 11: 35–44.

42. Caesar JC, Macdonald AD (1984) Shoot development in *Betula papyrifera*. V. effect of male inflorescence formation and flowering on long shoot development. Can J Bot 62: 1708–1713. 10.1139/b84-231. http://dx.doi.org/10.1139/b84-231.

43. Macdonald AD, Mothersill DH, Caesar JC (1984) Shoot development in *Betula papyrifera*. III. long-shoot organogenesis. Can J Bot 62: 437–445. 10.1139/b84-066. Available: http://dx.doi.org/10.1139/b84-066.

44. Macdonald AD, Mothersill DH (1983) Shoot development in *Betula papyrifera*. I. short-shoot organogenesis. Can J Bot 61: 3049-3065. 10.1139/b83-342. Available: http://www.nrcresearchpress.com/doi/abs/10.1139/b83-342.

45. Anonymous (2014) Netherlands phenological observation network (transl.). Environmental Systems Analysis Group, Wageningen University, The Netherlands. Available: www.natuurkalender.nl

46. Kasprzyk I (2009) Forecasting the start of *Quercus* pollen season using several methods - the evaluation of their efficiency. Int J Biometeorol 53: 345–353. 10.1007/s00484-009-0221-8. Available: http://dx.doi.org/10.1007/s00484-009-0221-8.

47. Tyler G (2001) Relationships between climate and flowering of eight herbs in a Swedish deciduous forest. Annals of Botany 87: 623–630. 10.1006/anbo.2001.1383.

48. Rasmussen A (2002). The effect of climate change on the birch pollen season in Denmark. Aerobiologia 18: 253–265.

49. Ranta H, Hokkanen T, Linkosalo T, Laukkanen L, Bondestam K, et al. (2008). Male flowering of birch: Spatial synchronization, year-to-year variation and relation of catkin numbers and airborne pollen counts. Forest Ecology and Management 255: 643–650.

11

Patterns and Variability of Projected Bioclimatic Habitat for *Pinus albicaulis* in the Greater Yellowstone Area

Tony Chang*, Andrew J. Hansen, Nathan Piekielek

Department of Ecology, Montana State University, Bozeman, Montana, United States of America

Abstract

Projected climate change at a regional level is expected to shift vegetation habitat distributions over the next century. For the sub-alpine species whitebark pine (*Pinus albicaulis*), warming temperatures may indirectly result in loss of suitable bioclimatic habitat, reducing its distribution within its historic range. This research focuses on understanding the patterns of spatiotemporal variability for future projected *P.albicaulis* suitable habitat in the Greater Yellowstone Area (GYA) through a bioclimatic envelope approach. Since intermodel variability from General Circulation Models (GCMs) lead to differing predictions regarding the magnitude and direction of modeled suitable habitat area, nine bias-corrected statistically down-scaled GCMs were utilized to understand the uncertainty associated with modeled projections. *P.albicaulis* was modeled using a Random Forests algorithm for the 1980–2010 climate period and showed strong presence/absence separations by summer maximum temperatures and springtime snowpack. Patterns of projected habitat change by the end of the century suggested a constant decrease in suitable climate area from the 2010 baseline for both Representative Concentration Pathways (RCPs) 8.5 and 4.5 climate forcing scenarios. Percent suitable climate area estimates ranged from 2–29% and 0.04– 10% by 2099 for RCP 8.5 and 4.5 respectively. Habitat projections between GCMs displayed a decrease of variability over the 2010–2099 time period related to consistent warming above the 1910–2010 temperature normal after 2070 for all GCMs. A decreasing pattern of projected *P.albicaulis* suitable habitat area change was consistent across GCMs, despite strong differences in magnitude. Future ecological research in species distribution modeling should consider a full suite of GCM projections in the analysis to reduce extreme range contractions/expansions predictions. The results suggest that restoration strageties such as planting of seedlings and controlling competing vegetation may be necessary to maintain *P.albicaulis* in the GYA under the more extreme future climate scenarios.

Editor: Ben Bond-Lamberty, DOE Pacific Northwest National Laboratory, United States of America

Funding: This work was supported by the National Aeronautics and Space Administration Applied Sciences Program (Grant 10-BIOCLIM10-0034); Funder URL: http://www.nasa.gov (AJH TC). It also received support from the National Science Foundation Experimental Program to Stimulate Competitive Research (EPSCoR) Track-I EPS-1101342 (INSTEP 3); Funder URL: http://www.nsf.gov/div/index.jsp?div = EPSC (NBP TC); and the North Central Climate Science Center (G13AC00392-G-8829-1); Funder URL: http://www.doi.gov/csc/northcentral/index.cfm (AJH NBP). The funders had no role in study design, data collection and analysis, decision to publish, or preparation of the manuscript.

Competing Interests: The authors have declared that no competing interests exist.

* Email: tony.chang@msu.montana.edu

Introduction

Over the next century, it is expected that most of North America will experience climate changes related to increased concentrations of anthropogenic greenhouse gas emissions and natural variability [1]. At regional scales these changes are highly variable and can result in areas of increased mesic, xeric, or even hydric habitat conditions relative to present day. These shifting climates in turn also transform the suitable habitat for individual species that may result in changes in species composition and dominant vegetation types.

Whitebark pine (*Pinus albicaulis*) is a native conifer of the Western U.S. that is considered a keystone species in the sub-alpine environment. It provides a food source for animals such as the grizzly bear (*Ursus arctos*), red squirrel (*Tamiasciurus hudsonicus*), and Clark's nutcracker (*Nucifraga columbiana*) [2]. It also serves the ecosystem functions of stabilizing soil, moderating snow melt and runoff, and facilitating establishment for other species [2,3]. Whitebark pine has experienced a notable decline in

the past two decades within the U.S. Northern Rockies due to high rates of infestation from the mountain pine beetle (*Dendroctonus ponderosae*) and infections from white pine blister rust (*Cronartium ribicola*), resulting in an 80% mortality rate within the adult population [4–7]. Given the potential loss of important ecosystem functions that whitebark pine contribute to the landscape under this mortality event, there is an emphasis to understand the climate characteristics of its habitat to identify the restoration strategies and locations that may aid the persistence of the species under future climates.

One method of understanding species response to climate change is through bioclimate niche modeling, which has become a common practice for assessing potential vegetation shifts under new environmental conditions [8–13]. Ecological niche theory proposes there exists some range of bioclimatic conditions within which a species can persist [14]. In bioclimatic niche modeling, the realized niche is modeled by empirical relationships between the presence or absence of a species and the associated abiotic, and

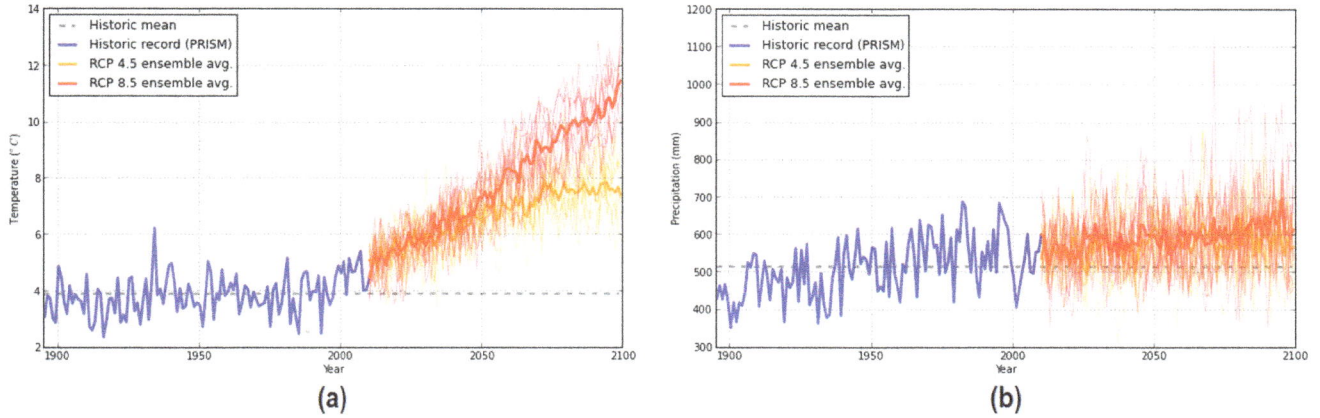

Figure 1. Historic and projected climates variables for the GYA from 1895–2099 under RCP 4.5 and 8.5 scenarios. Light shaded orange and red lines represent individual GCMs for RCP 4.5 and 8.5 respectively. Bold lines represent GCM ensemble average. (a) Mean annual temperature (b) Mean annual precipitation.

sometimes biotic, variables that describe the niche space. Bioclimatic models assume that species are in equilibrium with their environment and that the current abiotic relationships reflect a species environmental preferences which may be retained into the future [15,16]. At macro scales, bioclimatic approaches have demonstrated success at predicting current distributions of species [17,18]. Most bioclimatic models do not explicitly consider the many additional ecological factors that ultimately influence a

species distribution such as dispersal, disturbance, or biotic interaction. Thus the approach does not predict where a species will actually occur in the future, but rather it predicts locations where climatic conditions will be suitable for the species.

Bioclimatic niche methodology has demonstrated utility in modeling historic ranges of species for conservation and management applications. By modeling the present day suitable habitat and then projecting those habitats into the future, bioclimatic

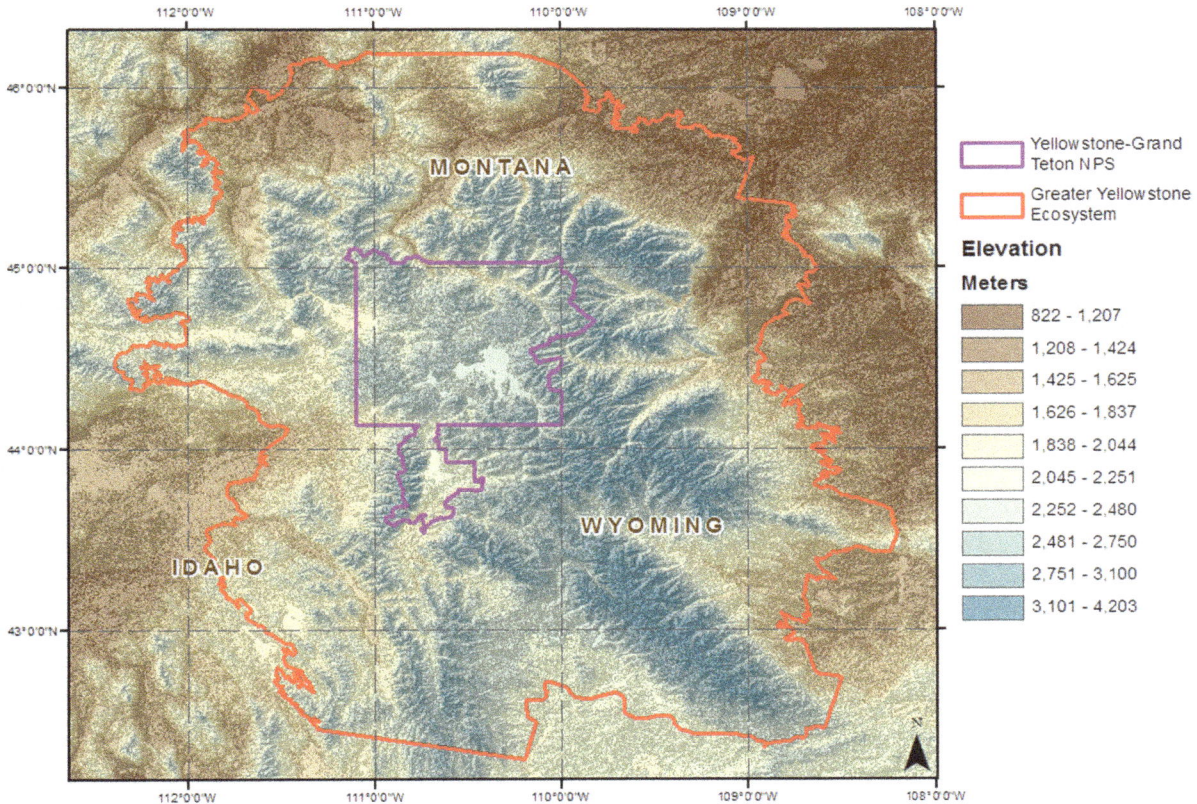

Figure 2. The Greater Yellowstone Area, representing an area of 150,700 km² with an elevational gradient from 522–4,206 m.

Table 1. General Circulation Models for analysis.

Name	Institute	Country
CESM1-CAM5	National Center for Atmospheric Research	US
CCSM4	National Center for Atmospheric Research	US
CESM1-BGC	National Center for Atmospheric Research	US
CNRM-CM5	Centre National de Recherche Meteorologiques	FR
HadGEM2-AO	Met Office Hadley Centre Climate Programme	UK
HadGEM2-ES	Met Office Hadley Centre Climate Programme	UK
HadGEM2-CC	Met Office Hadley Centre Climate Programme	UK
CMCC-CM	Centro Euro-Mediterraneo per Cambiamenti Climatici	ITA
CanESM2	Canadian Centre for Climate Modelling and Analysis	CAN

Selection of AR5 GCMs that represent historic climate in the U.S. Pacific Northwest region for future bioclimate habitat modeling.

niche models can serve as the first step filter for conservation action plans, such as mapping suitable species reintroduction sites or habitat reserve selection [19–21]. For *P.albicaulis*, McLane and Aitken [22] utilized bioclimate niche models to successfully implement experimental assisted migration on persisting climate habitat in British Columbia. Additionally, models of *Pinus flexis*, a closely related species of five needle pine, have been used to evaluate management options in Rocky Mountain National Park [23]. Given these examples, an effort to model and projected suitable climate habitat for *P.albicaulis* within a regional domain can provide valuable insight to land resource managers.

In this study, we present a bioclimatic habitat model for *P.albicaulis* within the Greater Yellowstone Area (GYA). Although *P. albicaulis* has a range-wide distribution that is split into two broad sections, one along Western North America: the British Columbia Coast Range, the Cascade Range, and the Sierra Nevada; and the other section in the Intermountain West that covers the Rocky Mountains from Wyoming to Alberta [2,24]; the GYA was selected as the primary geographic modeling domain for three reasons: 1) evidence that the *P. albicaulis* sub-population in the GYA is genetically distinct from other regional populations with different climate tolerances [25]; 2) the high regional investment in *P. albicaulis* conservation in the area [6]; 3) the high density of climate stations within the region. Climate within the GYA is highly heterogenous due to complex topography, and sharp elevational gradients. Current knowledge of the region expects climate to shift towards increased mean annual temperatures and earlier spring snowmelt [26,27]. This shift is expected to have an impact on the total suitable habitat area for *P. albicaulis*. Modeling at a regional scale can provide a finer resolution spatially explicit description of the bioclimatic envelope of *P. albicaulis* in the GYA.

Here we also present an opportunity to investigate the effect of future climate variability on projected species distributions. In 2013, the World Climate Research Programme Coupled Model released the new generation General Circulation Model (GCM) projections through the Coupled Model Intercomparison Project Phase 5 (CMIP5) [28]. These new GCM projections also include four possible climate futures are modeled with each GCM under the Representative Concentration Pathways (RCP) of greenhouse gas/aerosol. These RCP scenarios designate four different levels of radiative forcing (2.6, 4.5, 6.0 and 8.5 W/m^2) that may occur by the year 2099 [29]. In practice, research of future species suitable climate generally use a small suite of GCM/RCP combinations to project future climate [8,11,30]. However, internal variability in

these GCMs that arise from modeled coupled interactions among the atmosphere, oceans, land, and cryosphere can result in atmospheric circulation fluctuations that are characteristic of a stochastic process [31]. Such intrinsic atmospheric circulation variations from model structure induce regional changes in air temperature and precipitation on the multi-decadal time scale [31]. For the GYA specifically, this GCM variability has been observed with mean annual temperatures projected to increase by $2-9°C$ and mean annual precipitation to change by -50 to $+225$ mm (Fig. 1). This suggests that magnitude and direction of projected species distributions at a regional scale can vary depending on the GCM selected and the modeled species response to more xeric or mesic future climate conditions [32].

To summarize, this study presents a bioclimatic niche model for *P. albicaulis* based on historic climate observations and field sampling of *P. albicaulis* presence and absence. Using this modeled bioclimate envelope, projections of future total climate suitable habitat area under nine GCMs and two RCP scenarios will be measured. Since different GCMs may project a diverging spectrum of climates, it is expected that measures of total suitable habitat will reduce with varying degrees of area loss. It is also expect that number and size of continous patches of *P. albicaulis* habitat will reduce due to the limited available number of sub-alpine areas distributed within the landscape. This research provides an analysis of the variability of biotic response under a large suite of GCMs to provide managers/researchers with a measure of the uncertainty associated with future species distribution models. Furthermore, this analysis explicitly describes the spatial patterns of bioclimatic niches for *P. albicaulis* to gain a better understanding of topographic characteristics, such as elevation, on suitable habitat. Changes in these spatial patterns are examined through quantifying landscape patch dynamic that may result from GCM projections to understand the species trends for persistence on the landscape.

Methods

Study area

The GYA, which includes Yellowstone National Park, Grand Teton National Park, and a number of state and federally managed forests, is a mid- to high-latitude region in the Northern Rocky Mountains of western North America. Conifers are dominant in the range, with forest types composed of *Pinus contorta*, *Abies lasiocarpa*, *Pseudotsuga menziesii*, *Pinus albicaulis*, *Juniperus scopulorum*, *Pinus flexis* and *Picea engelmannii*, although the deciduous hardwood *Populus tremuloides*, is also wide spread. Plateaus and lowlands are dominated by species of *Artemisia tridentata* and open grasslands of mixed composition. The GYA study area encompasses 150,700 km^2 with an elevational gradient from 522–4,206 m that represents 14 surrounding mountain ranges (Fig. 2).

Data

Biological data. Field observations of *P. albicaulis* adult presences and absences were compiled from three data sources. First, 2,545 observations from the Forest Inventory and Analysis (FIA) program were assembled. FIA plots are located on a regular

gridded sampling design with one plot at approximately every 2,500 forested hectares, with swapped and fuzzed exact plot locations within 1.6 km to protect privacy [33]. Gibson et al. [34] found that model accuracy to not be dramatically affected by data fuzzing, but to provide the most spatial accuracy, this study culled FIA field points where measured elevation were > 300 m different from a 30 m USGS DEM [35]. To capitalize on additional field observations of *P. albicaulis* within the study area, and because false absences are one of the most problematic data issues in constructing bioclimatic niche models [36]; supplementary points were drawn from the Whitebark/Limber Pine Information System (WLIS) [37], and long-term monitoring plots established by the

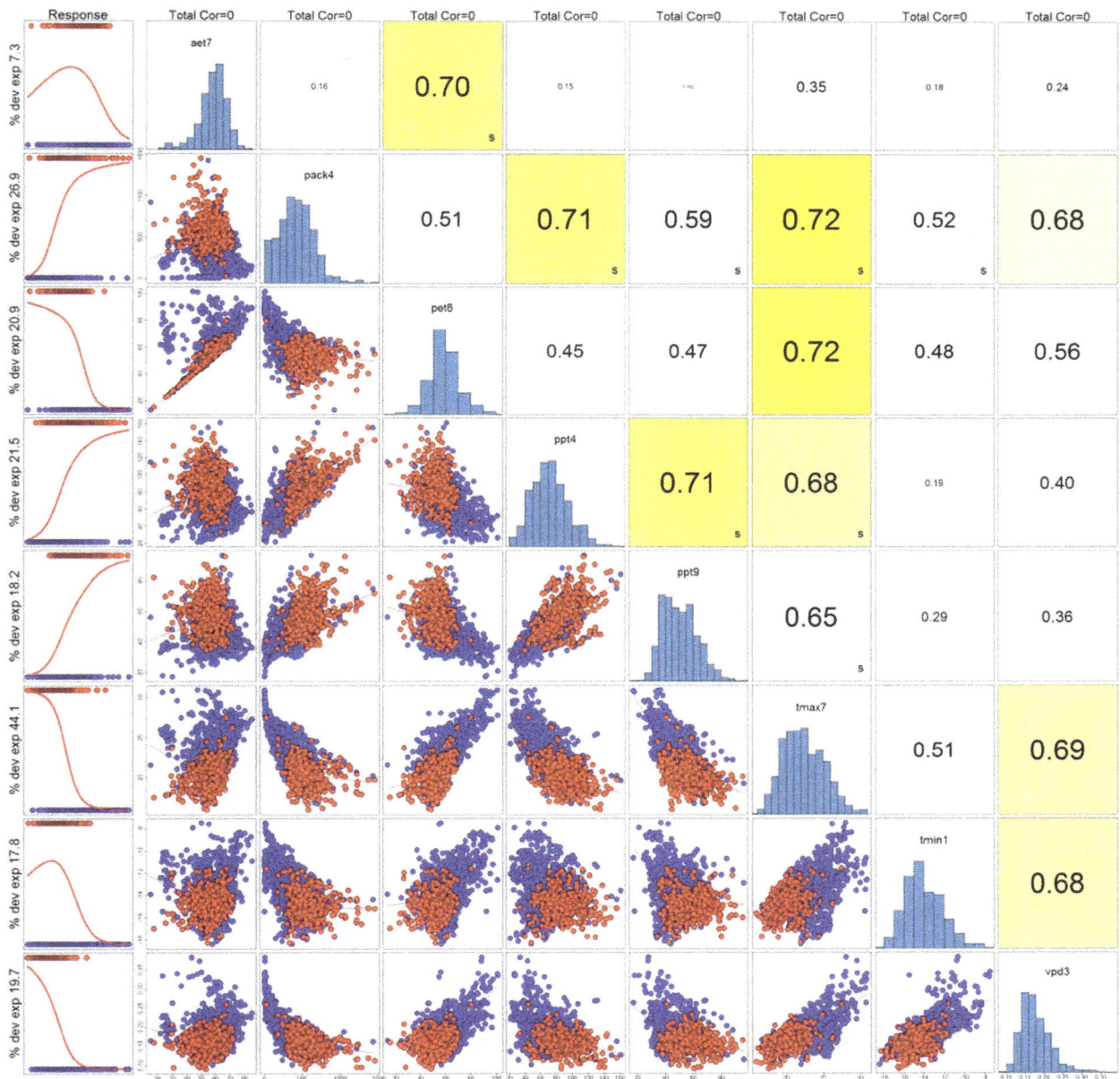

Figure 3. Selected predictor variables based on Principal Component Analysis and a maximum correlation filter of ≤ 0.75. Scatter plots represent one-to-one covariate plots where red points represent *P. albicaulis* presence, and blue points represent absence from field data. Far-left columns display logistic-regression of covariates from Generalized Additive Modeling using the Software for Assisted Habitat Modeling (SAHM [59]).

Table 2. Bioclimatic predictor variable list.

Code	Predictor Variable
tmin1	Minimum Temperature January
vpd3	Vapor Pressure Deficit March
ppt4	Precipitation April
pack4	Snow Water Equivalent April
tmax7	Maximum Temperature July
aet7	Actual Evapotranspiration July
pet8	Potential Evapotranspiration August
ppt9	Precipitation September

Final predictor variable set for Random Forest modeling. All variables were calculated as a 30-year climate mean from 1950–1980.

National Park Service Greater Yellowstone Inventory and Monitoring Network (GYRN) [38]. The presences in these two additional datasets were collocated within predictor pixels of FIA absence to correct for false absences. In doing so, only one *P. albicaulis* presence or absence record was associated per predictor pixel, thereby avoiding issues associated with sampling bias that are common when building bioclimate niche models with data from targeted surveys [39]. This compilation of data represents an effort for "completeness" as described by Kadmon et al. [40] and Franklin [36], to capture all climate conditions where a species does exist. New data sources added 119 *P. albicaulis* presences that would have been missed by using FIA data alone, for a total of 938 presences and 1,633 absences.

"Adult" class *P. albicaulis* were selected for modeling based on a recorded diameter at breast height (DBH) > 20 cm. *P. albicaulis* within the Central Montana are reported to reach 100 years of age at approximately 8–12 m in height with DBHs between 15–20 cm

[41]. Given previous silvicultural studies, it was assumed that 20 cm DBH *P. albicaulis* represent adult class individuals for the GYA, with potential to reproduce [24]. Furthermore, this study focused on adult size class due to difficulties distinguishing younger age class *P. albicilus* from *P. flexis*.

Historic climate data. Climate inputs for modeling were acquired from the 30-arc-second (∼800 m) monthly Parameter-elevation Regressions on Independent Slopes Model (PRISM), a derived product that interpolates local station measurements across a continuous grid [42]. PRISM data includes monthly average minimum temperatures (T_{min}), maximum temperature (T_{max}), mean temperature (T_{mean}), and mean precipitation (**Ppt**). All monthly data were averaged for the temporal extent of 1950–1980 for bioclimatic niche model fitting. The 1950–1980 temporal extent was selected for modeling since: 1) a sufficient density of weather stations were operating by 1950 to provide a reasonable network; 2) evidence of anthropogenic warming that begins in the

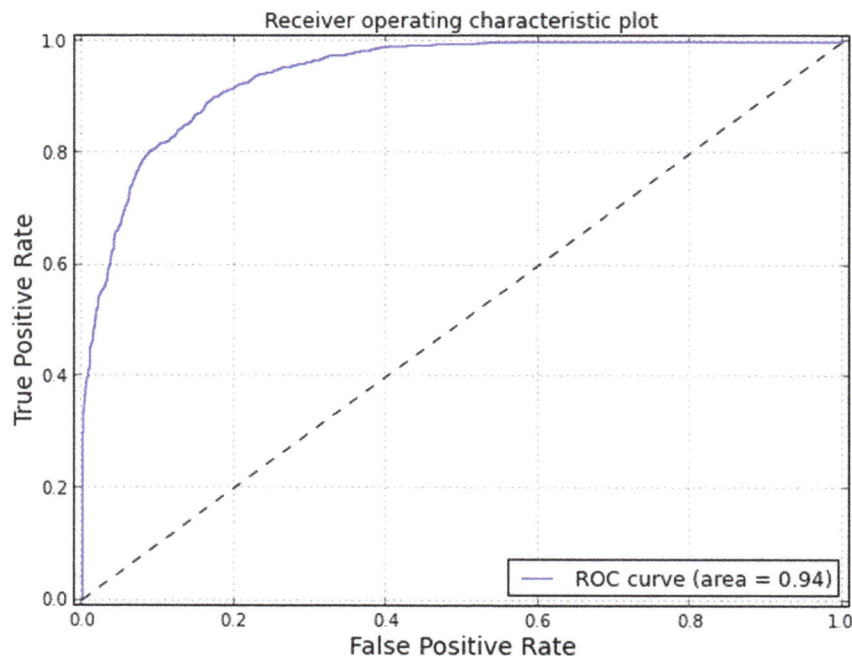

Figure 4. Area under curve for the receiver operating characteristic plot suggests adequate performance from the Random Forest modeling.

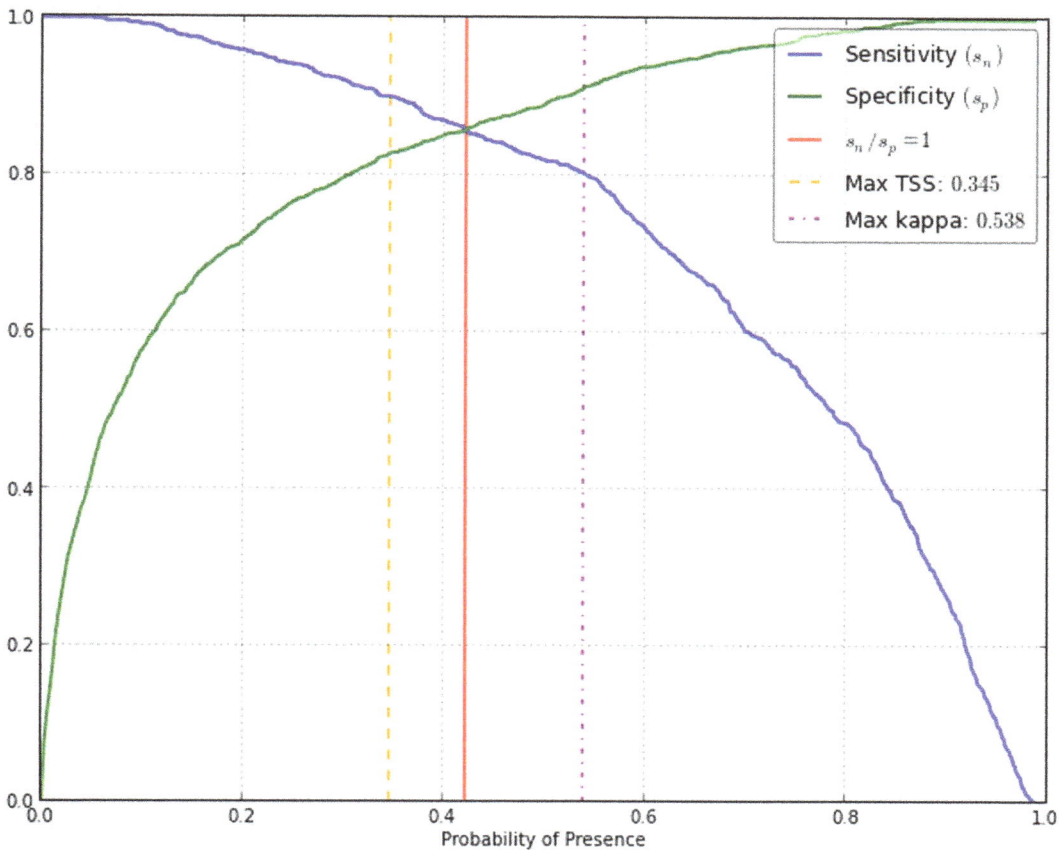

Figure 5. Threshold for probability of presence of 0.421 determined at the intersection of true positive rate (TPR) and true negative rate (TNR). Equivalent TPR and TNR, displayed a compromise between the maximum true skill statistic (TSS : 0.345) and maximum Kappa statistic (κ : 0.538).

late 1980s; 3) trees old enough to bear seeds today likely established under a similar climates to the 1950–1980 period.

Water balance. A Thornthwaite-based dynamic water balance model was used to estimate a number of variables that include actual evapotranspiration (AET) and potential evapotranspiration (PET) [43–45]. The model required only monthly mean temperatures, dew point temperatures, and precipitation (see Text S1). Water was stored as soil moisture or in surface snowpack, with the excess taking the form of evaporated vapor or loss through seepage/runoff. In addition to the climatic variables, latitude and physical characteristics of the soil were required to define water holding capacity. Soil attributes assigned by the Soil Survey Geographic (STATSGO) datasets were allocated from the Natural Resource Conservation Service at a 30-arc-second

resolution to determine soil water holding capacity and estimates for soil depth [46]. All water balance variables, which include PET, AET, soil moisture, vapor pressure deficit (vpd), and snow water equivalent (pack), were averaged by month over 1950–1980 to match with historic climate data for bioclimate model fitting.

GCM data. The general circulation model (GCM) experiments conducted under CMIP5 for the Intergovernmental Panel on Climate Change Fifth Assessment Report provided future projected climate data sets for assessing the effects of global climate change. Using a Bias-Correction Spatial Disaggregation (BCSD) approach, an archive of statistically down-scaled CMIP5 climate projections for the conterminous United States at 30-arc-second spatial resolution was assembled by the NASA Center for Climate Simulation NEX-DCP30 [47]. For this analysis, a subset of the

Table 3. Confusion matrix from out-of-bag analysis.

		Validation data set	
		Presence	**Absence**
Model	Presence	763 (81.9%)	169 (13.1%)
	Absence	176 (10.9%)	1437 (89.1%)

Random Forest tree estimators displays higher OOB specificity than sensitivity. Area Under Curve (AUC) value of 0.94 suggests model has high predictive capacity for projecting future suitable bioclimate habitat.

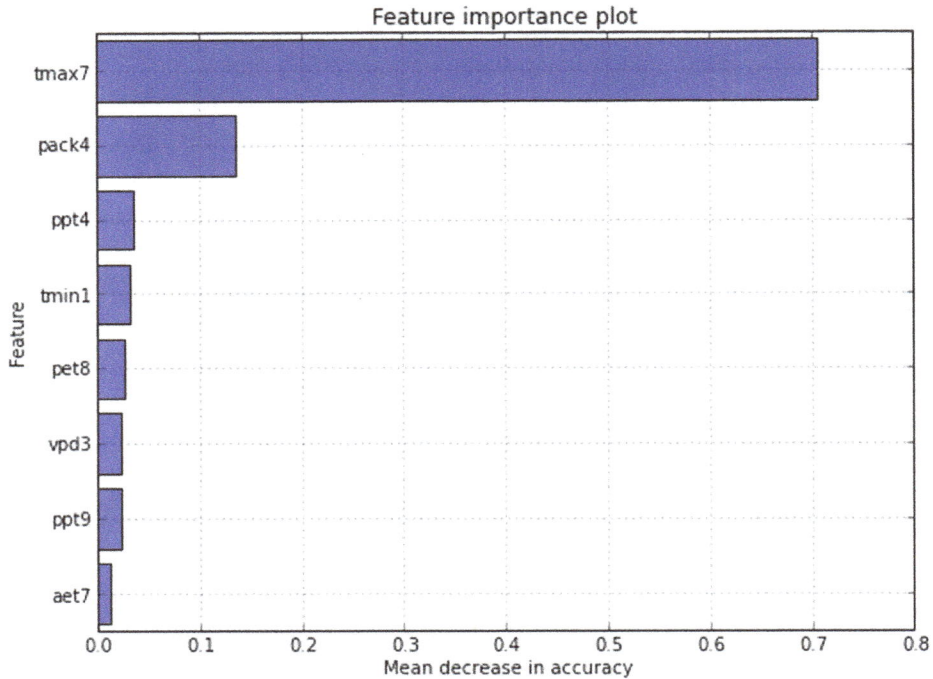

Figure 6. Random Forest out-of-bag variable importance plots find removal of maximum temperatures for July and April snow water equivalent to create the greatest reducing in model accuracy.

total GCM models available from NASA were selected that best represent the Northwestern US. Rupp et al. [48] recently presented an analysis of GCM performance versus the observed historic climate in the U.S. Pacific Northwest under 18 specified climate metrics. In their analysis, Rupp et al. ranked GCMs for accuracy using an empirical orthogonal function (EOF) analysis of the total normalized error compared to reference data. This analysis selected models with a normalized error score <0.5 as a threshold to cull the full suite of GCMs to the top nine models. These GCMs were used to project modeled *P. albicaulis*

distributions into the future (Table 1). Two RCP scenarios were selected to understand effects of differing carbon futures under climate change from 2010 to 2099. RCP 4.5 was the first, representing increased radiative forcing until stabilization of greenhouse emissions between 2040–2050 and total radiative forcing of 4.5 W/m^2 by 2099. RCP 8.5 was the second, representing the "business as usual" scenario, with uncontrolled radiative forcing increasing with stabilization of 8.5 W/m^2 by 2099 [49,50].

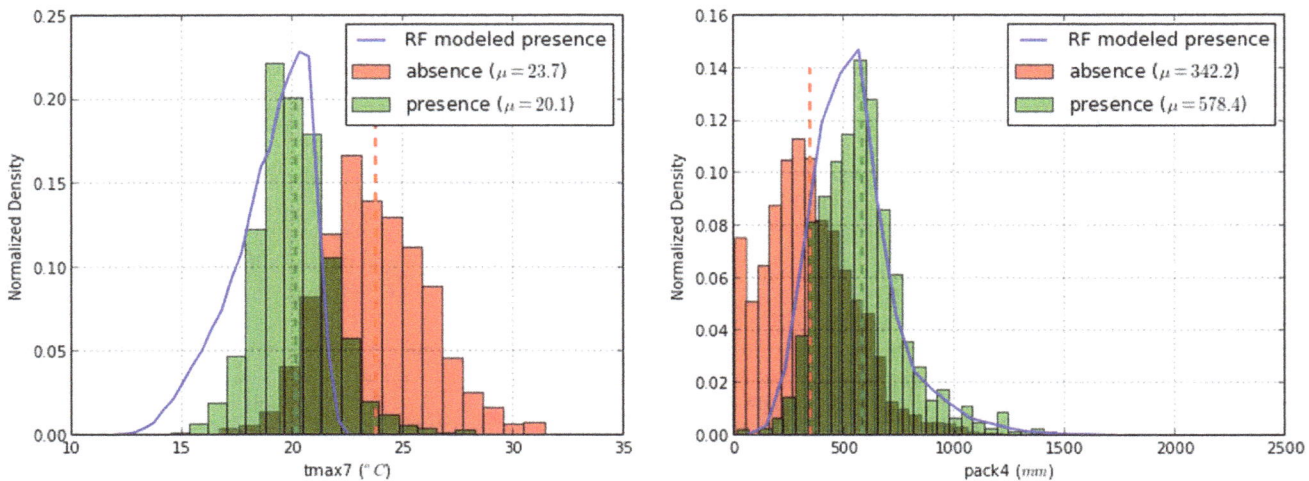

Figure 7. Modeled binary presence for *P.albicaulis* under 1980–2010 mean July maximum temperatures and mean April snow water equivalent bioclimate variables shows agreement with field presence data. Dotted lines designate climate means for corresponding *P. albicaulis* field points. Blue lines represent the distribution of Random Forest modeled presence within the GYA.

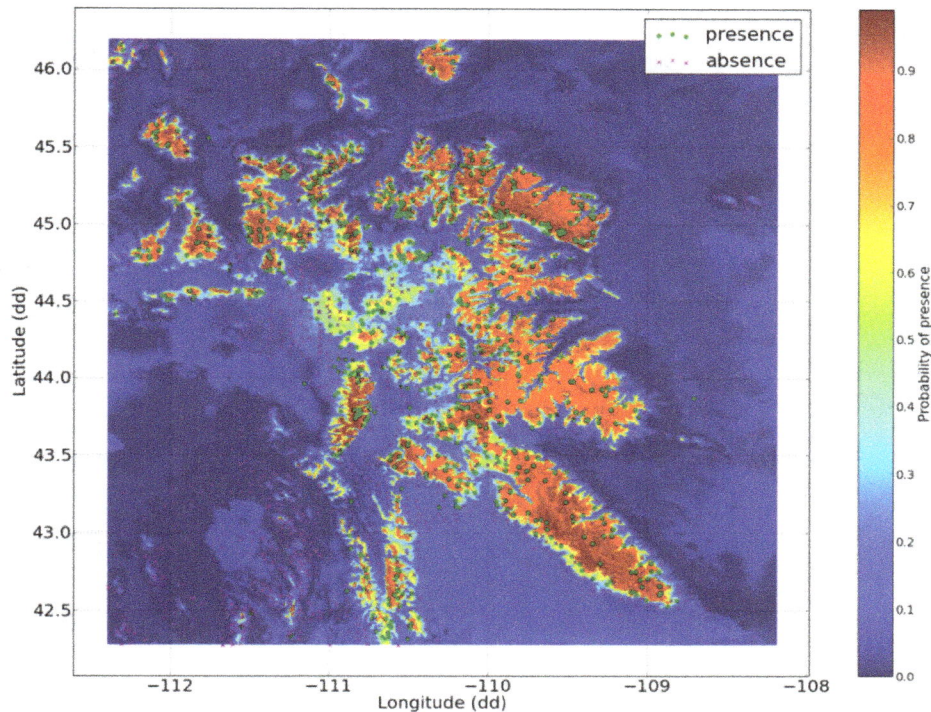

Figure 8. Probabiliy presence for *P.albicaulis* ≥ **20 cm DBH within the GYA for the 2010 climate period.**

Modeling methods

A random forest (RF) [51] algorithm was used to create a bioclimate niche model of *P. albicaulis* in the GYA. Random forest is an ensemble learning technique that generates independent random classification trees using a subset of the total predictor variables and classifies a bootstrap random subsample of the data. These trees are aggregated and a majority vote over all trees in the random forest defines the resulting response class. This method of random trees with subsampling ensures a robust ensemble classification reducing overfitting and collinearity issues, especially with a large number of trees [9,51–53]. The python programming language (Python 3.3) and the Scikit-Learn library was used to fit the random forest model and predict current habitat niche, with parameters for number of trees ($n_{estimators} = 1000$), number of variables ($max_{features} = 4$), and node size ($min_{samplesleaf} = 20$) [54].

First pass filtering of environmental covariates was performed using Principal Component Analysis (PCA) to generate proxy sets [55–57]. After initial list was constructed, an additional filter was imposed on the variables with a 0.75 maximum correlation threshold to avoid collinearity issues (Fig. 3) [55]. Physiologically relevant variables to *P. albicaulis* presence were given precedence in final culling in cases of correlation above the specified maximum threshold. The final variable list selected were tmin1, vpd3, ppt4, pack4, tmax7, aet7, pet8, ppt9 (Table 2). The Software for Assisted Habitat Modeling (SAHM) was used to visualize correlations with the pairs function embeded in the VisTrails scientific workflow management system [58,59].

Model evaluation was performed under a variety of methods. An out-of-bag (OOB) error estimate was calculated by comparing the modeled probability of presence using approximately two-thirds of the field data, while withholding a subset of the remainder. Accuracy was evaluated by calculating: 1) the sensitivity, representing the true positive rate (TPR), 2) the specificity, representing the true negative rate (TNR), 3) the receiver operator characteristic curve (AUC). Importance of a specific predictor variable was calculated by examination of the increase in prediction error within the OOB sample when the predictor variable was permuted while others were held constant [54,60]. The rate of prediction error with permutation of a specified variable can be interpreted as the level of dependence of presence or absence response to that variable [61].

Projections for *P. albicaulis* were computed using 30 year moving climate averages for the period from 2010–2099 for both RCP 4.5 and 8.5 climate scenarios. Changes of suitable habitat area were determined using a binary classification of expected presence and absence. Binary class assignment was made under a probability of presence threshold where the ratio of sensitivity and specificity equalled 1. This method ensured an equal ability of the model to detect presence and absence. The Kappa and True Skill Statistic (TSS) were also calculated to observe how sensitivity and specificity responded under differing probability thresholds [62]. Survey plots predicted as suitable under climatic conditions in 2010 served as a reference for projections. The presence classifications were evaluated as the amount of suitable habitat changed over time, confined within specified elevational limits. To account for the need for a minimum patch size, total number of patches and median sizes using the an eight-neighbor rule (see Text S1) for patch identification were tracked over time [63].

Results

Model evaluation

The random forest model displayed an out-of-bag (OOB) error rate of 16.1% with greater errors of commission (13.1%) than omission (10.9%) (Table 3). The AUC was 0.94, displaying high

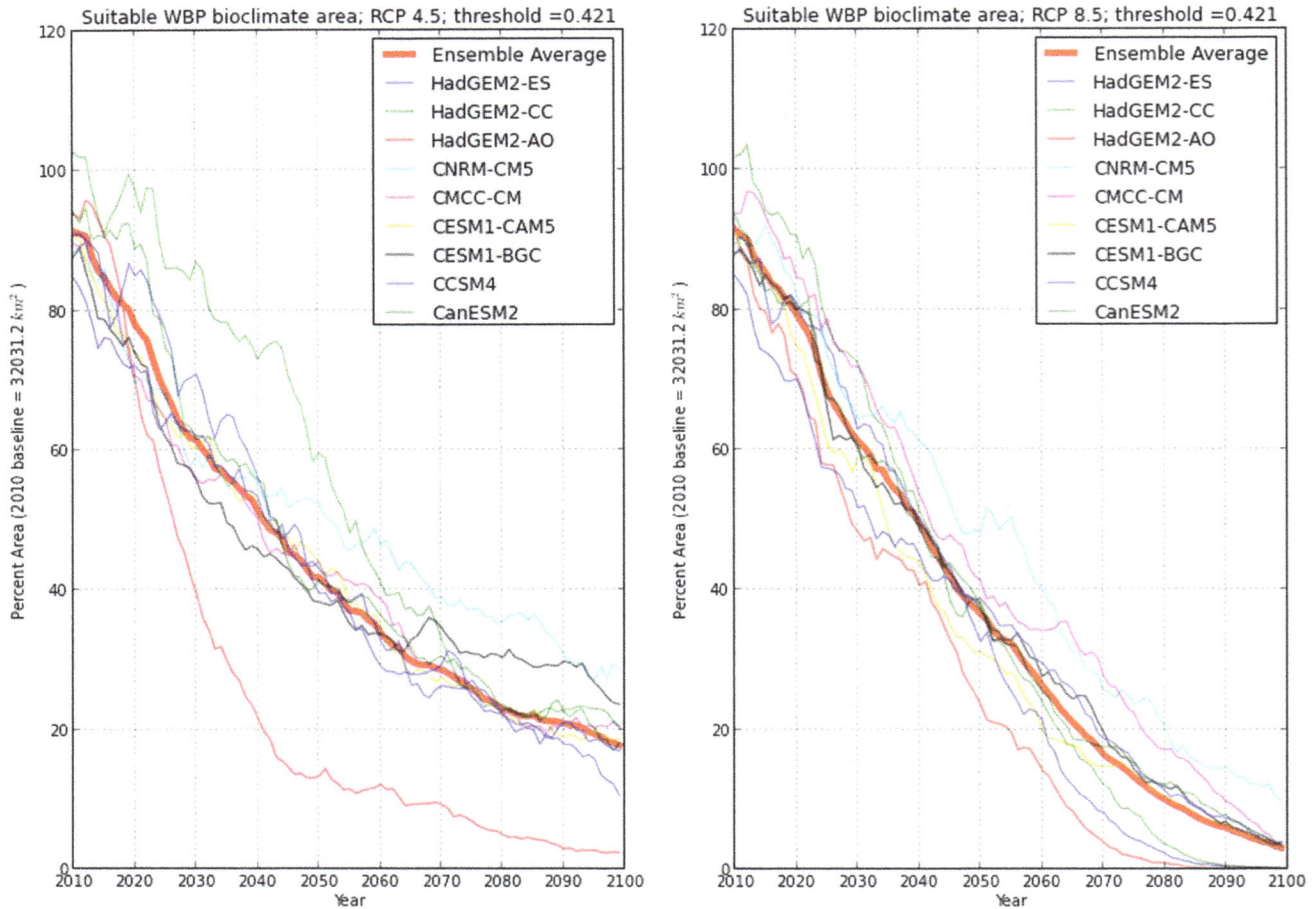

Figure 9. Bioclimate projections for *P.albicaulis* for 2010 to 2099 under 30-year moving averaged climates.

specificity and sensitivity (Fig. 4). Threshold probability of presence for a binary classification was selected at 0.421 (i.e where sensitivity = specificity). A probability threshold where TPR and TNR were equal was compared to the maximum Kappa statistic (0.538) and the maximum True Skill Statistic (TSS) (0.345) and found to be a compromise between the diagnostics (Fig. 5).

Estimates of variable importance plots revealed that permutation of maximum temperatures of summer months from all random trees resulted in a large drop in mean accuracy for distinguishing presence and absence of *P. albicaulis* (0.706 decrease in mean accuracy). This was followed by spring time snowpack (0.137 decrease in mean accuracy) (Fig. 6). Histogram plots of July maximum temperatures and April snowpack provided evidence of discrimination for presence and absence that are consistent with the modeled probability of presence for the year of calibration (Fig. 7).

Spatially explicit probability plots for the 2010 climate displayed highest probability of presence values within the ≥2500 m mountain ranges of the GYA in agreement with studies employing aerial imagery and remote sensing [4,5] (Fig. 8). Assuming that the modeled suitable bioclimate for *P. albicaulis* remains similar in the next century, the model demonstrated capacity to predict probable future *P. albicaulis* suitable habitat under projected climate conditions.

Model projections

Under both RCP 4.5 and 8.5, there was a predicted steady reduction of suitable bioclimate habitat for *P. albicaulis* over the course of this century, with RCP 8.5 displaying steeper declines than RCP 4.5 (Fig. 9). Under the RCP 4.5 and 8.5 scenarios, suitable habitat shifts from 100–85% to 2–29% by 2099, and 100–85% to 0.04–10% by 2099 respectively (Table 4).

CNRM-CM5, CMCC-CM, and CESM1-BGC projections showed the highest probabilities for suitable habitat area at the end of the century, while HadGEM2-AO, HadGEM2-ES, and HadGEM2-CC indicated the lowest probabilities. The standard deviations per year for both RCPs progressively decreased over time (Fig. 10). Among climate scenarios, standard deviations for both RCPs display low variability for the first five projection years and a rapid increase of variability peaking at 2043. For RCP 4.5, high variability existed primarily due to differing climate projections by models HadGEM2-AO and HadGEM2-CC, resulting in uncertainties in probabilities of presence fluctuating between 8 and 15% until 2068, after which variability was between 6–8%. Under RCP 8.5, standard deviations between GCMs were consistently lower than RCP 4.5. Regardless of the GCM, by 2079 the areas of suitable habitat converged to similar values.

Spatially explicit mapping of probability surfaces presented similar contractions of *P. albicaulis* habitat suitability toward the

Table 4. Projected binary *P. albicaulis* presence area within GYA to 2099.

Ensemble Average RCP 4.5	2010	2040	2070	2099
Area (km²)	29250.9	16381.2	9151.1	5685.9
	(27134–32858)	(6918–23359)	(2962–12477)	(763–9194)
% Total Threshold Area*	91.3	51.1	28.6	17.8
	(85–103)	(22–73)	(9–39)	(2–29)
Mean Elevation (m)	2875.7	3020.2	3128.0	3217.9
	(2842–2895)	(2938–3182)	(3055–3297)	(3114–3471)
2.5 Percentile Elevation (m)	2356.3	2494.2	2595.0	2691.5
	(2320–2376)	(2433–2656)	(2506–2758)	(2571–3041)
97.5 Percentile Elevation (m)	3521.9	3603.5	3677.8	3734.6
	(3507–3530)	(3551–3701)	(3636–3783)	(3673–3905)
Ensemble Average RCP 8.5	2010	2040	2070	2099
Area (km²)	29259.3	15746.0	5271.5	960.0
	(27188–32604)	(12985–19581)	(1247–8850)	(13–3105)
% Total Threshold Area*	91.3	49.2	16.5	3.0
	(85–102)	(40–61)	(4–28)	(0–10)
Mean Elevation (m)	2874.7	3022.5	3225.5	3470.5
	(2845–2893)	(2974–3061)	(3116–3412)	(3255–3749)
2.5 Percentile Elevation (m)	2353.1	2492.2	2691.3	3001.5
	(2322–2369)	(2436–2547)	(2553–2934)	(2622–3401)
97.5 Percentile Elevation (m)	3522.1	3605.7	3739.5	3908.7
	(3508–3530)	(3576–3631)	(3677–3866)	(3775–4063)

Summary of projection outputs under RCP 4.5 and 8.5 climate scenarios displays loss of bioclimate habitat from 2010 to 2099 (low and high probability of presence GCM summaries displayed in parentheses). Projections into 2099 under all 9 GCMs suggest rapid loss of suitable bioclimate habitat to below 70% of the current modeled distribution and shifts towards the limited high elevation zones (>3000 m). *(Percent threshold areas calculated from the 2010 PRISM reference probabilities of presence.)*

upper elevation zones of the GYA that included the Beartooth Plateau and Wind River Ranges (Fig. 11). This implied that rapid warming may lead to conditions outside of the *P. albicaulis* niche in lower elevation areas, and limiting the species to the alpine zones. Elevational analysis of cells within threshold presence probabilities over time observed mean elevations of suitable bioclimates shifting from 2,875 to 3,218 m and 2,875 to 3,470 m for RCP 4.5 and 8.5 respectively. By 2099, ensemble averaged GCM projections displayed over 70% loss of habitat under both scenarios.

P. albicaulis patches from the 2010 baseline observed 202 patches with median patch size of ∼180 km². Projected patch dynamics analysis denoted a quadratic relationship of patch size over time. Patch dynamics displayed a slow increase in number of *P. albicaulis* patches to a maximum at 2074 and 2057 for RCP 4.5 and 8.5 respectively, followed by a decreasing trend. RCP 4.5 patch numbers were more sporadic, displaying fluctuations across the time period compared to RCP 8.5 associated with the greater interannual climate variability amongst GCM models. Median patch size saw a steady decrease from 72–65 km² to 21–8 km² for RCP 4.5 and 8.5 respectively, for the projection period, suggesting habitat loss through fragmentation (Fig. 12).

Discussion

In this analysis, the spatiotemporal patterns for *P. albicaulis* distributions were assessed under nine climate models and two emissions scenarios. Bioclimate modeling of *P. albicaulis* illustrat-

ed that presence and absence were strongly separated by summer temperatures and spring snowpack. This was in agreement with empirical findings of *P. albicaulis* presence in cool summertime environments where July temperatures range between 4–18°C [64]. Concordantly, these cool summer regions were synonymous with late snow melt, supporting snowpack as an important feature in distinguishing presence and absence.

Future projections by all nine GCMs suggested a contraction in suitable *P. albicaulis* climate area by the end of the century to <30% of current conditions. This was consistent with the results from various other research using either niche models or hybrid process models, predicting similar amounts of *P. albicaulis* contraction [8,9,65]. Variability among projected suitable habitat areas under differing GCMs decreased as all projected maximum temperatures increased above 1°C from the 100 year historic mean. This pattern of warming convergence occurred earlier for the GCMs under the RCP 8.5 scenario than those under RCP 4.5, resulting in the observed low variability of *P. albicaulis* suitable habitat area under RCP 8.5. Despite temperature variability remaining relatively constant amongst GCMs within a RCP, once mean annual temperatures increased beyond *sim* 1°C from the historic average, all bioclimatic habitat models exhibited a pattern of contracting total area and variability. These results lead to the conclusion that explicit selection of a GCM to model under may not necessarily matter for *P. albicaulis* bioclimatic niche modeling studies, especially if the direction of change is solely of concern. However, if investigation of the magnitude of change is relevant,

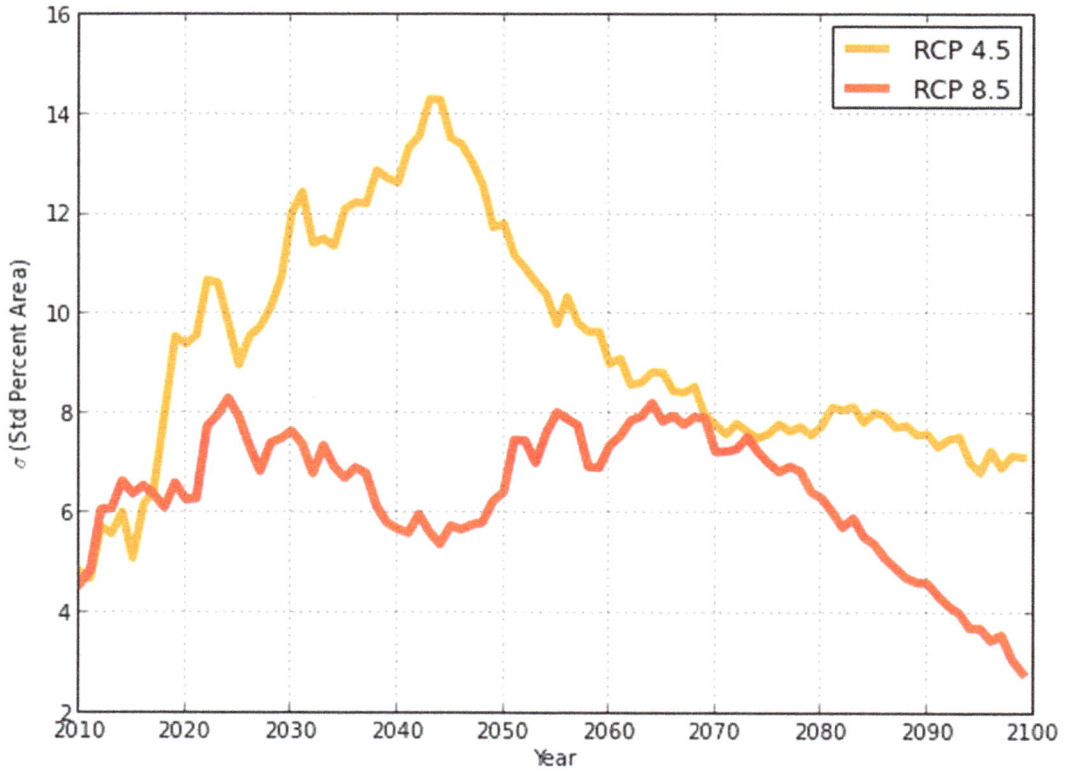

Figure 10. Evaluation of the standard deviation σ for percent suitable habitat area by RCP scenario.

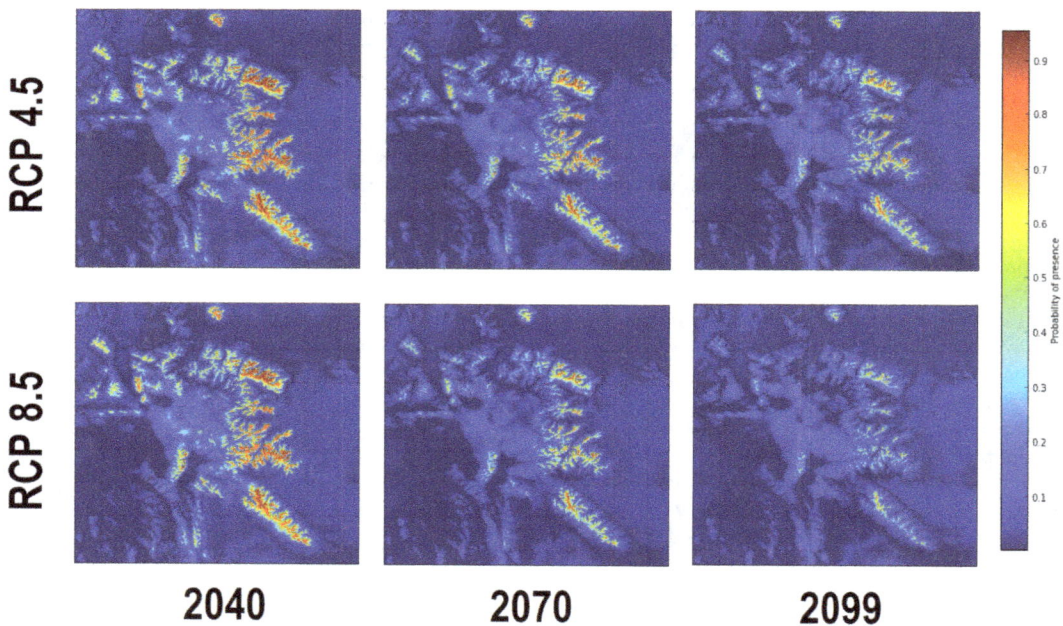

Figure 11. Spatially explicit probabilty surfaces for 2040 to 2099 suggest contraction of suitable bioclimatic habiatat for *P.albicaulis* into the ≥2500 m elevation zones.

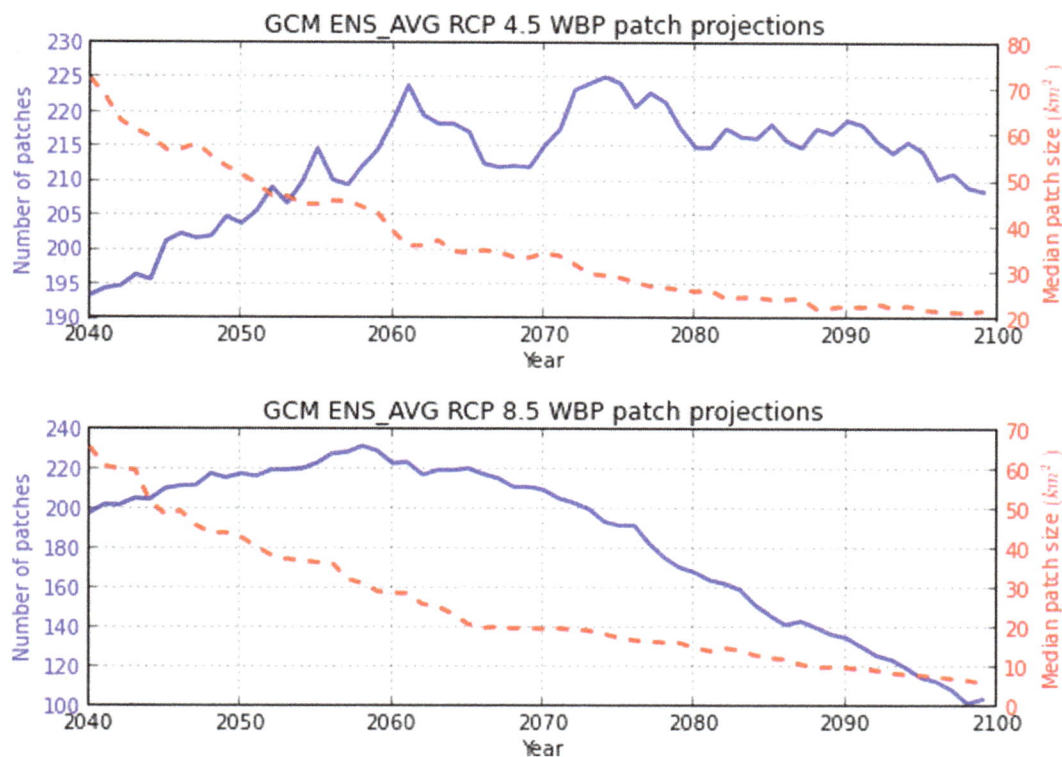

Figure 12. Patch dynamics of modeled *P.albicaulis*. Time series of *P.albicaulis* patch projections for number of patches and median patch size to 2099.

then GCM selection may directly influence the projected total suitable habitat area. This can be observed with RCP 4.5 habitat projection models differing by as much as 27% total suitable habitat area by the year 2099. Therefore arbitrary selection of a GCM for future projection modeling is likely inappropriate since it could lead to overly optimistic/pessimistic results for the species of concern.

Temporal patch dynamic analysis present an increase in fragmentation of the larger *P. albicaulis* suitable habitats over the next five decades, suggested through an increase in the total number of continuous patches but decreases in median size. This was followed by a contraction of small patches until they were almost absent from the system. Remaining habitat patches were smaller and less prevalent on the landscape by the end of the century. Reduced habitat patch size and density may reduce the likelihood for *N. columbiana* to disperse successful germinating seed caches, due to the limited size and area of suitable patch space. If changing climate habitats result in mortality within adult patches, genetic diversity may be lost resulting in a population bottleneck, thus reducing the robustness of the species to adapt to future disturbances. Experimental trials of P. albicaulis survival and fecundity under warmer and drier conditions outside the currently known range would provide greater confidence of the species ability to persist under future change. Limited analysis on seedling environmental conditions would also elucidate spatially explicit dispersal ranges and greater understanding of probable ranges for future establishment and survivorship.

Projected distributions of persistent *P. albicaulis* patches displayed a strong trend towards contraction into high elevation zones. Physiologically, there does not appear to be any upper elevation limit for *P. albicaulis* in the GYA. *P. albicaulis* in the

region has been reported to survive in absolute temperatures as low as $-36°C$ [64]. Lab experiments performed on *Pinus cembra*, a related five-needle pine residing in similar climates, were able to endure cold temperature extremes as low at $-70°C$ without cellular tissue damage [66]. Considering the current absolute minimum temperatures the species resides in and cold tolerance of its relatives suggests that *P. albicaulis* treeline in the GYA are not limited by lower temperatures. Controlled laboratory experimentation on *P. albicaulis* tolerances to temperatures would greatly improve this physiological understanding of cold tolerance.

Elevational habitat constriction do not imply that *P. albicaulis* will be completely gone from the region, but merely the loss of suitable climate habitat. Currently pre-established adult age class individuals will likely persist, since projected conditions of increased temperatures and CO_2 concentrations physiologically indicate increased growth rates of *P. albicaulis* [67]. Furthermore, micro-refugia sites may exist in the GYA that support *P. albicaulis* survival into the future, but were failed to have been modeled due to the coarseness of 30-arc-second climate data resolution. Since this bioclimatic envelope modeling approach was parameterized by the realized niche from in-situ data, it was difficult to determine if lower elevation limits are driven by warmer climate conditions or competition for light, water, or nutrients [15,17]. For example, lower treeline limits for *P. albicaulis* maybe driven primarily by competitive exclusion from late seral species *A.lasiocarpa*, *P.contorta*, and *P.engelmannii*. This follows from paleoecological pollen records of competitor migration during the Early Holocene (9000–5000 yr B.P), when climate conditions were warmer and drier. Longer growing seasons allowing competitors to invade likely drove *P. albicaulis* communities +500 m in elevation [68–70]. If future climate conditions become analogous to this Early

Holocene period, invasion of competitor species will likely contract *P. albicaulis* habitat to the limited high elevation zones of the GYA, specifically the Beartooth Plateau region and Wind River range [71].

Conclusion

This analysis examined the future of *P. albicaulis* suitable climate in the GYA and explicitly addressed the question of distribution variability under 9 representative GCMs and 2 emission scenarios. Increases in temperature within the GYA will likely result in a high level of contraction of suitable climate habitat for *P. albicaulis* over the next century. This contraction was consistent for all GCM projections, with approximately 20% uncertainty in total probable area. This analysis recommends that care be taken for species distribution modeling in future studies during the selection of GCMs due to their relevance for magnitudes of change. GCM ensemble averaging may be a solution to this issue, however it should be noted that averaging should take place after an individual GCM is projected in order to maintain interannual variability.

Although other studies have examined *P. albicaulis* species distribution models [8,65,72], this study is a step forward through its focus on relevant regional scale design, expansive local datasets, inclusion of high resolution climate and dynamic water balance variables, and selective projection under the latest AR5 GCMs. It is reiterated that the bioclimate niche model approach has high utility for understanding habitat conditions through correlative relationships with environmental variables, however, it may fail to explicitly model competitive exclusion, disturbance, phenotypic plasticity, and other complex interactions that are vital in determining a species' actual presence as it experiences changes in climate [15,17,73,74]. These unmodeled factors create uncertainties suggesting that this modeling effort does not identify the full potential climatic range of *P. albicaulis* in the future.

Uncertainties also exist regarding new suitable climates that may occur outside the current species range. Despite most rangewide studies confirming our results of total suitable habitat area reduction, there is potential for previously unsuitable habitat to become available under future climate change in the Northern regions [8,22,65]. Caution is therefore advised to individuals interpreting these findings. Changing climate will inevitably result in impacts on biomes and community structures. As such, mitigation and adaptation for potential futures are vital to conservation of climate sensitive species [75]. Future research that combines bioclimatic niche modeling with a mechanistic based disturbance, dispersal, and competition model will likely provide greater insight to the potential range of *P. albicaulis* in a climate changing world [76,77]. It would furthermore provide insight towards informing management options for restoration that may include controlled fire, selected thinning of competitor species, or assisted migration.

Acknowledgments

We are indebted for the insightful reviews from Richard Waring, William Monahan, and Tom Olliff. Many thanks to Marian and Colin Talbert, and Mark Greenwood for the software and statistical consultation.

Author Contributions

Conceived and designed the experiments: TC AJH NBP. Performed the experiments: TC NBP. Analyzed the data: TC. Contributed reagents/materials/analysis tools: TC AJH NBP. Wrote the paper: TC AJH NBP.

References

1. Intergovernmental Panel on Climate Change (2007) Fourth Assessment Report: Climate Change 2007: The AR4 Synthesis Report. Geneva: IPCC.
2. Tomback DF, Arno SF, Keane RE (2001) Whitebark pine communities: ecology and restoration. Island Press.
3. Callaway RM (1998) Competition and facilitation on elevation gradients in subalpine forests of the Northern Rocky Mountains, USA. Oikos 82: pp. 561–573.
4. Macfarlane WW, Logan JA, Kern W (2012) An innovative aerial assessment of greater yellowstone ecosystem mountain pine beetle-caused whitebark pine mortality. Ecological Applications.
5. Jewett JT, Lawrence RL, Marshall LA, Gessler PE, Powell SL, et al. (2011) Spatiotemporal relationships between climate and whitebark pine mortality in the greater yellowstone ecosystem. Forest Science 57: 320–335.
6. Logan JA, Macfarlane WW, Willcox L (2010) Whitebark pine vulnerability to climate-driven mountain pine beetle disturbance in the greater yellowstone ecosystem. Ecological Applications 20: 895–902.
7. Logan JA, Bentz BJ (1999) Model analysis of mountain pine beetle (coleoptera: Scolytidae) seasonality. Environmental Entomology 28: 924–934.
8. Rehfeldt GE, Crookston NL, Sáenz-Romero C, Campbell EM (2012) North American vegetation model for land-use planning in a changing climate: a solution to large classification problems. Ecological Applications 22: 119–141.
9. Rehfeldt GE, Crookston NL, Warwell MV, Evans JS (2006) Empirical analyses of plant-climate relationships for the western United States. International Journal of Plant Sciences 167: 1123–1150.
10. Thuiller W (2004) Patterns and uncertainties of species' range shifts under climate change. Global Change Biology 10: 2020–2027.
11. Iverson LR, Prasad AM, Matthews SN, Peters M (2008) Estimating potential habitat for 134 eastern US tree species under six climate scenarios. Forest Ecology and Management 254: 390–406.
12. Guisan A, Theurillat JP, Kienast F (1998) Predicting the potential distribution of plant species in an alpine environment. Journal of Vegetation Science 9: 65–74.
13. Busby J (1988) Potential impacts of climate change on Australias flora and fauna. Commonwealth Scientific and Industrial Research Organisation, Melbourne, FL, USA.
14. Hutchinson GE (1957) Concluding remarks. Cold Spring Harbor Symposia on Quantitative Biology 22: 415–427.
15. Austin M (2007) Species distribution models and ecological theory: a critical assessment and some possible new approaches. Ecological Modelling 200: 1–19.
16. Austin M (2002) Spatial prediction of species distribution: an interface between ecological theory and statistical modelling. Ecological Modelling 157: 101–118.
17. Pearson RG, Dawson TP (2003) Predicting the impacts of climate change on the distribution of species: are bioclimate envelope models useful? Global Ecology and Biogeography 12.
18. Willis KJ, Whittaker RJ (2002) Species diversity–scale matters. Science 295: 1245–1248.
19. Araújo MB, Cabeza M, Thuiller W, Hannah L, Williams PH (2004) Would climate change drive species out of reserves? an assessment of existing reserve-selection methods. Global Change Biology 10: 1618–1626.
20. Ferrier S (2002) Mapping spatial pattern in biodiversity for regional conservation planning: where to from here? Systematic Biology 51: 331–363.
21. Pearce J, Lindenmayer D (1998) Bioclimatic analysis to enhance reintroduction biology of the endangered helmeted honeyeater (lichenostomus melanops cassidix) in Southeastern Australia. Restoration Ecology 6: 238–243.
22. McLane SC, Aitken SN (2012) Whitebark pine (pinus albicaulis) assisted migration potential: testing establishment north of the species range. Ecological Applications 22: 142–153.
23. Monahan WB, Cook T, Melton F, Connor J, Bobowski B (2013) Forecasting distributional responses of limber pine to climate change at management-relevant scales in Rocky Mountain National Park. PloS ONE 8: e83163.
24. Arno SF, Hoff RJ (1989) Silvics of whitebark pine (pinus albicaulis). Intermountain Research Station GTR-INT-253.
25. Mahalovich MF, Hipkins VD (2011) Molecular genetic variation in whitebark pine (pinus albicaulis engelm.) in the inland west. In: Keane RE, Tomback DF, Murray MP, Smith CM, editors, The future of high-elevation, five-needle white pines in Western North America: Proceedings of the High Five Symposium. 28–30 June 2010; Missoula, MT. Proceedings RMRS.
26. Pederson GT, Gray ST, Ault T, Marsh W, Fagre DB, et al. (2011) Climatic controls on the snowmelt hydrology of the Northern Rocky Mountains. Journal of Climate 24: 1666–1687.

27. Westerling AL, Hidalgo HG, Cayan DR, Swetnam TW (2006) Warming and earlier spring increase western US forest wildfire activity. Science 313: 940–943.

28. Taylor KE, Stouffer RJ, Meehl GA (2012) An overview of CMIP5 and the experiment design. Bulletin of the American Meteorological Society 93.

29. Hibbard KA, van Vuuren DP, Edmonds J (2011) A primer on representative concentration pathways (RCPs) and the coordination between the climate and integrated assessment modeling communities. CLIVAR Exchanges 16: 12–13.

30. Lutz JA, van Wagtendonk JW, Franklin JF (2010) Climatic water deficit, tree species ranges, and climate change in Yosemite National Park. Journal of Biogeography 37: 936–950.

31. Deser C, Phillips AS, Alexander MA, Smoliak BV (2014) Projecting North American climate over the next 50 years: Uncertainty due to internal variability. Journal of Climate 27: 2271–2296.

32. Beaumont LJ, Hughes L, Pitman A (2008) Why is the choice of future climate scenarios for species distribution modelling important? Ecology Letters 11: 1135–1146.

33. Smith WB (2002) Forest inventory and analysis: a national inventory and monitoring program. Environmental Pollution 116: S233–S242.

34. Gibson J, Moisen G, Frescino T, Edwards Jr TC (2014) Using publicly available forest inventory data in climate-based models of tree species distribution: Examining effects of true versus altered location coordinates. Ecosystems 17: 43–53.

35. Gesch D, Oimoen M, Greenlee S, Nelson C, Steuck M, et al. (2002) The national elevation dataset. Photogrammetric engineering and remote sensing 68: 5–32.

36. Franklin J (2009) Mapping species distributions: spatial inference and prediction. Cambridge University Press.

37. Lockman IB, DeNitto GA, Courter A, Koski R (2007) WLIS: The whitebark-limber pine information system and what it can do for you. In: Proceedings of the conference whitebark pine: a Pacific Coast perspective. US Department of Agriculture, Forest Service, Pacific Northwest Region, Ashland, OR. Citeseer, pp. 146–147.

38. Jean C, Shanahan E, Daley R, DeNitto G, Reinhart D, et al. (2010) Monitoring white pine blister rust infection and mortality in whitebark pine in the Greater Yellowstone Ecosystem. Proceedings of the future of high-elevation five-needle white pines in Western North America: 28–30.

39. Edwards Jr TC, Cutler DR, Zimmermann NE, Geiser L, Moisen GG (2006) Effects of sample survey design on the accuracy of classification tree models in species distribution models. Ecological Modelling 199: 132–141.

40. Kadmon R, Farber O, Danin A (2003) A systematic analysis of factors affecting the performance of climatic envelope models. Ecological Applications 13: 853–867.

41. Weaver T, Dale D (1974) Pinus albicaulis in central Montana: environment, vegetation and production. American Midland Naturalist: 222–230.

42. Daly C, Gibson WP, Taylor GH, Johnson GL, Pasteris P (2002) A knowledge-based approach to the statistical mapping of climate. Climate Research 22: 99–113.

43. Thornthwaite C (1948) An approach toward a rational classification of climate. Geographical Review 38: 55–94.

44. Thornthwaite C, Mather J (1955) The water balance. Publication of Climatology 8.

45. Dingman S (2002) Physical hydrology. Prentice Hall.

46. National Resources Conservation Service (2014) Available: http://soildatamart.nrcs.usda.gov. Accessed 2013 Apr 3.

47. Thrasher B, Xiong J, Wang W, Melton F, Michaelis A, et al. (2013) Downscaled climate projections suitable for resource management. Eos, Transactions American Geophysical Union 94: 321–323.

48. Rupp DE, Abatzoglou JT, Hegewisch KC, Mote PW (2013) Evaluation of CMIP5 20th century climate simulations for the Pacific Northwest USA. Journal of Geophysical Research: Atmospheres 118: 10–884.

49. Gent PR, Danabasoglu G, Donner LJ, Holland MM, Hunke EC, et al. (2011) The community climate system model version 4. Journal of Climate 24: 4973–4991.

50. Moss RH, Babiker M, Brinkman S, Calvo E, Carter T, et al. (2008) Towards new scenarios for analysis of emissions, climate change, impacts, and response strategies.

51. Breiman L (2001) Random forests. Machine Learning 45: 5–32.

52. Roberts DR, Hamann A (2012) Method selection for species distribution modelling: are temporally or spatially independent evaluations necessary? Ecography 35: 792–802.

53. Lawrence RL, Wood SD, Sheley RL (2006) Mapping invasive plants using hyperspectral imagery and breiman cutler classifications (randomforest). Remote Sensing of Environment 100: 356–362.

54. Pedregosa F, Varoquaux G, Gramfort A, Michel V, Thirion B, et al. (2011) Scikit-learn: Machine learning in Python. Journal of Machine Learning Research 12: 2825–2830.

55. Dormann CF, Elith J, Bacher S, Buchmann C, Carl G, et al. (2013) Collinearity: a review of methods to deal with it and a simulation study evaluating their performance. Ecography 36: 027–046.

56. Booth GD, Niccolucci MJ, Schuster EG (1994) Identifying proxy sets in multiple linear regression: an aid to better coefficient interpretation. Research paper INT.

57. Tabachnick B, Fidell LS (1989) Using multivariate statistics, 1989. Harper Collins Tuan, PD A comment from the viewpoint of time series analysis Journal of Psychophysiology 3: 46–48.

58. Freire J (2012) Making computations and publications reproducible with vistrails. Computing in Science & Engineering 14: 18–25.

59. Morisette JT, Jarnevich CS, Holcombe TR, Talbert CB, Ignizio D, et al. (2013) Vistrails SAHM: visualization and workflow management for species habitat modeling. Ecography 36: 129–135.

60. Liaw A, Wiener M (2002) Classification and regression by randomforest. R news 2: 18–22.

61. Cutler DR, Edwards Jr TC, Beard KH, Cutler A, Hess KT, et al. (2007) Random forests for classification in ecology. Ecology 88: 2783–2792.

62. Allouche O, Tsoar A, Kadmon R (2006) Assessing the accuracy of species distribution models: prevalence, kappa, and the true skill statistic (tss). Journal of Applied Ecology 43: 1223–1232.

63. Turner MG, Gardner RH, O'Neill RV (2001) Landscape ecology in theory and practice: pattern and process. Springer.

64. Weaver T (2001) Whitebark pine and its environment. In: Tomback DF, Arno SF, Keane RE, editors, Whitebark pine communities: ecology and restoration, Washington D.C, USA: Island Press.

65. Waring RH, Coops NC, Running SW (2011) Predicting satellite-derived patterns of large-scale disturbances in forests of the pacific northwest region in response to recent climatic variation. Remote Sensing of Environment 115: 3554–3566.

66. Sakai A, Larcher W (1987) Frost survival of plants. Responses and adaptation to freezing stress. Springer-Verlag.

67. Chapin III FS, Chapin MC, Matson PA, Vitousek P (2011) Principles of terrestrial ecosystem ecology. Springer.

68. Whitlock C, Shafer SL, Marlon J (2003) The role of climate and vegetation change in shaping past and future fire regimes in the Northwestern US and the implications for ecosystem management. Forest Ecology and Management 178: 5–21.

69. Whitlock C (1993) Postglacial vegetation and climate of Grand Teton and southern Yellowstone national parks. Ecological Monographs: 173–198.

70. Bartlein PJ, Whitlock C, Shafer SL (1997) Future climate in the Yellowstone national park region and its potential impact on vegetation. Conservation Biology 11: 782–792.

71. Tausch RJ, Wigand PE, Burkhardt JW (1993) Viewpoint: plant community thresholds, multiple steady states, and multiple successional pathways: legacy of the quaternary? Journal of Range Management: 439–447.

72. Bell DM, Bradford JB, Lauenroth WK (2014) Early indicators of change: divergent climate envelopes between tree life stages imply range shifts in the western united states. Global Ecology and Biogeography 23: 168–180.

73. Keane B, Tomback D, Davy L, Jenkins M, Applegate V (2013) Climate change and whitebark pine: Compelling reasons for restoration. Whitebark Pine Ecosystem Foundation Whitepaper.

74. Guisan A, Thuiller W (2005) Predicting species distribution: offering more than simple habitat models. Ecology Letters 8: 993–1009.

75. Keane RE, Tomback DF, Aubry CA, Bower EM, Campbell CL, et al. (2012) A range-wide restoration strategy for whitebark pine (pinus albicaulis): General technical report. USDA FS, Rocky Mountain Research Station RMRS-GTR-279: 108.

76. Mathys A, Coops NC, Waring RH (2014) Soil water availability effects on the distribution of 20 tree species in Western North America. Forest Ecology and Management 313: 144–152.

77. Morin X, Thuiller W (2009) Comparing niche-and process-based models to reduce prediction uncertainty in species range shifts under climate change. Ecology 90: 1301–1313.

12

Mean Annual Precipitation Explains Spatiotemporal Patterns of Cenozoic Mammal Beta Diversity and Latitudinal Diversity Gradients in North America

Danielle Fraser[1,2]*, Christopher Hassall[1,3], Root Gorelick[1,4,5], Natalia Rybczynski[1,2]

1 Department of Biology, Carleton University, Ottawa, Ontario, Canada, **2** Palaeobiology, Canadian Museum of Nature, Ottawa, Ontario, Canada, **3** School of Biology, University of Leeds, Leeds, United Kingdom, **4** Department of Mathematics and Statistics, Carleton University, Ottawa, Ontario, Canada, **5** Institute of Interdisciplinary Studies, Carleton University, Ottawa, Ontario Canada

Abstract

Spatial diversity patterns are thought to be driven by climate-mediated processes. However, temporal patterns of community composition remain poorly studied. We provide two complementary analyses of North American mammal diversity, using (i) a paleontological dataset (2077 localities with 2493 taxon occurrences) spanning 21 discrete subdivisions of the Cenozoic based on North American Land Mammal Ages (36 Ma – present), and (ii) climate space model predictions for 744 extant mammals under eight scenarios of future climate change. Spatial variation in fossil mammal community structure (β diversity) is highest at intermediate values of continental mean annual precipitation (MAP) estimated from paleosols (~450 mm/year) and declines under both wetter and drier conditions, reflecting diversity patterns of modern mammals. Latitudinal gradients in community change (latitudinal turnover gradients, aka LTGs) increase in strength through the Cenozoic, but also show a cyclical pattern that is significantly explained by MAP. In general, LTGs are weakest when continental MAP is highest, similar to modern tropical ecosystems in which latitudinal diversity gradients are weak or undetectable. Projections under modeled climate change show no substantial change in β diversity or LTG strength for North American mammals. Our results suggest that similar climate-mediated mechanisms might drive spatial and temporal patterns of community composition in both fossil and extant mammals. We also provide empirical evidence that the ecological processes on which climate space models are based are insufficient for accurately forecasting long-term mammalian response to anthropogenic climate change and inclusion of historical parameters may be essential.

Editor: Alistair Robert Evans, Monash University, Australia

Funding: D. Fraser was supported by a Natural Science and Engineering Research Council of Canada (NSERC) postgraduate scholarship, a Fulbright Traditional Student Award, a Mary Dawson Pre-Doctoral Fellowship grant, an Ontario Graduate Scholarship (OGS), and a Koningstein Scholarship for Excellence in Science and Engineering. C. Hassall was supported by an Ontario Ministry of Research and Innovation Postdoctoral Fellowship. R. Gorelick was supported by an NSERC Discovery Grant (#341399). N. Rybczynski was supported by an NSERC Discovery Grant (#312193). The funders had no role in study design, data collection and analysis, decision to publish, or preparation of the manuscript.

Competing Interests: The authors have declared that no competing interests exist.

* Email: danielle_fraser@carleton.ca

Introduction

Terrestrial species from all major taxonomic groups show dramatic changes in richness and diversity across the landscape [1]. One of the fundamental goals in ecology is therefore to ascertain why there are more species in some places than in others. A satisfactory answer would identify and disentangle the drivers of biodiversity at all spatial scales, from the microhabitat to the globe, as well as explain changes through time. Attempts to provide such an answer have produced many studies of species richness patterns and community composition in extant organisms [1–8]. Prime examples are the numerous studies of latitudinal richness gradients (LRGs), which have been observed in many terrestrial groups including angiosperms, birds, mammals, insects and other invertebrates. The best supported hypotheses show that richness declines toward the poles in correlation with reductions in precipitation, temperature, and net primary productivity [9]. Correlation of global climate with animal richness over the past

65 Ma, specifically a decline in richness as climates cooled, similarly supports a link between diversity and climate [10–12]. However, of the spatial and temporal dimensions of diversity, spatial patterns of community differences ("β diversity") are infrequently studied despite considerable variation on both local and regional scales [2,13,14] and their influential role in the structuring of continental-scale richness patterns including LRGs [3,4].

β diversity has been defined most broadly as the differentiation in community composition (i.e. the species that make up the community) among regions or along environmental gradients [15]. Similar to LRGs, β diversity generally declines from the tropics to the poles in correlation with climate [2]. However, temporal changes in β diversity remain poorly studied despite their potential power for illuminating the drivers of past and present richness patterns and importance in modern conservation [16–18]. This study therefore tests the hypothesis that climatic influences on mammalian β diversity apply equally to temporal patterns, i.e. that the underlying ecological processes are "ergodic" (dynamic processes that are the same in both time and space).

The mid to late Cenozoic (36 Ma to present) has been a time of dramatic mammalian diversity change, shaped in part by the transition from the productive ice-free ecosystems of the early to mid Cenozoic to the more temperate glaciated ecosystems of the late Cenozoic. Under these changing climatic conditions, mammalian communities show dramatic reductions in richness, changes in community composition, and morphology [10,19–24]. The most dramatic changes occurred at high latitudes, where ecosystems transitioned from *Metasequoia* forests during the early to mid Cenozoic [25,26] to boreal-type forests during the later Cenozoic and to modern tundra [27]. Associated with Cenozoic climate change, were changes in latitudinal climate gradients; overall, the intensity of latitudinal climate gradients increased toward the present, reflecting disproportionate polar cooling due to the formation of permanent Arctic glaciation [28,29]. We therefore predict that latitudinal diversity gradients increased in strength under cooler, less productive environmental conditions just as modern LRGs are steeper in temperate than in tropical regions. Further, we predict that β diversity declined under cooler, less productive environmental conditions just as modern β diversity declines toward the poles [2,7].

Quaternary (2.6 Ma to present) climates have been cool relative to the majority of the late Cenozoic. Recently, however, high latitudes have experienced disproportionate increases in annual temperature (up to 2°C to date), increases in plant primary productivity, and loss of large areas of perennial ice under anthropogenic global warming [30]. Flora and fauna have responded through shifts in phenology [31], *in situ* evolution [32], and, in some cases, extinction [33]. However, perhaps the most often recorded response is the climatically-correlated pattern of extirpations and colonization that manifest as shifts in the location of a species' geographic range. Distributional studies over ecological timescales (<100 yrs) have recorded dramatic poleward range shifts and expansions for a wide range of terrestrial taxa in response to northern warming [34,35]. Projections (i.e. Special Report on Emissions Scenarios) for the next 100 years predict levels of global warming similar to the middle Miocene (+6°C) − a time of reduced or absent perennial Arctic glaciation [36,37] − or warmer (+11°C for the most extreme case; Table S1). We therefore expect continued range expansion, extinction, evolution, and community level changes among North American animals and plants.

A common approach to predicting the long-term outcomes of climate change for terrestrial organisms is climate space modeling (CSM). CSMs use distributional information and climate data to project species ranges into the future, usually under the assumption of no evolution and without adjustment for dispersal differences among species [38–40]. Rapid evolutionary changes on very short timescales and high degrees of variation in dispersal ability under climate change have been observed across a wide range of organisms [34,39,41], therefore CSMs are unlikely to generate accurate forecasts of climate change response. The fossil record, which encompasses many disparate environments and climates, might serve as record of a natural experiment by which ecological hypotheses can be tested in the temporal dimension. Fossil collections are a rich historical record of response to various climatic events that can be incorporated into predictive models, and mammals, in particular, are an excellent group for testing the generality of ecological hypotheses because they have an extensive Cenozoic fossil record. However, studies of extinct organisms have focused largely on richness [12,22,23,42,43] or morphology [44],

with limited focus on community composition [20,22]. Because changes in biological communities are not always associated with changes in richness, spatiotemporal patterns of community composition may be better indicators of climate change response [13,18].

We propose that integrating the study of fossil, modern, and projected spatiotemporal patterns of community composition i) allows for the testing of ecological principles in the temporal dimension, ii) provides the most complete picture of diversity responses to climate change, and iii) enables evaluation of the performance of commonly employed CSMs. Our approach of combining the study of fossil, modern, and projected diversity patterns provides novel insights into the ecological and evolutionary processes that drive continental patterns of biodiversity in space and time.

Methods

Data collection and preparation

We downloaded occurrences for modern North American mammals from NatureServe Canada. The extant mammal dataset included 744 species after the exclusion of a small number of unreadable or corrupted files [45]. We restricted our study of fossil mammals to the late Eocene through Pleistocene, thus avoiding the confounding effects of the early Paleogene mammal radiation. We partitioned the fossil mammal occurrence data by North American Land Mammal Age (NALMA) subdivisions because they delineate relatively temporally stable community assemblages and allowed us to obtain a nearly continuous sequence of mammal community change without large intervening gaps. Using NALMA subdivisions leads to time averaging of mammal communities and to differences in sampling (i.e. intensity, geographic coverage etc.) among time periods. However, we use a statistical approach to reduce these biases, described below. We based the dates for all NALMA subdivisions on Woodburne (2004). Further, we combined data for the entire Clarendonian and excluded for the Whitneyan, late Late Hemphillian, and early Chadronian due to poor sampling (Table 1).

We downloaded fossil mammal occurrence data for the Eocene, Oligocene, Pliocene, and Pleistocene from the the Paleobiology Database using the Fossilworks Gateway (fossilworks.org) in July and August, 2012, using the group name 'mammalia' and the following parameters: time intervals = Cenozoic, region = North America, paleoenvironment = terrestrial (primary contributor: John Alroy; literature sources summarized in Appendix S1). We downloaded Miocene mammal occurrence data from the Miocene Mammal Mapping Project in March 2011 [46] using the NALMA subdivision as our search criterion. For all analyses, with the exception of the Miocene, we used paleolatitudes and paleolongitudes. We chose to use MIOMAP for the Miocene data because it is the most complete Miocene dataset. However, MIOMAP does not provide paleo-coordinates. Fortunately, there are only small differences between modern and Miocene latitudes for the downloaded localities. We removed all taxa with equivocal species identifications (e.g. *Equus* sp.) unless they were the only occurrence for a genus. We assumed all occurrences of open nomenclature (e.g. *Equus* cf. *simplicidens*) were correct identifications.

We did not use latitudinal grids for fossil or extant mammals as in previous studies of latitudinal richness gradients [1,47] because our study is focused on community composition. We therefore do not need to clump localities by spatial proximity to employ rarefaction methods. In addition, the uneven spatial distribution of fossil localities makes the use of a grid method impractical. Instead,

Table 1. Summary of sampled North American Land Mammal Age (NALMA) subdivisions.

Epoch	NALMA subdivision	Age Range (Ma)	Midpoint Age (M)	Number of species	Number of fossil localities	Area (km^2)
Pleistocene	Rancholabrean	0.25–0.011	0.1305	222	180	176615.9
Pliocene	Irvingtonian II	0.85–0.25	0.55	189	94	144745.5
Pliocene	Irvingtonian I	1.72–0.85	1.285	102	37	60361.4
Pliocene	Blancan V	2.5–1.72	2.11	165	130	125042.6
Pliocene	Blancan III	4.1–2.5	3.3	183	163	122839.5
Pliocene	Blancan I	4.9–4.1	4.5	85	66	140433.4
Miocene	Early late Hemphillian	6.7–5.9	6.3	68	46	20108.2
Miocene	Late early Hemphillian	7.5–6.7	7.1	63	55	29446.7
Miocene	Early early Hemphillian	9–7.5	8.25	65	47	31455.8
Miocene	Clarendonian	12.5–9	10.75	104	90	36139.8
Miocene	Late Barstovian	14.8–12.5	13.6	195	194	33789.1
Miocene	Early Barstovian	15.9–14.8	15.5	150	168	51753.3
Miocene	Late Hemingfordian	17.5–15.9	16.7	100	83	25478.4
Miocene	Early Hemingfordian	18.8–17.5	18.15	107	105	45531.3
Miocene	Late late Arikareean	19.5–18.8	19.15	108	123	38307.2
Oligocene/Miocene	Early late Arikareean	23.8–19.5	21.65	71	67	37892.2
Oligocene	Late early Arikareean	27.9–23.8	25.85	95	65	20927.8
Oligocene	Early early Arikareean	30–27.9	28.95	116	124	15382.3
Oligocene	Late Orellan	33.1–32	32.55	38	36	17725.7
Oligocene	Early Orellan	33.7–33.1	33.4	88	130	5579.8
Eocene	Middle Chadronian	35.7–34.7	35.3	88	37	10349.7

we created taxon-by-locality occurrence matrices for extant and fossil mammals at the species taxonomic level excluding *Homo sapiens* [20,22]. In all cases, taxa and localities with fewer than two occurrences were removed from the dataset. Final numbers of localities and species are summarized in Table 1.

To make direct comparisons with modern mammals, we created occurrence matrices for extant mammals by pseudo fossil localities, which were generated using an iterative procedure in R with the maptools, sp, gpclib, ggplot2, rgeos, and MASS packages [48–54] (contact corresponding author for R code). To generate pseudo fossil localities and to ensure that we created pseudo fossil localities with the same spatial distributions as the fossil localities, we fit frequency distributions (normal, gamma, or β) to fossil localities for each NALMA subdivision (Fig. S1). We then generated point samples based on the frequency distributions and the number of fossil localities from which we created occurrence matrices (taxon-by-pseudo locality), repeating the procedure 100 times for each NALMA sub-age for a total of 2100 occurrence matrices. Fossil localities do not record the entire community and so show reduced richness compared to the actual communities (however, note that time averaging also increases richness at fossil localities). Further, most fossil localities, unless intensively screen washed, are biased against small species. Therefore, we also intentionally tested for the effects of sampling bias by removing 25%, 50%, and 75% of species from the extant mammal occurrence matrices for a total of 6300 occurrence matrices. Further, we tested for the effects of body mass bias by 25%, 50%, and 75% of species smaller than 5 kg for a total of 6300 occurrence matrices.

Climate space models

To create climate space models, we sampled the ranges of extant North and South American mammals at a series of 5066 points corresponding to a 1° grid (which we only used to project mammal occurrences under climate change models, but not to calculate biodiversity). Due to the focus on North America, we omitted any species with southern hemisphere ranges that did not cross the equator (n = 602; Table S2). We also excluded rare species (present in <20 cells) for which accurate species distribution models could not be generated (n = 361), leaving 706 species for the climate change projections. We extracted mean annual and winter (December, January, February) temperature and mean annual precipitation data from Climate Wizard (www.climatewizard.org) for the period of 1951–2006 and the following SRES scenarios and time periods: B1 2050s, A1b 2050s, A1b 2080s, A2 2050s, and A2 2080s [55] (Table S1). Each of these projections is based on an ensemble of 16 global circulation models [56]. However, to ensure that we sampled a range of potential warming, we also extracted the ensemble lowest B1 2050s projection (hereafter "B1 2050s low") and the ensemble highest A2 2080s projection (hereafter "A2 2080s high"). This gave a range of warming in North America from 1.49°C (B1 2050s low) to 6.78°C (A2 2080s high, see Table S1 for the full range).

We modeled species' ranges with the BIOMOD package in R using generalized linear models, generalized boosted models, classification tree analysis, artificial neural networks, surface range envelopes, flexible discriminant analysis, multiple adaptive regression splines, and random forests [57] (contact corresponding author for R code). We then used these models to make consensus forecasts for each of the projections described above, as well as current climate to evaluate the performance of the models. We tested model performance using area under the receiver operating curve (AUC), true skill statistic (TSS), and proportion correct classification (PCC, Fig. S2). Species and generic presences were determined across the 1° latitude-longitude grid to give presence or absence in each location at each time and SRES scenario.

Using the projections described above, we created pseudo localities, as before. From this, we created occurrence matrices as described above. We repeated this process 100 times for each projection for a total of 16,800 occurrence matrices.

Latitudinal turnover gradients (LTGs) and β diversity

We calculated β diversity as the change in mammalian communities across the North American landscape using multivariate dispersion and the Jaccard index for each NALMA subage, for modern mammals, and for the climate projections [58]. We calculated Euclidean distances from the centroid for localities using the R package vegan [59]. Larger distances from the centroid indicate greater spatial community turnover and thus higher β diversity. We did not regress the Jaccard index values against distance, as has been used for modern species [2] because we have found such an approach to be highly influenced by species-area relationships.

To estimate ancient, modern, and projected LTG strength for North American mammals, we calculated the amount of community change with latitude using detrended correspondence analysis (DCA; an ordination technique) in the vegan R package [59]. We used explained variance (R^2; how much of the variation in community change is explained by latitude) as a measure of LTG strength [13]. High values of explained variance indicate strong LTGs [60]. We did not compute latitudinal richness gradients because sampling bias (e.g. loss of taxa, body mass bias) is too great (Fraser, D. unpub.).

Sampling bias control

Although we have chosen methods that minimize the effects of sampling bias, we still used multiple methods to control for the non-independence of β diversity from the number of localities, the geographic area sampled, and the number of sampled taxa. We used three approaches. Firstly, we used a re-sampling approach wherein we sub-sampled (without replacement) each NALMA 100× using a standardized number of localities (thirty) and limited to localities occurring between 30°and 50° North latitude. We also re-sampled the extant mammal ranges under various conditions of bias (taxonomic bias through the removal of 25%, 50%, 75% of taxa and body mass bias where we removed 25%, 50%, and 75% of species with a body mass lower than 5 kg) as above to test for direct causality of sampling bias. We also used a method of detrending whereby we regressed LTG strength and β diversity against statistically significant sampling bias metrics and further analyzed the residuals from the model. Finally, we used multivariate linear models to simultaneously account for the model variance explained by sampling and biological phenomena. The last multivariate method is similar to [61] and [62] (also addressed in [63]) who combine the predictive properties of models of biodiversity change and taphonomic bias.

Correlation with climate

We tested for correlations of β diversity and LTG strength with stable oxygen isotopes from benthic foraminifera ($\delta^{18}O$ ‰) [64,65], mean annual precipitation estimated from paleosols [66], number of localities, sampling area (km^2), number of species, latitudinal range (degrees), and length of the sampled interval (Ma) of the fossil localities using generalized least squares and using an autocorrelation structure of order one (corAR1) to account for

temporal autocorrelation in R [67,68]. Best fit models were selected using automated model selection in the MuMIn R package [69] and the Akaike Information Criterion (ΔAIC).

Results

Fossil mammal β diversity showed considerable variation with the warmest intervals (late Eocene, mid-late Oligocene, mid Miocene, and mid Pliocene), but showing generally higher β diversity than with cooler intervals (early Oligocene, late Miocene) (Fig. 1C). The best fit model includes mean annual precipitation (MAP squared), length of the NALMA subdivision, and number of taxa, which together accounts for 67% of model variance (Table 2). β diversity is statistically significant for all three predictors (p<0.05). Residual β diversity is significantly explained by MAP only (Table 2; Fig. 2B). Re-sampling did not alleviate the effects of sampling bias; re-sampled β diversity is significantly explained by MAP-squared, number of taxa, and NALMA subdivision length (Table 2). The remainder of the manuscript will discuss the results from the analyses of raw and residual β diversity only.

Mammalian latitudinal turnover gradients (LTGs) are weak prior to the late Miocene (Fig. 1D). Raw LTG strength (i.e. not detrended) peaks during late Miocene (Hemphillian) and late Pleistocene (Rancholabrean) (Fig. 1D). The best fit model includes mean annual precipitation (MAP) [66], number of taxa, area (km^2) and an the interaction of area and the number of taxa, which explains 47% of the model variance (Table 2; Fig. 2C). LTG strength of late Cenozoic mammal species is statistically significantly explained by all four metrics (p<0.001; Table 2). Residual LTG strength is significantly explained only by MAP (p<0.05; Table 2; Fig. 2D). As above, re-sampling did not alleviate the effects of sampling bias on LTG strength (Table 2). In other words, even accounting for variables that describe potential sources of bias, a climatic variable (MAP) still explains a significant proportion of the variance.

β diversity is much lower for extant mammals than for extinct mammals (Fig. 3A). LTG strength for extant mammals is also greater than for early to mid Cenozoic fossil mammals, but similar to the values for the late Miocene and Pleistocene (Fig. 3B). Extant mammal β diversity shows a slight decrease under incomplete sampling and a slight increase under body-mass–bias sampling (Fig. 3A), but the change is much smaller than observed for fossil mammals. LTG strength does not appear to be significantly affected by the sample size reduction.

Our forecast models (which showed a strong fit to modern mammalian distributions, see Fig. S2A–C) show a slight increase in β diversity for extant mammals (Fig. 3C), but no substantial change in LTG strength compared to the present (Fig. 3D).

Discussion

Spatiotemporal patterns of β diversity remain poorly studied despite being potentially very useful in conservation biology [17,18,70] and linkage to well-studied biogeographic phenomena such as latitudinal richness gradients [4]. Using an extensive analysis of past and present mammalian communities, we demonstrate that, over the past 36 Ma, spatiotemporal patterns of mammal community composition have varied by orders of magnitude in North America. Specifically, Cenozoic spatial turnover of mammal communities is explained by continental mean annual precipitation (MAP) (Fig. 2A–B), broadly supporting predictions drawn from published studies of modern terrestrial organisms [2,70,71] and our predictions outlined above.

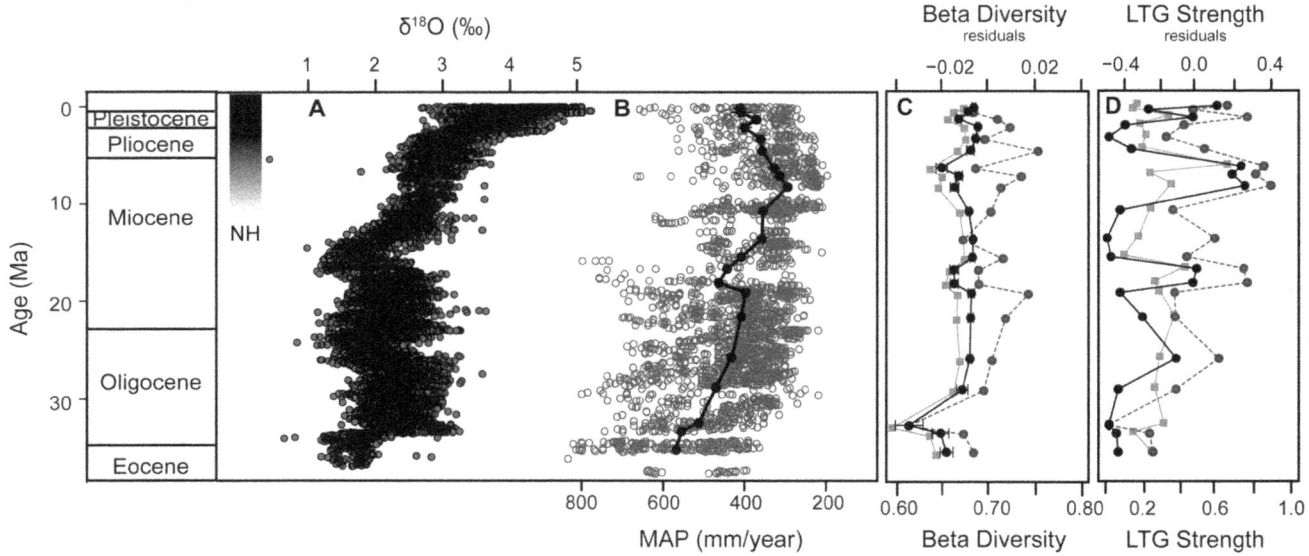

Figure 1. Mid to late Cenozoic trends of (A) $\delta^{18}O$ (‰) from benthic foraminifera (Zachos et al. 2008), (B) mean annual precipitation estimated from paleosols (Retallack, 2007), (C) β diversity of North American mammal species measured using multivariate dispersion (average distance from the centroid), and (D) strength of latitudinal turnover gradients (LTGs) measured as gradient strength for North American fossil mammals. Black lines are raw values, gray lines are residuals from significant sampling bias predictors, and gray dashed lines are re-sampled. Standard errors for re-sampled data are too small to display.

Contemporary ecological theory predicts that mammal diversity either declines monotonically with productivity or shows a unimodal pattern, declining with both low and high productivity [1,2,70,72]. Further, stronger latitudinal diversity gradients are associated with cooler, less productive environments [71] and steeper latitudinal climate gradients [1,70]. Both sets of predictions assume that changes in climate, productivity, and seasonality influence rates of origination and extinction [72,73], niche breadths [74], as well as the carrying capacity of the ecosystem

[75], all factors that change the spatial turnover of terrestrial faunas [70]. Specifically, terrestrial organisms in low latitude, high productivity environments show low rates of speciation and extinction [73], high β diversity [2,76], and weak or absent latitudinal diversity gradients [71]. In contrast, high latitude organisms show high rates of speciation and extinction [73], low β diversity [2,76], and strong latitudinal diversity gradients [71]. Evolutionary history also plays a role in determining rates of spatial community turnover. Modern tropical organisms show

Table 2. Results of best fit generalized least squares models relating β diversity and latitudinal turnover gradient (LTG) strength to mean annual precipitation from paleosols (Retallack, 2007), $\delta^{18}O$ (‰) from benthic forams (mm/year; Zachos et al. 2001; 2008), length of North American Land Mammal Age subdivision, number of taxa sampled, sampling area (km^2), and number of fossil localities.

Dependent Variable	Parameters of Best Fit Model	Variance explained by model (%)	t value	p
Beta Diversity	Mean annual precipitation (quadratic)	66.51	−3.25	0.005
	Length of NALMA subdivision		2.43	0.027
	Number of taxa		5.30	<0.001
Beta Diversity Residuals	Mean annual precipitation (quadratic)	26.48	−3.50	0.002
Beta Diversity Re-sampled	Mean annual precipitation (quadratic)	66.04	−2.39	0.029
	Length of NALMA subdivision		2.51	0.023
	Number of taxa		5.47	<0.001
Latitudinal Turnover Gradient Strength (LTGs)	Mean annual precipitation (quadratic)	46.76	−5.65	<0.001
	Area		−4.62	<0.001
	Number of taxa		−4.36	<0.001
	Area : Number of taxa		4.85	<0.001
LTG Residuals	Mean annual precipitation (linear)	37.48	−3.79	0.001
LTG Re-sampled	Number of taxa	28.59	−2.55	0.020

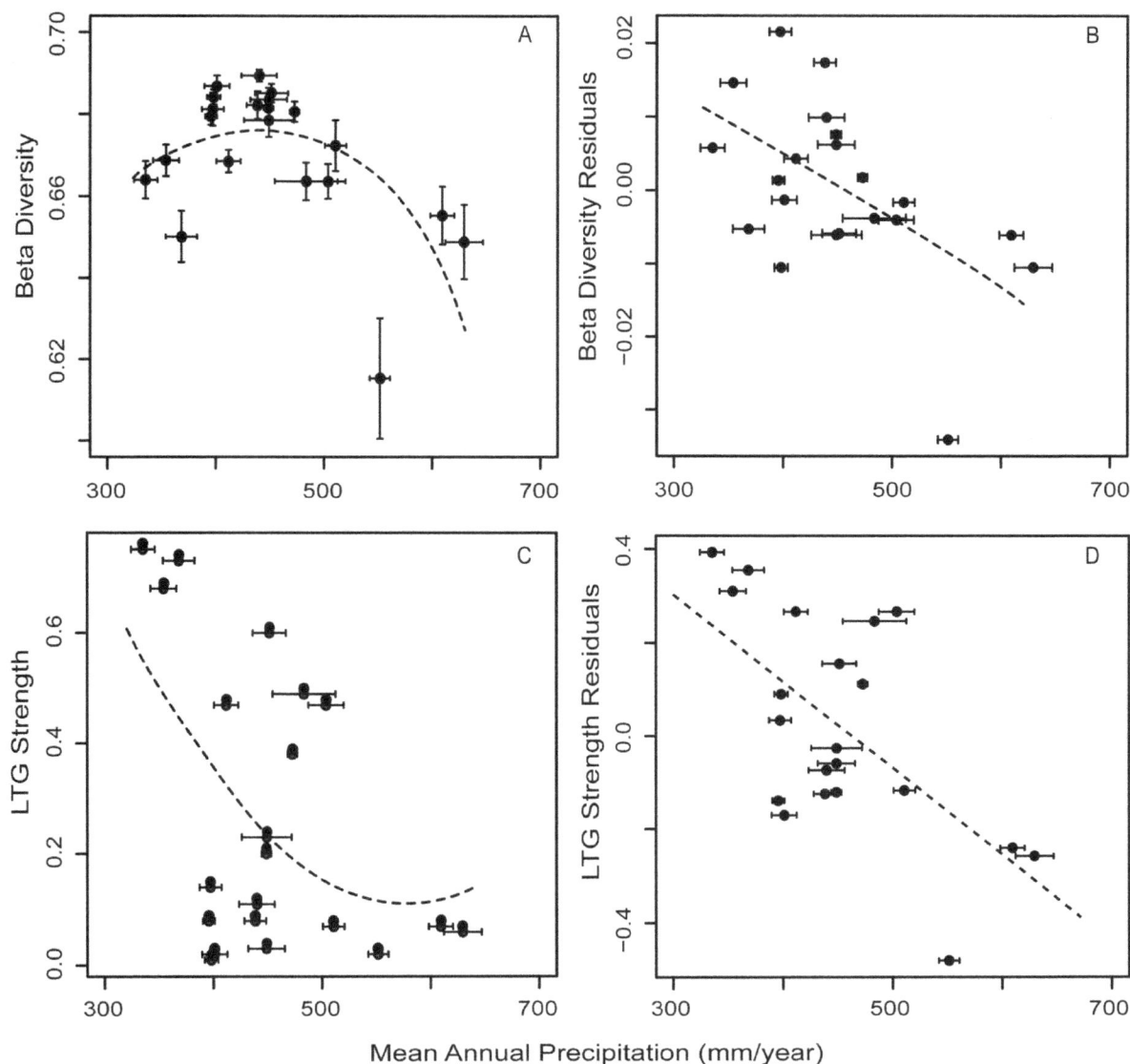

Figure 2. Relationship of mean annual precipitation estimated from paleosols (Retallack, 2007) with North American fossil mammal (A) raw β diversity ($R^2 = 0.43$), (B) residual beta diversity ($R^2 = 0.26$) and (C) raw latitudinal turnover gradient (LTG) strength ($R^2 = 0.25$), and (D) residual LTG strength ($R^2 = 0.37$).

faster turnover than their temperate counterparts regardless of the rate of environmental change [70]. Spatial and, by extension, temporal patterns of β diversity are the result of a mosaic of ecological and evolutionary processes.

Cenozoic fossil mammal β diversity peaked at intermediate values of mean annual precipitation and declined under both drier and wetter conditions (MAP; ~450 mm per year; Fig. 2B), showing a similar shape to latitudinal diversity curves for modern mammals [71]. Mammal β diversity was similarly lowest during periods of relative cooling, including the early Oligocene and late Miocene, coincident with declining atmospheric CO_2 [77–80] and, in the latter case, the expansion of ice sheets in the Northern Hemisphere [27,36], strengthening of thermohaline circulation [27,37,81–84], and transition from C_3 to C_4 dominated ecosystems at middle latitudes [66,85,86]. Declining β diversity during the late Miocene is also coincident with increased maximum body

mass [87], an ecologically relevant characteristic linked to lower ecosystem energy [88,89]. Water is a key component in photosynthesis and therefore net primary productivity (NPP) and MAP are correlated at a global scale, showing an asymptotic relationship [90]. Our results therefore suggest that putatively lower energy ecosystems (e.g. early Oligocene, late Miocene) supported more spatially homogenous mammal faunas than putatively higher energy ecosystems (e.g. late Eocene, mid Miocene, mid Pliocene). Temporal changes in fossil mammal β diversity (this study) are therefore conceptually similar to spatial patterns observed in extant mammals.

Early Oligocene mammals had lower β diversity than expected based on MAP (Fig. 1C; Fig. 2A). The early Oligocene is associated with rapid global cooling [64] and expansion of open grassy ecosystems [91], which may have resulted in lower ecosystem energy. However, our taxonomic sample is the poorest

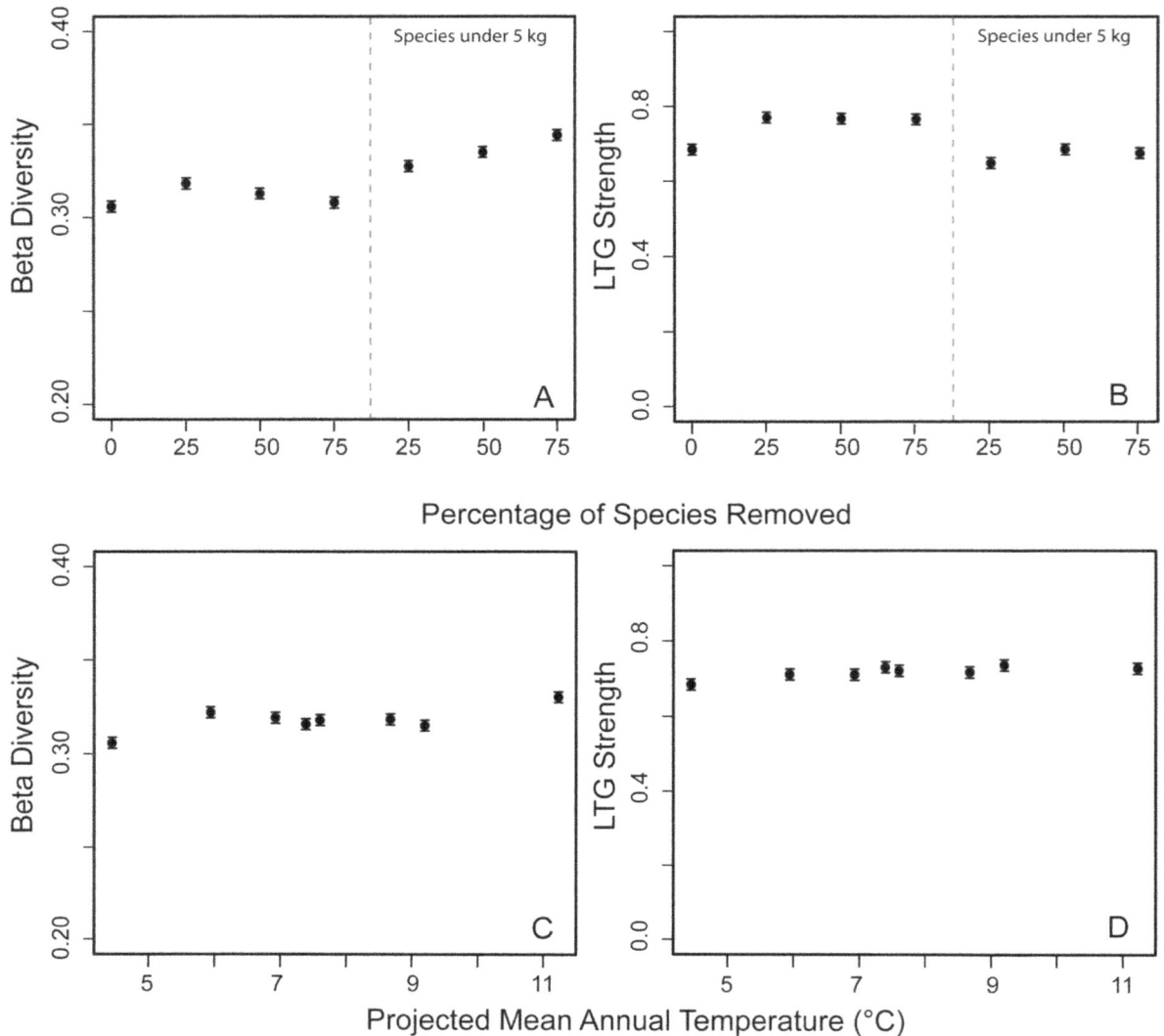

Figure 3. (A) β diversity (distance from centroid) and (B) latitudinal turnover gradients (LTG) strength of extant North American mammals under incomplete taxonomic sampling (removal of 25, 50, and 75% of species in sample) and body mass bias (removal of 25, 50, 75% of species smaller than 5 kg) and (C) β diversity (distance from centroid) and (D) latitudinal turnover gradients (LTG) strength of extant North American mammals under several International Panel on Climate Change scenarios (Special Reports on Emissions Scenarios).

for the early Oligocene; number of taxa is a significant predictor of fossil mammal β diversity (Table 2), suggesting some variation in preservation of species among NALMA subdivisions. Rarefied diversity also shows little change from the late Eocene to the early Oligocene [10]. However, our incomplete sampling trials show that removing even 75% of species reduces β diversity by a negligible amount (Fig. 3A), suggesting that at least some (but not all) of the observed decline in early Oligocene β diversity may have been climatically driven.

The magnitude of the latitudinal turnover gradient (LTG) for fossil mammals shows a temporally cyclic pattern that increases in amplitude during the late Cenozoic as well as a general trend toward stronger LTGs (Fig. 1D), coincident with the formation of ice on Svalbard at ~15 Ma and perennial Arctic sea ice at ~14 Ma, declining atmospheric CO_2 [37], and declining terrestrial MAP (Fig. 2B). Specifically, LTGs are strongest when precipitation is lowest (putatively lower productivity environments)

and weakest at when precipitation is highest (putatively high productivity environments; Fig. 2B), similar to modern mammals that show weak or absent latitudinal diversity gradients in the tropics and strong diversity gradients at mid to high latitudes [71]. Climate gradients are steeper at mid to high latitudes in North America due to the albedo of high latitude glaciation. Northern glaciation is an important means by which solar radiation is reflected from high latitudes, resulting in cool, low productivity Arctic environments [92,93]. Mammal communities are sorted along a latitudinal axis according to their climatic tolerances and the process of abiotic filtering, whereby taxa meet the limits of their environmental tolerances and are excluded from communities farther north [94]. Although late Miocene sea and land ice thickness and extent were reduced compared to the modern, increasing northern albedo and strengthening of thermohaline circulation are coincident with that strengthening of mammal

LTGs during the late Miocene (25–60% stronger than for any preceding NALMA; Fig. 1D) [27,81–84].

At first glance, the Pliocene appears to be anomalous because the magnitude of the mammalian LTG declines dramatically (60–70% reduction in the magnitude of the LTG; Fig. 1D). However, evidence from fossil deposits on Ellesmere Island show that approximately 3.5 Ma the Pliocene Arctic was ~14–22°C warmer than present [83,95,96] with an associated reduced volume of Arctic sea ice [27,82]. Pliocene Arctic warming is similarly coincident with reduced richness gradients of marine zooplankton [81]. The Pliocene might therefore be the "exception" that proves the rule.

Under modern global warming, Arctic winter temperatures have increased at a greater rate than at southern latitudes [97]. Long-term projections suggest boosts in high latitude net primary productivity due to increasing nitrogen fertilization and increases in mean annual precipitation of 100–150 mm per year or 5–20% at middle to high latitudes [98]. From our analyses of fossil North American mammals and published studies of beta diversity [18], we therefore expect weakened climate gradients and thus weakened LTGs due to northward range shifting, and, in the long-term, declining β diversity under the influence of modern anthropogenic climate change. β diversity decline may be facilitated by the homogenization of communities due to any of the following (note the lack of mutual exclusivity): i) extinction of species with small geographic ranges and replacement with wide-ranging species, ii) evolution toward larger range sizes within species, and, iii) invasion by wide-ranging species even without the extinction of residents [18]. However, our climate space models that are based on SRES scenarios corresponding to absolute mean annual temperatures of 4.4–11.2°C (averaged across North and South America) did not show changes in mammal LTGs or β diversity (Fig. 3C–D). We suggest that climate space models (CSMs) are unlikely to accurately forecast the outcomes of anthropogenic climate change for modern mammals because current CSM algorithms do not incorporate microevolutionary, macroevolutionary, or ecological processes, such as niche shifts, niche creation, and differences in dispersal abilities that are inherent in the response of animals to climate change. However, even on modern ecological timescales, rapid evolutionary changes and niche shifts have been observed in native and invasive populations [41], and this local adaptation complicates the prediction of range shifts. On longer timescales, taxa adapt to new climates and the processes of speciation and extinction help form new terrestrial communities. Without the explicit inclusion of evolutionary parameters and historical data for the taxa of interest, we are unlikely to accurately predict long-term changes in terrestrial biodiversity patterns.

We have shown here that macroecological patterns of North American mammal community composition varied considerably over the past 35 million years in response to changes in global climate change and Arctic glaciation (Fig. 1C–D). Furthermore, our comparison of fossil evidence with climate-space forecast models (CSMs) suggests that CSMs (in which species are modeled to simply track climate variables) may distort the degree of community composition change we should expect in the future. A unifying ecological theory relating diversity to climate must address both the spatial and temporal dimensions of diversity, as well as both richness and community composition. However, studies of organismal richness are far more common than studies of community composition (β diversity), despite the importance of the latter in conservation and their vast potential for contributing to our understanding of the processes underlying modern biodiversity. Studying the community composition of fossil animals represents a new frontier in paleontological research with potential to truly inform modern conservation.

Supporting Information

Figure S1 Maps of North America showing the distribution of fossil localities for all sampled North American Land Mammal Age subdivisions.

Figure S2 Model fit statistics for climate space models of extant North American mammals. Model performance was tested using area under the operating curve (A; AUC), the true skill statistics (B; TSS), and the proportion of correct classification (C).

Table S1 Summary of Special Emissions Report Scenarios (SERs) to which we fit climate models for extant mammalian species.

Table S2 List of mammalian taxa included and excluded from the species distribution models.

Appendix S1 Sources for the majority of mammal occurrence data downloaded from the Fossilworks database.

Acknowledgments

We thank John P. Hunter for a thorough review of this paper. Further, we thank John Alroy for his substantial contributions to the fossil data used in this analysis, accessed via his Fossilworks website, and his detailed review of the paper. We would also like to thank two anonymous reviewers, D. Currie, M. Clementz, M. Churchill, R. Haupt, J. Hoffmann, and E. Lightner for reviewing earlier versions of this manuscript, as well as L. Fahrig and S. Kim for constructive comments on this project.

Author Contributions

Conceived and designed the experiments: DF CH NR. Performed the experiments: DF. Analyzed the data: DF. Contributed reagents/materials/analysis tools: CH RG. Contributed to the writing of the manuscript: DF. Manuscript copyediting: CH RG NR.

References

1. Hawkins BA, Field R, Cornell HV, Currie DJ, Guegan JF, et al. (2003) Energy, water, and broad-scale geographic patterns of species richness. Ecology 84: 3105–3117.

2. Qian H, Badgley C, Fox DL (2009) The latitudinal gradient of beta diversity in relation to climate and topography for mammals in North America. Global Ecology and Biogeography 18: 111–122.

3. Condit R, Pitman N, Leigh EG Jr, Chave J, Terborgh J, et al. (2002) Beta-diversity in tropical forest trees. Science 295: 666–669.

4. Baselga A, Lobo JM, Svenning JC, Aragón P, Araújo MB (2012) Dispersal ability modulates the strength of the latitudinal richness gradient in European beetles. Global Ecology and Biogeography 21: 1106–1113.

5. Engle VD, Summers JK (1999) Latitudinal gradients in benthic community composition in Western Atlantic estuaries. Journal of Biogeography 26: 1007–1023.

6. Condamine FL, Sperling FAH, Wahlberg N, Rasplus JY, Kergoat GJ (2012) What causes latitudinal gradients in species diversity? Evolutionary processes and ecological constraints on swallowtail biodiversity. Ecology Letters 15: 267–277.

7. Currie DJ, Fritz JT (1993) Global patterns of animal abundance and species energy use. Oikos 67: 56–68.

8. Currie DJ, Francis AP, Kerr JT (1999) Some general propositions about the study of spatial patterns of species richness. Ecoscience 6: 392–399.

9. Mittelbach GG, Schemske DW, Cornell HV, Allen AP, Brown JM, et al. (2007) Evolution and the latitudinal diversity gradient: speciation, extinction and biogeography. Ecology Letters 10: 315–331.

10. Figueirido B, Janis CM, Pérez-Claros JA, Renzi MD, Palmqvist P (2012) Cenozoic climate change influences mammalian evolutionary dynamics. Proceedings of the National Academy of Sciences USA 109: 722–727.

11. Sepkoski JJ (1998) Rates of speciation in the fossil record. Philosophical Transactions of the Royal Society of London B 353: 315–326.

12. Mayhew PJ, Bell MA, Benton TG, McGowan AJ (2012) Biodiversity tracks temperature over time. Proceedings of the National Academy of Sciences USA 109: 15141–15145.

13. Kent R, Bar-Massada A, Carmel Y (2011) Multiscale analyses of mammal species composition-environment relationship in the contiguous USA. PLoS One 6: e25440.

14. Legendre P, Borcard D, Peres-Neto PR (2005) Analyzing beta diversity: partitioning the spatial variation of community composition data. Ecological Monographs 75: 435–450.

15. Whittaker RJ, Willis KJ, Field R (2001) Scale and species richness: towards a general, hierarchical theory of species diversity. Journal of Biogeography 28: 453–470.

16. Soininen J (2010) Species turnover along abiotic and biotic gradients: patterns in space equal patterns in time? BioScience 60: 433–439.

17. Hassall C, Hollinshead J, Hull A (2012) Temporal dynamics of aquatic communities and implications for pond conservation. Biodiversity and Conservation 21: 829–852.

18. Dornelas M, Gotelli NJ, McGill B, Shimadzu H, Moyes F, et al. (2014) Assemblage time series reveal biodiversity change but not systematic loss. Science 344: 296–299.

19. Janis CM, Damuth J, Theodor JM (2000) Miocene ungulates and terrestrial primary productivity: where have all the browsers gone? Proceedings of the National Academy of Sciences USA 97: 7899–7904.

20. Atwater AL, Davis EB (2011) Topographic and climate change differentially drive Pliocene and Pleistocene mammalian beta diversity of the Great Basin and Great Plains provinces of North America. Evolutionary Ecology Research 13: 833–850.

21. Finarelli JA, Badgley C (2010) Diversity dynamics of Miocene mammals in relation to the history of tectonism and climate. Proceedings of the Royal Society of London, Series B 277: 2721–2726.

22. Davis EB (2005) Mammalian beta diversity in the Great Basin, western USA: palaeontological data suggest deep origin of modern macroecological structure. Global Ecology and Biogeography 14: 479–490.

23. Barnosky AD, Hadly EA, Bell CJ (2003) Mammalian response to global warming on varied temporal scales. Journal of Mammalogy 84: 354–368.

24. Barnosky AD (2005) Effects of Quaternary climatic change on speciation in mammals. Journal of Mammalian Evolution 12: 247–264.

25. Eberle J, Fricke H, Humphrey J (2009) Lower-latitude mammals as year-round residents in Eocene Arctic forests. Geology 37: 499–502.

26. Eberle JJ, Fricke HC, Humphrey JD, Hackett L, Newbrey MG, et al. (2010) Seasonal variability in Arctic temperatures during early Eocene time. Earth and Planetary Science Letters 296: 481–486.

27. Polyak L, Alley RB, Andrews JT, Brigham-Grette J, Cronin TM, et al. (2010) History of sea ice in the Arctic. Quaternary Science Reviews 29: 1757–1778.

28. Clementz MT, Sewall JO (2011) Latitudinal gradients in greenhouse seawater $\delta^{18}O$: evidence from Eocene sirenian tooth enamel. Science 332: 455–458.

29. Micheels A, Bruch A, Mosbrugger V (2009) Miocene climate modelling sensitivity experiments for different CO_2 concentrations. Palaeontologia Electronica 12: 5A.

30. Post E, Forchhammer MC, Bret-Harte MS, Callaghan TV, Christensen TR, et al. (2009) Ecological dynamics across the Arctic associated with recent climate change. Science 325: 1355–1358.

31. Primack RB, Ibáñez I, Higuchi H, Lee SD, Miller-Rushing AJ, et al. (2009) Spatial and interspecific variability in phenological responses to warming temperatures. Biological Conservation 142: 2569–2577.

32. Bradshaw WE, Holzapfel CM (2006) Evolutionary response to rapid climate change. Science 312: 1477–1478.

33. Parmesan C (2006) Ecological and evolutionary responses to recent climate change. Annual Review of Ecology and Systematics 37: 637–639.

34. Chen IC, Hill JK, Ohlemüller R, Roy DB, Thomas CD (2011) Rapid range shifts of species associated with high levels of climate warming. Science 333: 1024–1026.

35. Parmesan C, Yohe G (2003) A globally coherent fingerprint of climate change impacts across natural systems. Nature 421: 37–42.

36. Foster GL, Lunt DJ, Parrish RR (2009) Mountain uplift and the threshold for sustained Northern Hemisphere glaciation. Climate of the past discussions 5: 2439–2464.

37. Foster GL, Lear CH, Rae JWB (2012) The evolution of pCO2, ice volume and climate during the middle Miocene. Earth and Planetary Science Letters 341–344: 243–254.

38. Lawler JJ, White D, Neilson RP, Blaustein AR (2006) Predicting climate-induced range shifts: model differences and model reliability. Global Change Biology 12: 1568–1584.

39. Hoffmann AA, Sgró CM (2011) Climate change and evolutionary adaptation. Nature 470: 479–485.

40. Thuiller W, Münkemüller T, Lavergne S, Mouillot D, Mouquet N, et al. (2013) A road map for integrating eco-evolutionary processes into biodiversity models. Ecology Letters 16: 94–105.

41. Lavergne S, Mouquet N, Thuiller W, Ronce O (2010) Biodiversity and climate change: integrating evolutionary and ecological responses of species and communities. Annual Review of Ecology, Evolution, and Systematics 41: 321–350.

42. Sepkoski JJ (1997) Biodiversity: past, present, and future. Journal of Paleontology 71: 533–539.

43. Rose PJ, Fox DL, Marcot J, Badgley C (2011) Flat latitudinal gradient in Paleocene mammal richness suggests decoupling of climate and biodiversity. Geology 39: 163–166.

44. Secord R, Bloch JI, Chester SGB, Boyer DM, Wood AR, et al. (2012) Evolution of the earliest horses driven by climate change in the Paleocene-Eocene thermal maximum. Science 335: 959–962.

45. Patterson BD, Ceballos G, Sechrest W, Tognelli MF, Brooks T, et al. (2007) Digital distribution maps of the mammals of the Western Hemisphere, version 3.0. NatureServe, Arlington, Virginia, USA.

46. Carrasco MA, Kraatz BP, Davis EB, Barnosky AD (2005) Miocene mammal mapping project (MIOMAP). University of California Museum of Paleontology.

47. McCoy ED, Connor EF (1980) Latitudinal gradients in the species diversity of North American mammals. Evolution 34: 193–203.

48. Lewin-Koh NJ, Bivand R (2008) maptools package version 0.8–16.

49. Pebesma EJ, Bivand RS (2005) Classes and methods for spatial data in R. R News 5.

50. Bivand RS, Pebesma EJ, Gomez-Rubio V (2008) Applied spatial data analysis with R. New York: Springer.

51. Peng RD (2007) The gpclib package version 1.5–5.

52. Wickham H (2009) ggplot2: elegant graphics for data analysis. New York: Springer.

53. Bivand R, Rundel C (2012) rgeos: interface to geometry engine version 0.2–16.

54. Venables WN, Ripley BD (2002) Modern and applied statistics with S. New York: Springer.

55. Nakicenovic N, Swart R (2000) Emissions scenarios: a special report of Working Group III of the Intergovernmental Panel on Climate Change. Cambridge: Cambridge University Press.

56. Girvetz EH, Zganjar C, Raber GT, Maurer EP, Kareiva P, et al. (2009) Applied climate-change analysis: the climate wizard tool. PLoS One 4: e8320.

57. Thuiller W, Georges D, Engler R (2012) BIOMOD: Ensemble platform for species distribution modeling. Ecography 32: 369–373.

58. Anderson MJ, Ellingsen KE, McArdle BH (2006) Multivariate dispersion as a measure of beta diversity. Ecology Letters 9: 683–693.

59. Oksanen J, Blanchet FG, Roeland Kindt PL, Minchin PR, O'Hara RB, et al. (2012) Package vegan version 2.0–7.

60. Tuomisto H, Ruokolainen K (2006) Analyzing and explaining beta diversity? understanding the targets of different methods of analysis. Ecology 87: 2697–2708.

61. Benson RBJ, Mannion PD (2012) Multi-variate models are essential for understanding vertebrate diversification in deep time. Biology Letters 8: 127–130.

62. Mannion PD, Upchurch P, Carrano MT, Barrett PM (2011) Testing the effect of the rock record on diversity: a multidisciplinary approach to elucidating the generic richness of sauropodomorph dinosaurs through time. Biological Reviews 86: 157–181.

63. Benton MJ, Dunhill AM, Lolyd GT, Marx FG (2011) Assessing the quality of the fossil record: insights from vertebrates. In: A. J McGowan and A. B Smith, editors. Comparing the geological and fossil records: implications for biodiversity studies. London: Geological Society of London. 63–94.

64. Zachos JC, Dickens GR, Zeebe RE (2008) An early Cenozoic perspective on greenhouse warming and carbon-cycle dynamics. Nature 451: 279–283.

65. Zachos J, Pagani M, Sloan L, Thomas E, Billups K (2001) Trends, rhythms, and aberrations in global climate 65 Ma to present. Science 292: 686–693.

66. Retallack GJ (2007) Cenozoic paleoclimate on land in North America. Journal of Geology 115: 271–294.

67. Development core team R (2012) R: A language and environment for statistical computing. Vienna, Austria: Foundation for Statistical Computing.

68. Dornelas M, Magurran AE, Buckland ST, Chao A, Chazdon RL, et al. (2013) Quantifying temporal change in biodiversity: challenges and opportunities. Proceedings of the Royal Society B 280: 1–10.

69. Barton K (2013) Multi-model inference package 'MuMIn' version 1.10.0 (http://cran.r-project.org/web/packages/MuMIn/MuMIn.pdf).

70. Buckley LB, Jetz W (2008) Linking global turnover of species and environments. Proceedings of the National Academy of Sciences USA 105: 17836–17841.

71. Currie DJ (1991) Energy and large-scale patterns of animal- and plant-species richness. American Naturalist 137: 27–49.

72. VanderMeulen MA, Hudson AJ, Scheiner SM (2001) Three evolutionary hypotheses for the hump-shaped productivity–diversity curve. Evolutionary Ecology Research 3: 379–392.

73. Weir JT, Schluter D (2007) The latitudinal gradient in recent speciation and extinction rates of birds and mammals. Science 315: 1574–1576.

74. Vázquez DP, Stevens RD (2004) The latitudinal gradient in niche breadth: concepts and evidence. American Naturalist 164: E1–E19.

75. Buckley LB, Davies J, Ackerly DD, Kraft NJB, Harrison SP, et al. (2010) Phylogeny, niche conservatism and the latitudinal diversity gradient in mammals. Proceedings of the Royal Society, Series B 277: 2121–2138.

76. Qian H, Xiao M (2012) Global patterns of the beta diversity energy relationship in terrestrial vertebrates. Acta Oecologica 39: 67–71.

77. Franks PJ, Beerling DJ (2009) Maximum leaf conductance driven by CO$_2$ effects on stomatal size and density over geologic time. Proceedings of the National Academy of Sciences USA 106: 10343–10347.

78. DeConto RM, Pollard D, Wilson PA, Pälike H, Lear CH, et al. (2008) Thresholds for Cenozoic bipolar glaciation. Nature 455: 652–657.

79. Tripati AK, Roberts CD, Eagle RA (2009) Coupling of CO$_2$ and ice sheet stability over major climate transitions of the last 20 million years. Science 326: 1394–1397.

80. Zhang YG, Pagani M, Liu Z, Bohaty SM, DeConto R (2013) A 40-million-year history of atmospheric CO$_2$. Philosophical Transactions of the Royal Society, Series A 371: 1–20.

81. Yasuhara M, Hunt G, Dowsett HJ, Robinson MM, Stoll DK (2012) Latitudinal species diversity gradient of marine zooplankton for the last three million years. Ecology Letters 15: 1174–1179.

82. Haywood AM, Valdes PJ, Sellwood BW, Kaplan JO, Dowsett HJ (2001) Modelling middle Pliocene warm climates of the USA. Palaeontologia Electronica 4: 1–21.

83. Ballantyne AP, Greenwood DR, Damsté JSS, Csank AZ, Eberle JJ, et al. (2010) Significantly warmer Arctic surface temperatures during the Pliocene indicated by multiple independent proxies. Geology 38: 603–606.

84. Ballantyne AP, Rybczynski N, Baker PA, Harington CR, White D (2006) Pliocene Arctic temperature constraints from the growth rings and isotopic composition of fossil larch. Palaeogeography, Palaeoclimatology, Palaeoecology 242: 188–200.

85. Fox DL, Honey JG, Martin RA, Peláez-Campomanes P (2012) Pedogenic carbonate stable isotope record of environmental change during the Neogene in the southern Great Plains, southwest Kansas, USA: Oxygen isotopes and paleoclimate during the evolution of C$_4$-dominated grasslands. Geological Society of America Bulletin 124: 431–443.

86. Strömberg CAE, McInerney FA (2011) The Neogene transition from C$_3$ to C$_4$ grasslands in North America: assemblage analysis of fossil phytoliths. Paleobiology 37: 50–71.

87. Smith FA, Boyer AG, Brown JH, Costa DP, Dayan T, et al. (2010) The evolution of maximum body size of terrestrial mammals. Science 330: 1216–1219.

88. Freckleton RP, Harvey PH, Pagel M (2003) Bergmann's rule and body size in mammals. American Naturalist 161: 821–825.

89. Blackburn TM, Gaston KJ, Loder N (1999) Geographic gradients in body size: a clarification of Bergmann's rule. Diversity & Distributions 5: 165–174.

90. Del Grosso S, Parton W, Stohlgren T, Zheng D, Bachelet D, et al. (2008) Global potential net primary production predicted from vegetation class, precipitation, and temperature. Ecology 89: 2117–2126.

91. Jacobs BF, Kingston JD, Jacobs LL (1999) The origin of grass-dominated ecosystems. Annals of Missouri Botanical Garden 86: 590–643.

92. Alexeev VA, Langen PL, Bates JR (2005) Polar amplification of surface warming on an aquaplanet in "ghost forcing" experiments without sea ice feedbacks. Climate Dynamics 24: 655–665.

93. Holland MM, Bitz CM (2003) Polar amplification of climate change in coupled models. Climate Dynamics 21: 221–232.

94. Soininen J, McDonald R, Hillebrand H (2007) The distance decay of similarity in ecological communities. Ecography 30: 3–12.

95. Csank AZ, Tripati AK, Patterson WP, Eagle RA, Rybczynski N, et al. (2011) Estimates of Arctic land surface temperatures during the early Pliocene from two novel proxies. Earth and Planetary Science Letters 304: 291–299.

96. Rybczynski N, Gosse JC, Harington CR, Wogelius RA, Hidy AJ, et al. (2013) Mid-Pliocene warm-period deposits in the High Arctic yield insight into camel evolution. Nature Communications 4: 1–9.

97. Kaplan JO, Bigelow NH, Prentice IC, Harrison SP, Bartlein PJ, et al. (2003) Climate change and Arctic ecosystems: 2. Modeling, paleodata-model comparisons, and future projections. Journal of Geophysical Research 108: 1–17.

98. Oechel WC, Vourlitis GL (1994) The effects of climate change on land-atmosphere feedbacks in arctic tundra regions. Trends in Ecology and Evolution 9: 324–329.

13

Molecular Phylogeography and Population Genetic Structure of *O. longilobus* and *O. taihangensis* (*Opisthopappus*) on the Taihang Mountains

Yiling Wang, Guiqin Yan*

College of Life Sciences, Shanxi Normal University, Linfen, China

Abstract

Historic events such as the uplift of mountains and climatic oscillations in the Quaternary periods greatly affected the evolution and modern distribution of the flora. We sequenced the *trnL–trnF*, *ndhJ-trnL* and ITS from populations throughout the known distributions of *O. longilobus* and *O. taihangensis* to understand the evolutionary history and the divergence related to the past shifts of habitats in the Taihang Mountains regions. The results showed high genetic diversity and pronounced genetic differentiation among the populations of the two species with a significant phylogeographical pattern ($N_{ST} > G_{ST}$, $P < 0.05$), which imply restricted gene flow among the populations and significant geographical or environmental isolation. Ten chloroplast DNA (cpDNA) and eighteen nucleus ribosome DNA (nrDNA) haplotypes were identified and clustered into two lineages. Two corresponding refuge areas were revealed across the entire distribution ranges of *O. longilobus* and at least three refuge areas for *O. taihangensis*. *O. longilobus* underwent an evolutionary historical process of long-distance dispersal and colonization, whereas *O. taihangensis* underwent a population expansion before the main uplift of Taihang Mountains. The differentiation time between *O. longilobus* and *O. taihangensis* is estimated to have occurred at the early Pleistocene. Physiographic complexity and paleovegetation transition of Taihang Mountains mainly shaped the specific formation and effected the present distribution of these two species. The results therefore support the inference that Quaternary refugial isolation promoted allopatric speciation in Taihang Mountains. This may help to explain the existence of high diversity and endemism of plant species in central/northern China.

Editor: Igor Mokrousov, St. Petersburg Pasteur Institute, Russian Federation

Funding: The study was financially supported by Natural Science Foundation of Shanxi Normal University (ZR1106), Natural Science Foundation of Shanxi Province, China (2011011031-2) and High-tech Industrialization project of Shanxi Province, China (20110014). The funders had no role in study design, data collection and analysis, decision to publish, or preparation of the manuscript.

Competing Interests: The authors have declared that no competing interests exist.

* Email: guiqiny@dns.sxnu.edu.cn

Introduction

During the last three million years, repeated glacial and interglacial cycles have greatly affected the landform and fauna of the total earth [1–3]. Historical ecological and biogeographic factors are often considered to have played significant roles in shaping global biodiversity by influencing regional differences in speciation, extinction, and migration [4–6]. The genetic structures of the current species record the simultaneous consequences of two fundamental processes: population dynamics in response to past geological or climatic changes, and lineage sorting within a species under natural selection. Thus, knowledge of the evolutionary history of many plant species is central to the identification of divergence and speciation processes [3,7–8].

Phylogeographical analyses can provide an understanding of how paleo-environmental changes in landscape and climate have influenced species distributions and population demography [9]. Such analyses play an important role in understanding the evolutionary history of species with changes [3]. Over the last decade, molecular research on the evolutionary history of plant species, both in China and throughout East Asia, has been conducted with reference to past climatic oscillations [10–18]. One of the results of such research is the proposal by some authors that

glacial refugia were maintained in both the northern and the southern regions, or at different spatial-temporal scales in China, during these glacial periods. These refugia are thus suggested to have acted as sites for subsequent range expansion during the interglacial (or postglacial) periods [12–14,19–20]. Some plant species have extended their distribution from southwestern China to central/northern China [21]. Reduction of genetic diversity is expected under a scenario of rapid postglacial expansion, as has been found in northern Europe and America [2]. Because central/northern China is believed not to have been glaciated during the Quaternary [2,22–23], it is an open question whether species experienced past northern expansion and southern retreat during the Quaternary climatic oscillations.

Taihang Mountains locates on 35°19′–40°51′N and 113°10′–115°48′E. As a natural boundary mountain of the east, southeast of Shanxi province and Hebei and Henan provinces, it is the important mountain range and geographical boundary for eastern China. The northern part of Taihang Mountains is higher than its southern part. Most areas of Taihang Mountains are higher than 1200 meters. The uplift of Taihang Mountains, followed by the formation of high mountains and deep valleys within the plateau [24–25], was one of the most important geological events. Several lines of evidence suggest that the rapid uplift of Taihang

Mountains took place after the late Pleistocene [26–27]. Rich sources of species, most of which are endemic, are found in this region [24]. The high species richness in Taihang Mountains has led to the hypothesis that this region is a distribution and diversity center for many plant genera [24].

Molecular techniques have provided powerful tools for studying the phylogeography or migratory footprints of species [7–9,28–29]. Maternally inherited chloroplast DNA lineages in natural populations often display geographical structure [9], which is useful for deciphering the evolutionary history of species. The cpDNA markers are thought to be the most appropriate candidates because of their slow evolution and lack of recombination [30]. Nevertheless, the joint use of molecular markers derived from different genomes provides a more complete description of population structure and insights into population history and dynamics, particularly for comparisons of maternally inherited organelle and bi-parentally inherited nuclear markers [31]. Several studies about inter-specific and intra-specific divergence from China, using different molecular markers, have played an important role in discovering phylogeographical patterns in East Asian flora. This is particularly true for the mechanisms of plant speciation, along with the production of the high plant biodiversity and endemism found in East Asian flora [12–13,15,20,32]. However, most of the studied species from China have been trees [13,33–34]. Herbaceous vegetation has received much less attention [16]. Therefore, it would be of great interest to study the phylogeography of an herbaceous species to understand the evolution and modern distribution of the vegetation, especially in Taihang Mountains of central/northern China.

Opisthopappus longilobus Shih and *Opisthopappus taihangensis* (Ling) Shih belong to *Opisthopappus* (Asteraceae). The genus *Opisthopappus* is endemic to China, and its wild distribution is mainly restricted to Taihang Mountains across the provinces of Shanxi, Hebei, and Henan, occurring on the slopes at an elevation of about 1000 m or in the cracks of the steep cliffs [35]. *O. longilobus* is mainly distributed in the province of Hebei, whereas *O. taihangensis* is found in the provinces of Shanxi and Henan [35–36]. Both species are diploid (2n = 18) [25]. A strict morphological differentiation occurred between *O. longilobus* and *O. taihangensis*. The leaf blade of *O. longilobus* is smooth and subcylindrical, and most stem and leaves are pinnatifid, with one pair of bracteal leaf. *O. taihangensis*, apprised pubescent on both surfaces of leaf blade, stem, and leaves, is bipinnatifid, with no bracteal leaf [35–36]. *O. longilobus* and *O. taihangensis* have been listed among the Class II State-Protected Endangered Plant Species [36] due to the narrow distribution range, unique ecological environment, and serious artificial plucking.

O. longilobus and *O. taihangensis* are an ideal candidate for investigating the influences of Quaternary climate change in Taihang Mountains for the following reasons: (i) Geographically, *O. longilobus* and *O. taihangensis* are distributed only in Taihang Mountains, including Shanxi, Hebei, and Henan provinces. This geographic range may be advantageous for enabling investigation into whether the glacial refuges were maintained in Taihang Mountains during the last glacial maximum, or earlier cold periods. (ii) *O. longilobus* and *O. taihangensis* have intra-specific and endemic taxa in China that display visible phenotypic differences, which are important in exploring the incipient speciation caused by isolation provided by a glacial refuge. Finally, (iii) *O. longilobus* and *O. taihangensis* are an herbaceous plant and sensitive to environmental changes. In research on the effects of climate oscillations on plant evolution, it can represent the phylogeography of an herb, and can serve as a model to reveal

the historical and evolutionary processes of such plants since the period of Quaternary climate change.

Thus, our objectives included the following: (1) explore the phylogeographical structure and the evolutionary history of *O. longilobus* and *O. taihangensis*; (2) elucidate the effects of the uplift of Taihang Mountains and the Quaternary climatic oscillations on the genetic structure, divergence and distribution of *O. longilobus* and *O. taihangensis*; (3) infer the possible refugia of *O. longilobus* and *O. taihangensis* across their whole distribution range.

Materials and Methods

Ethics statement

This study was conducted in accordance with all People's Republic of China laws. No specific permits were required for the described field studies because all researchers collecting the samples had introduction letters from the College of Life Science, Shanxi Normal University, Linfen.

Plant sampling

Thirteen populations of *Opisthopappus* were investigated in this study, covering the entire distribution area on Taihang Mountains. The samples comprised five populations of *O. longilobus* and eight populations of *O. taihangensis* (Table 1, Figure 1). Each population included 10 to 25 individuals that were collected at least 5 m apart. Healthy leaves were collected and dried in silica gel until total DNA was extracted. In each of the 13 populations, the ecological factors of each location were determined and recorded (Table 1).

DNA extraction, amplification and sequencing

Total DNA was extracted from the silica gel–dried leaves using the modified 2×CTAB procedure [37]. DNA quality was checked by electrophoresis on 0.8% agarose gels, and DNA concentration was determined using an Eppendorf biophotometer protein nucleotide analyzer (Eppendorf China Ltd., Beijing, Germany). The DNA samples were diluted to 10 ng·μL^{-1} and stored at 20°C for subsequent use.

The *trn*L-*trn*F sequences were amplified using *trn*L and *trn*F [38].The primers *ndh*J and *trn*L [39] were used to amplify the *ndh*J-*trn*L sequences. ITS sequences were then amplified using the primers ITS4 and ITS5 [40]. A polymerase chain reaction (PCR) was then performed in a 25 μL volume, with 50 ng plant DNA, 2×MasterMix (0.2 mM dNTPs, 3 mM MgCl$_2$, 1×PCR buffer, and 0.1 unit Taq DNA polymerase), and 0.6 mM of both forward and reverse primers. The PCR parameters for all amplification programs of ITS and cpDNA were as follows: 4 min of pre-denaturation at 94°C, followed by 34 cycles of 30 s of denaturation, 40 s of annealing (49.2°C for ITS or 58.8°C for *trn*L-*trn*F or 59.8°C *ndh*J-*trn*L), 1 min 20 s of elongation at 72°C, and a final elongation step of 7 min at 72°C. PCR products were purified using the Wizard PCR Preps DNA purification system (Promega, Madison, WI, USA) following the manufacturer's instructions. Cycle sequencing reactions were conducted using the purified PCR product, AmpliTaq DNA polymerase, and fluorescent BigDye terminators. The sequencing products were analyzed using an ABI Prism 310 DNA sequencer (Applied Biosystems Inc., Foster City, CA, USA).

Data analysis

The sequences were edited manually based on the chromatograms and aligned by CLUSTAL X [41] and then adjusted manually. Inserts and indels within all cpDNA and nrDNA sequences were firstly treated as a single character resulting from

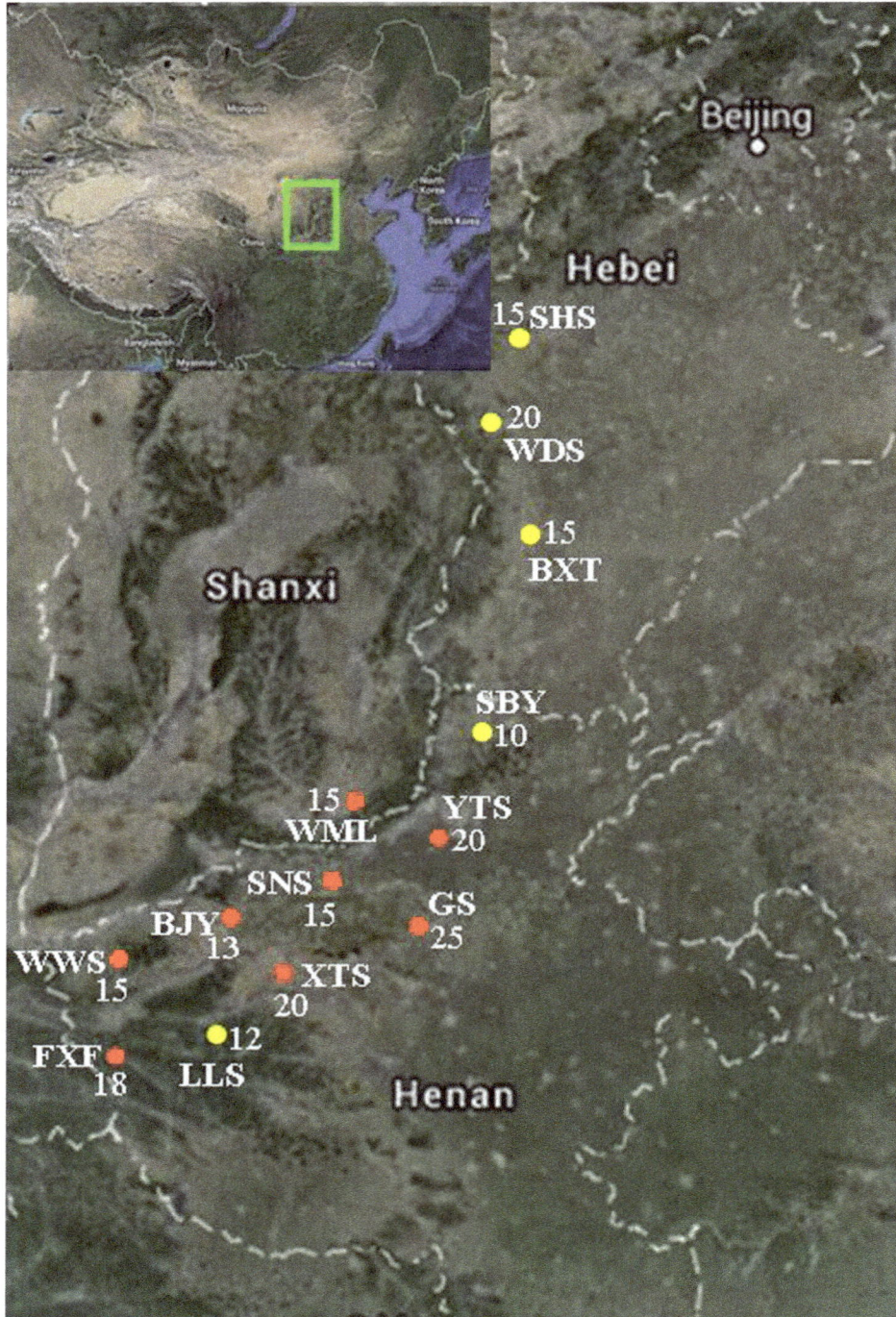

Figure 1. Map of sample sites for *Opisthopappus* **(Asteraceae) on the Taihang Mountains.** Location details are given in Table 1, and locality numbers correspond to those in Table 1. Yellow circle dots represent the populations of *O. longilobus*; red circle dots represent the populations of *O. taihangensis*.

Table 1. Location of the sampled *Opisthopappus* populations and the estimated diversity indexes.

Population		Geographic origin	Sample size	Latitude	Longitude	Number hapelotypes cpDNA (ITS)	Haplotype diversity cpDNA (ITS)	Nucleotide diversity cpDNA (ITS)
O. longilobus	WDS	Wudangshan, Hebei	20	113°47'	36°57'	6(7)	0.854(0.819)	0.00194(0.00240)
	BXT	Beixiangtang, Hebei	15	114°09'	36°02'	1(3)	0.000(0.556)	0.00000(0.00130)
	SHS	Shuanghuangshan, Hebei	15	113°57'	36°54'	2(3)	0.500(0.833)	0.00119(0.00391)
	LLS	Linlvshan, Henan	12	113°02'	34°43'	1(5)	0.000(0.857)	0.00000(0.00196)
	SBY	Shibanya, Henan	10	113°32'	36°03'	1(1)	0.000(0.000)	0.00000(0.00000)
Species level						8(15)	0.838(0.908)	0.00175(0.00537)
O.taihangensis	GS	Guanshan, Henan	25	113°34'	34°45'	1(3)	0.000(0.714)	0.00000(0.00251)
	YTS	Yuntaishan, Henan	20	113°25'	35°25'	2(2)	0.500(0.400)	0.00089(0.00059)
	XTS	Xiantaishan, Henan	20	113°49'	34°11'	2(1)	0.600(0.000)	0.00107(0.00000)
	SNS	Shennongshan, Henan	15	112°44'	35°16'	1(3)	0.000(0.618)	0.00000(0.00155)
	FXF	Foxifeng, Henan	18	112°62'	35°11'	1(3)	0.000(0.607)	0.00000(0.00162)
	WWS	Wangwushan, Henan	15	112°26'	35°38'	1(3)	0.000(0.700)	0.00000(0.00176)
	BJY	Baijiayan, Henan	13	112°92'	35°46'	2(2)	0.667(0.667)	0.00040(0.00293)
	WML	Wangmangling, Shanxi	15	113°35'	35°41'	2(2)	0.533(0.667)	0.00063(0.00098)
Species level						5(6)	0.562(0.558)	0.00073(0.00158)

one mutation. Haplotype diversity (h) and nucleotide diversity (π) were calculated for each population (h, π) and at the species level (h_d, π_d) using DNAsp [42]. The program PERMUT [44] was used to calculate the within-population diversity (h_S), total diversity (h_T), geographical total haplotype diversity (V_T), geographical average haplotype diversity (V_S), level of population differentiation at the species level (G_{ST}), and an estimate of population subdivisions for phylogenetically ordered alleles (N_{ST}). We further tested the phylogeographical structure at the species range between (N_{ST}) and (G_{ST}) by using the U-statistic. An N_{ST} value higher than the G_{ST} value indicates that closely related haplotypes are observed more often in a given geographical area than would be expected by chance [43].

To quantify the genetic differentiation partitioned among different groups and total genetic variance, analyses of molecular variance (AMOVA) were carried out using ARLEQUIN [44], with 1000 random permutations to test for significance of partitions. The spatial genetic structure of haplotypes was analyzed by spatial analysis of molecular variance using SAMOVA [45]. This program uses a simulated annealing approach to define groups of populations (K) that are geographically homogenous and maximally differentiated from each other. In this analysis of haplotypes, K varied from 2 to 13, with each simulation starting from 100 random initial conditions. An F_{CT} index of genetic differentiation among initial K groups was computed, followed by an iterative simulated annealing process to obtain the optimal configuration of groups and final F_{CT} values. The simulated annealing process was repeated 1000 times. The configuration with the largest F_{CT} value among the 100 tested was retained as the best grouping of populations.

Genealogical relationships between haplotypes were inferred from a maximum parsimony median-joining network calculated in NETWORK 4.5.0.2 [46]. To complement the results of NETWORK, TCS 1.21 [47] was also used to construct haplotype relationships. Phylogenetic analyses were conducted for the nrDNA and cpDNA sequence data using maximum likelihood (ML) and Bayesian inference (BI), respectively.

The time of the most recent common ancestor (TMRCA) of all haplotypes was estimated via a Bayesian approach implemented in *BEAST (Star BEAST) [48–49] using a GTR+I substitution model. The best substitution model was determined according to the Akaike Information Criterion (AIC) in jModeltest [50]. These models were applied in ML, BI, and subsequent *BEAST analyses. Bayesian inference was performed using MrBayes 3.1.2 [51]. Two independent runs of Metropolis-coupled Markov chain (MCMC) analysis were executed, each including one cold chain and three incrementally heated chains that started randomly in the parameter space. Two independent runs of 10^8 generations were carried out, with sampling at every 1000 generations. The first 25% of sampled trees were discarded as burn-in, and the remaining trees were used to construct a Bayesian consensus tree. The convergence of chains was checked using Tracer 1.5 [52]. The remaining trees were pooled to estimate the posterior probabilities (PPs).

No fossil records or biogeographic events isolating distinct populations are available to calibrate a cpDNA-IGS substitution rate for *O. longilobus* and *O. taihangensis*. Therefore, for our combined chloroplast non-coding regions, we assumed minimum and maximum values of a range of average mutation rates reported for synonymous sites of plant chloroplast genes [i.e., 1.2 and 1.7×10^{-9} substitutions per site per year (s/s/y)] [53]. These rates were then used for estimating TMCRA in *BEAST under a relaxed molecular clock assumption.

To detect whether population groups experienced recent population expansion and satisfied the assumption of neutrality, a mismatch distribution analysis was performed using the DNAsp program. Tajima's D and Fu and Li's D* were calculated for the entire genus and groups of populations. The statistical significance of D and D* was estimated with coalescent simulations as implemented in this program [54–55]. To further infer demographic processes, the null hypotheses of a spatial expansion and of a pure demographic expansion were tested in ARLEQUIN by comparing observed and expected distributions of pairwise sequence differences (mismatch distributions). For each expansion model, goodness-of-fit was tested using the sum of squared differences (SSD) and Harpending's [56] raggedness index (HRag).

Finally, isolation by distance (IBD) for both nrDNA and cpDNA data was tested based on pairwise geographical and genetic distances (F_{ST}) with the Isolation by Distance Web Service [57] (http://ibdws.sdsu.edu/~ibdws/), running 10,000 Mantel permutations.

Results

Patterns of variability in cpDNA and ITS sequences

The *trn*L-*trn*F and *ndh*J-*trn*L intergenic spacers varied from 800 to 866 bp and from 790 to 835 bp, respectively. The *trn*L-*trn*F intergenic spacer, which exhibits considerable nucleotide polymorphism, is characterized by 10 substitutions and a 4 bp insertions and deletions (Table S1). The 4 bp insertions and deletions region was mainly observed in *O. longilobus* populations. The *ndh*J-*trn*L spacer was characterized by one indel, three substitutions, and a 10 bp insertions and deletions (Table S1). The total cpDNA combined matrix comprised 1701 sites, of which 13 positions were variable and 18 harbored gaps. Ten haplotypes (Figure 2A) were observed by combining cpDNA data: eight haplotypes were within *O. longilobus* populations and five were within *O. taihangensis* populations. The haplotype diversity (h) ranged from 0 to 0.854, and the nucleotide diversity (π) from 0 to 0.00194 in the 5 populations of *O. longilobus* (Table 1). At species level, $h_d = 0.838$ and $\pi_d = 0.00175$ for *O. longilobus*. Within 8 *O. taihangensis* populations, h ranged from 0 to 0.667, π from 0 to 0.00107, h_d was 0.562 and π_d was 0.00073. The highest haplotype diversity was observed in population WDS from *O. longilobus* and BJY from *O. taihangensis* (Table 1).

The full ITS sequences, including ITS1+5.8S+ITS2, were 682 bp in length, comprising of 254 bp and 224 bp for ITS1 and ITS2, respectively. Eighteen haplotypes (Figure 2B) and 17 polymorphic sites, the latter consisting of 15 parsimony informative sites and 2 singleton variable sites, were detected at ITS sequences. Two populations, SBY and XTS, were monomorphic, whereas the remaining populations were polymorphic (Table 1). The haplotype diversity (h) ranged from 0 to 0.857 and the nucleotide diversity (π) from 0 to 0.00391 in the 5 populations *O. longilobus*. Within 8 *O. taihangensis* populations, the haplotype diversity h ranged from 0 to 0.714 and the nucleotide diversity π from 0 to 0.00293. For *O. longilobus*, $h_d = 0.908$, $\pi_d = 0.00537$; for *O. taihangensis* populations, $h_d = 0.558$, $\pi_d = 0.00158$ (Table 1).

Genetic diversity and population structure

A genetic diversity analysis revealed high genetic diversity harbored in *O. longilobus* and *O. taihangensis* populations. The parameters h_T, V_T, h_S, and V_S were 0.908, 0.916, 0.669, and 0.581, respectively, for *O. longilobus* based on cpDNA sequences. By ITS sequences, $h_T = 0.944$, $V_T = 0.937$, $h_S = 0.824$, and $V_S = 0.889$ within *O. longilobus* populations. For *O. taihangensis*

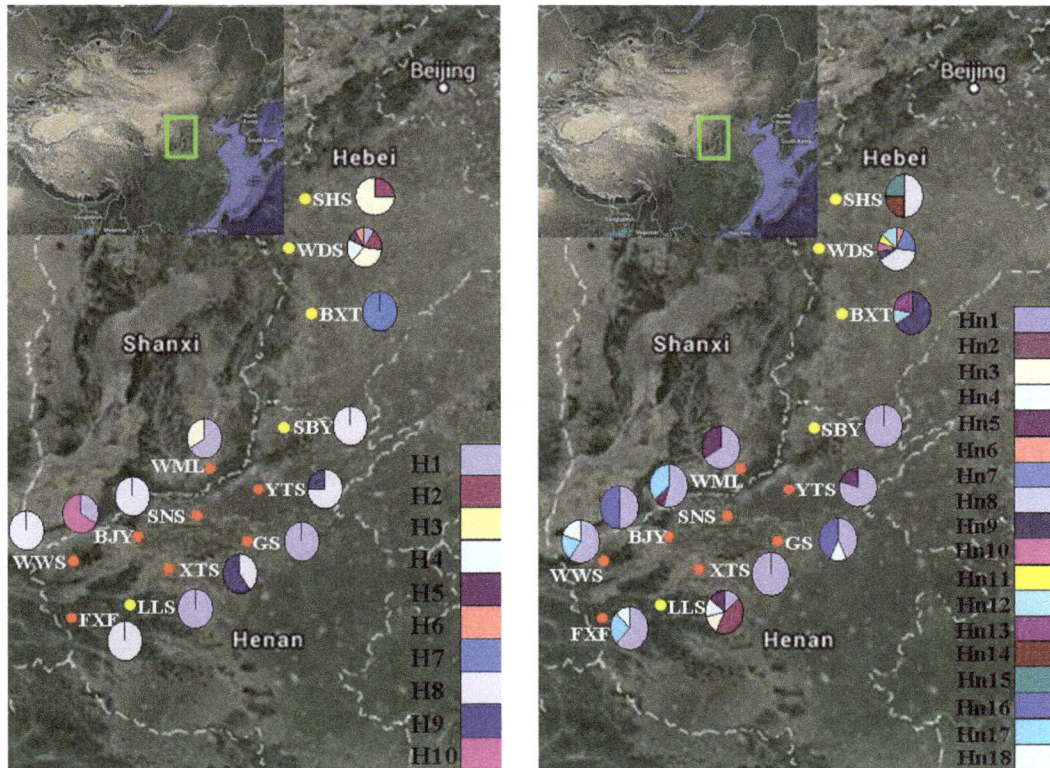

Figure 2. Geographical distribution of 10 cpDNA (A) and 18 (B) nrDNA haplotypes. The pie charts reflect the frequency of haplotype occurrence in each population. Haplotype colours correspond to those shown in panel.

populations, $h_T = 0.828$, $V_T = 0.833$, $h_S = 0.531$, and $V_S = 0.455$ by cpDNA sequences; $h_T = 0.806$, $V_T = 0.810$, $h_S = 0.704$, and $V_S = 0.592$ with nrDNA sequences.

Based on ITS sequences, phylogenetic analyses were carried (Figure S1). All studied populations made up a single monophyletic lineage, indicating that all of these divergent populations were derived within *Opisthopappus* itself, rather than the result of introgression by other species. Spatial genetic analyses of cpDNA and nrDNA haplotypes in all 13 populations using SAMOVA indicated that F_{CT} increased to a maximal value of 0.921 when $K = 2$ (K, the number of groups). Thus, the division of all the 13 sampled populations approximately into two groups is appropriate.

AMOVA analysis revealed that 60.58% and 67.70% ($P<0.01$) of the total variation was due to differences among the cpDNA and nrDNA populations for all studied populations, respectively (Table 2). When the populations were grouped by taxonomic variety, the cpDNA showed 27.28% variation among populations within species ($P<0.01$; Table 2). Similarly, for the nrDNA, AMOVA revealed that 26.68% of the variation was attributed to differences among populations within the species ($P<0.01$; Table 2). There was a significant ($P<0.01$) variation between species, suggesting that the two separately distributed groups have a genetic differentiation. In addition, 60.71% of the total cpDNA variation and 68.24% of the total nrDNA variation existed among *O. longilobus* populations (Table 2). The genetic structure of *O. taihangensis* populations showed a similar trend with that of *O. longilobus* populations, with most of the genetic differentiation occurring among populations (Table 2).

Results of the Mantel test showed a significant correlation between nrDNA data differentiation among populations and the natural logarithm of the geographical distances throughout the sampled range of *O. longilobus* ($r = 0.442$; $P = 0.001$) and *O. taihangensis* ($r = 0.385$; $P = 0.001$). Likewise, a significant correlation was detected among populations from cpDNA data of *O. longilobus* ($r = 0.401$; $P = 0.001$) and *O. taihangensis* ($r = 0.399$; $P = 0.001$). This indicates a significant correlation between genetic differentiation and geographical distance, implying strong IBD.

Haplotype and phylogeographical structure

In this study, similar topologies were obtained from the two different network approaches used to infer the relationships among the cpDNA and nrDNA haplotypes. Only the median-joining network is shown. The haplotype H8 was the most frequent haplotype in cpDNA data, occurring in 6 of the 13 populations. Three haplotypes, H1, H3, and H8, were shared by *O. longilobus* and *O. taihangensis* populations (Table 1 and Figure 3). In nrDNA data, Hn1 was the most frequent haplotype, occurring in 10 of the 13 populations (Figure 3 and Figure 4). The haplotypes Hn1, Hn4, and Hn5 were shared by *O. longilobus* and *O. taihangensis* populations (Table 1 and Figure 4). The distributions of haplotypes that were restricted to a single population were extremely skewed. Four cpDNA and nine nrDNA haplotypes were found in *O. longilobus* single populations. Only one cpDNA haplotype was identified for *O. taihangensis* populations.

Based on cpDNA data, we estimated a G_{ST} value (0.649, $P<$ 0.05), which is significantly smaller than the N_{ST} value (0.761). When the nrDNA was examined, the G_{ST} value (0.253) also

Table 2. Analyses of Molecular Variance (AMOVA) based on the nrDNA (ITS) and cpDNA sequences.

Source of variation	d.f	Sum of squares	Variance components	Variation percentage	Fixation indices
ITS(all population)					
Among populations	19	105.101	1.218	67.70	$F_{ST} = 0.677$ ($P<0.01$)
Within populations	178	44.179	0.581	32.30	
ITS (species)					
Among species	1	54.511	1.113	48.16	$F_{CT} = 0.482$($P<0.01$)
Among populations within species	18	50.590	0.617	26.68	$F_{ST} = 0.748$($P<0.01$)
Within populations	178	44.179	0.581	25.16	$F_{SC} = 0.515$($P<0.01$)
ITS(O. longilobus)					
Among populations	12	44.767	1.499	68.24	$F_{ST} = 0.682$ ($P<0.01$)
Within populations	76	23.022	0.698	31.76	
ITS(O. taihangensis)					
Among populations	14	21.157	0.492	90.04	$F_{ST} = 0.900$ ($P<0.01$)
Within populations	95	5.823	0.054	9.96	
ITS (SMOVA species)					
Among species	7	57.023	2.456	50.21	$F_{CT} = 0.502$($P<0.01$)
Among populations within species	12	55.147	2.078	28.34	$F_{ST} = 0.786$($P<0.01$)
Within populations	178	40.563	1.789	21.45	$F_{SC} = 0.497$($P<0.01$)
cpDNA(all population)					
Among populations	19	78.452	2.187	60.58	$F_{ST} = 0.606$ ($P<0.01$)
Within populations	178	50.234	1.023	39.42	
cpDNA (species)					
Among species	1	60.231	1.586	46.59	$F_{CT} = 0.466$($P<0.01$)
Among populations within species	18	58.427	0.789	27.28	$F_{ST} = 0.738$($P<0.01$)
Within populations	178	45.126	0.567	26.13	$F_{SC} = 0.533$($P<0.01$)
cpDNA (O. longilobus)					
Among populations	12	29.313	1.032	60.71	$F_{ST} = 0.607$ ($P<0.01$)
Within populations	76	19.364	0.668	39.29	
cpDNA (O. taihangensis)					
Among populations	14	21.601	0.467	68.64	$F_{ST} = 0.686$ ($P<0.01$)
Within populations	95	9.183	0.214	31.36	
cpDNA (SMOVA species)					
Among species	7	56.273	1.756	52.34	$F_{CT} = 0.523$($P<0.01$)
Among populations within species	12	55.012	1.111	27.55	$F_{ST} = 0.799$($P<0.01$)
Within populations	178	41.603	0.654	22.36	$F_{SC} = 0.498$($P<0.01$)

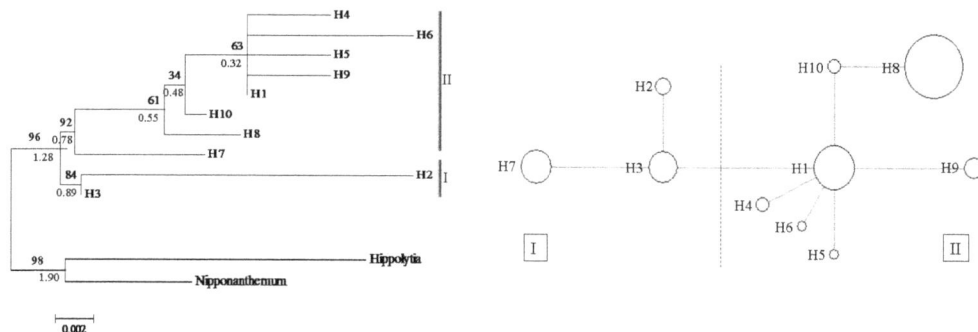

Figure 3. The evolutionary relationships among cpDNA haplotypes. (A) NJ phylogenetic tree for the 10 cpDNA haplotypes. Numbers above branches are support values from bootstrap resampling/Bayesian inference. (B) Median-joining network. Sizes of the circles are proportional to the overall frequency of the haplotypes in the entire sample of all species.

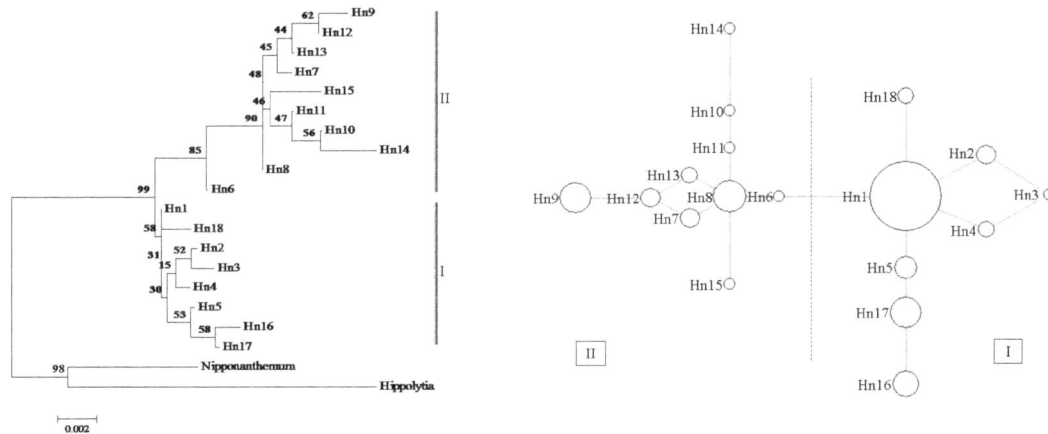

Figure 4. The evolutionary relationships among nrDNA haplotypes. (A) NJ phylogenetic tree for the 18 nrDNA haplotypes. Numbers above branches are support values from bootstrap resampling. (B) Median-joining network. Sizes of the circles are proportional to the overall frequency of the haplotypes in the entire sample of all species.

differed significantly from the N_{ST} value (0.641). A significant phylogeographic structure was indicated by both cpDNA and nrDNA data ($N_{ST}>G_{ST}$, $P<0.05$).

The phylogeographical relationships of the 10 cpDNA haplotypes and 18 nrDNA haplotypes were assessed under Neighbor-joining analyses (NJ) and Bayesian inferences drawn using *Hippolytia* and *Nipponanthemum* as outgroups (Figure 3 and Figure 4). Similar topology to phylogeographical relationships indicated a single phylogenetic split. All haplotypes were clustered into two lineages: I and II (Figure 3 and Figure 4). Lineage I contained some populations of *O. longilobus* with a high bootstrap support. The remaining haplotypes formed lineage II, containing some *O. longilobus* populations and all *O. taihangensis* populations, also with a high bootstrap support. This unrooted network of cpDNA haplotypes is consistent with the strict consensus tree produced by NJ and Bayesian inference and furthermore displays the relationship of interior (ancestral) and tip (derived) haplotypes (Figure 3). For lineage I (Figure 3), the haplotype H3 located in the center and fixed in three populations, being an ancestral haplotype. In clade II, H1 was shared by five populations. It appears to be an ancestral haplotype distributed in the center of clade II. The haplotype H8 was located in the tip of the network despite though contained a higher frequency of haplotypes. In the nrDNA haplotype network (Figure 4), Hn8 was ancestral for the lineage I; Hn1 was the predominant and widespread one in the clade II.

To test whether the current expansion had occurred in different populations, we calculated the frequency distribution of pairwise nucleotide differences among *O. longilobus* populations and *O. taihangensis* populations respectively. The resultant mismatch distribution consisted of multiple multimodal curves (Figure S2). Neutrality tests revealed nonsignificant positive values in considering all populations (Table 3). The observed mismatch distributions of cpDNA haplotypes were multimodal for all groups except for *O. taihangensis*. The mismatch distributions based on nrDNA haplotypes were present a similar pattern with cpDNA data (Table 3). Thus, a recent population expansion for *O. longilobus* is unlikely. *O. taihangensis* perfectly fits the expected expansion distribution (Figure 3).

Divergence times were estimated using the program *BEAST. The *ESS* values were from 150 to 240 for all the nodes discussed as follows. All sampled haplotypes of *Opisthopappus* coalesced at

about 1.90 Ma (95% highest posterior density, HPD) (0.50–1.72), which indicated the origin of *Opisthopappus* during the late Tertiary periods. The divergence times, estimated at approximately 1.28 Ma (95% HPD) (0.70–2.34) between *O. longilobus* and *O. taihangensis*, are shown in the haplotype phylogenic tree (Figure 3A). *O. longilobus* and *O. taihangensis* thus appear to have diverged from each other during the early Pleistocene.

Discussion

Taxonomic implications

At the morphological character levels, the evidences that illustrated *O. taihangensis* as a separate species from *O. longilobus* were only leaf form and involucres. So, these two species were once regarded as a species. In this study, the analyses of the combined cpDNA and nrDNA matrix support the monophyly of two clades (Figure 3A, Figure 4A, and Figure S1). The phylogenetic relationships among all populations show an overall congruence with the result of the haplotype network (Figure 3B and Figure 4B). All haplotypes are split into two major groups according to cpDNA and nrDNA data. The populations from two distinct clades approximately exhibited specific geographical distributions in Taihang Mountains. SAMOVA analysis identified two approximately defined groups corresponding to *O. longilobus* and *O. taihangensis* populations at every level of divergence. Within DNA sequences of cpDNA and ITS, the total of thirty polymorphic sites harbored in populations. Unique haplotypes found in *O. longilobus* and *O. taihangensis* populations respectively. Moreover, a significant variation based on AMOVA occurred between *O. longilobus* and *O. taihangensis* ($F_{STcpDNA}=0.482$ and $F_{STnrDNA}=0.466$, Table 2). By used SRAP and cpSSR markers [58–59], AMOVA indicated that a significant genetic differentiation between *O. longilobus* and *O. taihangensis*. Accordingly, our molecular results support that *O. longilobus* and *O. taihangensis* were regarded as two species rather than a species. That result is consistent with the point of Hu [25]. The divergence times between *O. longilobus* and *O. taihangensis* (Figure 3A) indicate that genetic divergence between two species occurred in early Pleistocene. This split was related to the ecological and geographical habitat changes resulting from climate oscillations during the Quaternary glacial periods and complicated topography with the uplift of Taihang Mountains.

Table 3. Parameters of mismatch distribution analysis.

	Tajima's D	P	Fu's Fs	P	Fu and Li's D*	P	Fu and Li's F*	P	SSD	P	H_{RAG}	P
ITS data												
O. longilobus	0.085	>0.10	-3.988	0.012	0.283	>0.10	0.257	>0.10	0.111		0.461	<0.05
O. taihangensis	0.483	>0.10	-0.787	0.157	1.002	>0.10	0.984	>0.10	0.106		0.350	<0.05
cpDNA data												
O. longilobus	0.017	>0.10	0.271	0.182	0.012	>0.10	0.015	>0.10	0.168		0.424	<0.05
O. taihangensis	0.261	>0.10	0.495	0.217	1.095	>0.10	0.977	>0.10	0.254		0.806	<0.05

Geographic patterns of genetic diversity and structure

The high levels of genetic diversity were observed in *O. longilobus* and *O. taihangensis* populations (Table 1). These results are similar to previous research on herbaceous plants in East Asia and significantly exceed the average value of 0.67 for the 170 plant species documented [8,12,19,31,60]. In *O. longilobus*, the WDS population obtained the most haplotypes and highest genetic diversity, followed by the SHS population. This suggests the presence of two genetic diversity centers for *O. longilobus*, i.e., the regions where WDS and SHS populations are located. For *O. taihangensis*, the BJY and WML populations, which obtained the high genetic diversity, were genetic diversity centers for *O. taihangensis*. Significant genetic divergence and a highly structure were illustrated in all studied populations (Figure 3, Figure 4, and Table 2). For *O. longilobus* ($F_{STcpDNA} = 0.607$, $F_{STnrDNA} = 0.682$) and *O. taihangensis* ($F_{STcpDNA} = 0.686$, $F_{STnrDNA} = 0.805$), the results of our analysis suggested that significant variation was detected from among populations rather than from variation within groups (Table 2). Moreover, very high genetic differentiation in *O. longilobus* and *O. taihangensis* greatly exceeding the mean value ($G_{ST} = 0.22$) reported by Nybom [61].

Species genetic diversity and structure can be affected by life history traits (e.g., life cycle, breeding system, pollination mechanism) and environmental effects (e.g., geographical range, climate, topography) [61–62]. *O. longilobus* and *O. taihangensis* possess a sexual reproductive mode and a relatively long life cycle as perennial herbs, all of which are traits associated with low total genetic diversity [63–64]. However, high genetic diversity was observed in our study, and this diversity appears to result from high among-population genetic variation. It has been widely documented that the dispersal of pollen and seeds in plant species is strongly linked to the development of a population genetic structure in bi-parentally inherited (nuclear-encoded) and maternally inherited (cytoplasmic) genetic markers [12,31,61,62,64]. Generally, outcrossing taxa with high seed dispersal capacity retain the majority of both types of genetic marker variation within populations, whereas selfing (and/or asexual) taxa with restricted seed dispersal allocate the majority of such variation among populations. *O. longilobus* and *O. taihangensis* are self-compatible plant species [25]. These two species, therefore, might be expected to have a heterogeneous distribution of both cpDNA and nrDNA variation among populations. Taken at face value, the above results may suggest that restricted gene flow via pollen and seed has resulted in significant population genetic differentiation in *O. longilobus* ($N_{mcpDNA} = 0.12$; $N_{mnrDNA} = 0.10$) and *O. taihangensis* ($N_{mcpDNA} = 0.19$; $N_{mnrDNA} = 0.14$). The physiographic complexity of Taihang Mountains and the deeply carved valleys and ravines between the inhabit areas probably would imposes significant barriers to gene flow among populations of *O. longilobus* and *O. taihangensis*. The genetic divergence between the two lineages occurred at 1.28 Ma, corresponding to the early Pleistocene. This divergence falls within the glaciations during the Taku Glaciation (1.05–1.20 Ma). The glaciers and/or extremely low temperature in the mountains might also have created barriers to gene flow between geographically isolated populations, which therefore promoted the divergence. For both marker systems, we generally observed a good fit of these populations to an IBD model. When coupled with high levels of population subdivision, such conditions strongly suggest that geographical isolation had a larger historical role in extant population structure compared with limited pollen and seed flow alone. The habitats of *O. longilobus* and *O. taihangensis* are either rocky or massifs, and these species grow discontinuously in different habitats along the altitudinal gradient from 700 to 1500 m, in which the variable micro-surrounding, complex topography, and great altitudinal variability of the region might promote the high degree of genetic population differentiation [65–66]. Limited migration coupled with a heterogeneous environment could promote local adaptation and fixation of different alleles in *O. longilobus* and *O. taihangensis* populations.

Phylogeographical structure and inference of demographic history

Both *O. longilobus* and *O. taihangensis* populations have a significant phylogeographic structure. The current distribution of haplotypes could be a result of past climatic changes related to the advance/retreat of glaciations [11]. During the Quaternary periods, climatic oscillations resulted in repeated drastic environmental changes, which further caused massive range shifts of most plants and animals, leading to accumulated genetic differences and particular phylogeographic patterns [2–3].

Our analyses suggested that *Opisthopappus* originated during the Pliocene. Before the Pleistocene, the current range of Taihang Mountains would have been covered by grassland with roughly similar geographic conditions [67]. As herbs, *O. longilobus* and *O. taihangensis* can disperse following grassland migration. On this basis, we suspect that *O. longilobus* and *O. taihangensis* were once distributed throughout Taihang Mountains and its adjacent region. With uplift of Taihang Mountains [27] and climate fluctuation of Quaternary glacial periods, *O. longilobus* and *O. taihangensis* were fragmented and forced into the refuge areas. The flora would migrate and expand during interglacial and glacial periods. This migration and expansion resulted in ecological displacement, which might have resulted in populations situated at different spatial-temporal scales. Different habitats might enhance the isolation of plant populations and prevent gene flow, which in turn could lead to incipient allopatric speciation or further allopatric speciation [12,18,19,68,69]. During the Quaternary periods, the paleovegetation in Taihang Mountains repeatedly appear the replacement of grasslands and forests cycles [67]. Those multiple replacements likely fragmented and isolated the habitat of herbs. Therefore, the refuge areas for *O. longilobus* and *O. taihangensis* would have remained separated and fragmented at different spatial-temporal scales for a long time. Moreover, dramatic geological changes induced by the Taihang Mountains uplift contributed to the extremely complicated landscape, which likely acted as a major driving force for the separation of the *O. longilobus* and *O. taihangensis*. Divergent selection between populations in contrasting environments, longer temporal and spatial separation and fragmentation would accelerate their differentiation, and eventually promote allopatric speciation.

Following this divergence, *O. longilobus* and *O. taihangensis* underwent strikingly different histories. For *O. longilobus*, long-distance dispersal and colonization may be a major historical process. *O. longilobus* was divided into different lineages around 0.89 Ma, which fell within the interglacial stage between the Taku Glaciation and Lushan Glaciation. The paleovegetation grasslands of Taihang Mountains [67] might be regarded as a corridor connecting the currently isolated populations. Populations of *O. longilobus* in Hebei province may have extended across this corridor and reached Henan province. Haplotypes H1 and Hn1 located in populations LLS and WDS or LLS and SBY imply this long dispersal (Table 2). The 10 bp insertions of *ndh*J-*trn*L occurring in population LLS also supported this viewpoint (Table S1). Subsequently, with the advent of inter-glaciations or post-glaciations, the climate changed, which led to the paleovegetation of Taihang Mountains experiencing a transition from grasslands to forests. The corridor for *O. longilobus* migration gradually disappeared, which would have resulted in isolation of the

population [11,12,68]. Additionally, we did not detect any population expansion due to the clear multimodel mismatch distribution (Figure S2). The complex geological conditions of Taihang Mountains would limit *O. longilobus* population expansion on a macro scale and thus restricting *O. longilobus* to sustain in situ during the Quaternary periods, perhaps by moving upwards and downwards in their mountain ranges. For *O. taihangensis*, a pattern of population expansion was observed, because the observed mismatch distribution for cpDNA and nrDNA haplotypes was unimodal (a pattern consistent with expansion) (Figure S2). This expansion was dated to be 0.78 Ma, which corresponding the interglacial stage between the Taku Glaciation and Lushan Glaciation. The above results suggest an expansion consistent with a range expansion and re-colonization before the mainly uplift of Taihang Mountains. With the uplift of Taihang Mountains, large geographical distances and significant topographical barriers could restrict gene flow and limit any large-scale population expansion [12,70]. The cpDNA sequences characteristic of the *trn*L-*trn*F and *ndh*J-*trn*L intergenic spacers further confirmed this hypothesis (Table S1).

Crandall and Templeton [71] suggest that ancestral haplotypes are located in a central position within the haplotype phylogeographical network and that potential refugia are usually characterized by widely dispersed ancestral haplotypes and other tip and unique haplotypes. Geographic areas displaying increased levels of genetic diversity are thus good candidates in the search for past refuges [8,34]. These regions are characterized by relatively stable ecological conditions during environmental fluctuations, which foster the accumulation of genetic diversity [72]. Most *O. longilobus* individuals or populations gathered in lineage I (Figure 3 and Figure 4). Haplotypes H3 and Hn8 are located in the center of the haplotype network. And Haplotype H3 is detected in three populations (WDS, SHS and WML), Hn8 in two populations (WDS and SHS) (Figure 2). WDS and SHS are located in the southern part of Hebei province and have higher haplotype diversity (Table 1). Thus, we suggest that the WDS and SHS populations' regions were two major potential refugia for *O. taihangensis*. These regions may therefore have been characterized by a relatively mild climate, potentially containing microclimatic environments able to impart relative stability to a range of habitats. Lineage II consisted of all populations of *O. taihangensis* and some individuals of *O. longilobus*. Haplotypes H1 and Hn1, which are located in the center of the haplotype network, were the most dominant haplotypes found in five and ten populations, respectively. Moreover, some populations in these haplotypes (H1 and Hn1) had higher haplotype diversity (Table 1, e.g., BJY, WML, and YTS). Thus, we suggest that there were multiple small refuges located in Shanxi and Henan provinces during the last glacial maximum or earlier cold periods for *O. taihangensis*. The above populations of *O. taihangensis*, distributed in southern Taihang Mountains, would have been provided with a complex and stable environment. Thus, the potential refugia area may have been located in those regions.

However, our comprehensive sampling showed that assessing the species-level relationships in our study is complicated by the fact that not all cpDNA and nrDNA haplotypes are species-specific. Incongruence between the species boundaries and the genealogy of their cpDNA and nrDNA sequences is illustrated by the sharing of haplotypes (e.g., H1 and Hn8) between the two species. On the other hand, several haplotypes were found within a single species (Table 1 and Figure 2). Haplotype distribution does not strictly follow species circumscription. Moreover, an association of haplotypes with geographically circumscribed regions rather than with taxonomic boundaries is a phenomenon observed in *O. longilobus* and *O. taihangensis*. The sharing of haplotypes and lack of reciprocal monophyly might be explained by the persistence of ancestral polymorphisms during speciation events and/or exchange of genes by inter-specific hybridization (Table S1 and Figure S1). These two species have similar habitats and the same period of flowering [25], and possible visitation by the same pollinators in the absence of physiological or genetic reproductive barriers might have enabled hybridization events. Nonetheless, further work on chromosomes and the sequencing of additional genomic regions are needed to make more in-depth conclusions about the hybrid status of these taxa.

Conclusions

In summary, *O. longilobus* and *O. taihangensis* have a high level of genetic diversity. The following phylogeographical structure and historical scenario of *O. longilobus* and *O. taihangensis* can also be drawn. The vicariance following the uplift of Taihang Mountains and a transition of paleovegetation of Taihang Mountains mainly shaped the present distribution of these two species. The above findings imply that multiple small refugia could have been maintained in the Taihang Mountains regions, which allowed *O. longilobus* and *O. taihangensis* to persist in situ and maintain sizable haplotype and nucleotide diversity at the lineage-wide scale during the glaciations. The current distribution pattern of *O. longilobus* and *O. taihangensis* refuges is similar to that of other East Asian plants (e.g., *Cathaya argyrophylla*, *Picea crassifolia*, *Sinopodophyllum hexandrum*, *Lagochilus Bunge* ex Bentham) [8,20,73], with multiple refuges sustained across their distribution ranges. Furthermore, complex topography rendered

these refuge areas both isolated and fragmented in different geographic units. Geographical and ecological isolation likely restricted seed and pollen migration ability. The isolation and fragmentation further promoted the intra-specific split. These results support the conclusion that the high plant diversity and endemism found in central/northern China and throughout eastern Asia has mainly resulted from allopatric speciation due to the complex topography of Mountains and allopatric fragmentation during the late Tertiary and Quaternary periods.

Supporting Information

Figure S1   Phylogenetic analyses based on ITS sequences.

Figure S2   Pairwise mismatch distribution analyses (MDAs) for *O. taihangensis* (A, B) and *O. longilobus* (C, D) populations inferred from cpDNA and ITS sequences.

Table S1   The cpDNA sequences characteristic of *trn*L-*trn*F and *ndh*J-*trn*L intergenic spacers.

Author Contributions

Conceived and designed the experiments: GQY YLW. Performed the experiments: YLW. Analyzed the data: YLW. Contributed reagents/materials/analysis tools: YLW. Wrote the paper: YLW. Revised the manuscript: GQY.

References

1. Avise JC, Walker D, Johns GC (1998) Speciation durations and Pleistocene effects on vertebrate phylogeography. Proceedings of the Royal Society of London Series B Biological Sciences 265: 1707–1712.
2. Hewitt GM (2000) The genetic legacy of the Quaternary ice ages. Nature 405: 907–913.
3. Hewitt GM (2004) Genetic consequences of climatic oscillations in the Quaternary. Philosophical Transactions of the Royal Society of London Series B Biological Sciences 359: 183–195.
4. Willis KJ, Whittaker RJ (2002) Species diversity-scale matters. Science 1245–1248.
5. Turchetto-Zolet AC, Cruz F, Vendramin GG, Simon MF, Salgueiro F, et al. (2012) Large-scale phylogeography of the disjunct Neotropical tree species *Schizolobium parahyba* (Fabaceae-Caesalpinioideae). Molecular Phylogenetics and Evolution 65: 174–182.
6. Xu JW, Chu KH (2012) Genome scan of the mitten crab *Eriocheir sensu stricto* in East Asia: Population differentiation, hybridization and adaptive speciation. Molecular Phylogenetics and Evolution 64: 118–129.
7. Cao MM, Jin YT, Liu NF, Ji WH (2012) Effects of the Qinghai–Tibetan Plateau uplift and environmental changes on phylogeographic structure of the *Daurian Partridge* (Perdix dauuricae) in China. Molecular Phylogenetics and Evolution 65: 823–830.
8. Meng HH, Zhang ML (2013) Diversification of plant species in arid Northwest China: Species-level phylogeographical history of *Lagochilus Bunge* ex Bentham (Lamiaceae). Molecular Phylogenetics and Evolution 68: 398–409.
9. Avise JC (2000) Phylogeography: The History and Formation of Species. Harvard University Press, Cambridge.
10. Yu G, Chen X, Ni J, Cheddadi R, Guiot J, et al. (2000) Palaeovegetation of China: a pollen databased synthesis for the mid-Holocene and last glacial maximum. Journal of Biogeography 27: 635–664.
11. Zhang YH, Volis S, Sun H (2010) Chloroplast phylogeny and phylogeography of *Stellera chamaejasme* on the Qinghai-Tibet Plateau and in adjacent regions. Molecular Phylogenetics and Evolution 57: 1162–1172;
12. Zhao YP, Qi ZC, Ma WW, Dai QY, Li P, et al. (2013) Comparative phylogeography of the *Smilax hispida group* (Smilacaceae) in eastern Asia and North America – Implications for allopatric speciation, causes of diversity disparity, and origins of temperate elements in Mexico. Molecular Phylogenetics and Evolution 68: 300–311.
13. Xie KQ, Zhang ML (2013) The effect of Quaternary climatic oscillations on *Ribes meyeri* (Saxifragaceae) in northwestern China. Biochemical Systematics and Ecology 50:39–47.
14. Tian B, Liu RR, Wang LY, Qiu Q, Chen KM, et al. (2009) Phylogeographic analyses suggest that a deciduous species (Ostryopsis davidiana Decne.,

Betulaceae) survived in northern China during the Last Glacial Maximum. Journal of Biogeography 36: 2148–2155.
15. Jia DR, Liu TL, Wang LY, Zhou DW, Liu JQ (2011) Evolutionary history of an alpine shrub *Hippophaë tibetana* (Elaeagnaceae): allopatric divergence and regional expansion. Biological Journal of Linnean Society 102: 37–50.
16. Chen KM, Abbott RJ, Milne RI, Tian XM, Liu JQ (2008) Phylogeography of *Pinus tabulaeformis* Carr. (Pinaceae), a dominant species of coniferous forest in northern China. Molecular Ecology 17: 4276–4288.
17. Bai WN, Liao WJ, Zhang DY (2010) Nuclear and chloroplast DNA phylogeography reveal two refuge areas with asymmetrical gene flow in a temperate walnut tree from East Asia. New Phytologist 188: 892–901.
18. Harrison SP, Yu G, Takahara H, Prentice IC (2001) Palaeovegetation: diversity of temperate plants in East Asia. Nature 413: 129–130.
19. Qiu YX, Guan BC, Fu CX, Comes HP (2009) Did glacials and/or interglacials promote allopatric incipient speciation in East Asian temperate plants? Phylogeographic and coalescent analyses on refugial isolation and divergence in *Dysosma versipellis*. Molecular Phylogenetics and Evolution 51: 281–293.
20. Li Y, Stocks M, Hemmilä S, Källman T, Zhu HT, et al. (2010) Demographic histories of four spruce (Picea) species of the Qinghai-Tibetan Plateau and neighboring areas inferred from multiple nuclear loci. Molecular Biology and Evolution 27: 1001–1014.
21. Wu CY (1995) Vegetation of China. Science Press, Beijing.
22. Zhao SL, Li GG (1990) Desertization on the shelves adjacent China in the Later Pleistocene. Oceanologia et Limnologia Sinica 8: 289–298.
23. Liu KB (1988) Quaternary history of the temperate forests of China. Quaternary Sciences Reviews 7: 1–20.
24. Zhu LM (2008) Spider community structure in fragmented habitats of Taihang Mountain area, China. Thesis for Master' Degree, Hebei Univiersity.
25. Hu X (2008) Preliminary studies on inter-generic hybridization within *Chrysanthemum* in broad sense (III). Thesis for Master' Degree, Beijing Forestry University.
26. Gong MQ (2010) Uplifting process of southern Taihang Mountain in Cenozoic. Chinese Academy of Geological Science Thesis for Doctor' Degree.
27. Wu C, Zhang XQ, Ma YH (1999) The Taihang and Yan Mountains rose mainly in Quarteranary. Norht China Earthquake Sciences 17(3): 1–7.
28. Gao LM, Möller M, Zhang XM, Hollingsworth ML, Liu J, et al. (2007) High variation and strong phylogeographic pattern among cpDNA haplotypes in *Taxus wallichiana* (Taxaceae) in China and North Vietnam. Molecular Ecology 16: 4684–4698.
29. Zhou TH, Li S, Qian ZQ, Su HL, Huang ZH, et al. (2010) Strong phylogeographic pattern of cpDNA variation reveals multiple glacial refugia

for *Saruma henryi* Oliv. (Aristolochiaceae), an endangered herb endemic to China. Molecular Phylogenetics and Evolution 57: 176–188.

30. Wolfe KH, Li WH, Sharp PM (1987) Rates of nucleotide substitution vary greatly among plant mitochondria, chloroplast, and nuclear DNAs. Proceeding of the National Academy of Sciences of the United States of America 84: 9054–9058.

31. Petit RJ, Duminil J, Fineschi S, Salvini D, Vendramin GG (2005) Comparative organization of chloroplast, mitochondrial and nuclear diversity in plant populations. Molecular Ecology 14: 689–701.

32. Zhang RY, Song G, Qu YH, Alström P, Ramos R, et al. (2012) Comparative phylogeography of two widespread magpies: Importance of habitat preference and breeding behavior on genetic structure in China. Molecular Phylogenetics and Evolution 65: 562–572.

33. Zhang Q, Chiang TY, George M, Liu JQ, Abbott RJ (2005) Phylogeography of the Qinghai-Tibetan Plateau endemic *Juniperus przewalskii* (Cupressaceae) inferred from chloroplast DNA sequence variation. Molecular Ecology 14: 3513–3524.

34. Zhao C, Wang CB, Ma XG, Liang QL, He XJ (2013) Phylogeographic analysis of a temperate-deciduous forest restricted plant (*Bupleurum longiradiatum* Turcz.) reveals two refuge areas in China with subsequent refugial isolation promoting speciation. Molecular Phylogenetics and Evolution 68: 628–643.

35. Wang FZ, Tang J, Chen XQ, Liang SJ, Dai LK, et al. (1978) Liliaceae. In: Editorial Board of the Flora of China of the China Science Academy. Flora of China 73–74 p.

36. Ding BZ, Wang SY (1998) Flora of Henan. Henan Science & Technology Press, Zhengzhou.

37. Doyle JJ, Doyle JL (1987) A rapid DNA isolation procedure for small quantities of fresh leaf material. Phytochemistry Bulletin 19: 11–15.

38. Taberlet PT, Gielly L, Patou G, Bouvet J (1991) Universal primers for amplification of three noncoding regions of chloroplast DNA. Plant Molecular Biology 17: 1105–1109.

39. Vijverberg K, Bachmann K (1999) Molecular evolution of a tandemly repeated trnF (GAA) gene in the chloroplast genomes of *Microseris* (Asteraceae) and the use of structural mutations in phylogenetic analyses. Molecular Biology and Evolution 16(10): 1329–1340.

40. White TJ, Bruns T, Lee S, Taylor J (1990) Amplification and direct sequencing of fungal ribosomal RNA genes for phylogenetics. In: Innis MA, Gelfand DH, Shinsky JJ, White TJ (eds), PCR Protocols: A Guide to Methods and Applications. Academic Press, San Diego 315–322 p.

41. Thompson JD, Gibson TJ, Plewniak F, Jeanmougin F, Higgins DG (1997) The ClustalX windows interface: flexible strategies for multiple sequence alignment aided by quality analysis tools. Nucleic Acids Research 25: 4876–4882.

42. Rozas J, Sanchez-DelBarrio JC, Messeguer X, Rozas R (2003) DnaSP, DNA polymorphism analyses by the coalescent and other methods. Bioinformatics 19: 2496–2497.

43. Pons O, Petit RJ (1996) Measuring and testing genetic differentiation with ordered versus unordered alleles. Genetics 144: 1237–1245.

44. Excoffier L, Laval G, Schneider S (2005) Arlequin ver. 3.0: an integrated software package for population genetics data analysis. Evolutionary Bioinformatics Online 1: 47–50.

45. Dupanloup I, Schneider S, Excoffier L (2002) A simulated annealing approach to define the genetic structure of populations. Molecular Ecology 11: 2571–2581.

46. Polzin T, Daneshmand SV (2003) On Steiner trees and minimum spanning trees in hypergraphs. Operations Research Letters 31: 12–20.

47. Clement M, Posada D, Crandall KM (2000) TCS: a computer program to estimate gene genealogies. Molecular Ecology 9: 1657–1660.

48. Drummond AJ, Rambaut A (2007) BEAST: Bayesian evolutionary analysis by sampling trees. BMC Evolutionary Biology 7: 214.

49. Heled J, Drummond AJ (2010) Bayesian inference of species trees from multilocus data. Molecular Biology and Evolution, 27, 570–580.

50. Posada D (2008) JModelTest: phylogenetic model averaging. Molecular Biology and Evolution 25: 1253–1256.

51. Ronquist F, Huelsenbeck JP (2003) MrBayes 3: Bayesian phylogenetic inference under mixed models. Bioinformatics 19: 1572–1574.

52. Rambaut A, Drummond AJ (2009) Tracer v1.5. <http://tree.bio.ed.ac.uk/software/tracer/>.

53. Graur D, Li WH (200) Fundamentals of Molecular Evolution, second ed., Sinauer Associates Inc., Sunderland, Massachusetts.

54. Tajima F (1989) Statistical method for testing the neutral mutation hypothesis by DNA polymorphism. Genetics 123: 585–595.

55. Fu YX (1997) Statistical tests of neutrality of mutations against population growth, hitchhiking and background selection. Genetics 147: 915–925.

56. Harpending HC (1994) Signature of ancient population growth in a low-resolution mitochondrial DNA mismatch distribution. Human Biology 66: 591–600.

57. Jensen JL, Bohonak AJ, Kelley ST (2005) Isolation by Distance, Web Service. BMC Genet. 6, 13 (Version 3.16, <http://ibdws.sdsu.edu/>).

58. Wang YL (2013) Chloroplast microsatellite diversity of *Opisthopappus* Shih. Plant Systematic and Evolution 299: 1849–1858.

59. Wang YL, Yan GQ (2013) Genetic diversity and population structure of *Opisthopappus longilobus* and *Opisthopappus taihangensis* (Asteraceae) in China determined using sequence related amplified polymorphism markers. Biochemical Systematics and Ecology 49: 115–124.

60. Wang YL, Li X, Guo J, Guo ZG, Li SF, et al. (2010) Chloroplast DNA phylogeography of *Clintonia udensis* Trautv. & Mey. (Liliaceae) in East Asia. Molecular Phylogenetics and Evolution 55: 721–732.

61. Nybom H (2004) Comparison of different nuclear DNA markers for estimating intraspecific genetic diversity in plants. Molecular Ecology 13: 1143–1155.

62. Loveless MD, Hamrick JL (1984) Ecological determinants of genetic structure in plant populations. Annual Review of Ecology and Systematics 15: 65–95.

63. Hamrick JL, Godt MJW (1989) Allozyme diversity in plant species. In: Brown AHD, Clegg MT, Kahler AL, Weir BS (eds.), Plant Population Genetics. Breeding and Genetic Resources. Sinauer, Sunderland MA 43–63 p.

64. Hamrick JL, Godt MJW (1996) Effects of life history traits on genetic diversity in plant species. Philosophical Transactions of the Royal Society of London Series B Biological Sciences 351: 1291–1298.

65. Till-Bottraund I, Gaudeul M (2002) Intraspecific genetic diversity in alpine plants. In: Körner C, Spehn EM (eds.), Mountain Biodiversity: A Global Assessment. Parthenon Publishing, New York 23–34 p.

66. Liu L, Hao ZZ, Liu YY, Wei XX, Cun YZ, et al. (2014) Phylogeography of *Pinus armandii* and its relatives: heterogeneous contributions of geography and climate changes to the genetic differentiation and diversification of Chinese white pines. PLoS one 9(1): e85920.

67. Yang XL, Xu QH, Zhao HP (1999) The vegetation succession of Taihang Mountains during the late glaciations. Geography and Territorial Research 15(1): 81–88.

68. Qiu YX, Fu CX, Comes HP (2011) Plant molecular phylogeography in China and adjacent regions: tracing the genetic imprints of Quaternary climate and environmental change in the world's most diverse temperate flora. Molecular Phylogenetics and Evolution 59: 225–244.

69. Qian H, Ricklefs RE (2001) Diversity of temperate plants in East Asia–reply. Nature 413: 130.

70. Chou YW, Thomas PI, Ge XJ, LePage BA, Wang CN (2011) Refugia phylogeography of Taiwaniii East Asia. Journal of Biogeography 38: 1992–2005.

71. Crandall KA, Templeton AR (1993) Empirical tests of some predictions from coalescent theory with applications to intraspecific phylogeny reconstruction. Genetics 134: 959–969.

72. Gong W, Chen C, Dobeš C, Fu CX, Koch MA (2008) Phylogeography of a living fossil: Pleistocene glaciations forced *Ginkgo biloba* L. (Ginkgoaceae) into two refuge areas in China with limited subsequent postglacial expansion. Molecular Phylogenetics and Evolution 48:1094–1105.

73. Wang HW, Ge S (2006) Phylogeography of the endangered *Cathaya argyrophylla* (Pinaceae) inferred from sequence variation of mitochondrial and nuclear DNA. Molecular Ecology 15: 4109–4122.

14

A Spring Forward for Hominin Evolution in East Africa

Mark O. Cuthbert[1,2]*, Gail M. Ashley[3]

1 Connected Waters Initiative Research Centre, UNSW Australia, Sydney, NSW, Australia, **2** School of Geography, Earth and Environmental Sciences, University of Birmingham, Birmingham, United Kingdom, **3** Department of Earth and Planetary Sciences, Rutgers University, Piscataway, New Jersey, United States of America

Abstract

Groundwater is essential to modern human survival during drought periods. There is also growing geological evidence of springs associated with stone tools and hominin fossils in the East African Rift System (EARS) during a critical period for hominin evolution (from 1.8 Ma). However it is not known how vulnerable these springs may have been to climate variability and whether groundwater availability may have played a part in human evolution. Recent interdisciplinary research at Olduvai Gorge, Tanzania, has documented climate fluctuations attributable to astronomic forcing and the presence of paleosprings directly associated with archaeological sites. Using palaeogeological reconstruction and groundwater modelling of the Olduvai Gorge paleo-catchment, we show how spring discharge was likely linked to East African climate variability of annual to Milankovitch cycle timescales. Under decadal to centennial timescales, spring flow would have been relatively invariant providing good water resource resilience through long droughts. For multi-millennial periods, modelled spring flows lag groundwater recharge by 100 s to 1000 years. The lag creates long buffer periods allowing hominins to adapt to new habitats as potable surface water from rivers or lakes became increasingly scarce. Localised groundwater systems are likely to have been widespread within the EARS providing refugia and intense competition during dry periods, thus being an important factor in natural selection and evolution, as well as a vital resource during hominin dispersal within and out of Africa.

Editor: Richard G. Taylor, University College London (UCL), Canada

Funding: MOC was supported by funding from the European Community's Seventh Framework Programme [FP7/2007–2013] under grant agreement n °299091 (http://ec.europa.eu/research/mariecurieactions/). GMA received funding from the Spanish Ministry of Education and Science through the European project I+D HUM2007-6381507-63815 and under National Science Foundation grant(Sedimentary Geology and Paleobiology, EAR-1349651 to D.M. Deocampo and G.M. Ashley) (http://www.nsf.gov). The funders had no role in study design, data collection and analysis, decision to publish, or preparation of the manuscript.

Competing Interests: The authors have declared that no competing interests exist.

* Email: m.cuthbert@bham.ac.uk

Introduction

Hominin fossil discoveries in the last few decades have shown that humans evolved in Africa and then migrated in waves to other parts of the world, starting as early as 1.85 Ma [1]. Both marine and terrestrial records in the region point to a general increase in aridity during the Plio-Pleistocene [2–4] primarily due to progressive rifting in East Africa and associated tectonic uplift [5]. The climate during the last 6–7 million years when hominins evolved to modern humans was characterized by high variability [6,7]. In the tropics of Africa it was dominated by wet and dry cycles driven by orbital forcing and, in particular, the effect of precession on monsoon strength [8–11]. The climate is also likely to have varied on the millennial timescale due to tropical expressions of Dansgaard-Oeschger cycles [12], as well as on a decadal timescale governed by variations in El Nino Southern Oscillation (ENSO) and Indian Ocean Dipole (IOD) variations [13,14]. A major unknown connected with human evolution in this climatically turbulent environment [6] is the availability of resources, particularly freshwater. Recent research has suggested the importance of freshwater availability from some lakes for hominin survival and dispersal in the EARS [6,11,15,16]. Finlayson [17] even suggests that "the need for swift and efficient movement between ever shrinking sources of water" was *the* trigger for human evolution. However, many of the EARS lakes

are saline and were likely so in the past, thus their potability, during periods of hominin radiations in the last 5 Ma, is speculative [11,18]. Furthermore, streams which originate in arid and semi-arid areas are ephemeral in nature; perennial flows are only possible in large catchments where humid source areas upstream contribute consistent inflows via runoff and groundwater baseflow [19]. During periods of increased aridity, sources of riverine or potable lake water would therefore have been scarce in many parts of the EARS.

Groundwater is protected from evaporation and thus potentially provides a key alternative potable resource for sustaining life through drought periods in areas with variable rainfall [20]. Hence, springs and groundwater-fed habitats could have played a decisive role in the survival and dispersal of hominins during these times of known climate variability [6,15] when potable surface water was limited. The significance of groundwater in this context has, nevertheless, received little attention, although geological evidence for active freshwater springs has recently been linked to archaeological hominin remains at Olduvai Gorge [21,22] and the first fossil chimpanzee [23] elsewhere in the EARS.

Despite the geological evidence for the presence of springs, it is presently unclear how such spring discharges may have varied with climate. Our objectives are therefore to provide estimates for the temporal variability of spring flow at Olduvai Gorge, at around 1.8 Ma BP under different climate variability scenarios and to

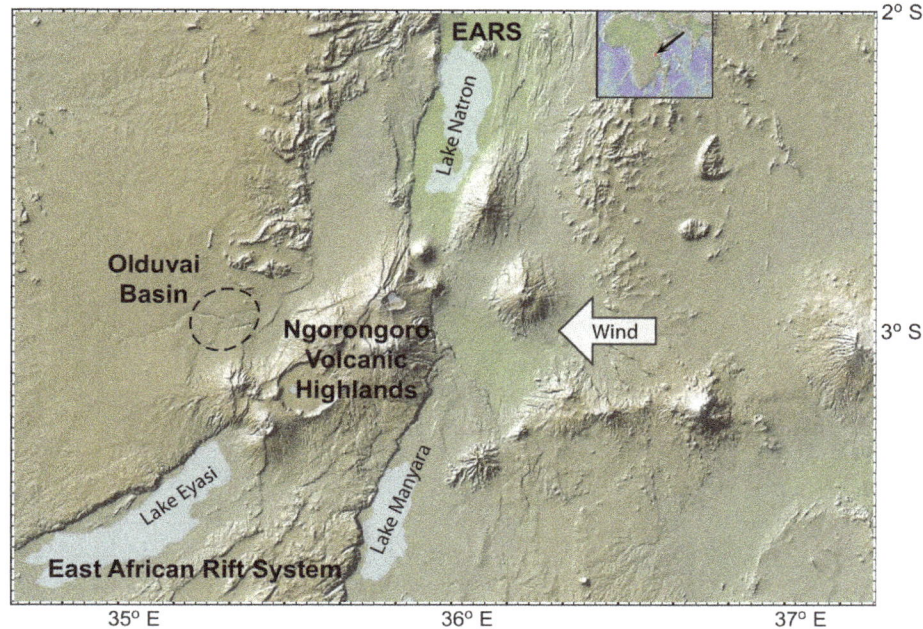

Figure 1. Location map. Insert shows with arrow the location of study area in eastern Africa. Map of the Northern Tanzanian Divergence Zone depicts the East African Rift System (EARS), containing Lake Natron (north), diverging around the Ngorongoro Volcanic Highland massif and splitting into two separate rift valleys (Lake Eyasi on west) and Lake Manyara (on east). Prevailing wind is from the east. Olduvai basin lies to the west of and in the rain shadow of Ngorongoro. (Map made by Sara Mana, http://www.geomapapp.org).

consider the wider implications of this new information in the context of current hypotheses about human evolution in the East African Rift System (EARS) (Figure 1, http://www.geomapapp. org).

Physical Setting

Geology

The Olduvai sedimentary basin was formed ~2.2 Ma years ago on the western margin of the EARS, at 3°S, in response to the growth of a large volcanic complex (Ngorongoro) (Figure 1). The region is in the Northern Tanzanian Divergent Zone, a prominent bifurcation in the main rift valley. Ngorongoro Volcanic Highland is a ~4,000 km^2 massif comprised of 8–10 eruptive centers of alkali magma compositions [24,25]. The paleo basin is an estimated 3,500 km^2 in area and bordered on the south and east by the volcanic highlands and on the west and north by metamorphic terrain. The preserved sedimentary record in the basin center is 30 km in diameter and composed of 100 meters of interbedded pyroclastics (tuffs), volcaniclastic sediments and minor limestones. A playa lake occupied the center of the basin during the first million years, seasonal rivers drained the periphery and an alluvial fan comprised of fine- and coarse-grained pyroclastics fringed the basin margin on east and south (Figure 2).

Olduvai Basin is ideal for studying paleohydrology. The tuffs are precisely dated using $^{40}Ar/^{39}Ar$ single crystal method [26] and can be traced through the basin using tephrocorrelation [27]. The variation in stratigraphic sequences was documented throughout the basin [25,28]. Field mapping identified the lithologic record of Milankovitch (precession) driven lake cycles [8,29] and the location of groundwater-fed environments. The paleoenvironmental reconstruction was based on sedimentary facies and structures [30,31], clay mineralogy [32,33], stable isotopes [21,34,35], and fossilized plant remains [31,36–38]. The association of spring and wetland deposits is documented in deposits 2 to 1 Ma in age [39–42].

Hydrology

The shallow Olduvai basin contained a saline-alkaline lake flanked to the east and south by freshwater wetlands [21]. Two rainy seasons currently provide around 550 mm/a of precipitation to the Olduvai lowland and around twice that in the highlands due to the elevation difference and rain-shadow effect [36]. This difference was also likely present in the past since the topographical contrast was similar. Annual rainfall data show substantial interannual and decadal variability for the 20th century (Figure 3). Around 1.8 Ma, wet-dry cycles driven by precession (21 ka) and modulated by eccentricity (100 ka) (orbital monsoon hypothesis [7]) were superimposed on a longer term drying trend [43]. There were 5 wet and 5 ½ dry cycles between 1.85–1.74 Ma [43] during which the seasonal variability in insolation, and thus also temperature, potential evapotranspiration and monsoon strength [8], would have fluctuated. Annual rainfall is thought to have varied from 250 to 700 mm/a during arid and wetter intervals respectively [44]. Shorter term (years to decades) variations also occurred [35] with centennial to millennial variability also likely. The strengthening of the Walker circulation around this time [13] is also likely to have led to more extreme climate variability on the annual to decadal timescale. This may have been particularly significant for providing an intensification of rainfall and therefore increased frequency of runoff-recharge events [14,45].

Our conceptual model of the hydrology of paleo-Olduvai is as follows. Higher rainfall and steep slopes on the flanks of the Ngorongoro Volcanic Highland caused significant runoff feeding the upper parts of a pyroclastic alluvial fan at the mountain front (Figures 1 and 2). This is likely to have occurred under intense rainfall conditions even when potential evapotranspiration in the region was high (likely >2000 mm/a [14,22]). Ephemeral stream

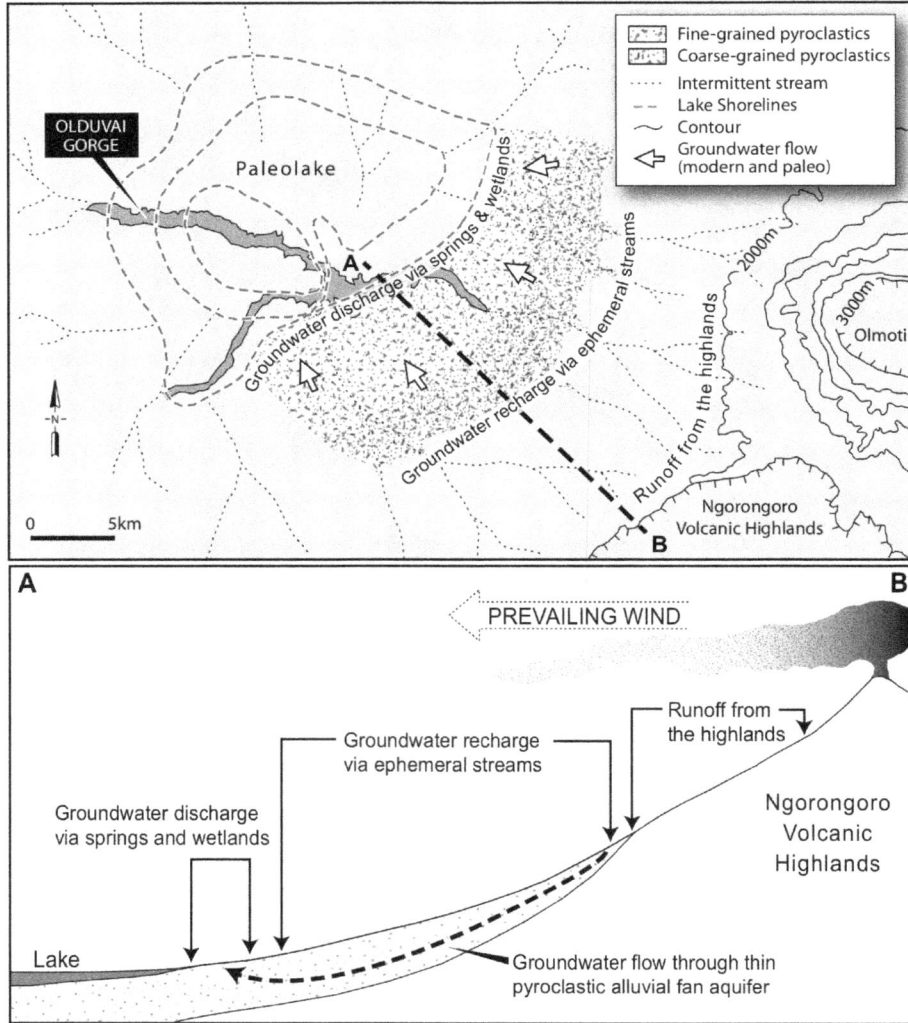

Figure 2. Paleoenvironmental reconstruction of Olduvai and schematic hydrogeological conceptual model (modified from [28]).

flow then led to 'indirect' groundwater recharge through the base of the streams into the fan deposits [43,46]. 'Diffuse' groundwater recharge across the wider region was unlikely given the predominant semi-arid climate conditions unless there were occasional very prolonged and intense rainfall events. Such events may have enabled soil moisture deficits to be overcome locally and/or triggered preferential flow pathways through the soils to enable water to escape the zone of evapotranspiration [14,47]. How far down the fan the ephemeral streams flowed during flow events would have been a complex interaction between antecedent moisture conditions, streambed permeability and the magnitude of the flow event [46]. However, initial groundwater mounding underneath the stream channels would have spread transversely into the fan deposits, at the same time as flowing longitudinally downslope to the lower parts of the fan. Flow-focussing via faults and the juxtaposition of lower permeability deposits on the fan margin led to local spring discharges and significant tufa deposition [21].

Groundwater Model

Model Formulation

The objective of the modelling is to provide the likely range of timescales for groundwater discharge recession for the paleo-Olduvai groundwater system. The models have been designed to contain, conceptually, the most important features of the flow system while being kept mathematically simple enough to enable analytical solutions to the problems to be found. This enables the large range of parameter uncertainty to be analysed without excessive computational effort while also making it possible to test a variety of different boundary conditions to cover the necessary groundwater recharge input scenarios.

Numerical models were developed based on analytical solutions to the following 1–D linearised Boussinesq equation for groundwater flow through a homogeneous and isotropic sloping aquifer [48] sketched conceptually in Figure 4:

$$\frac{\partial \eta}{\partial t} = \frac{k_0 \eta_0 cos\alpha}{n_e}\frac{\partial^2 \eta}{\partial x^2} + \frac{k_0 sin\alpha}{n_e}\frac{\partial \eta}{\partial x} \tag{1}$$

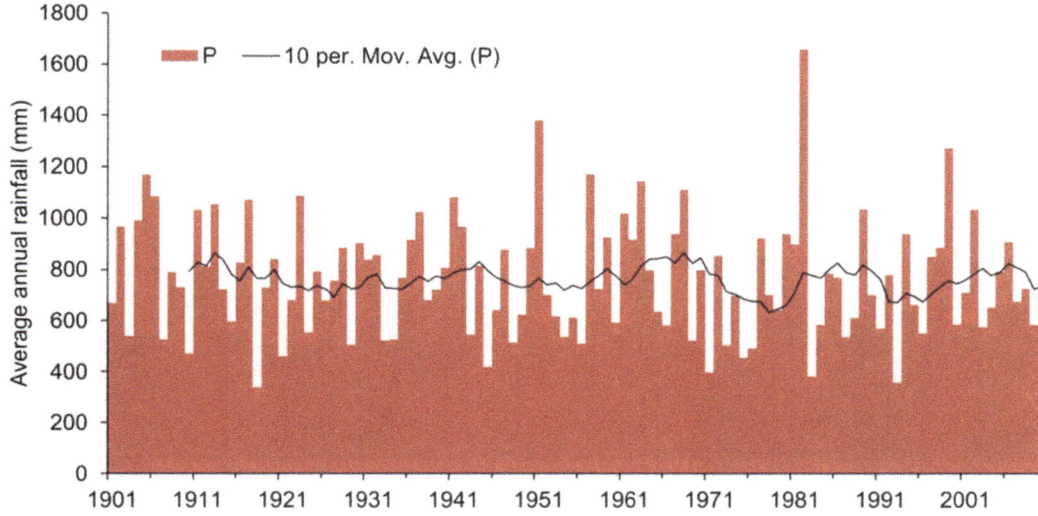

Figure 3. Gridded annual precipitation data (P) from the Global Precipitation Climatology Centre (GPCC) for grid square 2.75 deg S, 35.25 deg E in which Olduvai Gorge is situated. A 10 year moving average is also plotted. Data was accessed on 26/4/13from: http://iridl.ldeo.columbia.edu/SOURCES/.WCRP/.GCOS/.GPCC/.FDP/.version6/.

where t is time [T], x is distance along the aquifer base [L], η is hydraulic head [L], k_0 is aquifer hydraulic conductivity, n_e is aquifer specific yield [−], α is the slope of the aquifer [−], η_0 is the average water table height equal to $0.3D$ where D is the maximum saturated thickness of the aquifer [L].

It is assumed that the lateral spreading occurring away from the ephemeral streambeds during recharge events occurs at timescales orders of magnitude faster than the longitudinal drainage of the fan. This is reasonable since the groundwater response time is controlled by $L^2 n_e/T$ [49] and the lateral spreading occurs over a length scale (L) much smaller than the length of the fan longitudinally. Thus, groundwater recharge (I [LT^{-1}]) is assumed to be evenly distributed across the domain between $x = 0$ at the spring discharge point which is represented by a constant head boundary, and a groundwater divide represented by a no flow boundary at $x = B_x$ (Figure 4). The total recharge input is thus equal to IB where B is the horizontal distance between the constant head and no flow boundaries. Two boundary value problem scenarios are solved for the time variant groundwater discharge (q [L^{3}T^{-1}]) at $x = 0$ as follows.

Scenario 1: Sudden cessation of recharge. The first case considered is the sudden cessation of recharge after a wet period when the aquifer is at a steady state. An existing solution to equation (1) for the stated boundary conditions which proves very useful for the Olduvai case is as follows [48]:

$$q(t) =$$
$$-2B_x I \cos\alpha \sum_{n=1,2,3\ldots}^{\infty} \frac{z_n^2 \left[1 - 2\cos(z_n)\exp\left(\frac{Hi}{2}\right)\right] \exp\left[-\left(z_n^2 + \frac{Hi^2}{4}\right)t_+\right]}{\left(z_n^2 + \frac{Hi^2}{4} + \frac{Hi}{2}\right)\left(z_n^2 + \frac{Hi^2}{4}\right)} \quad (2)$$

with z_n being the nth root of $\tan(z) = 2z/Hi$ and:

$$Hi = \frac{B_x \tan\alpha}{\eta_0} \quad (3)$$

$$t_+ = \left[\frac{k_0 \eta_0 \cos\alpha}{n_e B_x^2}\right] t \quad (4)$$

Scenario 2: Periodic variation of recharge. No solution to equation (1) for the stated boundary conditions exists in the literature to calculate the outflow for the case of a periodic variation in recharge in a sloping aquifer. However, using Brutsaert's [48] solution for a unit response to a step change in recharge as follows:

$$u(t) =$$
$$-\frac{2k_0 \eta_0 \cos\alpha}{n_e B_x} \sum_{n=1,2,3\ldots}^{\infty} \frac{z_n^2 \left[1 - 2\cos(z_n)\exp\left(\frac{Hi}{2}\right)\right] \exp\left[-\left(z_n^2 + \frac{Hi^2}{4}\right)t_+\right]}{\left(z_n^2 + \frac{Hi^2}{4} + \frac{Hi}{2}\right)} \quad (5)$$

and then applying a convolution integral for a periodic variation around some average recharge (I_{av}), with:

$$I = I_{av}(1 - \cos(\omega t)) \quad (6)$$

leads to the following equation for the outflow at $x = 0$:

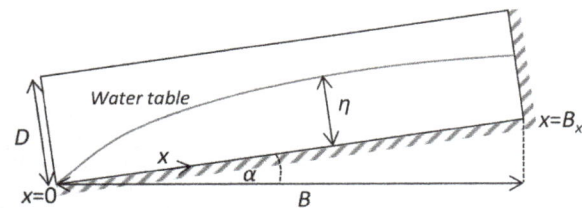

Figure 4. Simplified conceptual groundwater flow model for a sloping aquifer.

$$q(t) = \int_{-\infty}^{t} I_{av}(1 - \cos(\omega\tau))u(t-\tau)d\tau \qquad (7)$$

The solution to this integral is:

$$q(t) = I_{av} \sum_{n=1,2,3...}^{\infty} \left[\frac{A}{C^2 + \omega^2}(-\omega \sin(\omega t) + C\cos(\omega t)) - \frac{A}{C} \right] \qquad (8)$$

with

$$A = -\frac{2k_0\eta_0 cos\alpha}{n_e B_x} \left[\frac{z_n^2 \left[1 - 2\cos(z_n)\exp\left(\frac{Hi}{2}\right) \right]}{\left(z_n^2 + \frac{Hi^2}{4} + \frac{Hi}{2} \right)} \right] \qquad (9)$$

and

$$C = -\frac{k_0\eta_0 cos\alpha}{n_e B_x^2} \left(z_n^2 + \frac{Hi^2}{4} \right) \qquad (10)$$

Model outputs have been plotted as amplitude ratios (AR) and phase shifts (PS) normalised to the recharge input (i.e. using $q/(I_{av}.B)$) and the period of oscillation ($P = 2\pi/w$) respectively as follows:

AR = (amplitude of q)/(amplitude of $I_{av}.B$)

PS = time lag between maxima (or minima) in q, and maxima (or minima) in $I_{av}.B/P$

The AR is thus a measure of how much the outflow varies in comparison with the input signal; the PS is how much the output signal lags the input signal as a proportion of the input period. Smaller ARs mean the groundwater outflow is highly damped in comparison to the input recharge signal and this is normally accompanied by larger PS.

The analytical solution, Eq. 2–4, used for the first 'sudden cessation of recharge' scenario is a well-known published result [48]. Since the analytical solution for the periodic case was derived for this paper (Eq. 8–10) it was therefore tested against a published solution for the end member where the aquifer is not sloping [50,51] and agreed perfectly. Although it is unlikely that recharge varied exactly as a step function (Scenario 1) or exactly sinusoidally (Scenario 2), the two solutions have been chosen to provide plausible and contrasting end members of the possible recharge variability under likely variations in climate. The software Maple (v.17) was used for the model calculations; the model files and the model data used to produce Figures 5 to 7 are included as Dataset S1.

Model Parameterisation

The geometry of the alluvial fan and thus the length scale of the groundwater flow system (B) is relatively well constrained from the paleo-reconstruction and is likely to have been in the range 15 km $+/-5$ km. The slope angle (α) is likely to have been in the range $0.04 +/-0.01$.

While there are relatively few detailed studies on the hydraulic characterisation of pyroclastic materials, a literature review (Table 1) suggested [52–57] that a range of k_0 of 0.05 to 5 m/d is reasonable (for length scales of 10 s of km) in combination with

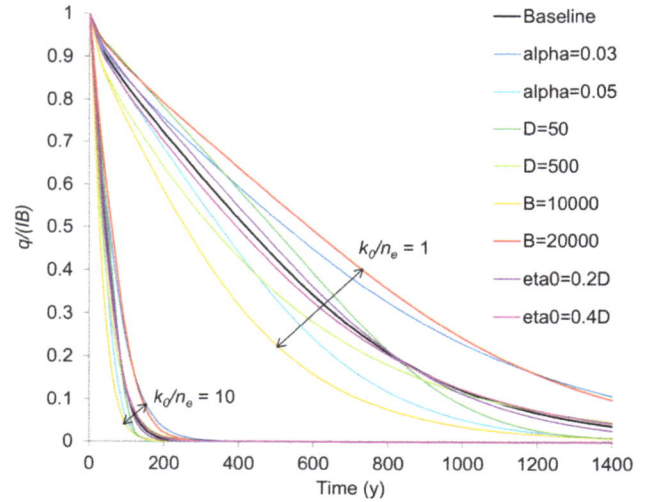

Figure 5. Geometric parameter sensitivity of spring flow recession using equation (2). Variation in recession rates is predominantly controlled by variations in the ratio of hydraulic conductivity to specific yield (k_0/n_e) and relatively insensitive to variations in the geometry, i.e. the spread *around* the baseline recession for the two k_0/n_e end members ($k_0/n_e = 1$ and $k_0/n_e = 10$) is much smaller than variation *between* the end members.

likely saturated thicknesses of <200 m for the fan materials to keep $T<1000$ m^2/d consistent with the literature values. Specific yield of porous materials is normally in the range 0.05 to 0.5.

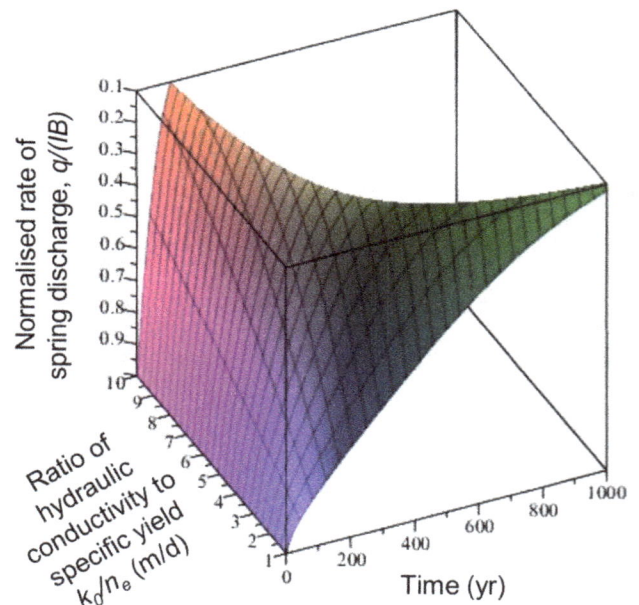

Figure 6. Rate of groundwater discharge (q, modelled using Eq. 2) at Olduvai springs normalised to the rate of groundwater recharge (IB) for the case of a sudden cessation of groundwater recharge after a period of steady state. Baseline geometry parameters have been used and hydraulic properties (k_0/n_e) were varied across the likely range. The vertical axis, $q/(IB)$, is clipped at 0.1 to illustrate the range of timescales by which the discharge recedes to 10% of its original value.

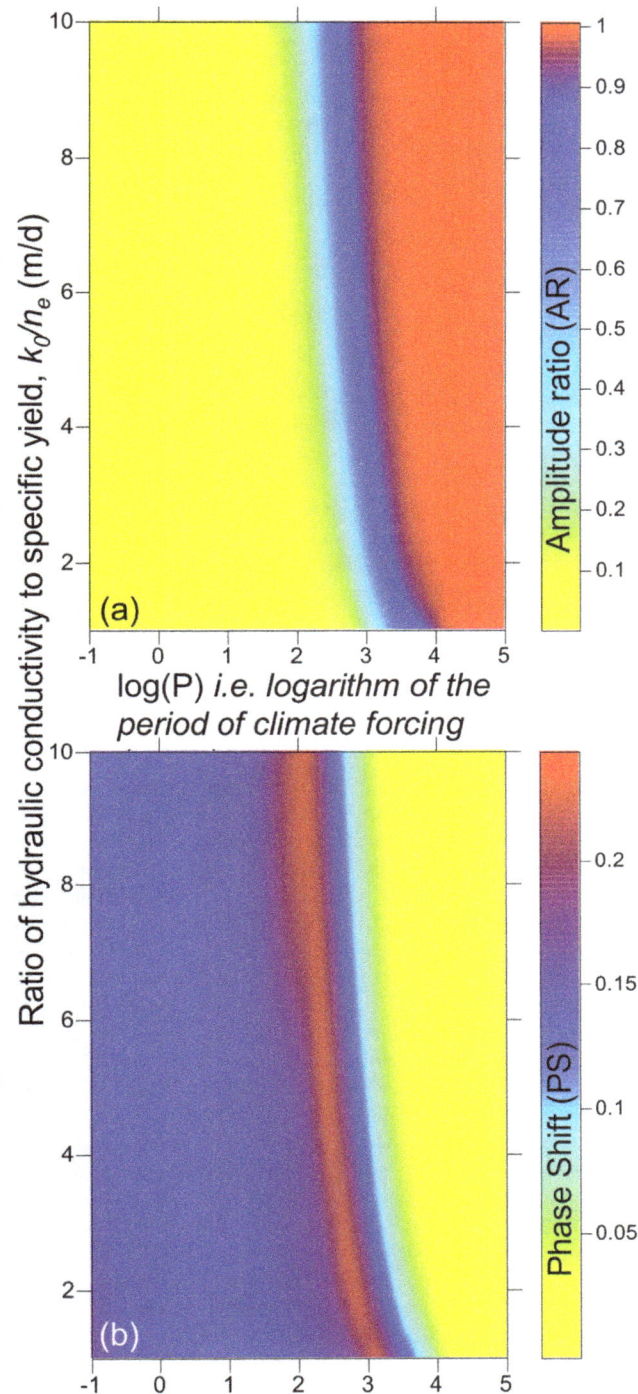

Figure 7. Modelled (a) amplitude ratio (AR) and (b) phase shift (PS) of the input groundwater recharge forcing relative to the spring discharge output on periods from 0.1 year to 100 000 years across a range of likely aquifer hydraulic properties (using Eq. 8–10). The transition to lower amplitude ratios and higher phase lags for periods lower than 100 to 1000 years implies greater buffering of the climate signal and increased potential resilience of the spring discharge to climate variability on these timescales.

However, covariation in porous materials of k_0 and n_e leads to a likely range of k_0/n_e between minimum $0.05/0.05 = 1$ and maximum $5/0.5 = 10$ which was used for all the models.

A sensitivity analysis for the geometric parameters used in the models (α, D, B and η_0) was carried out by altering each parameter across its likely range away from a baseline parameter set, and observing the change in the recession timescale using Eq. 2–4. The baseline parameters are as follows: $B = 15$ km, $\alpha = 0.04$, $\eta_0 = 0.3D$, $D = 200$ m with sensitivity ranges as defined in Figure 5. Of the geometric parameters, the rate of recession is most sensitive to the length of the flow system, B. However, it is clear that the sensitivity to the geometric parameters is insignificant in comparison to the sensitivity to the hydraulic diffusivity (ratio of k_0/n_e) (Figure 5). The results we present and discuss in the paper, despite having a fixed geometry, are thus illustrative of the likely range of hydrodynamic characteristics.

Results and Discussion

Limits of Spring Longevity at Olduvai

Even in the present day it is often impossible to estimate groundwater recharge rates to within an order of magnitude in semi-arid areas solely on the basis of known climate variables [58]. Based on the record of extensive tufa deposition in spring and associated wetland environments at Olduvai between 1.84 to 1.36 Ma [21,30] we can be sure that freshwater would have been flowing actively and thus available for direct consumption even, to some extent, during the driest periods of the precessional cycles. However, temporally, the geological record is discontinuous and it is not possible to deduce how persistent the spring flows may have been on the basis of the geology alone. Whereas it is not possible to be precise about the absolute rates of recharge and discharge, the groundwater models we have developed here, based on our paleoenvironmental reconstruction and hydrological conceptual model, are able to realistically estimate the plausible range of variation in flows as well as their persistence during dry periods. As would be expected intuitively, as aquifer hydraulic diffusivity increases, modelled spring flows become more responsive to the climate forcing and have characteristically shorter recessions (periods of declining flow). For the case of a sudden reduction of groundwater recharge during an extended drought period, across the likely range of hydraulic properties (k_0/n_e), spring discharge recedes to almost zero flow after approximately 150 to >1000 years (Figure 6). This suggests that the Olduvai springs and wetland would likely have continued to be supported by groundwater flow for a few 100 to approx. 1000 years even with no recharge.

For the case of a periodic variation in the groundwater recharge forcing, as the period increases the amplitude ratio (AR) increases indicating decreased damping, irrespective of the hydraulic diffusivity (Figure 7). The phase shift (PS) shows a more complex relationship with period which is more strongly dependent on the diffusivity. For 20 ka Milankovitch recharge cycles, the spring discharge lags recharge by a few hundred years and is almost identical in amplitude (Figure 7). Once the period is shorter than around 2 ka, discharge maxima and minima lag those of the forcing recharge signal by up to 400 years for the lower hydraulic diffusivity end member and are damped by as much as 50% of the magnitude of the recharge variation. This means that groundwater would have been available during the driest periods of millennial climate cycles, only diminishing significantly on the rising limb when potable surface water sources may have become more plentiful. For recharge oscillations on the order of 100 years or less there is almost complete damping of the input signal (Figure 7)

Table 1. Range of literature values for volcanically derived aquifers used to parameterise the groundwater models.

Reference	Range of hydraulic properties: hydraulic conductivity (k_0, m/d), transmissivity (T, m²/d) and specific yield (n_{e}, -)
Gleeson et al. [54]	$k_0 \approx 0.5$ for mid-range 'regional' volcanic values
Yihdego & Becht [57]	$k_0 \approx 0.1$ to 0.4 for reworked volcanics near Lake Naivasha EARS
Belcher et al. [53]	$k_0 \approx 0.04$ to 4 for Death Valley tuffs
Ayenew et al. [52]	$T \approx 27$ to 135 in Ethiopian EARS
Greco et al. [55]	$k_0 \approx 7$ for Italian shallow pyroclastic deposits
Moghaddam & Najib [56]	$k_0 \approx 1.05$ to 3.5, $T \approx 242$ to 858 for Iranian tuff

meaning that groundwater discharge would have varied by less than 20% during droughts over these timescales despite large changes in groundwater recharge.

Considering the superposition of climate variations occurring at different frequencies, these results indicate that, for multi-millennial and greater period variations, the wetland springs would only have provided a refugium if sufficient groundwater recharge still occurred at centennial or higher frequencies during the driest parts of longer term climate cycles. At Olduvai, this is likely to have been the case since there is stratigraphic evidence of some tufa deposition during dry parts of the 21 ka climate oscillation [22]. Sufficient, albeit likely sporadic, groundwater recharge must have been occurring despite the arid conditions. The recharge was made possible due to the focussing effect of runoff on the highland flank which caused indirect recharge through the base of ephemeral stream channels. This phenomenon is well known to occur in similar modern day arid settings [58–61].

Therefore, in combination, the geology and the modelling suggest that the groundwater system at Olduvai would have provided a freshwater resource throughout the precessional cycle even during long droughts occurring on decadal to multi-centennial timescales.

Wider availability of groundwater within the EARS

While Olduvai Gorge was the first archaeological locality at which geological evidence of groundwater discharge was recognised, other localities within the EARS have recently been identified [62,63]. Many semi-arid EARS settings in the present day have active groundwater systems [64,65] providing seeps and springs supporting vegetation, watering holes and lakes, and it is likely that they also would have done so in the past in similar settings and provided essential freshwater resources during dry periods to hominins and other animals. Many of the EARS lakes during the critical hominin radiation periods appear to have been saline [11,18] and we propose that groundwater springs may have provided important 'landscapes for evolution' [66] in the wider context of the EARS in addition to any surface water sources that may have also been potable. It has been suggested that that it took about 2 ka for the large EARS lakes to completely dry up [16]. However, during such dry periods it is possible that, while runoff in such catchments was not hydrologically effective enough at the basin scale to keep a lake wet, localised groundwater recharge could still feed spring systems and provide critical freshwater resources. The modelling results presented here show, for the first time, the likely timescale for the springs at Olduvai to remain active with no rainfall was in the range 100 s–1000 years, and geological evidence points to the fact that the springs were active during the driest times of the precessional cycle. Other sites in

similar settings in the EARS may have been more or less 'drought proof' in comparison to Olduvai. For example our model sensitivity analysis shows that increased length scales, decreased diffusivity, increased thickness or more gentle slopes in a groundwater system leads to increased buffering of the climate signal (Figure 5). Different catchment geometries and runoff characteristics would have also led to variations in recharge thresholds with varying aridity across the EARS. In catchments without steep slopes or increased rainfall at higher elevations, indirect recharge may not have occurred frequently, and recharge may only have been possible during periods of intense rainfall for example driven by particularly strong El Nino events [14].

Implications for Hominin Evolution and Dispersal

There is a broad coincidence between aridification of Africa (and the associated expansion of open woodlands and grasslands [2]) over the last 7 Ma, and the development of bipedalism. Increasing climate variability at about 2.5 Ma is associated with the first record of genus Homo, the first appearance of stone tools (evidence of technological capabilities) [67], the increase in cranial capacity [68], and eventually the migration of hominins out of Africa. The development of complex cognitive processes (e.g. language) and art [69] cannot be tied directly to climate. However, there have been several attempts to explain the role of past climate change on hominin evolution which focus on the importance of climate instability. For example, Potts' "variability selection" theory proposes that dramatic climatic shifts favoured animals that were truly generalists and could adapt to a wide range of environmental conditions [70]. More recently, in a variation on this theme, the "pulsed climate variability hypothesis" [71] argues for extreme wet-dry cycles, in particular the precession-driven appearance and disappearance of deep EARS lakes, driving hominin evolution. While around 2 Ma most EARS lakes dried up or became saline during arid precessional phases, a notable exception is that of Lake Turkana, due its extensive catchment in the Ethiopian Highlands, which may have provided a continuous "aridity-refugium" during the whole precessional cycle [72].

The major contribution we propose here to this debate is in showing how groundwater refugia could also have persisted during the driest parts of the precessional cycle. We hypothesise that as surface water sources became more scarce during a given precessional cycle, the only species to survive may have been those with adaptations for sufficient mobility to discover a new and more persistent groundwater source, or those already settled within home range of such a resource. Such groundwater refugia may have been sites for intense competition between hominin and other animal species and hence selective pressure favouring those who could maintain access to water, something for which there is no substitute. Furthermore we speculate that, during wetter periods, springs may have formed ways of 'bridging' longitudinal dispersal of hominins between larger freshwater bodies or rivers providing a critical resource during hominin migration within and out of Africa.

Thus we consider that, while the argument for the persistence of springs during arid periods is robust, further exploration is needed to test hypotheses as to how groundwater flow systems produced by the EARS played a significant role in the evolution and dispersal of humans in the region.

Acknowledgments

We are grateful to the Editor (Richard Taylor), Mark Maslin, and two anonymous reviewers for comments which improved the manuscript. The geological context is based on data collected under permits from the Tanzania Commission for Science and Technology and the Tanzanian Antiquities Department to TOPPP (The Olduvai Paleoanthropology and Paleoecology Project), PIs M. Domínguez-Rodrigo, H.T. Bunn, A.Z.P. Mabulla, and E. Baquedano.

Author Contributions

Analyzed the data: MOC GMA. Contributed reagents/materials/analysis tools: MOC. Contributed to the writing of the manuscript: MOC GMA. Conceptual model derivation: GMA MOC. Groundwater modelling: MOC.

References

1. Ferring R, Oms O, Agusti J, Berna F, Nioradze M, et al. (2011) Earliest human occupations at Dmanisi (Georgian Caucasus) dated to 1.85–1.78 Ma. Proceedings of the National Academy of Sciences of the United States of America 108: 10432–10436.
2. Cerling TE, Wynn JG, Andanje SA, Bird MI, Korir DK, et al. (2011) Woody cover and hominin environments in the past 6 million years. Nature 476: 51–56.
3. Feakins SJ, Levin NE, Liddy HM, Sieracki A, Eglinton TI, et al. (2013) Northeast African vegetation change over 12 my. Geology 41: 295–298.
4. Ségalen L, Lee-Thorp JA, Cerling T (2007) Timing of C4 grass expansion across sub-Saharan Africa. Journal of Human Evolution 53: 549–559.
5. Sepulchre P, Ramstein G, Fluteau F, Schuster M, Tiercelin J-J, et al. (2006) Tectonic uplift and Eastern Africa aridification. Science 313: 1419–1423.
6. Potts R (2012) Environmental and behavioral evidence pertaining to the evolution of early Homo. Current Anthropology 53: S299–S317.
7. Ruddiman WF (2013) Earth's Climate: past and future. New York: W. H. Freeman & Co Ltd. 445 p.
8. Ashley GM (2007) Orbital rhythms, monsoons, and playa lake response, Olduvai Basin, equatorial East Africa (ca. 1.85–1.74 Ma). Geology 35: 1091–1094.
9. Campisano C (2012) Milankovitch cycles, paleoclimatic change, and hominin evolution. Nature Education Knowledge 3: 5.
10. Deino AL, Kingston JD, Glen JM, Edgar RK, Hill A (2006) Precessional forcing of lacustrine sedimentation in the late Cenozoic Chemeron Basin, Central Kenya Rift, and calibration of the Gauss/Matuyama boundary. Earth and Planetary Science Letters 247: 41–60.
11. Trauth MH, Maslin MA, Deino AL, Strecker MR, Bergner AGN, et al. (2007) High- and low-latitude forcing of Plio-Pleistocene East African climate and human evolution. Journal of Human Evolution 53: 475–486.
12. Cruz FW, Burns SJ, Karmann I, Sharp WD, Vuille M, et al. (2005) Insolation-driven changes in atmospheric circulation over the past 116,000 years in subtropical Brazil. Nature 434: 63–66.
13. Ravelo AC, Andreasen DH, Lyle M, Lyle AO, Wara MW (2004) Regional climate shifts caused by gradual global cooling in the Pliocene epoch. Nature 429: 263–267.
14. Taylor RG, Todd MC, Kongola L, Maurice L, Nahozya E, et al. (2013) Evidence of the dependence of groundwater resources on extreme rainfall in East Africa. Nature Climate Change 3: 374–378.
15. Shultz S, Maslin M (2013) Early human speciation, brain expansion and dispersal influenced by African climate pulses. Plos One 8: e76750.
16. Trauth MH, Maslin MA, Deino AL, Junginger A, Lesoloyia M, et al. (2010) Human evolution in a variable environment: the amplifier lakes of Eastern Africa. Quaternary Science Reviews 29: 2981–2988.
17. Finlayson C (2014) The improbable primate: how water shaped human evolution: Oxford University Press. 202 p.
18. Deocampo DM, Cuadros J, Wing-Dudek T, Olives J, Amouric M (2009) Saline lake diagenesis as revealed by coupled mineralogy and geochemistry of multiple ultrafine clay phases: Pliocene Olduvai Gorge, Tanzania. American Journal of Science 309: 834–868.
19. Simmers I (2003) Understanding water in a dry environment: IAH International Contributions to Hydrogeology 23. Lisse, The Netherlands: Taylor & Francis. 341 p.
20. MacDonald AM, Bonsor HC, Dochartaigh BEO, Taylor RG (2012) Quantitative maps of groundwater resources in Africa. Environmental Research Letters 7: 024009.
21. Ashley GM, Dominguez-Rodrigo M, Bunn HT, Mabulla AZP, Baquedano E (2010) Sedimentary geology and human origins: a fresh look at Olduvai Gorge, Tanzania. Journal of Sedimentary Research 80: 703–709.
22. Ashley GM, Tactikos JC, Owen RB (2009) Hominin use of springs and wetlands: Paleoclimate and archaeological records from Olduvai Gorge (similar to 1.79–1.74 Ma). Palaeogeography Palaeoclimatology Palaeoecology 272: 1–16.
23. McBrearty S, Jablonski NG (2005) First fossil chimpanzee. Nature 437: 105–108.
24. Dawson JB (2008) The Gregory rift valley and Neogene-recent volcanoes of northern Tanzania: Geological Society, London, Memoirs, 33.
25. Hay RL (1976) Geology of the Olduvai Gorge: a study of sedimentation in a semiarid basin: Univ of California Press.
26. Deino AL (2012) Ar-40/Ar-39 dating of Bed I, Olduvai Gorge, Tanzania, and the chronology of early Pleistocene climate change. Journal of Human Evolution 63: 251–273.
27. McHenry LJ (2004) Characterization and correlation of altered Plio-Pleistocene tephra using a "multiple technique" approach: case study at Olduvai Gorge, Tanzania: Rutgers University.
28. Ashley GM, Hay R L (2002) Sedimentation in Continental Rifts: SEPM Special Publication 73 107–122.
29. Magill CR, Ashley GM, Freeman KH (2013) Water, plants, and early human habitats in eastern Africa. Proceedings of the National Academy of Sciences of the United States of America 110: 1175–1180.
30. Ashley GM, de Wet CB, Domínguez-Rodrigo M, Karis AM, O'Reilly TM, et al. (In Press) Freshwater limestone in an arid rift basin, a Goldilocks Effect. Journal of Sedimentary Research 84.
31. Liutkus CM, Ashley GM (2003) Facies model of a semiarid freshwater wetland, Olduvai Gorge, Tanzania. Journal of Sedimentary Research 73: 691–705.
32. Deocampo DM, Blumenschine RJ, Ashley GM (2002) Wetland diagenesis and traces of early hominids, Olduvai Gorge, Tanzania. Quaternary Research 57: 271–281.
33. Hover VC, Ashley GM (2003) Geochemical signatures of paleodepositional and diagenetic environments: A STEM/AEM study of authigenic clay minerals from an arid rift basin, Olduvai Gorge, Tanzania. Clays and Clay Minerals 51: 231–251.
34. Ashley GM, Dewet C, Karis AM, O'Reilly T, Baluyot RD (2013) Freshwater limestone in an arid rift basin, a Goldilocks effect. Geological Society of America Abstracts with Programs 45.
35. Liutkus CM, Wright JD, Ashley GM, Sikes NE (2005) Paleoenvironmental interpretation of lake-margin deposits using delta C-13 and delta O-18 results from early Pleistocene carbonate rhizoliths, Olduvai Gorge, Tanzania. Geology 33: 377–380.
36. Ashley GM, Barboni D, Dominguez-Rodrigo M, Bunn HT, Mabulla AZP, et al. (2010) Paleoenvironmental and paleoecological reconstruction of a freshwater oasis in savannah grassland at FLK North, Olduvai Gorge, Tanzania. Quaternary Research 74: 333–343.
37. Ashley GM, Barboni D, Dominguez-Rodrigo M, Bunn HT, Mabulla AZP, et al. (2010) A spring and wooded habitat at FLK Zinj and their relevance to origins of human behavior. Quaternary Research 74: 304–314.
38. Barboni D, Ashley GM, Dominguez-Rodrigo M, Bunn HT, Mabulla AZP, et al. (2010) Phytoliths infer locally dense and heterogeneous paleovegetation at FLK North and surrounding localities during upper Bed I time, Olduvai Gorge, Tanzania. Quaternary Research 74: 344–354.
39. Ashley GM (2001) Archaeological sediments in springs and wetlands. Sediments in Archaeological Contexts University of Utah Press, Salt Lake City, UT, USA: 183–210.
40. Ashley GM (2009) 50th Golden Anniversaries of Zinjanthropus discovery and the current establishments of Serengeti National Park and Ngorongoro Conservation Area Authority, Tanzania: Ministry of Natural Resources and Tourism, Dar es Salaam, Tanzania.
41. Ashley GM, Bunn HT, Delaney JS, Barboni D, Domínguez-Rodrigo M, et al. (2013) Paleoclimatic and paleoenvironmental framework of FLK North archaeological site, Olduvai Gorge, Tanzania. Quaternary International 322–323: 54–65.
42. Dominguez-Rodrigo M, Bunn HT, Mabulla AZP, Ashley GM, Diez-Martin F, et al. (2010) New excavations at the FLK Zinjanthropus site and its surrounding landscape and their behavioral implications. Quaternary Research 74: 315–332.
43. Ashley GM, Beverly EJ, Sikes NE, Driese SG (2014) Paleosol diversity in the Olduvai Basin, Tanzania: Effects of geomorphology, parent material, depositional environment, and groundwater on soil development. Quaternary International 322–323: 66–77.

44. Magill CR, Ashley GM, Freeman KH (2013) Ecosystem variability and early human habitats in eastern Africa. Proceedings of the National Academy of Sciences of the United States of America 110: 1167–1174.

45. Pool D (2005) Variations in climate and ephemeral channel recharge in southeastern Arizona, United States. Water resources research 41.

46. Hogan JF, Phillips FM, Scanlon BR (2004) Groundwater recharge in a desert environment: the southwestern United States: American Geophysical Union.

47. Cuthbert MO, Tindimugaya C (2010) The importance of preferential flow in controlling groundwater recharge in tropical Africa and implications for modelling the impact of climate change on groundwater resources. Journal of Water and Climate Change 1: 234–245.

48. Brutsaert W (2005) Hydrology: an introduction: Cambridge University Press.

49. Cuthbert M (2014) Straight thinking about groundwater recession. Water Resources Research 50: 2407–2424.

50. Cuthbert MO (2010) An improved time series approach for estimating groundwater recharge from groundwater level fluctuations. Water Resources Research 46: W09515.

51. Townley LR (1995) The response of aquifers to periodic forcing. Advances in Water Resources 18: 125–146.

52. Ayenew T, Demlie M, Wohnlich S (2008) Hydrogeological framework and occurrence of groundwater in the Ethiopian aquifers. Journal of African Earth Sciences 52: 97–113.

53. Belcher WR, Sweetkind DS, Elliott PE (2002) Probability distributions of hydraulic conductivity for the hydrogeologic units of the Death Valley regional ground-water flow system, Nevada and California: US Department of the Interior, US Geological Survey.

54. Gleeson T, Smith L, Moosdorf N, Hartmann J, Durr HH, et al. (2011) Mapping permeability over the surface of the Earth. Geophysical Research Letters 38: GL045565.

55. Greco R, Comegna L, Damiano E, Guida A, Olivares L, et al. (2013) Hydrological modelling of a slope covered with shallow pyroclastic deposits from field monitoring data. Hydrology and Earth System Sciences 17: 4001–4013.

56. Moghaddam AA, Najib MA (2006) Hydrogeologic characteristics of the alluvial tuff aquifer of northern Sahand Mountain slopes, Tabriz, Iran. Hydrogeology Journal 14: 1319–1329.

57. Yihdego Y, Becht R (2013) Simulation of lake-aquifer interaction at Lake Naivasha, Kenya using a three-dimensional flow model with the high conductivity technique and a DEM with bathymetry. Journal of Hydrology 503: 111–122.

58. Scanlon BR, Keese KE, Flint AL, Flint LE, Gaye CB, et al. (2006) Global synthesis of groundwater recharge in semiarid and arid regions. Hydrological Processes 20: 3335–3370.

59. Dunkerley DL (2008) Bank permeability in an Australian ephemeral dry-land stream: variation with stage resulting from mud deposition and sediment clogging. Earth Surface Processes and Landforms 33: 226–243.

60. Houston J (2002) Groundwater recharge through an alluvial fan in the Atacama Desert, northern Chile: mechanisms, magnitudes and causes. Hydrological Processes 16: 3019–3035.

61. Simmers I (1997) Recharge of phreatic aquifers in (semi-) arid areas.

62. Johnson CR, Ashley GM, De Wet CB, Dvoretsky R, Park L, et al. (2009) Tufa as a record of perennial fresh water in a semi-arid rift basin, Kapthurin Formation, Kenya. Sedimentology 56: 1115–1137.

63. Lee RKL, Owen RB, Renaut RW, Behrensmeyer AK, Potts R, et al. (2013) Facies, geochemistry and diatoms of late Pleistocene Olorgesailie tufas, southern Kenya Rift. Palaeogeography Palaeoclimatology Palaeoecology 374: 197–217.

64. Darling WG, Gizaw B, Arusei MK (1996) Lake-groundwater relationships and fluid-rock interaction in the east African rift valley: Isotopic evidence. Journal of African Earth Sciences 22: 423–431.

65. Olago D, Opere A, Barongo J (2009) Holocene palaeohydrology, groundwater and climate change in the lake basins of the Central Kenya Rift. Hydrological Sciences Journal-Journal Des Sciences Hydrologiques 54: 765–780.

66. Bailey GN, Reynolds SC, King GCP (2011) Landscapes of human evolution: models and methods of tectonic geomorphology and the reconstruction of hominin landscapes. Journal of Human Evolution 60: 257–280.

67. Semaw S, Harris J, Feibel C, Renne P, Bernor R, et al. (1997) The oldest archaeological sites with an early Oldowan Industry from the Gona River deposits of Ethiopia. Nature 385: 333–336.

68. Neubauer S, Hublin J-J (2012) The evolution of human brain development. Evolutionary Biology 39: 568–586.

69. McBrearty S, Brooks AS (2000) The revolution that wasn't: a new interpretation of the origin of modern human behavior. Journal of human evolution 39: 453–563.

70. Potts R (1998) Variability selection in hominid evolution. Evolutionary Anthropology: Issues, News, and Reviews 7: 81–96.

71. Maslin MA, Trauth MH (2009) Plio-Pleistocene East African pulsed climate variability and its influence on early human evolution. The First Humans–Origin and Early Evolution of the Genus Homo: Springer. 151–158.

72. Joordens JC, Vonhof HB, Feibel CS, Lourens LJ, Dupont-Nivet G, et al. (2011) An astronomically-tuned climate framework for hominins in the Turkana Basin. Earth and Planetary Science Letters 307: 1–8.

15

Satsurblia: New Insights of Human Response and Survival across the Last Glacial Maximum in the Southern Caucasus

Ron Pinhasi[1]*, **Tengiz Meshveliani**[2], **Zinovi Matskevich**[3], **Guy Bar-Oz**[4], **Lior Weissbrod**[4], **Christopher E. Miller**[5], **Keith Wilkinson**[6], **David Lordkipanidze**[2], **Nino Jakeli**[2], **Eliso Kvavadze**[7], **Thomas F. G. Higham**[8], **Anna Belfer-Cohen**[9]*

1 Earth Institute and School of Archaeology, University College Dublin, Dublin, Ireland, 2 Georgian State Museum, Department of Prehistory, Tbilisi, Georgia, 3 Israel Antiquities Authority, Jerusalem, Israel, 4 Zinman Institute of Archaeology, University of Haifa, Haifa, Israel, 5 Institute for Archaeological Sciences, and Senckenberg Centre for Human Evolution and Paleoenvironment, University of Tübingen, Tübingen, Germany, 6 Department of Archaeology, University of Winchester, Winchester, United Kingdom, 7 Institute of Paleobiology, National Museum of Georgia, Tbilisi, Georgia, 8 Oxford Radiocarbon Accelerator Unit, Research Laboratory for Archaeology & the History of Art, University of Oxford, Oxford, United Kingdom, 9 Institute of Archaeology, Hebrew University, Jerusalem, Israel

Abstract

The region of western Georgia (Imereti) has been a major geographic corridor for human migrations during the Middle and Upper Palaeolithic (MP/UP). Knowledge of the MP and UP in this region, however, stems mostly from a small number of recent excavations at the sites of Ortvale Klde, Dzudzuana, Bondi, and Kotias Klde. These provide an absolute chronology for the Late MP and MP–UP transition, but only a partial perspective on the nature and timing of UP occupations, and limited data on how human groups in this region responded to the harsh climatic oscillations between 37,000–11,500 years before present. Here we report new UP archaeological sequences from fieldwork in Satsurblia cavein the same region. A series of living surfaces with combustion features, faunal remains, stone and bone tools, and ornaments provide new information about human occupations in this region (a) prior to the Last Glacial Maximum (LGM) at 25.5–24.4 ka cal. BP and (b) after the LGM at 17.9–16.2 ka cal. BP. The latter provides new evidence in the southern Caucasus for human occupation immediately after the LGM. The results of the campaigns in Satsurblia and Dzudzuana suggest that at present the most plausible scenario is one of a hiatus in the occupation of this region during the LGM (between 24.4–17.9 ka cal. BP). Analysis of the living surfaces at Satsurblia offers information about human activities such as the production and utilisation of lithics and bone tools, butchering, cooking and consumption of meat and wild cereals, the utilisation of fibers, and the use of certain woods. Microfaunal and palynological analyses point to fluctuations in the climate with consequent shifts in vegetation and the faunal spectrum not only before and after the LGM, but also during the two millennia following the end of the LGM.

Editor: Nuno Bicho, Universidade do Algarve, Portugal

Funding: National Geographic-Global Exploration Fund funded this research from April 2013 to February 2014 (grant- GEFNE78-13). Additional support was receieved from European Research Council Starter (grant- ERC-2010-StG 263441). The funders had no role in study design, data collection and analysis, decision to publish, or preparation of the manuscript.

Competing Interests: The authors have declared that no competing interests exist.

* Email: ron.pinhasi@ucd.ie (RP); Anna.Belfer-Cohen@mail.huji.ac.il (ABC)

Introduction

The Southern Caucasus region played a major role in human evolution. During the past two decades research in this region has mainly focused on the Middle-Upper Palaeolithic (MP-UP) transition in the context of Neanderthal extinction and the first appearance of modern humans [1–4]. Less emphasis has been placed on charting human occupation during and right after the Last Glacial Maximum (LGM) at 24,000–18,000 calibrated years before present (ka cal. BP), a climate event which had a major demographic impact on human populations in Eurasia [5–6].

The interdisciplinary project reported on here focuses on fieldwork at Satsurblia cave, western Georgia (Fig. 1), a site discovered in 1975 by A. N. Kalandadze [7], and who

subsequently excavated it sporadically during 1976, 1985–88. Excavations were also carried out by K. Kalandadze in 1989–1992 [8], and by T. Meshveliani in 2008–2010.

Before the start of our campaigns current knowledge about UP occupation in Western Georgia was based predominantly on the study of Dzudzuana cave which has a UP archaeological sequence comprising three occupational episodes separated by millennia-long hiatuses: the lowermost UP phase, Unit D, dated to 34.5–32.2 ka cal. BP; followed by Unit C, dated to 27–24 ka cal. BP; and the latest UP phase, Unit B dated to 16.5–13.2 ka cal. BP [2]. It is unclear, however, whether the hiatuses are due to the absence of humans in the entire region or rather if they reflect the particular depositional and occupational history of this specific site. Analyses of the faunal assemblages from Dzudzuana, Ortvale Klde, and

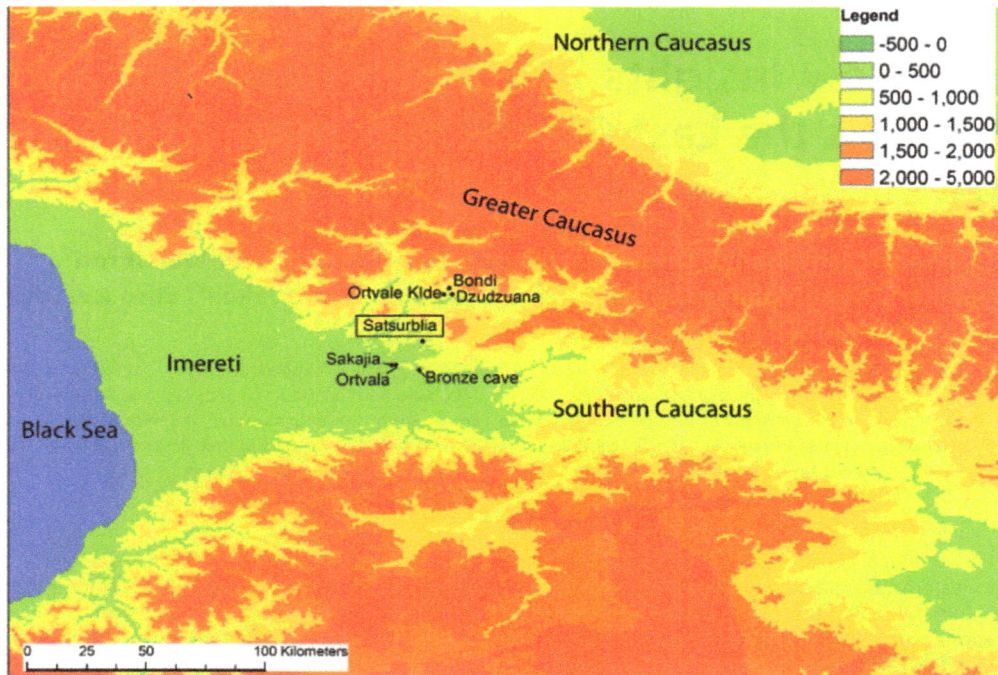

Figure 1. Location of Satsurblia Cave and other key sites with Upper Palaeolithic occupation in western Georgia.

Kotias Klde (the latter two sites yielded short UP sequences) indicate that hunting had focused on a few ungulate species with observed temporal variations in their relative proportions. Since there are limited differences in taphonomic history across the sites, it was postulated that the composition reflects variations in the availability of animal resources by season and period [2], [9–11].

In this paper we report on the record from Satsurblia and combine that with data from the sites of Dzudzuana, Ortvale Klde and Kotias Klde in order to develop a new regional Upper Palaeolithic archaeological sequence. This is followed by faunal data which provide information on the subsistence and behavior of humans during this period, and information about local palaeoclimate and palaeoecology in order to study the effect of climatic oscillations on human populations.

Stratigraphy and Radiocarbon Dating

The 2012–2013 excavations in Satsurblia were conducted in two areas. Area A is situated in the north-western part of the cave, near the entrance (squares R−T 20−24). Area B is in the rear of the cave (squares T−Z 4−7), adjacent to a trench excavated by K. Kalandadze in the early 1990s. Both areas revealed stratigraphic sequences comprising Pleistocene (Upper Palaeolithic) and Holocene (Eneolithic and more recent) deposits (Fig. 2).

Area A

The exposed stratigraphic sequence of the Area A so far includes five main lithological units (labelled A1–A5, with additional subdivisions, e.g. A4i–iii) corresponding to three main archaeological strata (labelled A/I, A/IIa, A/IIb) (Fig. 3). All the observed layers in Area A are associated with an extremely large boulder that is situated to the south of the area (in squares R-T 17−19), the upper face of which is exposed on the current surface of the cave, continuing to an as yet unexcavated depth (observed height of 1.5 meters).

Under the 'topsoil' (Unit A1), representing incipient pedogenesis within the upper levels of the underlying units, the deposits are formed mostly of open, matrix-supported pebble, cobble and boulder gravels, with occasional lenses of well-sorted silts and clays. This series of units (A2−A4ii) are grouped as the archaeological layer A/I and represent multiple episodes of sedimentation, occupation and roof collapse in the cave during the course of the Holocene.

Archaeological layer A/I yielded moderate quantities of finds – Eneolithic and (in the upper parts) Classical and Medieval pottery, as well as lithic items which seem mostly to derive from the earlier, Upper Palaeolithic levels. At least one discrete occupational event might be identified within Layer A/I (at an elevation of 4.30−4.40 m below the datum): a burnt, nearly horizontal floor-like surface consisting of yellowish brown clay with frequent granular to coarse sand-sized charcoal fragments ("Floor A"). Although there is a scarcity of cultural material directly associated with this feature, it is nevertheless cautiously interpreted as being of Eneolithic date.

The total thickness of the A/I layer varies widely: in the northwestern part of the excavated area (squares T22−23) it is ca. 80 cm (4.00 v4.80 m below the datum), while in the east (squares R20−23) it is ca. 40 cm (4.00−4.40 m below the datum). Consequently, the upper parts of the underlying Upper Palaeolithic Layer A/II (namely A/IIa), which are well preserved in the eastern part of the area, are truncated by erosion. In the northernmost part of the area (squares R-T24), a large depression with vertical walls (an excavated pit?) filled in with boulder-pebble-sized clasts is associated with Layer A/I. Its depth is at least 60 cm (4.90−5.45 m below the datum) and it continues to a yet unknown depth. A series flat boulder-sized horizontally laid stones (more than 1 m in diameter) is observed in the lower part of the pit. The feature contains Eneolithic pottery.

The Upper Palaeolithic archaeological layer (A/II) is associated with lithological units A4iii and A5. In the western section of the

Figure 2. Plan of Satsurblia Cave.

area, the upper part of these strata (lithological unit A4iii/ Archaeological layer A/IIa) comprises a brown fine cave earth of silt (mostly) and clay with moderate quantities of granular limestone clasts, occasional sub-angular limestone pebbles and granular-fine pebble-sized charcoal. Localized combustion features are manifested by a red coloration of the sediments. The

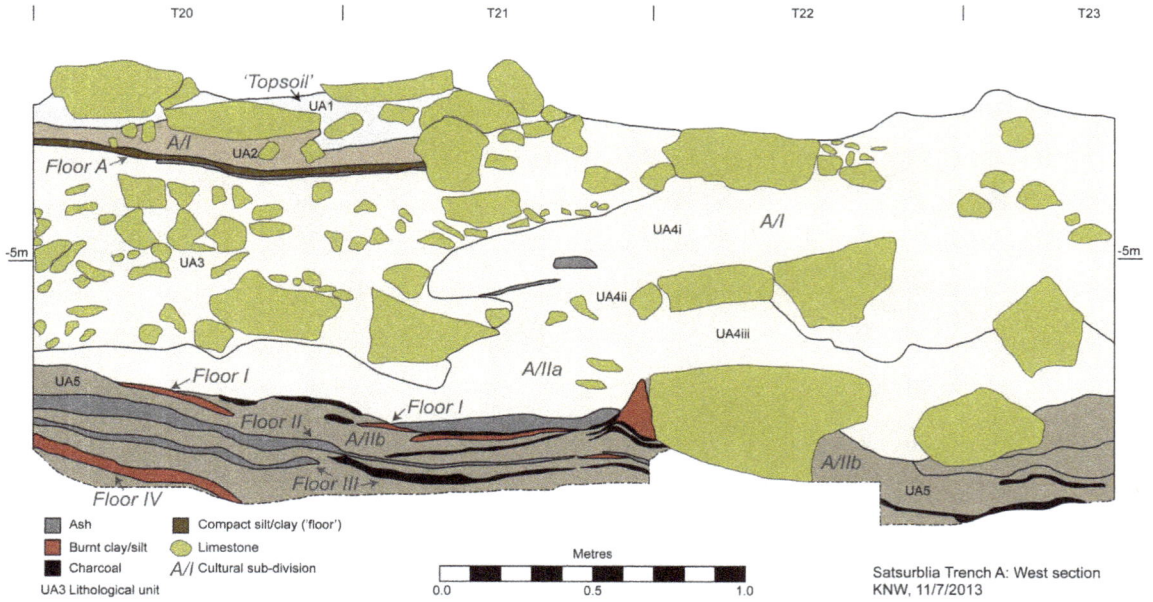

Figure 3. Transverse section of Area A, west.

layer is dated to 16,911−16,215 cal. BP (95.4% confidence interval, Fig. 4, Table S1).

Lithological unit A5 (archaeological layer A/IIb) comprises a complex of anthropogenic deposits including at least four discrete occupation episodes (Floors 1–4 identified in T-S 23−20), represented by compact surfaces of silt/clay. In the eastern part of the area (R23−20), where the upper part of the Upper Palaeolithic stratigraphy had not been truncated by later erosion,

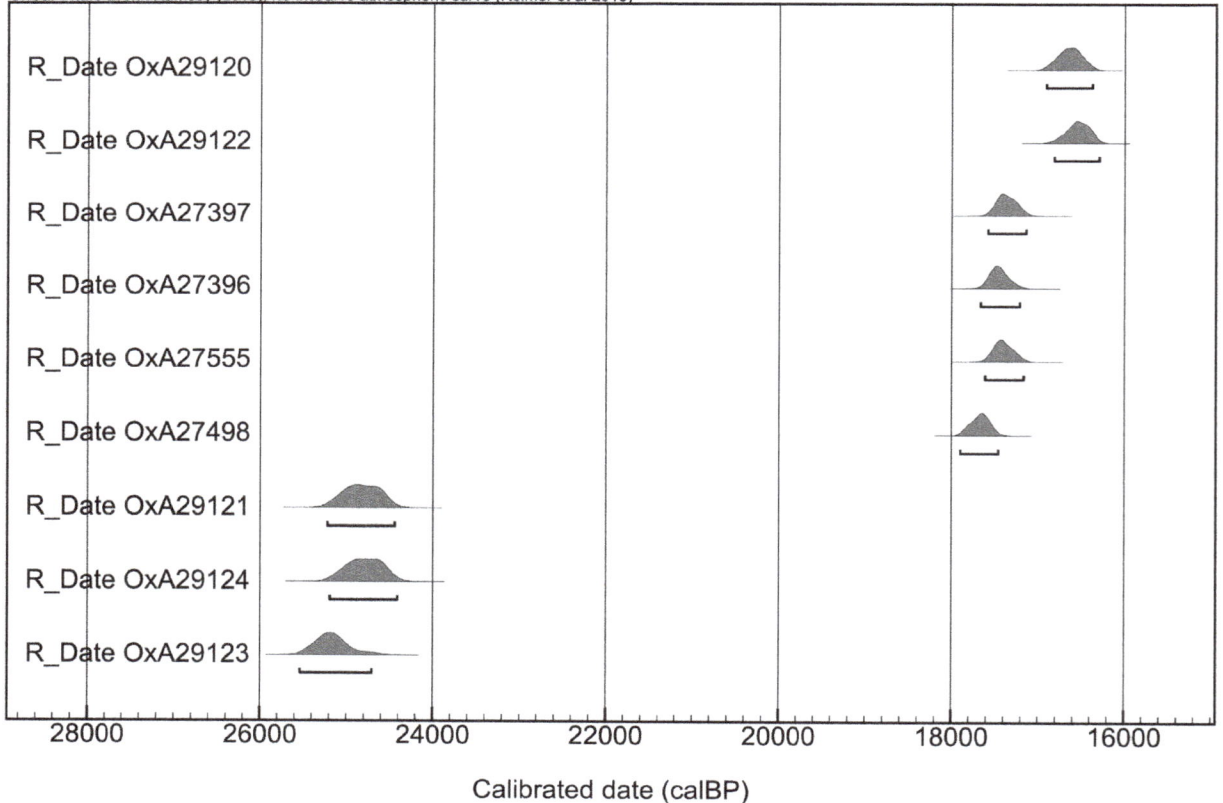

Figure 4. Absolute chronology of Units A/iia, A/IIb, B/II and B/III based on calibrated radiocarbon determinations.

the occupational surfaces with burnt material are also observed higher up, in the AII/a sub-layer. By the end of the 2013 season two of those surfaces (labelled "Floors B, and C") had been excavated. Most probably the lowermost of those surfaces correlate with Floors 1–3 of the western section (Layer A/IIb). The layer is dated to 17,895–17,140 cal. BP (95.4% confidence interval, Fig. 4, Table S1). The layer does not chronologically overlap, even at 95% confidence with dates obtained for A/IIa, and hence lends support to the hypothesis that the layers contain evidence of separate occupations. Nevertheless, there is no observed stratigraphic unconformity between the layers and hence it is plausible that future fieldwork and further dating might point to a continuity of occupation during the period 17.9–16.2 ka cal. BP.

The topography of the excavated 'floors' is generally irregular, with numerous shallow depressions, typically filled with charcoal, burnt pebbles, flints and bones. In one case (Floor 2, Layer A/IIb) a circular hearth constrained by medium-sized cobbles and covered by a thick layer of charcoal and ashes was identified in square T22.

Micromorphological analysis was carried out on (1) two block samples from the fireplace hearth in Floor 2, Layer A/IIb, square T22d, and (2) two block samples from Floor 1, Layer A/IIb, T23b. Information about sample preparation is provided in Text S1. Most sedimentary components are identical in all samples. The coarse components (those larger than silt, >62 μm), include sand-sized to medium pebbles (cm-sized) angular to sub-angular clasts of limestone, which are relatively common and likely derived from the cave walls and ceiling (éboulis) as well as occasional angular fragments of chipped fine-grained siliceous rock which are likely debitage from stone-tool production. Sand-sized fragments of burnt bone and finely comminuted fragments of charcoal and humified wood are dispersed evenly throughout the samples. Occasional rounded grains, 0.5 mm in size, of secondary carbonate, and with the appearance of "sinter" were also detected. Sinter is a precipitate that likely formed within the cave. The rounded form of these grains suggests that they are reworked. Sand to coarse silt sized grains of quartz are also present and might indicate an aeolian component to some of the sediments.

Another common coarse component in the samples are rounded aggregates of reddish clay and silt-sized grains of quartz. The size of the aggregates is variable: 100 μm-2 mm. Several of the aggregates exhibit oriented clays, as evidenced by granos-triated b-fabric. This characteristic is produced through rounding of the aggregates, likely during colluvial transport of the aggregates into the shelter. The aggregates display at least three different colors in Plane Polarised Light (PPL): yellowish brown, reddish, and dark reddish brown. The color difference may indicate different sources for the aggregates (Fig. 5). However, it may also indicate that the aggregates were heated in combustion features at different temperatures.

The fine fraction at Satsurblia is largely anthropogenic in the form of calcareous ashes.

A full description of the micromorphology of Floors 1 and 2, layer A/IIb, is provided in Text S1.

Area B

The stratigraphic sequence of Area B includes so far six lithological units (B1–B6) comprising three main archaeological layers (B/I, B/II, B/III) (Fig. 6). In a similar way to Area A all the stratigraphic units are associated with a coarse boulder (several meters long) exposed on the surface of the cave and extending to the depth of at least 2 m below the surface. Under the thin 'topsoil' layer (Unit B1), a massive pit-like feature (4 m in diameter) is observed in squares U-X 4–6, and it reaches a maximum depth of 1.4 m (4.00–5.40 m below the datum in squares X4–X5). The fill of the feature (Unit B3) is a clast-supported gravel of sub-angular limestone cobbles to granules in a brown silty clay matrix. Gravel particles are chaotically distributed and are densest towards the centre of the cut feature. The material is poorly sorted and also includes numerous Eneolithic artefacts, and is most probably a deliberate archaeological fill. On the top of the fill a weathered/pedogenically worked surface (Unit B2) is observed. The pit is associated with no well-defined floor levels, is archaeologically defined as Layer B/I and dates to the Eneolithic period.

The Eneolithic pit was cut into Upper Palaeolithic layers, which are elsewhere found directly under the present-day surface (in squares T4, Y–Z 4–8). The two uppermost Pleistocene lithological units (B4 and B5), comprise a single archaeological layer, B/II, which reaches a thickness of ca. 1 m (4.15–5.25 m below the datum) and is dated to 25,220–24,440 cal. BP (95.4% confidence interval, Fig. 4, Table S1). The upper lithological unit (B4) is yellow-reddish clay containing frequent granular limestone clasts, occasional sub-angular limestone pebbles and granular to pebble-sized charcoal particles. The latter decline in frequency to the east, but form two discrete charcoal-rich fine layers in the west. Unit B4 conformably overlies unit B5, which is a brown silt/clay with moderate granular limestone clasts and which also contains frequent charcoal fragments. Its darker colour in comparison to unit B4 is a result of the presence of finely divided charcoal. Layer B/II contains a moderate quantity of Upper Palaeolithic artefacts and animal bones.

Archaeological layer B/III, associated with lithological Unit B6, has at present been excavated to the thickness of 40–50 cm (i.e. to 5.70 m below the datum). The layer is composed of a brown diamict of moderate to frequent sub-angular limestone pebbles in a silt/clay and coarse sand matrix, structured as alternate bands of clast-rich and clast-poor strata. The layer is extremely rich in archaeological finds (bones and lithics). In its upper part, in square Y6 at the elevation of 5.25–5.30 m below datum, a circular installation, 40 cm. in diameter, built of cobbles and associated with large quantities of burnt material was discovered. The layer is dated to 25,535–24,408 cal. BP (95.4% confidence interval, Fig. 4, Table S1) and hence seems to chronologically overlaps with Layer B/II. However, it is important to note that the chronology for the archaeological layers in Area B is currently based on only three radiocarbon determinations. Additional determinations will provide better information about the absolute chronology of these archaeological layers.

Lithic Assemblages and Other Artefacts

The size and content of the lithic assemblages reflect the complexity and history of the various archaeological layers observed in the cave. In Area B the UP levels (layers B/II and B/III), were deeply cut by the Eneolithic Layer B/I and a noticeable level of mixture between the various assemblages was discerned. A similar phenomenon was observed in Area A where Layer A/I cut into the UP Layer A/II. Hence it is more acceptable to discuss for the archaeological units only general techno-typological trends. The overall picture is further distorted due to differences in samples sizes, especially for layers where samples comprise less than 100 tools (e.g., B/II with 87 items) as well as the high percentage among them of broken items (e.g., 85.4% of the tools in A/IIb), predominantly among the bladelet tools.

Table 1 provides the percentages of the major tool groups in the different archaeological units. Table S2 presents the debitage

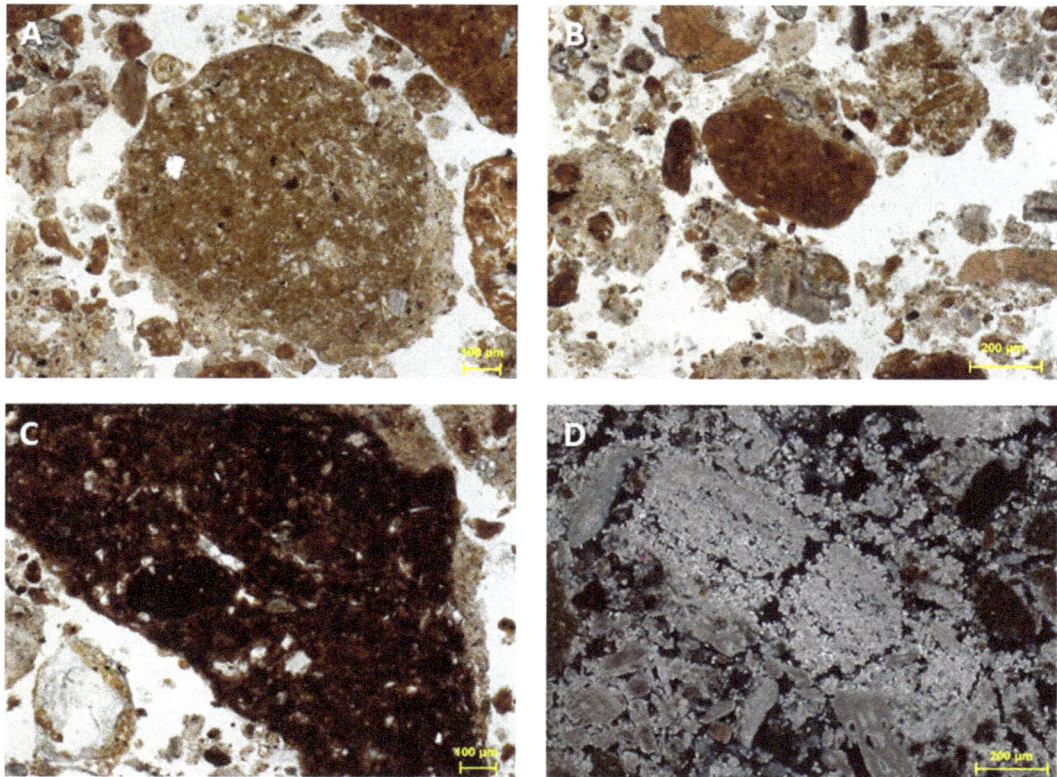

Figure 5. Photomicrographs of sediment thin sections from Satsurblia. At least three types of clay and silt aggregates were identified in the Satsurblia samples, based on color variation. Yellowish brown aggregates (A), reddish aggregates (B), and dark reddish brown aggregates (C). The color variation could indicate different sources, or, could be a result of differential heating of the aggregates in combustion features. The fine material at Satsurblia contains a high proportion of calcareous ash rhombs (D). A,B,C are from sample SAT-12–48 and photographed in plan polarized light (PPL). D is from sample 12–46 and photographed in cross polarized light (XPL).

counts including debris and cores. Since these are but preliminary observations, we discuss in detail only the tool types we consider as diagnostic. Still one should observe that as always in the Palaeolithic industries of the region (e.g. [12]), the endscrapers consistently account for more than 10% of the assemblages (Fig. 7:1−3), outnumbering the burins (which vary in frequencies from 2.6% to 19.2%) (Fig. 7:4−6), with the exception of the B/III assemblage, which points to a clear difference between the upper and lower UP entities (and see below). The same can be said concerning the *pièce esquillée* 'tool' type which numbers diminish from the earlier to the later assemblages (Fig. 7:7−8). In addition to these tools the assemblages also comprise some awls and borers,

notches and denticulates, retouched flakes (mostly ad-hoc) and varia.

The assemblages of Layer A/I and Layer B/I are quite late in the sequence (13.3−13.1 ka cal. BP) and it is clear that human occupations at those times dug into the earlier layers and consequently introduced into underlying levels some elements that are evidently intrusive, e.g., the few lunates and triangles in the A/II assemblages (and a single triangle in B/II). There are some lithics which firmly anchor A/I and B/I typologically to post Palaeolithic times: two lekalla/fishtail items in A/I [13−14] and a polished axe in B/I, not to mention the pottery retrieved therein (a detailed account is in preparation). When examining the lithic

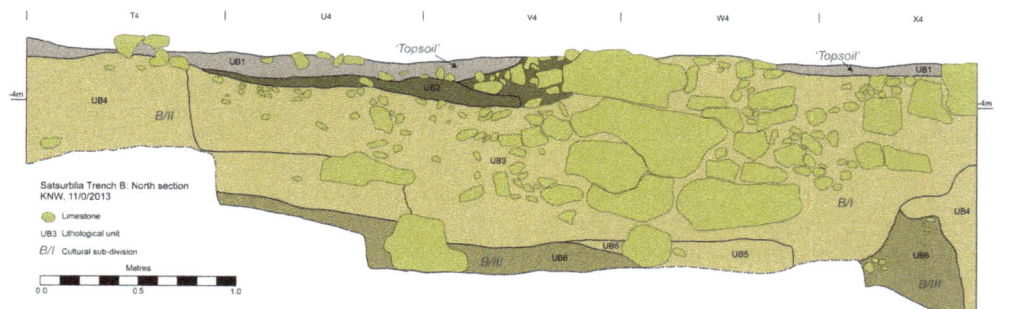

Figure 6. Transverse section of Area B, north.

Table 1. Percentages of various tool types in Satsurblia, by unit.

tool types	A/I		A/IIa		A/IIb		B/I		B/II		B/III	
	N	%	N	%	N	%	N	%	N	%	N	%
endscraper	17	14.7	60	24.5	76	11.6	40	16.8	16	18.4	17	10.9
burin	3	2.6	19	7.8	25	3.8	34	14.3	7	8.0	30	19.2
composite	0	0.0	0	0.0	3	0.5	1	0.4	2	2.3	5	3.2
blade, backed	3	2.6	7	2.9	7	1.1	3	1.3	2	2.3	0	0.0
blade, backed & trunc.	1	0.9	2	0.8	6	0.9	0	0.0	1	1.1	0	0.0
blade, truncated	0	0.0	3	1.2	8	1.2	3	1.3	1	1.1	2	1.3
blade, retouched	1	0.9	3	1.2	10	1.5	6	2.5	4	4.6	11	7.1
bladelet, backed	22	19.0	83	33.9	333	50.6	56	23.5	10	11.5	22	14.1
bladelet, backed & trunc.	6	5.2	6	2.4	36	5.5	9	3.8	5	5.7	3	1.9
bladelet, truncated	0	0.0	2	0.8	5	0.8	4	1.7	1	1.1	0	0.0
bladelet, retouched	25	21.6	18	7.3	24	3.6	33	13.9	9	10.3	38	24.4
bladelet, shouldered	1	0.9	1	0.4	7	1.1	1	0.4	0	0.0	0	0.0
flakes, truncated	0	0.0	0	0.0	1	0.2	1	0.4	0	0.0	0	0.0
flakes, retouched	2	1.7	2	0.8	2	0.3	4	1.7	3	3.4	5	3.2
fragment, backed	2	1.7	9	3.7	21	3.2	2	0.8	2	2.3	1	0.6
fragment, retouched	6	5.2	3	1.2	16	2.4	5	2.1	5	5.7	5	3.2
Gravette pt.	0	0.0	0	0.0	2	0.3	2	0.8	0	0.0	0	0.0
microgravette pt.	2	1.7	7	2.9	33	5.0	6	2.5	2	2.3	4	2.6
geometrics: lunate & triangle	1	0.9	4	1.6	1	0.2	1	0.4	1	1.1	0	0.0
rectangle	3	2.6	3	1.2	4	0.6	1	0.4	4	4.6	1	0.6
notches & denticulates	1	0.9	2	0.8	2	0.3	2	0.8	2	2.3	0	0.0
awls and borers	2	1.7	3	1.2	11	1.7	7	2.9	2	2.3	1	0.6
p. esquillee	2	1.7	0	0.0	4	0.6	2	0.8	2	2.3	4	2.6
other and varia	16	13.8	8	3.3	21	3.2	15	6.3	6	6.9	7	4.5
Total	**116**	**100.0**	**245**	**100.0**	**658**	**100.0**	**238**	**100.0**	**87**	**100.0**	**156**	**100.0**

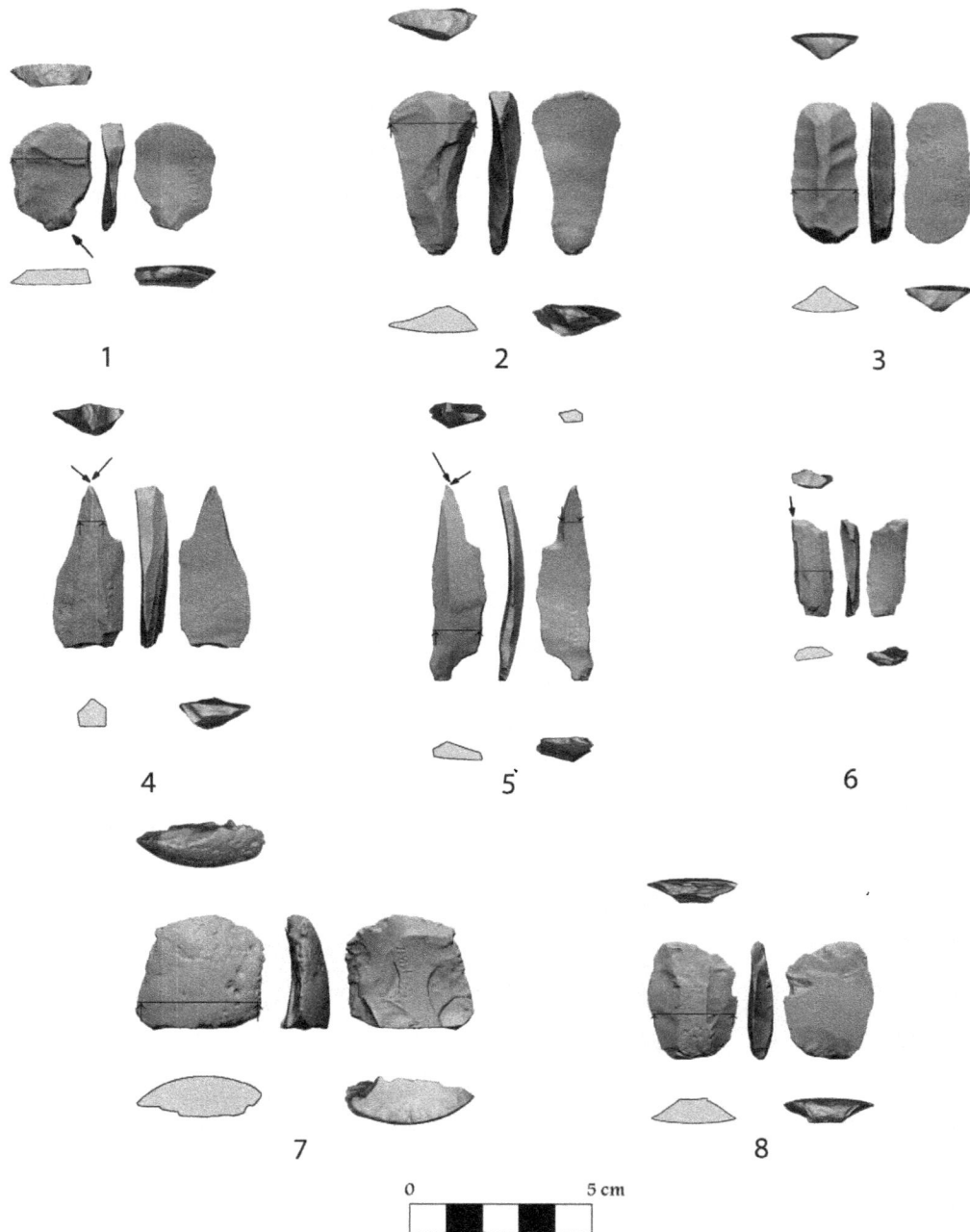

Figure 7. Tools from the UP layers, Satsurblia. 1–3 endscrapers; 4–6 burins; 7–8 piece esquielles.

assemblage of A/IIa vs. A/IIb there is an observed bottom to top (A/IIb to A/IIa) decline in the percentages of microgravettes and backed and truncated items and a corresponding increase in the number of retouched bladelets (Table 1).

The UP techno-typological entity best represented is one comprising microgravettes varieties (Fig. 8:4, 6–8), with rare occurrence of Gravette points (Fig. 8:5), and some truncated varieties including rectangles (Fig. 8:1–3) (though the numbers are low, see Table 1). It seems that these components are mostly observed in the assemblages of Layers A/IIa-b and B/II. Backed and truncated (straight) bladelets, which commonly represent broken rectangles, comprise in layer A/I –5.2%, layer A/IIa

3.3%, layers A/IIb - 6.2%, layer B/I –5.5%, layerB/II –6.9% and layer B/III - 3.2%. The frequencies of microgravette are 1.2% in A/I, 2.9% in A/IIa, 5.0% in A/IIb, 2.5% in B/I, 2.3% in B/II and 2.6% in B/III. The production technology is quite simple consisting of unipolar (the majority) and bipolar bladelet cores. It seems that most of the raw material was of poor quality, and there are many discarded core endeavours. Cores indicating intensive use are small and exhausted, indicating that the available good quality raw material was exploited to the utmost. Most of the flaked material is flint of various colours and quality with very rare obsidian pieces, including tools which comprise: in layer A/I, 3 out of 116 (3/116), in A/IIa–4/245 in A/IIb–7/658, in B/I–5/238;

B/II–3/87, and in B/III–3/156. Based on the debitage analysis (Table S2) it is clear that some degree lithic production took place on site though the percentages of artefacts (debitage and tools, excluding the chips and chunks) per core greatly vary: in Layer A/I there are 680 artefacts per 12 cores (57:1), in layer A/IIa–2391 artefacts per 26 cores (92:1), in layer A/IIb–5316 artefacts per 12 cores (443:1); and in Area B – B/I - 1796 artefacts per 51 cores (35:1), in B/II–592 artefacts per 6 cores (99:1) and in B/III–1841 artefacts per 15 cores (123:1).

Microgravette industries, with rare Gravette points, were reported from recent excavations in the region (e.g., Dzudzuana B, [2], yet those represent a different (later) facies as they lack the other dominant component evident in Satsurblia A/II and B/II assemblages, namely the truncated items, in particular the rectangles. Another illustrative example from farther afield is the recently published assemblage from the site of Kalavan I (Armenia) which also represents a microgravette industry, but without the rectangles and the truncated items [15]. Going through the literature and material from past excavations, only the illustrations of finds from the UP layers at Gvardjilas Klde point to the existence of such a particular late UP variant in the region, similar to that from Satrusblia (and see figures in [16–19]). It should be emphasized that those rectangles are elongated items, different from the later Mesolithic varieties in size and retouch ([20], [21], as well as possessing ventral retouch, most probably for thinning and hafting (see Fig. 8: 4, 6, 8). Unfortunately, the site of Gvardjilas Klde was excavated in 1916 and again in 1953 by antiquated methods which preclude detailed comparisons and the material available for study comprises several assemblages lumped together (per.obs., and see the cautious statement by Golovanova et al. 2014:215: *"Undoubtedly, the Epipaleolithic* [Late UP] *material dominates in both these not securely excavated collections."* Still it is of interest to note that the two dates from the Gvardjilas Klde site [22], 15,960±120 (OxA 7855) and 15,010±110 (OxA 7856), uncal. BP, when calibrated, provide a calibrated interval falling between 19,560–17,955 cal. BP, (calibrated using OxCal 4.2 confidence interval of 95.4%) which hence falls in the time interval preceding human occupation in Satsurblia in Floors 2 and 3 (see Fig. 4).

It is interesting to note that the morphological difference between the rectangles and the microgravettes is expressed gradually, with many specimens exhibiting intermediate morphological characteristics. In the microgravette, in addition to the straight backed/retouched lateral edge ending in a point, there is also a truncation at the opposite end (mostly proximal but sometimes distal) which is rounded, ventrally retouched or minutely flaked that continues obliquely to the opposite lateral edge. In the rectangle the straight backed/retouched lateral edges end with two straight truncations created by regular retouch. Yet quite frequently there is also some ventral 'treatment' which actually makes the truncation part thinner. There are, as with the microgravettes, some intermediate types, of a truncation on one end and a rounded butt on the other and either can be inverse or dorsal. For example, there are bladelets, backed and straight, that are truncated distally with a pointed base, with inverse retouch on the opposite lateral edge. The spectrum also varies from pointed, intensely retouched Gravette points (though the latter are rare, see Table 1) to a variety in which both ends are rounded and with a ventral treatment. Indeed, basal treatment is consistent in all the tool types mentioned above. In the microgravette varieties it is at the pointed end or the base, for thinning. In the truncated items it also appears either as thinning of the base, or inversely at the truncation, or both.

The ratio of the backed *vs.* retouched bladelets (all varieties) does show a clear trend of change from dominance of simple retouch to backing. Thus the percentages in A/IIa and A/IIb show a dominance of backing vs. retouch - 36.3% *vs.* 7.3% and 51.1% *vs* 3.6% respectively. The picture differs in Area B in which the proportions of backed vs. retouched bladelets show a clear trend of a change from B/III-15.4% (backed) *vs.* 25% (retouched), through B/II–17.2% *vs.* 10.3%; to B/I–27.3% *vs.*13.9%.

Besides the lithic artefacts discussed herein, we have also recovered pottery sherds, and 41 worked bone pieces, most of them polished bone fragments and horn cores, but also some points, awls and needles, as well as an 'incised' item (B/III). Other ornaments include two perforated stalagmite pendants (Layer A/IIb) and a perforated and polished bovid tooth (Layer B/I–II). Of interest are small lumps and crumbs of yellow, red and brown ochre recovered from all UP layers. These finds will be further discussed in future reports.

Palaeobotany

Palynological investigations of the archaeological deposits of Satsurblia cave have been conducted since 2007 [23–24]. Results show that the sediments are rich in both pollen and other organic remains of non-palynological character. Those include wood cells, spores of various fungi, microscopic remains of insects and other arthropods, and textile fibers.

More than 40 soil samples originating from various strata of the cave were analysed. The analysis has shown that climatic factors played a major part in the occupational history of the cave. Humans inhabited the cave mostly during warm and dry climatic phases. During humid and cold periods there was some standing water within the cave, at least in Area A, as is evident from remains of algae found in the samples [24].

A comparison of the pollen spectra of Floors 1–4 (Fig. S3) shows some variation as pollen of broad-leafed plants, including warm adapted species such as walnut is found only in Floors 1 (Fig. 9) and 2 suggesting fluctuations in the spectra of plants in the region near the cave. The pollen of Floor 1 is dominated by pine, Floor 2 by hazel, Floor 3 is very poor in pollen, while Floor 4 is also dominated by hazel. At the time of occupation of Floors 3 and 4, pollen of warm adapted plants is absent. Charred parenchyma cells of pine have been found, however, in all the four floor samples. It is possible that during formation of Floors 3 and 4 pine forest was growing in the vicinity of the cave. In the lower layers of the cave, besides the cells of pine, there were remains of pine needles [25] which, similarly to wood, cannot be transported over significant distances and therefore might serve as an useful indicator of pine forests near the cave. Moreover, the samples containing the remains of needles and pine wood cells comprise also pine pollen [24].

Additionally, large quantities of phytoliths and starch grains of cereals (Fig. S4), and the presence of Poaceae in the pollen spectra point to the importance of wild cereals in the diet of the Upper Palaeolithic inhabitants of the cave. A total of 40 flax textile fibers were recovered from all floors, including some dyed in blue and pink colours (Fig. S5).

Microfaunal Remains

Excavated deposits of the cave contain abundant remains of small rodents, shrews, hare and bats. It is apparent from the vertical distribution of the remains that depositional rates decreased during the formation of living floors within the stratigraphic sequence, although sample size does not allow detailed consideration of fluctuations in the microfaunal assem-

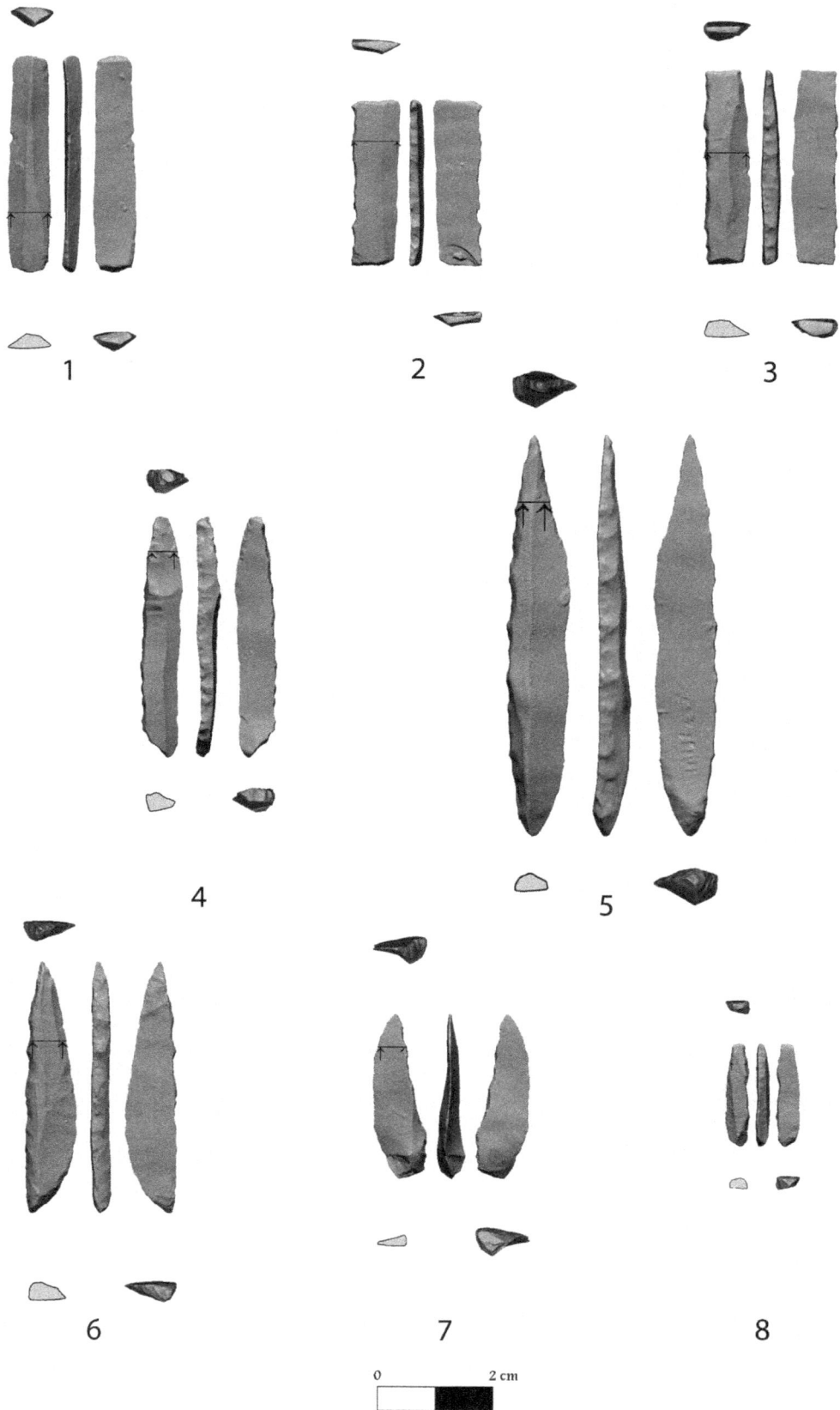

Figure 8. Tools from the UP layers, Satsurblia. 1–3 rectangles; 4, 6–8 microgravette varieties; 5 Gravette point.

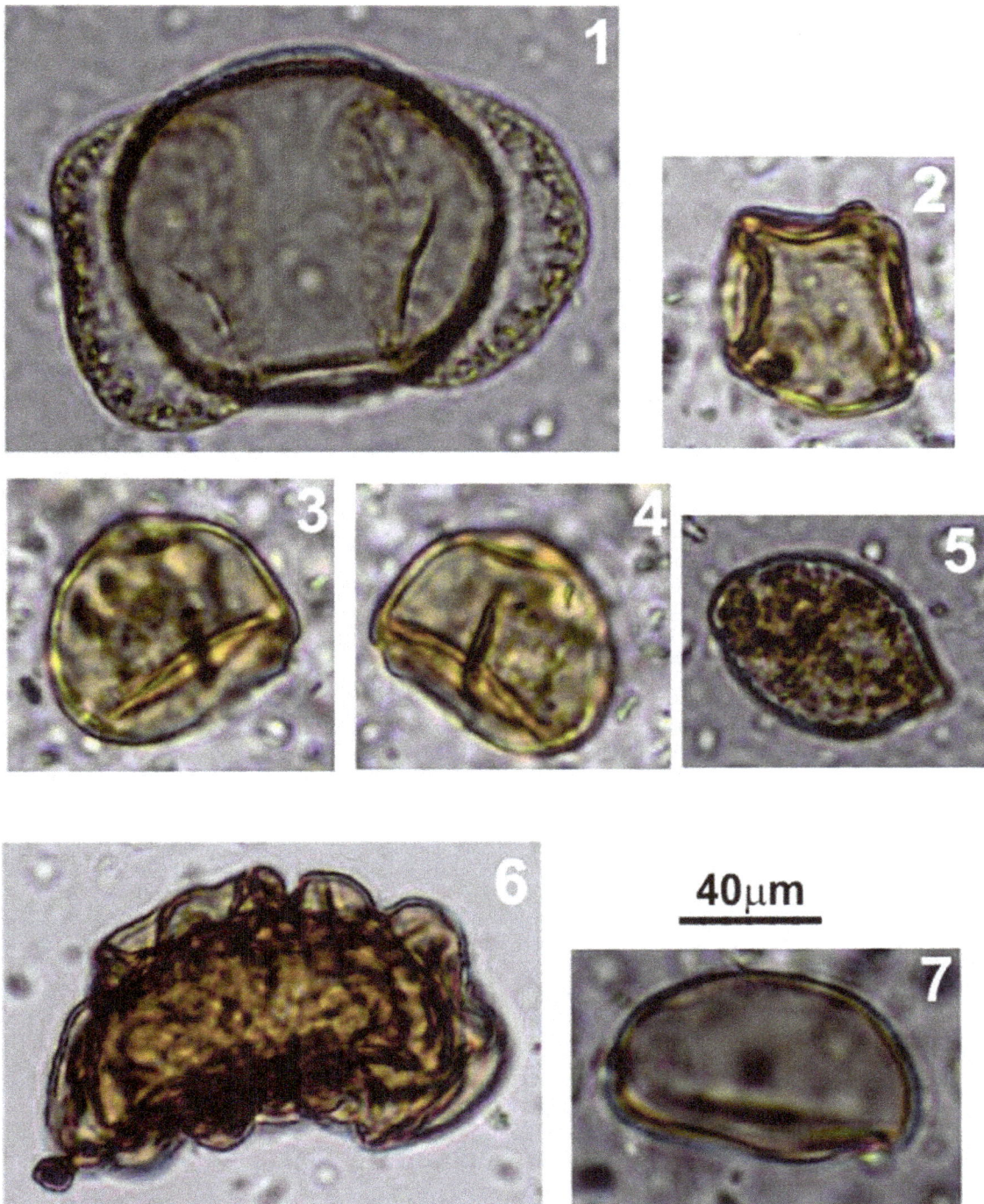

Figure 9. Pollen and spores, Floor 1: 1– pine (*Pinus*); 2– alder (*Alnus*); 3– hazel (*Corylus*); 5– not identified non-pollen paynomorph; 6– *Asplenium* fern; 7– *Polypodiaceae* (ferns).

blage within the stratigraphic layers. Initial identification of the material to general taxonomic categories at the family to genus levels is shown in Table S3 based on the molar teeth (n = 522). The data show that the assemblage is dominated by a number of species of small Arvicolinae voles and the large-bodied ciscaucasian hamster *Mesocricetus raddei*. Additional important taxa include *Apodemus* sp. wood mice, the water vole *Arvicola terrestris* and the mole vole *Ellobius* sp. Less frequent remains belong to the rodent families Sciuridae (squirrel) and Gliridae (dormice), which are likely represented by two species, the small hamster *Cricetulus migratorius* and two sub-families of shrews, Soricinae (red-toothed shrews) and Crocidurinae (white-toothed shrews). A few bat (Chiroptera) remains and a single toothless mandible and isolated incisor of the small hare *Ochotona rufesence* were recovered as well.

The taxonomic composition generally indicates affinities with both southern and northern Caucasus communities of small mammals. The modern distribution of *Mesocricetus raddei*, a dry steppe species, extends from the eastern section of the northern flanks of the Greater Caucasus into the Russian Plain, only bordering Georgia in the northeast ([25], see also [26]). Its smaller-bodied congener, *M. brandti*, occupies the southern parts of the Caucasus region into eastern Georgia. The remains of *Ellobius* sp., also a steppe species, may represent either *Ellobius talpinus*, which today is distributed north of the Greater Caucasus in the Russian Plain [27] or the southern *E. lutescens* presently found only as far north as southern Armenia [28]. More pronounced southern biogeographic affinities are indicated by the presence of rare remains of the Afghan pika *Ochotona rufesence* with a modern fragmented distribution in Iran and adjacent regions of Central Asia [29]. These specimens occur in Layers B/II. In the southern Armenian highlands this species is a dominant component of the UP preceding the LGM [30].

Present day distribution ranges of other taxa in the assemblage extend to both southern and northern sections of the Caucasus and adjacent regions. In the lower elevation (300 m asl) assemblage from Satsurblia, the steppic hamsters *Mesocricetus* and *Cricetulus* and other Arvicolinae are not as dominant as in Middle Palaeolithic layers at Kudaro Cave (1,600 m asl) [31]. The absence of *Ellobius* at Kudaro Cave may be due to its much higher elevation although this taxon was present at an even higher elevation of 2,040 m asl at Hovk Cave in northern Armenia in upper layers dating between the early Upper Palaeolithic and the Holocene [32].

The composition of taxa is fairly consistent along the stratigraphic units with an absence of rare species in some of the layers, likely due to sample size variation. This indicates that taphonomic mechanisms, mainly predation, were similar throughout the sequence. However, detailed taphonomic analysis will be needed to establish this claim. Variation can be detected in the taxonomic abundances which are analyzed using correspondence analysis in Fig. 10. The units are clearly organized along a horizontal gradient (Axis 1 = 63.84%) from an emphasis on steppic species on the left side to forest or woodland species (*Apodemus* sp., Sciuridae, Gliridae and Crocidurinae) on the right. Steppic and likely colder and more arid conditions are indicated for Layers B/II and A/IIa-b. Layers A/I and B/I, which are younger (tentatively assigned to Terminal Pleistocene based on the above-mentioned data), fall on the right side of the plot indicating a more forested environment. Nonetheless, forest taxa do not become overly abundant even in these later layers (ca. 31% of molars in each layer) and Terminal Pleistocene forests may have been less dense than the present humid deciduous forest of western Georgia.

Macro-Faunal Remains

The large faunal assemblage from Area A comprises a total of 327 complete and fragmentary bone specimens that were identified to taxon (excluding elements that were identified only to body-size group). Of those, 246 specimens were associated with Layer A/I, 65 specimens Layer A/IIa and 16 specimens with Layer A/IIb (Fig. 11). The taxonomic composition of layers A/I and A/II is similar and dominated by large ungulates, mainly boar (Sus *scrofa*) and red deer (*Cervus elaphus*) which comprises more than half of the identified elements in these assemblages. Other ungulates represented include large bovids (*Bos primigenius* and/ or *Bison priscus*), tur (*Capra caucasica*) and roe deer (*Capreolus capreolus*).

In Area B the UP assemblage (Layer B/II) consists of only 37 bones that were identified to taxon, of these were mainly of red deer, and boar as well as two bovid bones five capra bones and one bone of *Ursus* sp.

Carnivores are represented mainly by the remains of brown bear (*Ursus arctos*; identification was based on morphological and size criteria of selected bones following [33–35]. Additionally, there are remains of wolf (*Canis* sp.) and small carnivores, including fox (*Vulpes vulpes*). None of the carnivore bones was found in articulation and their remains were randomly dispersed across the Upper Palaeolithic horizontal strata and were mixed with the remains of the ungulate taxa.

The Upper Palaeolithic bone assemblage includes also a few bones of medium-sized species and two fish vertebrae. Additionally, we report the remains of Eurasian beaver in Layers A/I and A/IIa (*Castor fiber*). This species was once widespread in the Caucasus [36−37] until local extinction at the end of the 19th century [38].

The boar remains are dominated by a high number of young individuals (at least 25% are under the age of 24 months; based on DP$_4$/M$_3$ ratio following [39]. In addition we note the presence of two neonate specimens at age of less than three months (based on [40]). Assuming that wild boar in the Caucasus give birth to their young in early spring (in March–April, according to Heptner et al. [41]), the remains represent animals killed during late spring–early summer. The red deer and tur remains are dominated by prime-aged individuals. Also the bear remains derived from a prime-age individual.

The bone assemblage exhibits excellent preservation as evidenced by the presence of a whole range of bone densities, including porous parts such as sternum fragments. Bone preservation does not seem to vary among taxa. The long bones showed minor signs of surface weathering, indicating rapid burial of finds and the cave's protective conditions. Traces of carnivore bone ravaging activities are few (n = 9) and are present on the remains of all ungulate taxa. Rodent gnaw marks are also present in low numbers (n = 3). It appears that scavenging animals had only secondary access to the food remains.

Preliminary analysis of breakage patterns and bone surface modification reveals that the dominant agents of bone accumulation and bone damage were the humans. Virtually all ungulate long bones were split open to obtain marrow, evident by the high ratio of fresh (green) fractures (over 80%, and following the typology of Villa and Mahieu, 1991 typology [42]). Butchery marks are observed on boar and cervid, representing all butchery and carcass processing stages (skinning, dismemberment, and filleting).

Discussion

During the past two decades renewed and new excavations of cave sites in western Georgia have placed a major focus on the timing and nature of the Middle-Upper Palaeolithic transition, Neanderthal extinction and the arrival of modern humans to this region [1–3], [11], [13].

Many of the excavated cave sites in Western Georgia, such as Ortvale Klde, Sakajia, and Ortvala did not provide post-LGM Upper Palaeolithic archaeological sequences. Our knowledge was therefore limited as regards human occupation in this region before, during and after the LGM. Fieldwork in Dzudzuana provided us with three well-dated UP occupational phases of which Dzudzuana C (27–24 ka cal. BP) precedes the LGM and Dzudzuana B (16.5–13.2 ka cal. BP) begins several millennia after the end of the LGM.

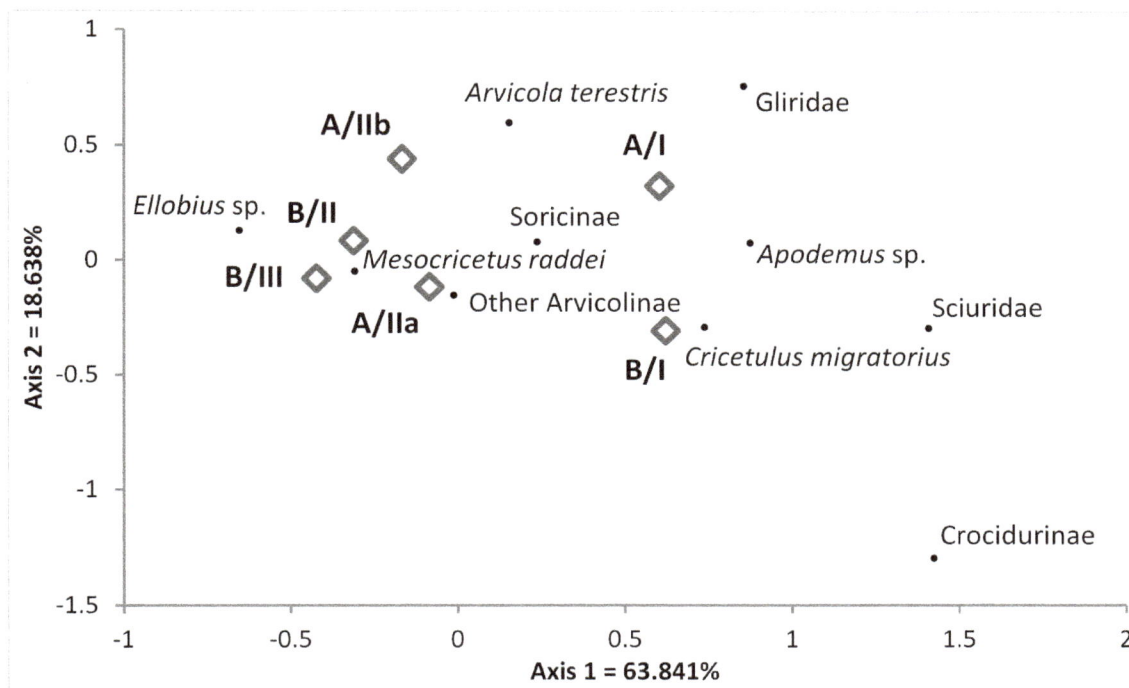

Figure 10. Correspondence analysis of abundances of micromammalian taxa in different stratigraphic units based on NISP data from molar teeth shown in Table S3.

The Upper Palaeolithic layers in Satsurblia cave provide new information about human occupation in the southern Caucasus during the period prior to (Layers B/II and B/III) and after (Layers A/IIa-b, A/I and B/I) the LGM. Layer B/II (25.2–24.4 cal. Ka BP) chronologically overlaps with Dzudzuana Unit C, while Layers A/IIa-b, A/I and B/I have a limited overlap with Dzudzuana Unit B (16.5–13.2 ka cal BP). We can therefore report with confidence the discovery and analysis of human occupation in western Georgia during the period spanning between 17.9–16.2 ka

cal. BP. The lithic analyses reveal that during this period, there existed a cultural (lithic) variant resembling the Eastern Epi-Gravettian, and similar to the industry reported from the site of Kalavan I, Armenia [15] dominated by bladelet tools, discrete among which are varieties of the microgravette point and truncated items. Moreover, besides rare occurrences of Gravette points, there is among the backed and truncated bladelets including a tool type that was not reported from earlier excavations in the region. This is the rectangle that differs from

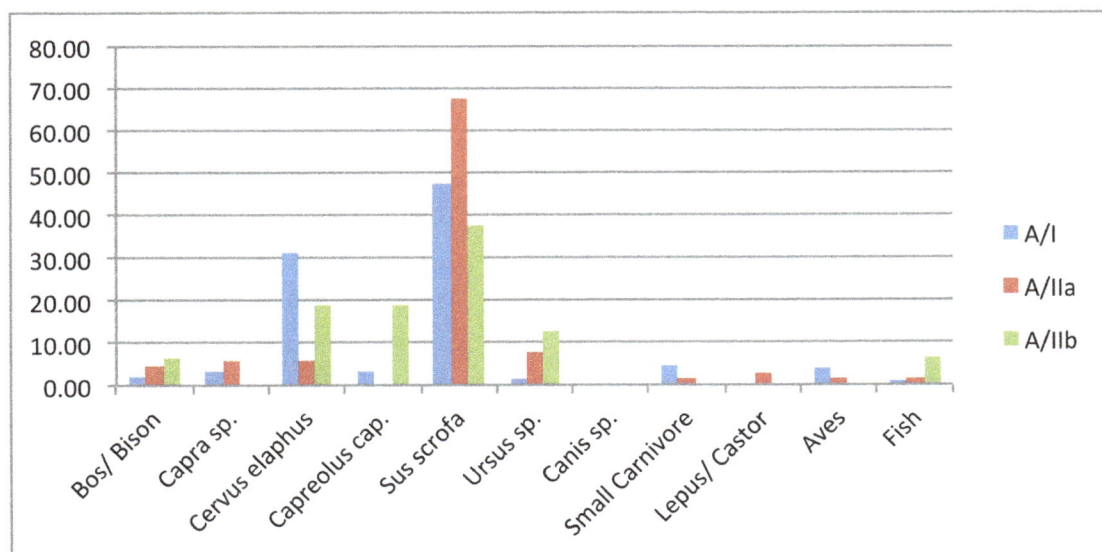

Figure 11. Faunal spectrum of identified species for UP Layers A/I, A/II (A/IIa and A/IIb).

the geometric trapeze-rectangle of the proceeding Mesolithic cultures in size, shape and retouch (Fig. 8:1–3). Additional typo-technological characteristics designate this assemblage as either a local or a temporal variant, hitherto not reported from this region.

It seems that the Gravette-microgravette-rectangles (the local Epi-Gravettian) industry identified in Satsurblia layers A/II and B/II illustrates the statement made by Bar-Yosef and co-authors [2], namely that the UP sequence reconstructed through the renewed excavations in the sites of Ortvale klde, Dzudzuana and Kotias Klde does not reflect the whole picture, and that we are still missing diachronic and synchronic techno-typological varieties which most probably existed in the region. It appears that the excavations at Satsurblia are providing some of the data to fill in those gaps in the Georgian UP sequence. Moreover, based on the nature of B/III assemblage we can cautiously state that the B/III assemblage represents a pre-Gravette/microgravette stage (in accord with the available dates) though we still cannot define its techno-typological characteristics in detail. Future excavations will both enlarge the samples of the assemblages discussed above as well as provide material from earlier times as we go deeper unto the occupation layers observed on site.

Another unique feature is the presence of 'floors' that are formed almost entirely as a result of anthropogenic processes. This is indicated by the micromorphological studies which point to the presence of reworked ashes and deposits, seemingly a result of hearth cleaning activities. This is further supported by the analyses of starch grains of wild grasses and the relatively high quantities of Flax fibers which were obtained from the 4 samples taken from Floors 1–4 (Layer A/IIb). These living surfaces therefore provide the opportunity to examine for the first time in the southern Caucasus, the study of intra-site spatial variation before and after the LGM.

The macro-faunal assemblage represents repeated seasonal visits by UP hunters targeting mainly forest-dwelling ungulates. *Sus scrofa* dominates in both Levels A/I and A/II (61.7% and 47.6%, respectively). The abundance of *Cervus elaphus* in Level A/II is 8.6% increasing to 31.3% in Level A/I. The common presence of prime-age prey implies deliberate hunting. In addition the remains of juvenile boar indicate that at least some of the hunting episodes took place during late spring - early summer. A similar pattern was observed for the assemblage from Layer B/II which is dominated by red deer and boar (73%–27/37 bones). Other UP and Early Mesolithic archaeo-faunal assemblages from the region are often dominated by open-landscape taxa (*Capra caucasica* and Bos/Bison in UP Dzudzuana Cave, both pre-LGM (Unit C and D) and post-LGM (Unit B) assemblages, and MP-UP Ortvale Klde; [9–11] (Fig. 12). The assemblage from the open-air UP site of Kalavan 1 in northern Armenia (dated to ca. 18–16 cal. BP [15] mainly consists of remains of open-landscape wild Caprinae (Ovis sp./Capra sp.). In all of the above-mentioned sites except Satsurblia there is a predominance of a single taxon suggesting a focus on specialised hunting, a pattern that differs from the one we report here for Satsurblia in which the faunal assemblage consists of a more diverse array of woodland dwelling ungulates. A somewhat similar pattern and composition of taxa was identified in Mesolithic Kotias Klde [43].

The different patterns of species abundances of hunted game when comparing Satsurblia to the above-mentioned Georgian UP sites reflect differences in the geographic location of the sites in relation to climate and vegetation zones. Satsurblia is located in a lower elevation part of western Georgia, well within the present day humid deciduous forest zone influenced from the Black Sea climatic regime. In contrast, the sites of Dzudzuana, Ortvale Klde and Bondi Caves are located 50 km east of Satsurblia, at a higher

elevation (~600–800 asl) region of Chiatura, in the foothills of the Greater Caucasus Mountains [2], [4]. The dominance of open-landscape taxa at these sites indicates either a preference of UP hunters for utilizing open landscapes combined with better access to such areas or a limited easterly shift of the climatic zones between the UP and present day.

Both the micro- and macro-faunas indicate a more dense distribution of woodland ecosystems associated with a climatic amelioration during the post-LGM human occupation of Satsurblia (i.e. between ca. 18–13 ka cal. BP). The microfaunal component overall points to an increase in woodland habitat when comparing the pattern in the pre-LGM (layers B/II and B/III) and post-LGM phases (layers A/IIa-b). This is supported by macrofaunal evidence for an increase in *Cervus elaphus* frequencies in the upper levels (AII-AI) and the persistent presence of pollen and wood cells of arboreal vegetation including *Pinus*, *Juglans* and *Corylus* throughout Layer AIIb. The existence of wetlands in the vicinity of the site is indicated by the presence of the Eurasian beaver, a key indicator species of wetland ecosystems, and the rodent water vole which occurs throughout the sequence (*Arvicola* sp.). A similar gradual forestation trend through post-LGM times and the final Pleistocene was recently detected in studies of micromammalian faunas in the Iberian Peninsula [44] and southern Italy [45]. It is suggested that in spite of the major biogeographic barrier of the Greater Caucasus Mountain Range, the region as a whole was permeable to biotic exchanges in different directions in relation to changing distributions of vegetation zones and climatic shifts up until relatively recent times of the final Pleistocene. Relative simplification of the vegetation cover during cold phases such as the LGM and subsequent Heinrich 1 event may have aided the movements of specific biotic elements, although not of entire communities, across steep mountainous sections of the Caucasus and provided the background for shifting patterns of human occupation of the region.

Conclusions

The study of the UP Layer in Satsurblia has revealed evidence of human occupation during the pre-LGM period (Area B, Layers B/II and B/III), 25.5–24.4 ka cal. BP, an interval broadly contemporaneous with part of the occupation in Dzudzuana C (27–24 ka cal. BP). The chronology of Layer AII/a and A/IIb indicate the presence of a new post-LGM phase (Area A, Layers A/IIa and A/IIb: 17.9–16.2 ka cal. BP). The latter provides new evidence for human occupation in this region more than a millennium prior to what was previously known based on the radiocarbon-based chronology of Dzudzuana B (16.5–13.2 ka cal. BP, [2]).

The lithic analyses reveal that during this post-LGM period, there existed a cultural (lithic) variant of the Eastern Epi-Gravettian, dominated by bladelet tools, discrete among which are varieties of the microgravette point. While microgravette industries, with rare Gravette points were reported from Dzudzuana B [2], those represent a later facies as they lack the other dominant component evident in Satsurblia A/II and B/II assemblages, namely the truncated items, and in particular the rectangles which were not reported from earlier excavations in the region.

The recovery of living floors and the presence of combustion features and hearths provide new information about the processing of wild cereals, the utilization of flax, and wood, as well as paleoenvironmental reconstruction based on palynological and micromorphological analyses. These indicate that during the

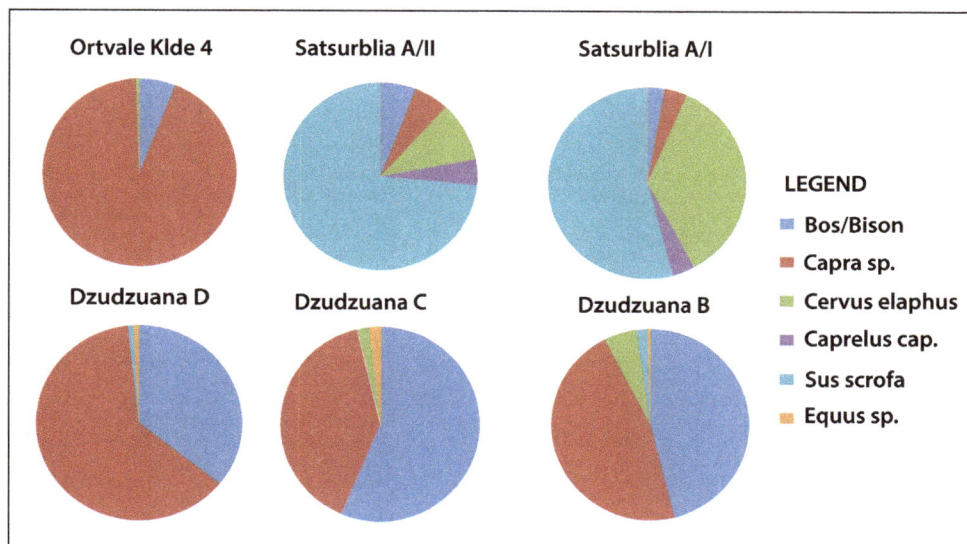

Figure 12. Regional pie charts of the main ungulate species in excavated UP assemblages.

formations of Floors 1–4 Layer A/IIb, the environment in the vicinity of the cave was predominantly temperate coniferous forest but with some evidence for pollen of deciduous taxa. Both the micro- and macro-faunas point to a relatively dense distribution of a wooded ecosystem during the post-LGM human occupation of Satsurblia (18–13 ka cal. BP).

The macrofaunal analysis indicates that subsistence focused on the hunting of wild boar and red deer as well as some wild goats and wild bovines. The Satsurblia UP assemblages differ from those reported for other sites with UP phases in the region in which hunting focused on bos/bison or wild goats. Future research will clarify whether these difference in the composition of UP macrofaunal assemblages from sites in this region reflect for the most part variations in the availability of animal resources by season and period [2], [9], [11], or also implies variations which are associated with cultural preferences.

The results of the campaigns in Satsurblia and Dzudzuana suggest that at present the most plausible scenario is one of a hiatus in the occupation of this region during the LGM (between 24.4–17.9 ka cal. BP). Future fieldwork will aim to assess earlier occupations and in particular to investigate whether the hiatus in occupation in Dzudzuana between Units D (34.5–32.2 ka cal. NP) and C (27–24 ka cal. BP) also occurs in Satsurblia, suggesting an additional regional (and potentially pan-regional) occupational hiatus.

Materials and Methods

Ethics

All necessary permits for the Satsurblia fieldwork and analyses of all sediments, human remains and artefacts, were obtained by Tengiz Meshveliani (Director of the excavation) and David Lordkipanidze (Direector Genral, Georgian Nationl Museum).

Supporting Information

Figure S1   Scan of sample SAT-12–46 T23b: Floor 1.

Figure S2   Scan of sample SAT-12–48 T22d: fireplace hearth on the 2nd floor.

Figure S3   Diagram of (a) non-pollen palynomorphs and (b) pollen and spores from samples taken from Floors 1–4 Layer A/IIb.

Figure S4   Cereal starch grain, Floor 1.

Figure S5   Flax fiber, fireplace installation, Floor 2.

Table S1   AMS determinations from Satsurblia measured at the ORA.

Table S2   Debitage counts by Layer, including debris and cores.

Table S3   Abundances of micromammalian taxa as numbers of identified specimens (NISP) and minimum numbers of individuals (MNI) by stratigraphic units and elevation within the stratigraphic sequence.

Text S1   Supporting information about the micromorphological analysis.

Acknowledgments

This paper is dedicated to the memory of Abesalom Vekua, a pioneer of Georgian paleontology, who passed away earlier this year. During our many years of working in the Georgian State museum we learned to appreciate Vekua's seminal work in the development of Georgian paleontology. His collaboration and assistance during our research was always inspiring.

The fieldwork in Satsurblia during 2013 was funded by a National Geographic-Global Exploration Fund, grant number GEFNE78-13. Additional support was obtained from the European Research Council Starter Grant- ERC-2010-StG 263441.

References

1. Adler DS, Bar-Yosef O, Belfer-Cohen A, Tushabramishvili N, Boaretto E, et al. (2008) Dating the demise: Neandertal extinction and the establishment of modern humans in the southern Caucasus. J Hum Evol 55: 817–833.
2. Bar-Yosef O, Belfer-Cohen A, Mesheviliani T, Jakeli N, Bar-Oz G, et al. (2011) Dzudzuana: an Upper Palaeolithic cave site in the Caucasus foothills (Georgia). Antiquity 85: 331–349.
3. Pinhasi R, Nioradze M, Tushabramishvili N, Lordkipanidze D, Pleurdeau D, et al. (2012) New chronology for the Middle Palaeolithic of the southern Caucasus suggests early demise of Neanderthals in this region. J Hum Evol 63: 770–780.
4. Tushabramishvili N, Pleurdeau D, Moncel MH, Agapishvili T, Vekua A, et al. (2012) Human remains from a new Upper Pleistocene sequence in Bondi Cave (Western Georgia). J Hum Evol 62: 179–185.
5. Gamble C, Davies W, Pettitt P, Hazelwood L, Richards M (2005) The archaeological and genetic foundations of the European population during the Late Glacial: Implications for "agricultural thinking". Camb Archaeol J 15: 193–223.
6. Soffer O, Gamble CE (1990) The World at 18,000 BP. High Latitudes, vol. 1. Unwin Hyman, London UK.
7. Kalandadze AN, Kalandadze KS (1978) Archaeological Research of Karstic Caves in Tskaltubo region (in Georgian, with Russian summary). Caves of Georgia: 116–136.
8. Kalandadze KS, Bugianishvili T, Ioseliani N, Jikia M, Kalandadze N (2004) Tskaltubo Expedition. The short reports of the archaeological expedition of 1989–1992. (in Georgian with Russian summary). Tbilisi.
9. Bar-Oz G, Belfer-Cohen A, Meshveliani T, Djakeli N, Bar-Yosef O (2008) Taphonomy and zooarchaeology of the Upper Palaeolithic cave of Dzudzuana, Republic of Georgia. Int J Osteoarch 18: 131–151.
10. Bar-Oz G, Adler DS (2005) Taphonomic history of the Middle and Upper Palaeolithic faunal assemblage from Ortvale Klde, Georgian Republic. Journal of Taphonomy 3: 185–211.
11. Adler DS, Bar-Oz G, Belfer-Cohen A, Bar-Yosef O (2006) Ahead of the Game. Middle and Upper Palaeolithic Behaviors in the Southern Caucasus. Curr Anthropol 47: 89–118.
12. Meshveliani T, Bar-Yosef O, Belfer-Cohen A (2004) The Upper Palaeolithic in western Georgia. In: Brantingham PJ, Kuhn SL, Kerry KW, (Eds), The Early Upper Paleolithic Beyond Western Europe. University of California Press, Berkeley, pp. 129–143.
13. Liubin VP (1966) Pervye svedeija o mezolite gornogo Kavkaza (Osetija) [First evidence concerning the Mesolithic of the Caucasus Highlands (Osetia)]. In: Gurnia NNeP, (Ed), U Istokov Drevnikh Kul'tur: Epokha Mezolita [At the Dawn of the Ancient Cultures: The Mesolithic Period]. Nauka, Materialy i issledovanija po arkheologii SSSR 126, Moscow, pp. 155–163.
14. Grigolia GK (1977) Centraluri Kolhetis neoliti Paluri [The Neolithic of Central Colchis: Paluri]. Tbilisi, (in Russian). Metsniereba.
15. Montoya C, Balasescu A, Joannin Sb, Ollivier V, Liagre Jrm, et al. (2013) The Upper Palaeolithic site of Kalavan 1 (Armenia): An Epigravettian settlement in the Lesser Caucasus. J Hum Evol 65: 621–640.
16. Golovanova LV, Doronichev VB, Cleghorn NE, Koulkova MA, Sapelko TV, et al. (2014) The Epipaleolithic of the Caucasus after the Last Glacial Maximum. Quat Int 337: 189–224.
17. Bader NO (1984) Pozdnij paleolit Kavkaza. In: Boriskovskij PI, (Ed), Paleolif SSSR. Nakua. Moscow, pp. 272–301.
18. Zamiatnin SN (1957) The Paleolithic of the Western Transcaucasus: Paleolithic Caves of Imeretia (In Russian). Collection of papers in Anthropology and Ethnography 17: 432–499.
19. Tushabramishvili DM (1960) Paleolithic Remains in Gvardjilas-klde Cave. Tbilisi: Tbilisi Nakua.
20. Gabunia M, Tsereteli L (2003) Mezolituri Kavkasshi [The Mesolithic in the Caucasus] (in Georgian with English summary). Dziebani 12: 5–30.
21. Meshveliani T, Bar-Oz G, Bar-Yosef O, Belfer-Cohen A, Boaretto E, et al. (2007) Mesolithic Hunters at Kotias Klde, Western Georgia: Preliminary results. Paleorient 33: 47–58.
22. Nioradze M, Otte M (2000) Paléolithique supérieur de Géorgie. L'Anthropologie (Paris) 104: 265–300.
23. Kvavadze E, Bar-Yosef O, Belfer-Cohen A, Boaretto E, Jakeli N, et al. (2011) Palaeoenvironmental change in the foothills of Imereti (western Georgia) during the Upper Palaeolithic Period according to new palynological data of cave material. Paper presented in the Conference: The View from the Mousntains. Bern, Switzerland, 21–27 July, 2011.
24. Kvavadze E, Meshveliani T, Jakeli N, Martkopishvili I (2011) Results of the palynological Investigation of the material taken in the cave Satsurblia in 2010. Proceedings of the Natural and Prehistoric Section, Georgian National Museum: Tbilisi, pp. 85–100.
25. Tsytsulina K (2008) Mesocricetus raddei. The IUCN Red List of Threatened Species. Version 2014.2.
26. Yiğit N, Çolak E, Gattermann R, Neumann K, Özkurt Ş, et al. (2006) Morphological and biometrical comparisons of Mesocricetus Nehring, 1898 (Mammalia: Rodentia) species distributed in the Palaearctic Region. Turkish Journal of Zoology 30: 291–299.
27. Tsytsulina K, Zagorodnyuk I (2008) Ellobius talpinus. The IUCN Red List of Threatened Species. Version 2014.2.
28. Kryštufek B, Shenbrot G (2008) Ellobius lutescens. The IUCN Red List of Threatened Species. Version 2014.2.
29. Smith AT, Boyer AF (2008) Ellobius lutescens. The IUCN Red List of Threatened Species. Version 2014.2.
30. Kandel AW, Gasparyan B, Bruch AA, Weissbrod L, Zardaryan D (2011) Introducing Aghitu-3, the first Upper Paleolithic cave site in Armenia. ARAMAZD 2: 7–23.
31. Gromov IM, Fokanov VA (1980) On remains of Upper Quaternary rodents from Kudaro 1 Cave. In: Ivanovka IK, Chernyakhovskij AG (Eds), The Kudaro Paleolithic Cave Sites in Southern Osetia. Moscow, pp. 79–89.
32. Pinhasi R, Gasparian B, Nahapetyan S, Bar-Oz G, Weissbrod L, et al. (2011) Middle Palaeolithic human occupation of the high altitude region of Hovk-1, Armenia. Quat Sci Rev 30: 3846–3857.
33. Kurten B (1958) Life and death of the Pleistocene cave bear: A study in paleoecology. Acta Zool Fenn 4: 1–59.
34. Bar-Oz G, Weissbrod L, Gasparian B, Nahapetyan S, Wilkinson K, et al. (2012) Taphonomy and zooarchaeology of a high-altitude Upper Pleistocene faunal sequence from Hovk-1 Cave, Armenia. J Arch Sci 39: 2452–2463.
35. Stiner MC (1998) Mortality analysis of Pleistocene bears and its paleoanthropological relevance. J Hum Evol 34: 303–326.
36. Baryshnikov GF (2002) Local biochronology of Middle and Late Pleistocene mammals from the Caucasus. Russ J Theriol 1: 61–67.
37. Vereshchagin NK (1967) The Mammals of the Caucasus: A History of the Evolution of the Fauna. Jerusalem: Israel Program for Scientific Translations.
38. Vereshchagin NK, Burchak-Abramovich NO (1958) History of distribution and opportunities of restoration of a beaver (Castor fiber L.) on Caucasus. Zool Zhurnal 37: 1874–1879.
39. Payne S, Bull G (1988) Components of variation in measurements of pig bones and teeth, and hte use of measurements to distinguish wild from domestic pig remains. Archaeozoologia 2: 27–76.
40. Amorosi T (1989) A Postcranial Guide to Domestic Neo-Natal and Juvenile Mammals - The Identification and Aging of Old World Species. BAR International Series 533, Oxford.
41. Heptner VG, Nasimovich AA, Bannikov AG (1989) Mammals of the Soviet Union. Vol. 1: Ungulates. Leiden: E.J. Brill.
42. Villa P, Mahieu E (1991) Breakage patterns of human long bones. J Hum Evol 21: 27–48.
43. Bar-Oz G, Belfer-Cohen A, Meshveliani T, Jakeli N, Matskevich Z, et al. (2009) Bear in mind: Bear hunting in the Mesolithic of the southern Caucasus. Archaeol Ethnol Anthropol Eurasia 37: 15–24.
44. Cuenca-Bescós G, Straus LG, González Morales MR, García Pimienta JC (2009) The reconstruction of past environments through small mammals: from the Mousterian to the Bronze Age in El Mirón Cave (Cantabria, Spain). J Arch Sci 36: 947–955.
45. López-García JM, Berto C, Colamussi V, Valle CD, Vetro DL, et al. (2014) Palaeoenvironmental and palaeoclimatic reconstruction of the latest Pleistocene–Holocene sequence from Grotta del Romito (Calabria, southern Italy) using the small-mammal assemblages. Palaeogeog Palaeoecol 409: 169–179.

Author Contributions

Conceived and designed the experiments: DL RP TM ZM ABC. Performed the experiments: RP TM ZM ABC GBO KW NJ TH. Analyzed the data: EK CM LW TM ZM ABC GBO KW NJ TH. Contributed to the writing of the manuscript: RP EK CM LW TM ZM ABC GBO KW NJ TH.

16

The Impact of 850,000 Years of Climate Changes on the Structure and Dynamics of Mammal Food Webs

Hedvig K. Nenzén[1]*, Daniel Montoya[2], Sara Varela[3]

1 Département des sciences biologiques, Université du Québec à Montréal, Montréal, Québec, Canada, **2** School of Biological Sciences, Life Sciences Building, University of Bristol, Bristol, United Kingdom, **3** Department of Ecology, Faculty of Science, Charles University, Prague, Czech Republic

Abstract

Most evidence of climate change impacts on food webs comes from modern studies and little is known about how ancient food webs have responded to climate changes in the past. Here, we integrate fossil evidence from 71 fossil sites, body-size relationships and actualism to reconstruct food webs for six large mammal communities that inhabited the Iberian Peninsula at different times during the Quaternary. We quantify the long-term dynamics of these food webs and study how their structure changed across the Quaternary, a period for which fossil data and climate changes are well known. Extinction, immigration and turnover rates were correlated with climate changes in the last 850 kyr. Yet, we find differences in the dynamics and structural properties of Pleistocene *versus* Holocene mammal communities that are not associated with glacial-interglacial cycles. Although all Quaternary mammal food webs were highly nested and robust to secondary extinctions, general food web properties changed in the Holocene. These results highlight the ability of communities to re-organize with the arrival of phylogenetically similar species without major structural changes, and the impact of climate change and super-generalist species (humans) on Iberian Holocene mammal communities.

Editor: Michael Hofreiter, University of York, United Kingdom

Funding: H.N. was supported by: UQAM, NSERC-Discovery, NSERC-Create, Canada Research Chairs. S.V. was supported by 'Support of establishment, development and mobility of quality research teams at the Charles University' (CZ.1.07/2.3.00/30.0022, European Science Foundation and Czech Republic). D.M. was supported by the European Commission (MODELECORESTORATION- FP7 Marie Curie Intra-European Fellowship for Career Development [301124]). The funders had no role in study design, data collection and analysis, decision to publish, or preparation of the manuscript.

Competing Interests: The authors have declared that no competing interests exist.

* Email: hedvig.nenzen@gmail.com

Introduction

Climate change is one of the major drivers affecting the diversity, composition, structure and functioning of ecological communities. Species respond in different ways to climate change which directly affects their persistence within the food web and, consequently, the composition and structure of the community [1,2]. Although evidence of the impacts of climate change mostly comes from studies of modern communities, life on Earth has experienced several climatic perturbations in the past and those changes impacted the composition and structure of ancient communities in similar ways. Understanding how ancient food webs responded to past climate provides information about how communities reorganize across time, and how food webs could respond to contemporary climate change [3,4].

Because the Quaternary fossil record is extensive and the climate changes during this period are relatively well understood, the examination of the Quaternary fossil record could be key to understanding long-term dynamics and structure of ancient biological communities [4]. Climate in the Quaternary is characterized by cyclic climatic changes, oscillating from cold, dry glacial scenarios to warm, wet interglacial scenarios [5]. Changes in annual mean temperature inferred from oxygen isotopes range from an increase of $5°C$ during the warm scenarios to a decrease of $-11°C$ in the extreme glacial periods [6]. This sequence of successive glacial and interglacial periods had direct and indirect effects on natural communities by forcing species to

migrate [4], eliminating species [7], introducing new species, and shaping several broad diversity patterns that we observe today (e.g. [8]).

In the Iberian Peninsula, fossil records show that the composition of species within communities changed during the Quaternary following climate changes (e.g. [9]). Whereas some species went extinct after inhabiting the Iberian Peninsula for long time periods (e.g. *Bos primigenius*, *Crocuta crocuta*), others remained for more than one million years (*Cervus elaphus*, *Sus scrofa*), and certain species appeared for short periods (*Cuon alpinus*, in the Middle Pleistocene; or *Coelodonta antiquitatis*, in the Last Glacial Maximum). Quaternary climate changes co-occurred with extinctions, migrations and the arrival of new species in the Iberian Peninsula that have likely impacted the structure of mammal communities living in this region.

In spite of the observed long-term species changes during the Quaternary, few studies have analysed long series of species turnovers (but see [10]), even fewer have investigated the structure of the mammal communities through time [7,11,12] and none have analyzed food web dynamics across the Quaternary. Analyzing food webs can help us to identify changes in biotic interactions, beyond counting the number of species present in the community [13,14]; these changes can alter the functioning of food webs and ecosystems [15], and affect their future stability and persistence [16,17]. However, food web approaches in paleoecological studies are scarce, and therefore we know little about how past climate changes have affected the long-term dynamics and food web structure of biological communities at regional scales, and to what extent these communities re-organize in response to these impacts (but see [18–21] for temporal food web studies). The reason for this paucity of studies is twofold. First, extending the approaches used in food web theory to ancient communities has been limited by the incompleteness or lack of fossil data. In particular there is little fossil evidence that can be used to establish trophic links, compared to highly-resolved modern networks where changes in feeding interactions can be directly observed (e.g. [22]).

Second, the information from one single fossil site is limited for studies at the community scale where several species are considered. Because of a number of biases that affect the fossil record [23], individual fossil sites often lack species from the regional species pool (the diversity of species at a spatial scale larger than the individual fossil site) and are not an appropriate source of data to construct ancient regional food webs. Here we overcome these limitations by using large-scale cumulative food webs, that are constructed by using fossil information from different sites, and are thus appropriate for comparing food webs over time or space [20]. Moreover, large-scale cumulative webs do not depend on single fossil site records, and minimize problems of undersampling or collection bias [23,24].

In this work, we assemble six large mammal communities in the Iberian Peninsula spanning 850,000 years (Pleistocene and Holocene time periods). To obtain a complete species pool for large Iberian mammals, communities were constructed using published data from 71 fossil sites. Thus, the results presented here describe how large mammal communities at a regional scale changed during the Quaternary climate changes. The aim of this study is to explore the long-term and community structure of a large mammal food web across 850,000 years of climate changes. We specifically address the following questions: (a) how have Quaternary mammal communities changed in terms of dynamics (species extinctions, immigrations, and turnover rates) and food web structure?, and (b) are the changes in the structure and dynamics of these communities associated with Quaternary climate changes?

Methods

Fossil data and food web construction

We assembled six different large mammal food webs across the Quaternary: one for the Early Pleistocene (EP; around 850,000 years before present), one for the Middle Pleistocene (MP; around 450 kyr BP), plus two food webs during the Late Pleistocene, one for the last interglacial maximum, LIM; 120 kyr BP, and one for the last glacial maximum, LGM; 21 kyr BP), and two food webs for the Holocene (H; 10 kyr BP and present, P). For each time, we identified the large mammals (>20 kg), including hominids, that were present using fossil records from several Iberian fossil sites (details about sites and references in Table S1 and figure S1 in File S1). The taxonomy has been revised and unified following [25]. The resulting communities are not observed local communities, but regional food webs constructed using data from 71 fossil sites within the Iberian Peninsula. Large mammals are highly mobile and we assume that their geographic ranges covered the entire Iberian Peninsula. We use the number of glacial cycles between each community, MIS/OIS boundaries (Marine Isotope Stages, or Oxygen Isotope Stages [26]) as a proxy for climatic changes. We assume that a higher number of cycles represent a higher number of recovery and reorganisation periods for the mammal communities, indicating that they experienced greater stress.

Next we determine the trophic links between the species present in each community (figure 1, figure S2 in File S1). This is the most challenging step since diet evidence of extinct species is rare and interactions cannot be observed in the field. Species eaten by Pleistocene carnivores are classically identified from indirect clues, such as cut marks, teeth marks or stone tools marks in the fossil bones [27]. There are methods to distinguish carnivorous vs. herbivorous species [28,29], and to identify ancient species diet requirements using stable isotopes [21] and DNA from coprolites [30]. Here, food web links were designated following three criteria: (i) spatio-temporal co-occurrence as shown by fossil record data (Tables S1, S2 in File S1), (ii) body-size relationships (Table S3 in File S1), and (iii) actualism (applying current species diet to infer past trophic links). The key role of body-size in determining feeding links and structuring food webs is broadly accepted [31]. We complement link information by using actualism, which can be applied to all Quaternary carnivores (e.g. human diet cannot be predicted by body weight alone). Actualism diets follow [32]. The networks analysed here are bipartite and consist of two trophic levels (predators and herbivores/prey).

Community and food web analysis

We explore community long-term dynamics by calculating the number of extinctions, immigrations and species turnover between communities. Extinction and immigration were estimated as the number of species that were lost or gained in each transition. We use Sørensen's dissimilarity index to estimate species turnover rates between the communities [33].

We also explore the possibility that extinction and immigration are non-random by using two plausible hypotheses based on phylogeny and body-size. For immigrating species, we test whether newly immigrated species are phylogenetically similar to the species that went extinct in the preceding time step. Following the phylogenetic conservatism hypothesis, new immigrants are more likely to replace extinct species if they are closely related within the evolutionary tree (because of functional similarity), a phenomenon that seems to be widespread across evolutionary time [34]. To test if species were replaced by closely related species, we count the number of times a species that went extinct was replaced by a species within the same genus (Table S2 in File S1). In the absence of a complete phylogeny this gives a simple measure of phylogenetic relatedness and functional similarity, as closely related species often fulfil the same function in an ecosystem [34–36]. Second, as large animals are likely to go extinct first [37,38], we test whether the body-size distribution of extinct species was random across time. For each food web we randomly remove the same number of species that were observed to go extinct in the next time step. After repeating this 10,000 times, we compare the mean body weight of the randomized and observed food webs.

To examine food web structure we use standard metrics: species richness, number of links and connectance (links/species^2) [39]. Two other food web properties were analysed: vulnerability (mean number of predators per prey), and generality (mean number of prey per predator) [40]. To further characterize generality we sort the number of links per prey and predator and plot the relationship for each food web. Because network structure is related to its stability, we analysed community stability by examining (i) the robustness of ancient food webs to species loss [41–43], and (ii) nestedness [44]. The robustness index measures the topological or structural stability of the food web by simulating how random removal of prey (or predators) induces secondary extinction among the predators (or prey) [43,45]. Robustness is measured as the area under the curve of the number of species

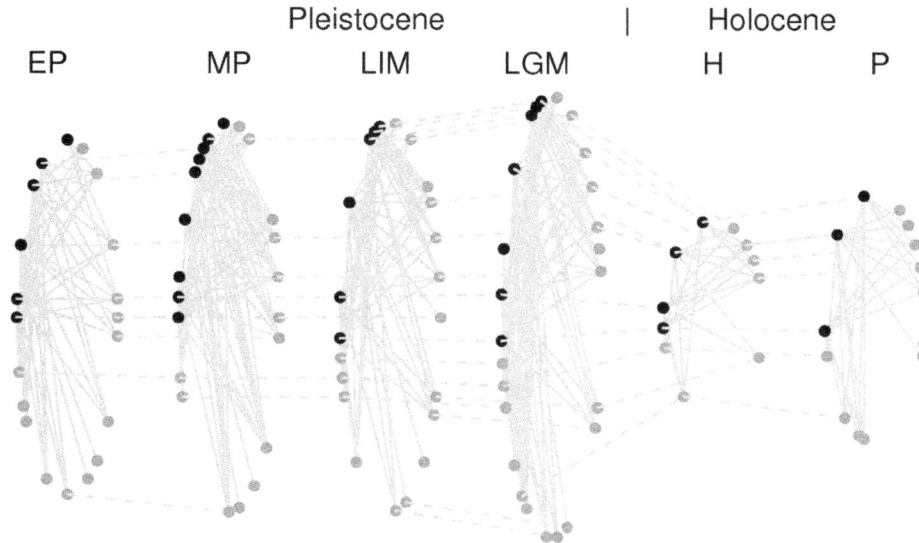

Figure 1. Large mammal food webs of the Iberian Peninsula during the Quaternary. Early Pleistocene (EP; 850,000 years before present, or 850 ky BP), Middle Pleistocene (MP; 450 kyr BP), Last Interglacial Maximum (LIM; 120 kyr BP), Last Glacial Maximum (LGM; 21 kyr BP), Early Holocene (H; 10 kyr BP) and present (P). Each species (black for predator, dark grey for prey) is a node, and each link indicates a trophic interaction. Horizontal links indicate that the species persists between communities.

being removed against the number of secondary extinctions, and ranges from 0 to 1, with high values representing more robust communities (the number of secondary extinctions is lower). We calculate community robustness for removal of both prey and predators individually. Nestedness measures the degree to which the diets of consumers are proper subsets of other, more generalist consumers. The nestedness algorithm used here is based on the nestedness temperature of the interaction matrix, and ranges from 0 which indicates high nestedness, to 100 which indicates no nestedness [46,47].

Finally, we examine the relationships between climate change, species turnover and food web properties. Controlled, replicated experiments cannot be conducted in paleoecological studies, but we can explore whether food web changes in ancient communities are associated to climatic changes in the past. We test if the number of glacial cycles between each community and the observed changes in species composition are related to changes in food web metrics, using the 15 unique pairwise comparisons between all six food webs. Given that non-linear responses of natural communities to climate changes are common (e.g. [48]), we assume that the relationship between the number of glacial cycles, species turnover and food web properties are non-linear, and for that reason we use the Spearman correlation index.

Results

Changes in Quaternary mammal food webs over the last 850,000 years

Pleistocene communities comprise 20–25 species >20 kg (figure 2a), a figure that falls within the range of values reported in comparative analyses of modern [49] and ancient food webs [20,21]. The complexity of these ancient webs is also similar to that observed in modern mammal food webs in Africa where megafauna did not go extinct in the Holocene [50].

During the Pleistocene the rates of extinction were high, (figure 3b, except the LIM-LGM transition), with one species disappearing every 20–36 kyr. Extinction analysis showed no

correlation between species' body-size and extinction probability (figure 3e). The number of new species entering Pleistocene communities equalled or exceeded the number of species that went extinct (figure 3c), with one new species entering the Pleistocene food webs every 12–36 kyr. This suggests that extinct species were replaced by new species at similar rates, and that the overall number of species in Pleistocene food webs did not change. Consistent with the phylogenetic conservatism hypothesis, a high proportion of these new species were phylogenetically similar to the extinct species (figure 3f). Despite the high turnover rates (figure 3d), the general diversity and food web properties of the Pleistocene communities remained relatively constant across almost 500 kyr.

The dynamic stability observed in Pleistocene mammal communities changed in the Holocene. Extinction rates dramatically increased (one extinct species every 830 years, figure 3b). These extinctions are non-random (figure 3e), and especially affect the least connected specialist predators (figure S3 in File S1) and species with large body-sizes (figure S4 in File S1). Immigration rate dropped (figure 3c), and, consequently, extinct species were not replaced by new ones. Intriguingly, no predator mammals entered into the food web during the Holocene and the few immigrant species were phylogenetically unrelated (figure 3f). These processes caused a reduction in the number of species at all trophic levels that affected food web structure in two ways: (i) connectance increased, and (ii) generality and vulnerability decreased. The observed differences in other food web properties between Holocene and Pleistocene communities may be a direct consequence of increases in connectance following a reduction in network size and the loss of specialist predators.

Regardless of the changes in community size between Pleistocene and Holocene communities, the robustness of large mammalian communities to secondary extinctions of both prey and predators remained constant across the Quaternary period (figure 2f). We found no correlation between connectance and robustness, nor between species richness and robustness across the Quaternary period (Spearman correlations, p>0.05). Nestedness

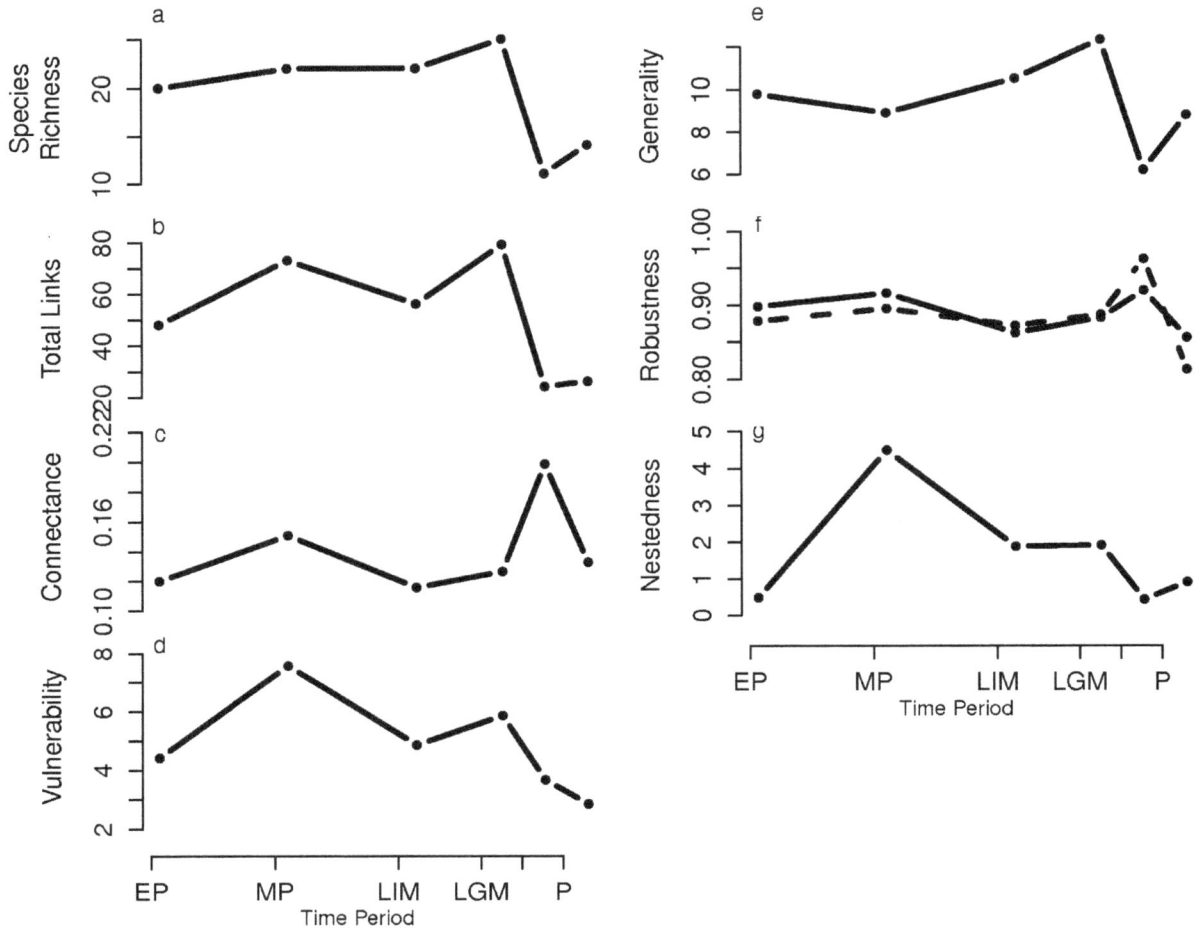

Figure 2. Food web properties at each time step. (a) Species richness (number of species), (b) total links (total number of links or trophic interactions between species), (c) connectance (links/species²), (d) vulnerability, (e) generality, (f) robustness against removal of prey (solid line) and predators (dotted line), (g) nestedness. The dates are the same as in figure 1. The x-axis is proportional to the time between each community.

patterns, on the other hand, have fluctuated but fall within high nestedness values (figure 2g).

The role of climate change in the dynamics and structure of Quaternary mammal food webs

Overall, turnover rates of Iberian mammal species across the Quaternary are associated with the pattern of glacial cycles (table 1). Quaternary climate changes are positively correlated with species turnover, extinction and immigration rates across all mammal communities. However, glacial-interglacial events show no significant relationship with food web properties, which suggests that climate changes across the Quaternary were not associated with changes in general structure of mammalian communities. We also assessed whether climate change is indirectly associated with food web properties through changes in species composition, yet no significant relationship was observed between species turnover rates and food web metrics (table 2). This suggests that the structure of the Quaternary mammal food webs is independent of observed species turnover rates, and that most of the changes in the food web properties between Pleistocene and Holocene mammal communities are associated with the loss of specialist predators and increases in connectance following a non-random reduction in network size between these two periods of time.

Discussion

Iberian large mammal communities have experienced important dynamical changes across the Quaternary. Extinction, immigration and turnover rates were highly correlated with climate changes in the last 850 kyr, yet food web properties were not significantly altered following these climate changes. Our results suggest that glacial-interglacial cycles are associated with changes in community dynamics, and highlights the ability of communities to re-organize with the arrival of phylogenetically similar species without major changes in food web properties. As a result, ancient mammal food webs were dynamically stable and able to recover from climatic perturbations even though many species went extinct.

Changes in large mammal food web structure and dynamics

Our results show differences in the dynamics and structure of Pleistocene *versus* Holocene mammal communities on the Iberian Peninsula. In the Pleistocene, extinctions were common and affected species irrespective of their body-size. Extinct species in the food web were replaced by newly arriving species that were closely related in the phylogenetic tree, so extinction and immigration rates were balanced. Consequently, food web

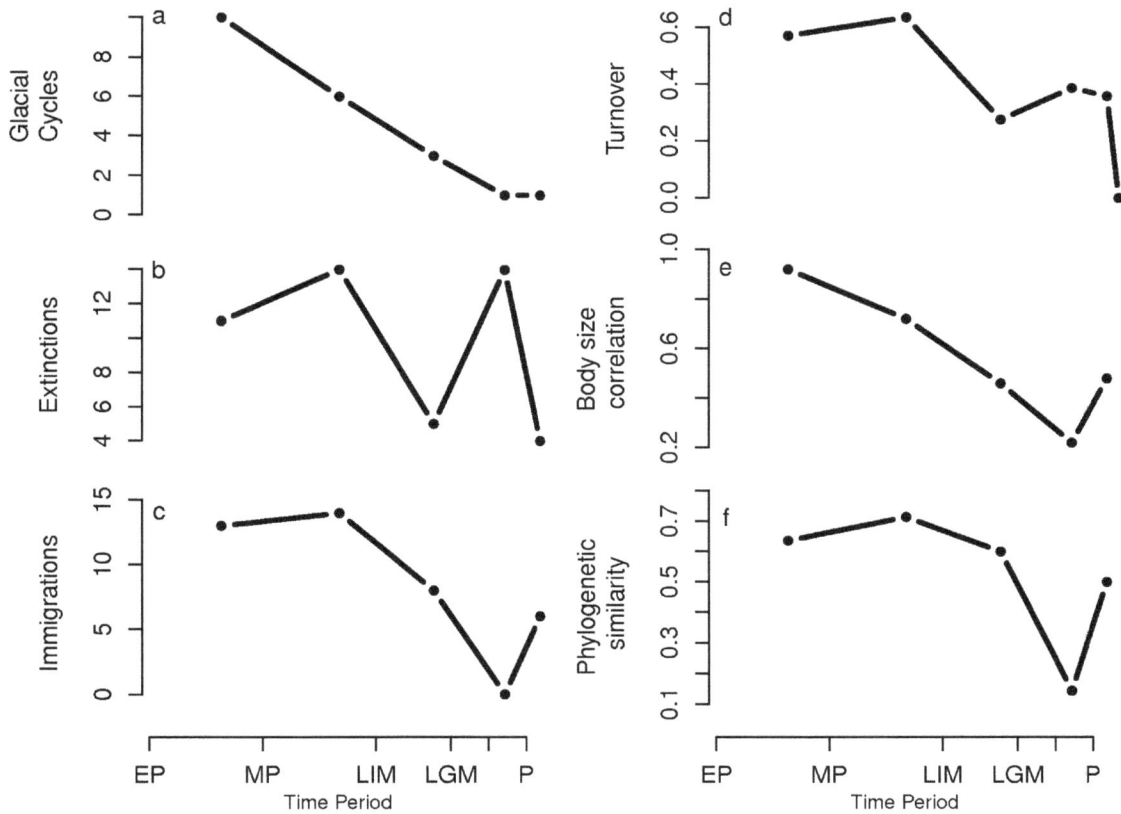

Figure 3. Food web structural changes at each time step. (a) Number of glacial cycles, (b) extinct species and (e) immigrating species, (f) species turnover, (g) correlation between observed and randomly expected body size distribution of extinct species and (f) phylogenetic similarity. The dates are the same as in figure 1.

properties remained unaltered and Pleistocene communities remained structurally stable despite the species identities. These results support the phylogenetic conservatism hypothesis [34], by which new immigrants are more likely to replace extinct species if they are closely-related within the evolutionary tree, as closely related species often fulfil the same function in the ecosystem. However, the extinction-immigration patterns of the Pleistocene are not reproduced in Holocene communities, which contracted as a result of higher extinction rates (56% of species went extinct from the early Pleistocene to Holocene) and low replacement rates. This reduction in community size and loss of specialist species increased

connectance, which in turn affected other properties of the food web.

After more than 800,000 years of relative stability, food web properties changed in two steps in the Holocene: species richness and generality first decreased and then increased. The first step was caused by the extinction of large mammals after the Last Glacial Maximum (21–10 kyr BP), when prey such as *Mammuthus primigenius* (woolly mammoth) or *Megaloceros giganteus* (Irish elk) went extinct, followed by large carnivores like *Panthera spelaea* (lion). This is consistent with the observation that modern mammalian communities are remnants of larger Pleistocene

Table 1. Spearman's rank correlation coefficient between climate change (number of glacial-interglacial cycles between scenarios) and species and food web structure parameters for all communities. ρ = Spearman correlation.

Climate change vs.		ρ	S	p
Species	Species turnover	0.92	47.54	<0.001
	Extinction	0.74	140.24	0.001
	Immigration	0.65	193.36	0.008
Food web structure	Species Richness	−0.15	654.77	0.585
	Number of food web links	−0.15	646.00	0.584
	Connectance	−0.24	692.00	0.397
	Predator/prey ratio	−0.07	598.00	0.812

Table 2. Spearman's rank correlation coefficient between species and food web structure parameters for all communities.

Species turnover vs.		rho	S	p
Food web structure	Species Richness	−0.06	594.33	0.828
	Number of links	−0.07	600.03	0.800
	Connectance	−0.34	752.17	0.210
	Predator/prey ratio	−0.11	623.05	0.689

mammal communities [51]. This reduction in community size is observed in contemporary mammal communities, and affected food web properties by reducing generality (fewer prey species) and increasing connectance. Theoretical studies suggest that spatially coupled food webs are especially sensitive to the loss of large animals [52], and there is evidence showing that the loss of megaherbivores such as mammoths had large impacts on ecosystem function [53,54]. Consequently, the loss of large mammals may have also affected the ecosystem functions provided by Holocene communities.

In the second step, during the Holocene only human-introduced herbivores, such as *Bos taurus* (cattle) or *Equus caballus* (horse) appeared as new species in the food webs. These introductions slightly increased the number of species and the mean number of prey species per predator, the latter being also a consequence of the existence of generalist predator species, especially the humans. In contrast to Pleistocene communities, the introduced species were not phylogenetically related to the extinct species, and this has likely impacted ecosystem functioning, as certain ecosystem functions may not have been completely replaced by phylogenetically related species.

Robustness and nestedness

Our analysis of robustness reveals across-time similarities in Pleistocene and Holocene communities. Despite the significant extinction and turnover rates across the Quaternary, and the reported changes in community size and food web connectivity, the robustness of mammal communities remained fairly constant for nearly one million years. On the other hand, all our mammal food webs present highly nested structures. Simulations have shown recently that nestedness is a destabilizing force in predator-prey food webs [55], and its absence in other ancient food webs may have promoted stability during the successive climatic changes that occurred in the past [21]. Although the combination of high robustness and high nestedness may have resulted in a neutral effect on stability, nested structures may have also acted as a buffer to secondary extinctions [56], so the interpretation of historical patterns of nestedness of Quaternary mammal communities is not straightforward.

The high nestedness reported in Quaternary mammal communities is explained by the presence of highly generalist predators. Both theoretical and empirical evidence show that super-generalist species become central nodes (most connected species) in the core of the nested community and may increase the overall nestedness [57]. Perturbation events usually favour generalist over specialist species, a pattern observed both in paleoecological [58] and contemporary communities [59], and this contributes to increase the generalist:specialist ratios. Quaternary mammal food webs were no exception to this rule and, as a result of the long-term dynamics following glacial-interglacial cycles, comprised highly generalist species. Specifically, the Holocene period experienced a population growth of the super-generalist anatomically modern

humans. Humans not only have a broad diet (figure S2 in File S1), but also introduced domesticated mammals into the food web, intensified the generality and contributed to the high connectivity of Iberian mammal food webs. Anatomically modern humans have been present in Spain since 42 kyr BP, but with a small population size [60]. Other hominids such as *Homo neanderthalensis* were present earlier, but our results suggest that they had a smaller impact on food webs compared to Holocene populations of *Homo sapiens*.

The high impact of omnivore humans on food webs has also been found in contemporary communities. A recent study explored a spatial (rather than temporal) gradient of human impacts in Serengeti food webs, average body mass and species richness also decreased with increasing human impact. Food webs tended to be more generalist following the human-induced extinction of the least-connected, most specialized species [61]. In contrast to Quaternary mammal communities, the Serengeti food web harbours more trophic levels and large predators, which results in a higher species number and lower connectance compared to our food webs.

Conclusion

Paleoecological community studies provide insight into the relationships between biological communities and perturbations such as climate change over long time scales. This work presents a first attempt to track temporal changes in food webs over long time periods. Although the spatial and temporal scales are coarse, the patterns revealed are significant. We found that large mammal communities in the Pleistocene were able to re-organize despite the high extinction rates reported in the last 850,000 years. However, our results indicate that Holocene mammal communities experienced changes that were related to the arrival of new species introduced in the communities by humans (cattle and game species). Future work should aim at investigating the actual changes in temperature at local scales and the duration of those cycles; this information could be used to make cross-comparisons of the species' extinction rates within and among continents, and their effects on ancient food webs, and to understand the mechanisms driving the likely spatial differences in the extinction rates.

Collectively, our results suggest differences between the Pleistocene and Holocene time periods in the structure and dynamics of large mammal food webs. These conclusions have consequences for understanding the current-day sixth extinction event, and to what extent modern communities will be able to re-organize during current climate change. Not only are current rates of change in climate high, but also they interact with other drivers of extinction such as habitat loss and overexploitation, and the feedbacks generated by these processes and biotic interactions may result in higher extinction rates than those expected only from climate change [62,63].

Supporting Information

File S1 Table S1, Reference sources used for determining the presence of species in each of the 71 sites. **Table S2,** The species present (indicated by 1) at each period. Weight categories: 1 = < 45 kg, 2 = 45–90 kg, 3 = 90–360 kg, 4 = 360–1000 kg, 5 = > 1000 kg. Phylogenetic replacement categories: 0 = Not replaced by a species in the same genus in the next time period, 1 = Replaced by a species in the next time period, 2 = Species still present. **Table S3,** Prey weight classes. This information, together with spatio-temporal co-occurrence and actualism (see main text) was used to establish the links between species in the food web. **Figure S1,** Geographic location on the Iberian Peninsula of the Quaternary fossil sites used for constructing the ancient food webs. **Figure S2,** Large mammal food webs of the Iberian Peninsula during the Quaternary. (a) Early Pleistocene (850,000 years before present, or 850 ky BP); (b) Middle Pleistocene (450 ky BP); (c) Last Interglacial Maximum (120 ky BP); (d) Last Glacial Maximum (21 ky BP); (e) Early Holocene (10 ky BP); and (f) Present. Each node (green for prey, red for predator) is a species, and each link indicates a trophic interaction.

Figure S3, Number of prey per predator in each time period, with species sorted in descending order. The time periods are the same as in figure 1. **Figure S4,** A random extinction experiment demonstrate that the distribution of the number of links, connectance and link density observed Holocene food web is not expected by chance. 1000 food webs have been created extracting randomly 11 species from last glacial maximum food web (the observed number of extinct species in the Holocene).

Acknowledgments

We thank J van der Made, D Storch, T Poisot, M Chevrinais, P Desjardins-Proulx, R Barrientos, A Nenzén and K Orford for their useful comments and advice.

Author Contributions

Conceived and designed the experiments: HN SV DM. Performed the experiments: HN SV DM. Analyzed the data: HN SV DM. Contributed reagents/materials/analysis tools: HN SV DM. Wrote the paper: HN SV DM.

References

1. Petchey OL, McPhearson P, Casey T, Morin P (1999) Environmental warming alters food-web structure and ecosystem function. Nature 402: 69–72. doi:10.1038/47023.

2. Lorenzen ED, Nogues-Bravo D, Orlando L, Weinstock J, Binladen J, et al. (2011) Species-specific responses of Late Quaternary megafauna to climate and humans. Nature 479: 359–364. doi:10.1038/nature10574.

3. Harnik PG, Lotze HK, Anderson SC, Finkel ZV, Finnegan S, et al. (2012) Extinctions in ancient and modern seas. Trends in Ecology & Evolution 27: 608–617. doi:10.1016/j.tree.2012.07.010.

4. Blois JL, Zarnetske PL, Fitzpatrick MC, Finnegan S (2013) Climate Change and the Past, Present, and Future of Biotic Interactions. Science 341: 499–504. doi:10.1126/science.1237184.

5. Kageyama M, Lainé A, Abe-Ouchi A, Braconnot P, Cortijo E, et al. (2006) Last Glacial Maximum temperatures over the North Atlantic, Europe and western Siberia: a comparison between PMIP models, MARGO sea–surface temperatures and pollen-based reconstructions. Quaternary Science Reviews 25: 2082–2102. doi:10.1016/j.quascirev.2006.02.010.

6. Jouzel J, Masson-Delmotte V, Cattani O, Dreyfus G, Falourd S, et al. (2007) Orbital and Millennial Antarctic Climate Variability over the Past 800,000 Years. Science 317: 793–796. doi:10.1126/science.1141038.

7. Ceballos G, Arroyo-Cabrales J, Ponce E (2010) Effects of Pleistocene environmental changes on the distribution and community structure of the mammalian fauna of Mexico. Quaternary Research 73: 464–473. doi:10.1016/j.yqres.2010.02.006.

8. Montoya D, Rodríguez MA, Zavala MA, Hawkins BA (2007) Contemporary richness of holarctic trees and the historical pattern of glacial retreat. Ecography 30: 173–182. doi:10.1111/j.0906-7590.2007.04873.x.

9. Agusti J, Oms O, Remacha E (2001) Long Plio-Pleistocene terrestrial record of climate change and mammal turnover in southern Spain. Quaternary Research 56: 411–418. doi:10.1006/qres.2001.2269.

10. Raia P, Piras P, Kotsakis T (2005) Turnover pulse or Red Queen? Evidence from the large mammal communities during the Plio-Pleistocene of Italy. Palaeogeography, Palaeoclimatology, Palaeoecology 221: 293–312. doi:10.1016/j.palaeo.2005.02.014.

11. Rodríguez J (2004) Stability in Pleistocene Mediterranean mammalian communities. Palaeogeography, Palaeoclimatology, Palaeoecology 207: 1–22. doi:10.1016/j.palaeo.2003.12.016.

12. Palombo MR, Raia P, Giovinazzo C (2005) Early-Middle Pleistocene structural changes in mammalian communities from the Italian peninsula. Geological Society, London, Special Publications 247: 251–262. doi:10.1144/GSL.SP.2005.247.01.14.

13. Jablonski D (2008) Biotic interactions and macroevolution: extensions and mismatches across scales and levels. Evolution 64: 715–739. doi:10.1111/j.1558-5646.2008.00317.x.

14. Bascompte J, Stouffer DB (2009) The assembly and disassembly of ecological networks. Philosophical Transactions of the Royal Society B: Biological Sciences 364: 1781–1787. doi:10.1098/rstb.2008.0226.

15. Thompson R, Brose U, Dunne JA (2012) Food webs: reconciling the structure and function of biodiversity. Trends in Ecology & Evolution 27: 689–697. doi:10.1016/j.tree.2012.08.005.

16. Paine RT (1966) Food Web Complexity and Species Diversity. The American Naturalist 100: 65–75.

17. Gross T, Blasius B (2008) Adaptive coevolutionary networks: a review. Journal of the Royal Society Interface 5: 259–271. doi:10.1098/rsif.2007.1229.

18. Roopnarine PD (2006) Extinction cascades and catastrophe in ancient food webs. Paleobiology 32: 1–19. doi:10.1666/05008.1.

19. Roopnarine PD (2010) Networks, Extinction and Paleocommunity Food Webs. Available: http://precedings.nature.com.ezproxy.its.uu.se/documents/4433/version/2. Accessed 4 June 2011.

20. Dunne JA, Williams RJ, Martinez ND, Wood RA, Erwin DH (2008) Compilation and network analyses of cambrian food webs. PLoS Biology 6: e102. doi:10.1371/journal.pbio.0060102.

21. Yeakel JD, Guimaraes PR, Bocherens H, Koch PL, Guimarães PR (2013) The impact of climate change on the structure of Pleistocene food webs across the mammoth steppe. Proceedings of the Royal Society B: Biological Sciences 280: 20130239. doi:10.1098/rspb.2013.0239.

22. Olesen JM, Stefanescu C, Traveset A (2011) Strong, Long-Term Temporal Dynamics of an Ecological Network. PLoS ONE 6: e26455. doi:10.1371/journal.pone.0026455.

23. Varela S, Lobo JM, Hortal J (2011) Using species distribution models in paleobiogeography: A matter of data, predictors and concepts. Palaeogeography, Palaeoclimatology, Palaeoecology 310: 451–463. doi:10.1016/j.palaeo.2011.07.021.

24. Martinez ND, Hawkins BA, Dawah HA, Feifarek BP (1999) Effects of Sampling Effort on Characterization of Food-Web Structure. Ecology 80: 1044–1055. doi:10.2307/177037.

25. Van der Made J (2005) La fauna del Pleistoceno europeo. In: Carbonell E, editor. Homínidos: las primeras ocupaciones de los continentes. Ariel. pp. 394–432.

26. Lisiecki LE, Raymo ME (2005) A Pliocene-Pleistocene stack of 57 globally distributed benthic δ18O records. Paleoceanography 20: PA1003. doi:10.1029/2004PA001071.

27. Selvaggio M (1994) Carnivore Tooth Marks and Stone Tool Butchery Marks on Scavenged Bones - Archaeological Implications. Journal of Human Evolution 27: 215–228. doi:10.1006/jhev.1994.1043.

28. Boesl C, Grupe G, Peters J (2006) A Late Neolithic vertebrate food web based on stable isotope analyses. International Journal of Osteoarchaeology 16: 296–315. doi:10.1002/oa.834.

29. Newsome SD, Collins PW, Rick TC, Guthrie DA, Erlandson JM, et al. (2010) Pleistocene to historic shifts in bald eagle diets on the Channel Islands, California. Proceedings of the National Academy of Sciences 107: 9246–9251. doi:10.1073/pnas.0913011107.

30. Bon C, Berthonaud V, Maksud F, Labadie K, Poulain J, et al. (2012) Coprolites as a source of information on the genome and diet of the cave hyena. Proceedings of the Royal Society B: Biological Sciences 279: 2825–2830. doi:10.1098/rspb.2012.0358.

31. Woodward G, Ebenman B, Emmerson MC, Montoya JM, Olesen JM, et al. (2005) Body size in ecological networks. Trends in Ecology & Evolution 20: 402–409. doi:10.1016/j.tree.2005.04.005.

32. Owen-Smith N, Mills MGL (2008) Predator-prey size relationships in an African large-mammal food web. The Journal of Animal Ecology 77: 173–183. doi:10.1111/j.1365-2656.2007.01314.x.

33. Baselga A, Orme CDL (2012) betapart: an R package for the study of beta diversity. Methods in Ecology and Evolution 3: 808–812. doi:10.1111/j.2041-210X.2012.00224.x.

34. Gómez JM, Verdú M, Perfectti F (2010) Ecological interactions are evolutionarily conserved across the entire tree of life. Nature 465: 918–921. doi:10.1038/nature09113.

35. Cattin M-F, Bersier L-F, Banasek-Richter C, Baltensperger R, Gabriel J-P (2004) Phylogenetic constraints and adaptation explain food-web structure. Nature 427: 835–839. doi:10.1038/nature02327.

36. Stouffer DB, Sales-Pardo M, Sirer MI, Bascompte J (2012) Evolutionary Conservation of Species' Roles in Food Webs. Science 1018: 1489–1492. doi:10.1126/science.1216556.

37. Voigt W, Perner J, Davis A, Eggers T (2003) Trophic levels are differentially sensitive to climate. Ecology 84: 2444–2453. doi:10.1890/02-0266.

38. Cardillo M, Mace GM, Jones KE, Bielby J, Bininda-Emonds ORP, et al. (2005) Multiple Causes of High Extinction Risk in Large Mammal Species. Science 309: 1239–1241. doi:10.1126/science.1116030.

39. Pascual M, Dunne JA (2005) Ecological Networks: Linking Structure to Dynamics in Food Webs. Oxford University Press.

40. Tylianakis JM, Tscharntke T, Lewis OT (2007) Habitat modification alters the structure of tropical host-parasitoid food webs. Nature 445: 202–205. doi:10.1038/nature05429.

41. Solé R V, Montoya JM (2001) Complexity and fragility in ecological networks. Proceedings of the Royal Society B: Biological Sciences 268: 2039–2045. doi:10.1098/rspb.2001.1767.

42. Dunne JA, Williams RJ, Martinez ND (2002) Food-web structure and network theory: The role of connectance and size. Proceedings of the National Academy of Sciences 99: 12917–12922. doi:10.1073/pnas.192407699.

43. Memmott J, Waser NM, Price MV (2004) Tolerance of pollination networks to species extinctions. Proceedings of the Royal Society B: Biological Sciences 271: 2605–2611. doi:10.1098/rspb.2004.2909.

44. Guimara PR, Ulrich W, Almeida-Neto M, Guimarães PR, Loyola RD, et al. (2008) A consistent metric for nestedness analysis in ecological systems: reconciling concept and measurement. Oikos 117: 1227–1239. doi:10.1111/j.0030-1299.2008.16644.x.

45. Burgos E, Ceva H, Perazzo RPJ, Devoto M, Medan D, et al. (2007) Why nestedness in mutualistic networks? Journal of Theoretical Biology 249: 307–313. doi:10.1016/j.jtbi.2007.07.030.

46. Rodríguez-Gironés MA, Santamaría L (2006) A new algorithm to calculate the nestedness temperature of presence–absence matrices. Journal of Biogeography 33: 924–935. doi:10.1111/j.1365-2699.2006.01444.x.

47. Dormann CF, Gruber B, Fruend J (2008) Introducing the bipartite Package: Analysing Ecological Networks. R news 8: 8–11.

48. Pope KS, Dose V, Da Silva D, Brown PH, Leslie CA, et al. (2013) Detecting nonlinear response of spring phenology to climate change by Bayesian analysis. Global Change Biology 19: 1518–1525. doi:10.1111/gcb.12130.

49. Williams RJ, Martinez ND (2000) Simple rules yield complex food webs. Nature 404: 180–183. doi:10.1038/35004572.

50. Baskerville EB, Dobson AP, Bedford T, Allesina S, Anderson TM, et al. (2011) Spatial Guilds in the Serengeti Food Web Revealed by a Bayesian Group Model. PLoS Computational Biology 7: e1002321. doi:10.1371/journal.pcbi.1002321.

51. Koch PL, Barnosky AD (2006) Late Quaternary Extinctions: State of the Debate. Annual Review of Ecology, Evolution, and Systematics 37: 215–250. doi:10.1146/annurev.ecolsys.34.011802.132415.

52. McCann KS, Rasmussen JB, Umbanhowar J (2005) The dynamics of spatially coupled food webs. Ecology Letters 8: 513–523. doi:10.1111/j.1461-0248.2005.00742.x.

53. Owen-Smith N (1987) Pleistocene Extinctions: The Pivotal Role of Mega-herbivores. Paleobiology 13: 351–362.

54. Gill JL, Williams JW, Jackson ST, Lininger KB, Robinson GS (2009) Pleistocene megafaunal collapse, novel plant communities, and enhanced fire regimes in North America. Science 326: 1100–1103. doi:10.1126/science.1179504.

55. Thébault E, Fontaine C (2010) Stability of ecological communities and the architecture of mutualistic and trophic networks. Science 329: 853–856. doi:10.1126/science.1188321.

56. Bastolla U, Fortuna MA, Pascual-García A, Ferrera A, Luque B, et al. (2009) The architecture of mutualistic networks minimizes competition and increases biodiversity. Nature 458: 1018–1020. doi:10.1038/nature07950.

57. Aizen MA, Morales CL, Morales JM (2008) Invasive mutualists erode native pollination webs. PLoS biology 6: e31. doi:10.1371/journal.pbio.0060031.

58. Sahney S, Benton MJ (2008) Recovery from the most profound mass extinction of all time. Proceedings of the Royal Society B: Biological Sciences 275: 759–765. doi:10.1098/rspb.2007.1370.

59. Estes J, Terborgh J, Brashares JS, Power ME, Berger J, et al. (2011) Trophic downgrading of planet Earth. Science 333: 301–306. doi:10.1126/science.1205106.

60. Voight BF, Adams AM, Frisse LA, Qian Y, Hudson RR, et al. (2005) Interrogating multiple aspects of variation in a full resequencing data set to infer human population size changes. Proceedings of the National Academy of Sciences of the United States of America 102: 18508–18513. doi:10.1073/pnas.0507325102.

61. De Visser SN, Freymann BP, Olff H (2011) The Serengeti food web: empirical quantification and analysis of topological changes under increasing human impact. The Journal of Animal Ecology 80: 484–494. doi:10.1111/j.1365-2656.2010.01787.x.

62. Brook BW, Sodhi NS, Bradshaw CJA (2008) Synergies among extinction drivers under global change. Trends in Ecology & Evolution 23: 453–460. doi:10.1016/j.tree.2008.03.011.

63. Zarnetske PL, Skelly DK, Urban MC (2012) Ecology. Biotic multipliers of climate change. Science 336: 1516–1518. doi:10.1126/science.1222732.

17

Origin and Expansion of the Yunnan Shoot Borer, *Tomicus yunnanensis* (Coleoptera: Scolytinae): A Mixture of Historical Natural Expansion and Contemporary Human-Mediated Relocation

Jun Lü[1,2❃], Shao-ji Hu[1,2❃], Xue-yu Ma[1,3], Jin-min Chen[4], Qing-qing Li[5], Hui Ye[1,2]*

1 Laboratory of Biological Invasion and Ecosecurity, Yunnan University, Kunming, 650091, China, **2** Yunnan Key Laboratory of International Rivers and Transboundary Eco-security, Yunnan University, Kunming, 650091, China, **3** School of Mathematics and Computer Science, Yunnan University of Nationalities, Kunming, 650031, China, **4** Laboratory for Conservation and Utilization of Bio-resources, Yunnan University, Kunming, 650091, China, **5** Life Science College, Yunnan Normal University, Kunming, 650092, China

Abstract

The Yunnan shoot borer, *Tomicus yunnanensis*, is a recently-discovered, aggressive pest of the Yunnan pine stands in southwestern China. Despite many bionomics studies and massive controlling efforts, research on its population genetics is extremely limited. The present study, aimed at investigating the origin and dispersal of this important forestry pest, analyzed the population genetic structure and demographic history using a mitochondrial *cox1* gene fragment. Our results showed that *T. yunnanensis* most likely originated from the Central-Yunnan Altiplano, and the divergence time analysis placed the origin approximately 0.72 million-years ago. Host separation and specialization might have caused the speciation of *T. yunnanensis*. Genetic structure analyses identified two population groups, with six populations near the origin area forming one group and the remaining six populations from western and eastern Yunnan and southwestern Sichuan comprising the other. Divergence time analysis placed the split of the two groups at approximately 0.60 million-years ago, and haplotype phylogenetic tree, network, as well as migration rate suggested that populations of the latter group were established via a small number of individuals from the former one. Migration analysis also showed a certain degree of recent expansion from southwestern Sichuan to eastern Yunnan. Our findings implied that *T. yunnanensis* underwent both historical expansion and recent dispersal. The historical expansion may relate to the oscillation of regional climate due to glacial and interglacial periods in the Pleistocene, while human-mediated transportation of pine-wood material might have assisted the relocation and establishment of this pest in novel habitats.

Editor: Wolfgang Arthofer, University of Innsbruck, Austria

Funding: The present study was supported by the National Natural Science Foundation of China (31360183), the Applied Basic Research Foundation of Yunnan Province (2013FA055), and the Applied Basic Research Foundation for Young Scientists of Yunnan Province (2014FD025). The funders had no role in study design, data collection and analysis, decision to publish, or preparation of the manuscript.

Competing Interests: The authors have declared that no competing interests exist.

* Email: yehui@ynu.edu.cn

❃ These authors contributed equally to this work.

Introduction

The Yunnan shoot borer, *Tomicus yunnanensis* Kirkendall & Faccoli (Coleoptera: Curculionidae: Scolytinae), is one of the most aggressive pest species in genus *Tomicus*, which has caused serious annual damage in up to 20,000 ha of the Yunnan pine, *Pinus yunnanensis* Franchet, since the major outbreak in southwestern China in the 1980s [1–3].

For the past two decades, *T. yunnanensis* has been confused with *T. piniperda* Linnaeus due to morphological resemblance, but genetic comparison between the southwestern Chinese population and the Eurasian *T. piniperda* suggested that the former should be treated as a previously undescribed *Tomicus* species [4]. Based on these findings, Kirkendall *et al.* (2008)

analyzed the morphological characters of all known *Tomicus* species and described this beetle for the first time [5].

The Yunnan pine, *P. yunnanensis*, is the only known host of *T. yunnanensis* to date [3]. This pine is the most important silvicultural tree in southwestern China due to its high tolerance to drought [6]. Therefore, the continuous infestation and constant outbreak of *T. yunnanensis* has brought severe damage to the environment in this region [7]. *T. yunnanensis* is endemic to southwestern China, and its distribution range is highly overlapped with the Yunnan pine [3]. The first population outbreaks of this beetle were recorded in the Central-Yunnan altiplano, and then in northern, western, and southern Yunnan in the following years [8–10]. Since 2000, the beetle has been reported in the Yunnan pine stands in southwestern Sichuan (i.e., Liangshan Prefecture) and western Guizhou (i.e., Panxian County) [11,12].

The origin of *T. yunnanensis* and the formation of its current distribution pattern are two interesting questions underpinning the infestation process of this pest. In an attempt to answer these questions, the present study used mitochondrial DNA data to analyze the genetic structure, population relationships, and demographic history of *T. yunnanensis*. The present study will elucidate the origin of the species, and understand its historical expansion routes. Meanwhile, the findings may also benefit the quarantine procedures to prevent further expansion of this pest.

Materials and Methods

Sample Collection

In total, 231 individuals were collected representing 12 populations of *Tomicus yunnanensis* (10 populations from Yunnan and two from Sichuan), with the sample size of the population from Yanshan being the lowest (10 individuals) and that of the population from Xiangyun being the greatest (23 individuals). The geographical coordinates and altitudes of sampling sites were recorded with a Garmin eTrex Vista GPS handset (Version 3.2; Garmin Ltd., Taiwan) (Table 1; Fig. 1).

Adult beetles were captured from egg galleries on trunks of *Pinus yunnanensis* in 2007 to 2012 and instantly preserved in 95% ethanol. In order to avoid sampling closely related individuals, for instance individuals sharing the same parents, at least five pine trees 100 m apart were chosen to collect the beetles, and the samples used in the present study were randomly taken from the mixed sample pool of each population.

An individual of *T. armandii* Li & Zhang and an individual of *T. piniperda* Linnaeus were chosen as outgroup due to their close phylogenetic lineage to *T. yunnanensis* (Ma *et al.* unpublished data). All samples were carefully identified under a Nikon SMZ1500 stereoscope (Nikon, Japan) following Kirkendall *et al.* (2008) [5] and Li *et al.* (2010) [13]. Selected samples were stored at −40°C in the Laboratory of Biological Invasion and Ecosecurity, Yunnan University until DNA extraction.

Ethics Statement

No specific permits were required in the present study for collection of this native forest insect. The authors confirm that the sampling sites were not privately owned or protected. This study did not involve any endangered or protected species.

Genomic DNA Extraction

In order to prevent contamination by parasitoids and fungi, each sample was washed twice with double distilled water, and the antennae, elytra, and abdomen were removed [4]. Samples were individually treated with 1 mL STE solution at room temperature for 24 h prior to DNA extraction to eliminate the residual ethanol and to rehydrate the tissue.

Endosymbiotons such as *Wolbachia* and the nuclear copies of mtDNA fragments (*Numts*) are the most common problems in phylogenetic research using mtDNA markers [14–17]. To minimize the possibility of such pitfall, the mesothoracic muscles containing high density of mitochondria were extracted for DNA extraction to reduce the possibility of obtaining *Numts* in PCR, and a *Wolbachia* search was also carried out to detect possible contamination (see below). Extraction protocol for genomic DNA followed the phenol-chloroform method described by Hu *et al.* (2013) [18]. Product DNA were quantified on an Eppendorf Biophotometer (Eppendorf AG, Hamburg, Germany), and preserved at −40°C in the same laboratory. The ≈100 ng/μL dilutions were used as templates in polymerase chain reactions (PCR).

PCR Amplification and Sequencing

For all samples, a ≈800 bp fragment of the *cox1* gene was amplified by PCR on a Biometra T-Professional Standard thermocycler (Biometra GmbH, Göttingen, Germany). The thermal profile consisted of an initial denaturation at 94°C for 3 min; followed by 35 cycles of denaturation at 94°C for 30 sec, annealing at 47°C for 1 min, and elongation at 72°C for 2 min; then a final elongation at 72°C for 10 min.

The PCR reaction was applied in a 25 μL volume system using the TaKaRa Ex *Taq* kit (TaKaRa Biotechnology Co., Ltd., Dalian, China), which contained 2.5 μL of 10× PCR buffer, 2.5 μL of $MgCl_2$ (25 mmol/L), 4.0 μL of dNTP mixture (2.5 mmol/L each), 0.5 μL of both the forward and reverse primers (20 μmol/L; Shanghai Sangon Biological Engineering Technology & Services Co., Ltd., Shanghai, China), 0.25 μL of Ex *Taq* polymerase (5 U/μL), and 1.0 μL of DNA template. The primers used to amplify the fragment were Cl-J-2183 (5′-CAA CAT TTA TTT TGA TTT TTT GG-3′) and T2-N-3014 (5′-TCC AAT GCA CTA ATC TGC CAT ATT A-3′) [19].

Table 1. Summary information of 12 sampling sites, arranged by ascending order of locality code.

Code	Locality	Coordinate	Alt./m	Time	Size
AN	Yunnan: Anning	24.97, 102.33	1,880	Jan. 2011	21
HL	Sichuan: Huili	26.30, 102.05	2,164	Aug. 2011	19
LL	Yunnan: Luliang	25.15, 103.66	1,883	Jan. 2011	21
MZ	Yunnan: Mengzi	23.46, 103.44	1,891	Jul. 2011	22
NH	Yunnan: Nanhua	25.09, 101.14	1,899	Mar. 2011	13
NL	Yunnan: Ninglang	27.38, 100.86	2,490	Aug. 2007	20
SL	Yunnan: Shilin	24.74, 103.41	1,903	Jan. 2011	20
XC	Sichuan: Xichang	27.49, 102.04	1,700	Aug. 2011	20
XY	Yunnan: Xiangyun	25.37, 100.61	2,257	Mar. 2011	23
YS	Yunnan: Yanshan	23.56, 104.31	1,566	Jul. 2011	10
YX	Yunnan: Yuxi	24.48, 102.59	1,703	Mar. 2011	20
ZY	Yunnan: Zhanyi	25.56, 104.01	2,132	Mar. 2011	22

In order to finally exclude the possibility of obtaining *Wolbachia* genes in the PCR products, a strict *Wolbachia* search was also carried out. Another set of PCR was applied to all samples for the *Wolbachia* surface protein (*wsp*) gene using primers *wsp* 81F (5′-TGG TCC AAT AAG TGA TGA AGA AAC-3′) and *wsp* 691R (5′-AAA AAT TAA ACG CTA CTC CA-3′) [20], and the PCR thermal profile and reaction system followed Braig et al. (1998) [20]. The 1% agarose gel electrophoresis was used to determine the presence and size of the *wsp* gene. The genomic DNA extracted from the abdomen of a *Wolbachia* infected female *Sogatella furcifera* (Horváth) (Hemiptera: Delphacidae) and a tetracycline-treated female *S. furcifera* were used as positive and negative controls respectively.

The PCR products of *cox1* gene were purified using TaKaRa Agarose Gel Purification Kit (version 2.0) (TaKaRa Biotechnology Co., Ltd.) and sequenced by Sangon Biological Engineering Technology & Services Co., Ltd. (Shanghai, China). Sequencing reactions were carried out in both forward and reverse directions on an ABI Prism 3730xl automatic sequencer (Applied Biosystems, Foster City, CA, USA).

Data Analyses

Sequence alignment. Raw sequences were proofread and aligned using Clustal W [21] in BioEdit 7.0.9 [22], and any sequence containing double peaks in the chromatograms was strictly excluded. The product sequences were checked by MEGABLAST against the genomic references (refseq_genomic) and nucleotide collection (nr/nt) in NCBI, as well as conceptual translation using the invertebrate mitochondrial criterion in MEGA 5.1 [23] to detect possible *Numts* (nuclear copies of mtDNA fragments). Also, a search for nonsynonymous mutations, in-frame stop codons, and indels was carried out to further minimize the existence of cryptic *Numts* [15,17]. Number of polymorphic sites and nucleotide composition were analyzed in MEGA 5.1. Haplotypes were defined by DnaSP 5.0 [24], and the

ratios of shared and private haplotypes were calculated respectively.

Genetic distance and diversity indices. Genetic distances and standard errors between populations were calculated with Kimura's two-parameter (K2P) model [25] in MEGA 5.1 with 1,000 iterations. Nei's average number of pairwise differences between populations [26], Nei's average number of pairwise differences within populations, pairwise Fst values, haplotype diversity (H) and nucleotide diversity (π) of each population, the degree of gene flow among populations (N_m) were calculated in Arlequin 3.11 [27]. The significance of Nei's average number of pairwise differences between populations and the pairwise Fst values were tested by 1,000 iterations for statistical significance, and the optimal gamma shape used in Arlequin 3.11 was estimated by jModelTest 0.1 [28,29].

Population genetic structure and phylogeny. A multidimensional scaling (MDS) plot [30] was drawn from the K2P distance in SPSS 17.0 (SPSS Inc., Chicago, IL, USA) to analyze the genetic relationship between the 12 populations of *T. yunnanensis*. To analyze the genetic structure of the 12 populations of *T. yunnanensis*, the spatial analysis of molecular variance (SAMOVA) was performed in SAMOVA 1.0 [31] by setting different numbers of groups (K) to the dataset. The optimal K value was selected when the F_{CT} value in the training results reached the plateau and no single population was assigned to any group. In the present study, the range of K was 2–6 in the training. Once the optimal K value was obtained, the result of SAMOVA analysis was tested by the analysis of molecular variance (AMOVA) [32] with 1,000 iterations in Arlequin 3.11. The above-mentioned genetic distances and the N_m value between the population groups were calculated using MEGA 5.1 and Arlequin 3.11. A median-joining haplotype network was constructed using Network 4.6 (Fluxus Engineering Ltd.) and categorized by population groups determined by the SAMOVA analysis.

A phylogenetic tree of haplotypes was reconstructed using the Bayesian Inference (BI) method using MrBayes 3.2.2 [33], with the

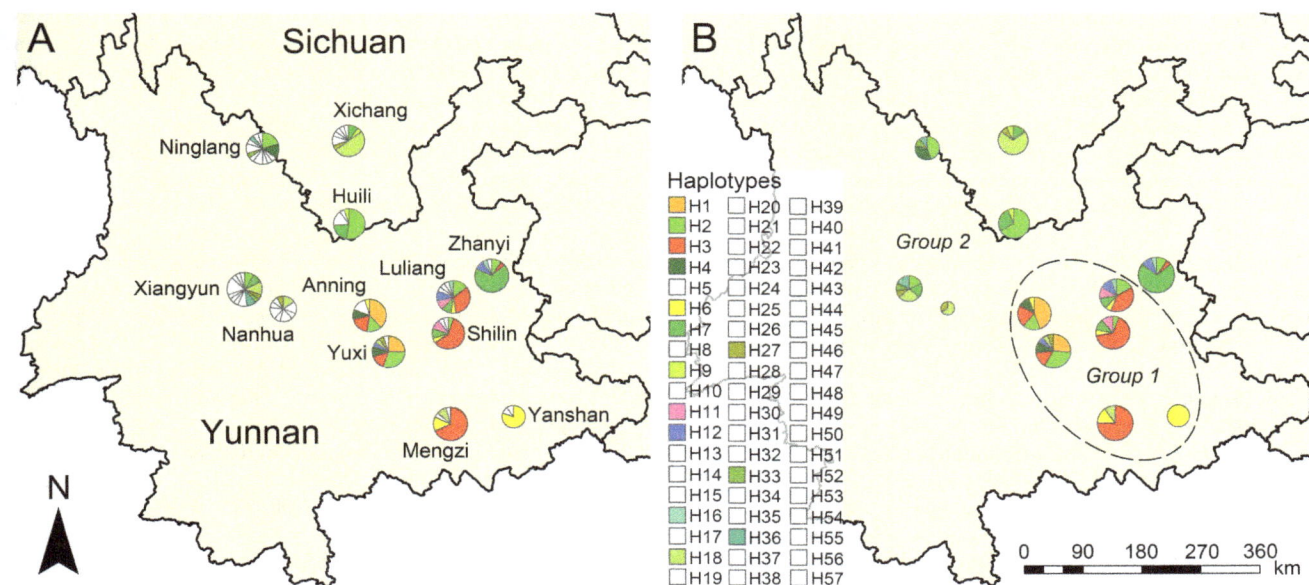

Figure 1. Haplotype distribution of the 12 populations of *T. yunnanensis*, (A) distribution of all haplotypes with private haplotypes marked in white and **(B)** distribution of only shared haplotypes and the tentative boundary (dash line) between the two population groups defined by SAMOVA, the circle sizes corresponding to the sample size of each population.

Table 2. Haplotype distribution and the number of *Wolbachia* infected samples (N_{Wol}) in the 12 populations of *T. yunnanensis*, the private haplotypes were underlined.

Population	Haplotypes	N_{Wol}
Anning	H1, H2, H3, H4, H5, H6	3
Huili	H2, H7, H8, H9, H10	0
Luliang	H2, H3, H6, H7, H11, H12, H13, H14, H15, H16	1
Mengzi	H3, H6, H17, H18, H19	2
Nanhua	H18, H20, H21, H22, H23, H24, H25, H26, H27	0
Ninglang	H2, H4, H28, H29, H30, H31, H32, H33, H34, H35, H36, H37, H38	0
Shilin	H2, H3, H6, H7, H11, H39, H40	1
Xichang	H7, H9, H18, H27, H41, H42, H43, H44, H45	0
Xiangyun	H2, H7, H18, H27, H33, H36, H46, H47, H48, H49, H50, H51, H52, H53	1
Yanshan	H6, H54, H55	0
Yuxi	H1, H2, H3, H4, H12, H18, H27, H56	0
Zhanyi	H2, H3, H7, H12, H16, H57	0

most appropriate nucleotide substitution model estimated by jModelTest 0.1 [28,29]. A 1,500,000 step of MCMC (Markov chain Monte Carlo) replications was applied for the Bayesian Inference in an attempt to obtain an average standard deviation of split frequency below 0.01, and a 25% burn-in criterion was used to summarize the tree [33]. The resultant tree was edited and annotated in FigTree 1.4 [34]. Molecular clock is an effective tool for inferring divergence time when fossil evidence is unavailable [35]. The divergence time was estimated using t_{MRCA} (time to most recent common ancestor) in BEAST 1.7.5 [36] with 10,000,000 steps of MCMC. The substitution rate for the t_{MRCA} inference was calibrated at 1.03% per million years for Polyphaga [37].

Demographic history and isolation-by-distance (IBD). In an attempt to analyze the demographic history of *T. yunnanensis*, the mismatch distributions [38] of each population as well as population groups were calculated in Arlequin 3.11 to detect past population expansion, as long-time stable populations show a multimodal curve while the expanded ones usually generate unimodal curves [39]. Significant sum squared deviations (SSD) and raggedness index were used to reject the rapid expansion model as expanded populations are expected to exhibit smaller values of SSD and raggedness index. Neutrality test of Tajima's D [40] and Fu's F_s [41] were also examined in Arlequin 3.11 with 1,000 iterations to test statistical significance, as the significant negative values indicate genetic hitchhiking, which might be caused by some events in demographic history (i.e., recent expansion, background selection, or genetic sweeping).

The software Migrate 3.5.1 was used to assess the degree of genetic exchange between populations by estimating the migration rates from one population to the another using both of the Bayesian inference and the maximum likelihood methods [42]. For Bayesian inference, three long chains were applied with 100,000,000 steps of MCMC each and 10,000 steps of burn-in; for maximum likelihood, 10 short chains and three long chains were applied with 1,000,000 steps and 100,000,000 steps, respectively. The combination of its analytic result and the previously obtained N_m values can be finally used to determine the extent of genetic exchange between populations.

The isolation-by-distance (IBD) is an effective tool for detecting the correlation between genetic and geographic distance. A null hypothesis was tested using the pairwise genetic differentiation Fst against the pairwise geographic distances (as well as the logarithm transformation) between populations. The approximate spherical distances between populations were measured between coordinates of each sampling locality in Google Earth. The test of IBD was performed using IBDWS 3.23 [43] with 10,000 iterations for Mantel test to determine the correlation between the genetic and geographic distances. The extent of this correlation was then determined by the regression of Fst against the geographic distance and its logarithm transformation.

Results

Sequence Variability and Haplotype Distribution

The proofread of all chromatograms did not find any double peaks. The *Wolbachia* search detected only eight infected samples scattered in five populations, and no sign of correlation between the infection rate and the molecular diversity indices was found (Fig. S1; Table 2). The alignment yielded a 720 bp *cox1* fragment, which correspond to the 2,247–2,966 bp portion in the mitogenome of *Drosophila melanogaster* Meigen. The sequence contained 41 polymorphic sites (5.7%), 28 informative parsimony sites (3.9%), and 13 singleton variation sites (1.8%). No indel and in-frame stop codon were detected. Based on current available analytical tools, no sign of *Numts* was detected. The nucleotide composition showed a high A-T bias of nucleotide usage comprising 69.3% of the total average nucleotide composition.

Fifty-seven haplotypes of the *cox1* sequence were defined based on polymorphic sites and designated in numeral order, among which 14 haplotypes (24.6%) were shared by at least two populations and the remaining 43 haplotypes (75.4%) were locally private. All haplotypes and outgroup were deposited in GenBank under the accession numbers JX448430–JX448477 (H1–H47 and *T. armandii*) and KC986943–KC986953 (H48–H57 and *T. piniperda*).

For each of the 12 populations of *T. yunnanensis*, the number of haplotypes varied greatly, with three haplotypes being the least in the population from Yanshan and 14 being the most abundant in the population from Xiangyun (Table 2; Table 3; Fig. 1). The haplotype distribution showed that 67.4% of the total number of private haplotypes was found in populations from Nanhua,

Table 3. The haplotype diversity (H), nucleotide diversity (π), number of haplotypes (h), ratio of shared haplotypes (R_{shr}), and ratio of private haplotypes (R_{prv}) of the 12 populations of *T. yunnanensis* and the two population groups defined by SAMOVA and the global dataset.

Population	H	π	h	R_{shr}	R_{prv}
Anning	0.805±0.059	0.00585±0.00338	6	0.83	0.17
Huili	0.684±0.092	0.00454±0.00273	5	0.60	0.40
Luliang	0.871±0.057	0.00704±0.00397	10	0.70	0.30
Mengzi	0.528±0.118	0.00334±0.00210	5	0.60	0.40
Nanhua	0.949±0.042	0.00578±0.00345	9	0.22	0.78
Ninglang	0.942±0.034	0.00397±0.00244	13	0.31	0.69
Shilin	0.642±0.118	0.00506±0.00299	7	0.71	0.29
Xichang	0.753±0.094	0.00443±0.00267	9	0.44	0.56
Xiangyun	0.949±0.026	0.00497±0.00292	14	0.43	0.57
Yanshan	0.378±0.181	0.00139±0.00115	3	0.33	0.67
Yuxi	0.847±0.051	0.00605±0.00349	8	0.88	0.13
Zhanyi	0.537±0.123	0.00485±0.00287	6	0.83	0.17
Group 1	0.830±0.027	0.00586±0.00325	22	0.41	0.59
Group 2	0.921±0.015	0.00596±0.00329	43	0.16	0.84
Global	0.922±0.009	0.00726±0.00390	57	0.25	0.75

Ningliang, Xichang, and Xiangyun. Meanwhile, five (62.5%) of the shared haplotypes in populations from Huili, Nanhua, Ninglang, Xichang, Xiangyun, and Zhanyi were also found in populations from Anning, Luliang, Mengzi, Shilin, and Yuxi.

Diversity Indices and Genetic Distances

The haplotype diversity (H) of the 12 populations of *T. yunnanensis* was 0.378–0.949, with the diversity index of the Yanshan population being the lowest (the sample size was also smaller) and that of the populations from Nanhua, Ninglang, and Xiangyun all exceeded 0.94 (Table 3). The nucleotide diversity (π) of the 12 populations varied from 0.00139 to 0.00704, with the index of the Yanshan population being the lowest and that of the Luliang population being the greatest (Table 3). The ratio of shared haplotypes (R_{shr}) also varied among the 12 populations, with that of the population from Nanhua being the lowest and that of the population from Yanshan being the highest (Table 3). The sum of squared deviations (SSD) were not significant in 11 populations but was flagged with 0.01-level significance for the population from Mengzi; however, no significant value of raggedness index was detected among all 12 populations (Table 3).

The Kimura two-parameter (K2P) distances, N_m values, pairwise Fst values, and Nei's average number of differences between populations were shown in Table S1 to S3.

Population Genetic Structure and Phylogeny

The SAMOVA analysis determined two population groups, with Anning, Luliang, Mengzi, Shilin, Yuxi, and Yanshan comprising one group (Group 1) and Huili, Nanhua, Ninglang, Xichang, Xiangyun, and Zhanyi forming the other (Group 2) (Fig. 1B). The AMOVA verification detected 32.16% of the total variation among the two population groups ($F_{CT} = 0.322$, $P < 0.01$), 13.35% among populations within groups ($F_{SC} = 0.197$, $P < 0.01$), and 54.49% within populations ($F_{ST} = 0.455$, $P < 0.01$). For the two groups, genetic variation among populations within Group 2 (24.81%) was greater than that within Group 1 (14.34%) (Table 4). Between the two population groups designated by SAMOVA analysis, the K2P distance was 0.011, pairwise Fst value was 0.344 ($P < 0.001$), Nei's average number of differences was 7.718 ($P < 0.001$), and the N_m value was 0.955. The multidimensional scaling (MDS) plot also divided the 12 populations of *T. yunnanensis* into two groups, with populations from

Table 4. Results of the AMOVA analysis.

Data division	Source of variation	d.f.	Variance component	% of variation	F	P
Global	Among groups	1	1.241 Va	32.16	$F_{CT}=0.322$	0.001
	Among populations within groups	10	0.515 Vb	13.35	$F_{SC}=0.197$	0.000
	Within populations	219	2.103 Vc	54.49	$F_{ST}=0.455$	0.000
Group 1	Among populations within groups	5	0.374 Va	14.34	$F_{ST}=0.143$	0.000
	Within populations	108	2.233 Vb	85.66	—	—
Group 2	Among populations within groups	5	0.653 Va	24.81	$F_{ST}=0.248$	0.000
	Within populations	111	1.978 Vb	75.19	—	—

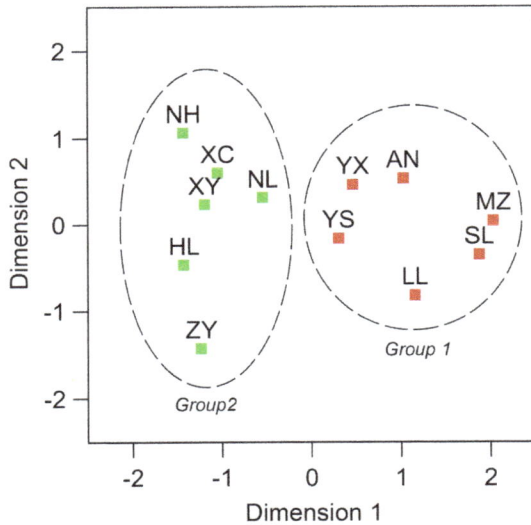

Figure 2. The multidimensional scaling (MDS) plot based on K2P distances of the 12 populations of *T. yunnanensis*, with tentative boundary (dash line) between two population groups.

Anning, Luliang, Mengzi, Shilin, Yanshan, and Yuxi forming one group (Group 1) and that from Huili, Nanhua, Ninglang, Xichang, Xiangyun, and Zhanyi comprising the other (Group 2) (Fig. 2).

The BI phylogenetic tree of the 57 haplotypes and two outgroups showed two somewhat clear clades. Clade a directly connected to the outgroups, was predominately occupied by haplotypes found in populations in Group 1 (marked in red), and clade b mostly represents haplotypes derived from populations in Group 2 (marked in green). There were seven haplotypes in the BI tree contained individuals from both of the two population groups (marked in blue) (Fig. 3A; Fig. S2). In general, the BI tree showed a trend that the derivation of haplotypes was from populations in Group 1 to those in Group 2. The t_{MRCA} inferred the divergence time between outgroups and clade a as approximately 0.72 ± 0.11 million years before present (Ma BP), and that between clade a and clade b as approximately 0.60 ± 0.10 Ma BP.

The median-joining haplotype network was clearly structured, even with three loops and five median vectors (Fig. 3B). The network showed a clear tendency of evolutionary process from Group 1 to Group 2, and it was noticeable that the middle portion of the network showed a highly admixed pattern of populations from the two population groups and intensively connected private haplotypes, while both ends of this network contained mostly the populations from either group (Fig. 3B).

Demographic History and IBD

Tajima's D values showed positive values in four populations from Anning, Huili, Luliang, and Yuxi, and negative values in the remaining eight populations as well as the global dataset and the populations in both groups. However, among all the negative values, only that of population from Yanshan was flagged a 0.05-level significance ($D = 1.74$, $P<0.05$) (Table 5).

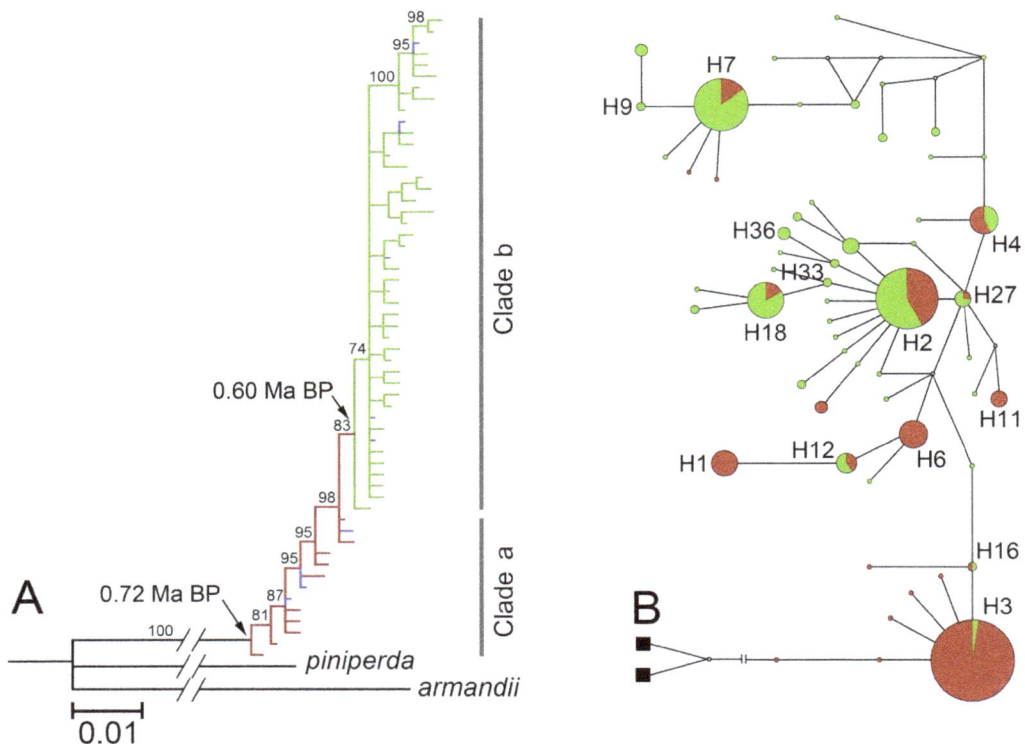

Figure 3. Population genetic structures of *T. yunnanensis*: (A) the Bayesian Inference (BI) tree of 57 haplotypes and outgroups (black terminal) with bootstrap values over 50 (see Fig. S2 for complete haplotype numbers and interior branch bootstrap values) and (B) the median-joining haplotype network denoted with shared haplotypes, the coloration of all three charts were in accordance with the group assignment of both of the SAMOVA analyses, red: Group 1, green: Group 2, blue: admixed.

Table 5. Statistics of neutrality test (Tajima's *D* and Fu's *F$_s$*) and mismatch distribution (SSD and raggedness index), with *P* values in parentheses.

Population	Tajima's *D*	Fu's *F$_s$*	SSD	Raggedness Index
Anning	1.78 (0.969)	2.17 (0.835)	0.048 (0.121)	0.138 (0.069)
Huili	2.10 (0.990)	2.09 (0.868)	0.083 (0.094)	0.166 (0.144)
Luliang	0.51 (0.715)	−0.65 (0.387)	0.018 (0.426)	0.037 (0.487)
Mengzi	−0.71 (0.276)	1.39 (0.797)	0.347 (0.000)	0.254 (0.975)
Nanhua	−0.03 (0.529)	−2.36 (0.101)	0.036 (0.112)	0.116 (0.064)
Ninglang	−0.56 (0.313)	−7.08 (0.001)	0.001 (0.906)	0.028 (0.762)
Shilin	−0.03 (0.517)	0.60 (0.650)	0.070 (0.327)	0.132 (0.370)
Xichang	−0.71 (0.273)	−1.54 (0.217)	0.051 (0.330)	0.131 (0.275)
Xiangyun	−0.21 (0.467)	−6.00 (0.004)	0.003 (0.727)	0.018 (0.863)
Yanshan	−1.74 (0.019)	0.48 (0.559)	0.039 (0.276)	0.285 (0.570)
Yuxi	0.11 (0.596)	0.31 (0.574)	0.046 (0.133)	0.094 (0.130)
Zhanyi	−0.08 (0.524)	1.61 (0.803)	0.140 (0.081)	0.322 (0.209)
Group 1	−0.41 (0.404)	−4.02 (0.113)	0.020 (0.225)	0.051 (0.127)
Group 2	−0.93 (0.184)	−25.70 (0.000)	0.005 (0.570)	0.016 (0.634)
Global	−0.68 (0.259)	−25.05 (0.000)	0.004 (0.536)	0.015 (0.438)

Fu's *F$_s$* values showed positive values in seven populations of *T. yunnanensis*, from Anning, Huili, Mengzi, Shilin, Yanshan, Yuxi, and Zhanyi. The values showed negative deviation in populations from Luliang, Nanhua, Ninglang, Xichang, and Xiangyun, and significant negative values were detected in populations from Ninglang ($F_s = -7.08$, $P<0.001$) and Xiangyun ($F_s = -6.00$, $P<0.01$). Neutrality tests showed significantly large negative values in the global dataset of the 12 populations ($F_s = -25.05$, $P<0.001$) and that of populations in Groups 2 ($F_s = -25.70$, $P<0.001$), while the test did not produce significant value of populations in Group 1 ($F_s = -4.02$, $P>0.1$) (Table 5).

Most of the sum squared deviation (SSD) were not significant except for that of the population from Mengzi (SSD $= 0.35$, $P<0.001$), and all raggedness indices were not significant in our analysis (Table 5).

The mismatch distribution analysis of the 12 populations of *T. yunnanensis* indicated multimodal pattern in the curves, but the results of the populations from Ninglang and Xiangyun showed a unimodal pattern (Fig. 4). The mismatch distribution curves of the global dataset of the 12 populations and that of the populations in Group 2 showed multimodal pattern in the observed mismatch distribution, but produced a unimodal pattern in the simulated mismatch distribution (Fig. 4A, C). However, the distribution curve of the populations in Group 1 did not show such a pattern (Fig. 4B).

Migration analysis showed that, with both the Bayesian inference and maximum likelihood methods, the effective population size (*θ*) of Group 2 was greater than that of Group 1 at all confident intervals (CI), and the migration rate (*M*) from Group 1 to Group 2 was all greater than that from Group 2 to Group 1 at all CI (Table 6). Large migration rates (>10,000) were detected between populations from Anning to Yuxi (168,291.0) and Zhanyi (10,976.6), Huili to Zhanyi (87,812.7), Luliang to Zhanyi (35,125.1), Mengzi to Shilin (10,384.4), Xichang to Yuxi (67,316.6), and Zhanyi to Yuxi (44,877.7) (Table S3).

Isolation-by-distance (IBD) was observed for the 12 populations of *T. yunnanensis* (Fig. 5). The correlation between genetic differentiation (*F*st) and the geographic distances was significantly positive (Mantel $r = 0.540$, $P = 0.001$, with 10,000 iterations), and the correlation between *F*st and the logarithm transformed geographic distances was also significantly positive (Mantel $r = 0.528$, $P = 0.000$, with 10,000 iterations). However, no significant positive correlation was found when testing IBD within each of the two population groups using either the geographic distance (for Group 1, Mantel $r = 0.227$, $P = 0.220$; for Group 2, Mantel $r = 0.405$, $P = 0.094$) or the logarithm transformed geographic distance (for Group 1, Mantel $r = 0.301$, $P = 0.146$; for Group 2, Mantel $r = 0.346$, $P = 0.140$).

Discussion

Origin of Tomicus yunnanensis

The distribution area of *T. yunnanensis* is currently confined in the Central-Yunnan Altiplano and the adjacent areas of western Yunnan, southwestern Sichuan, and western Guizhou [3]. This particular low-latitudinal area and the longitudinal range-gorge region in its west portion had become a hot spot of species' divergence and origination as a result of the impaction of the Indian plate and the rising of the Himalayas and the Tibetan Plateau [44,45]. Phylogeny of pine trees inferred that the Yunnan pine, *Pinus yunnanensis*, diverged in 4–3 Ma BP [46], while our analysis suggested that *T. yunnanensis* diverged from its closest ancestor, *T. piniperda*, in 0.72 ± 0.11 Ma BP (Fig. 3A). Moreover, the widespread Eurasian *T. piniperda* feeds on many taxa of Pinaceae [47,48], but *T. yunnanensis* only feeds only on *P. yunnanensis* [3,5]. It is interesting to note that another newly described shoot borer, *T. armandii*, which feeds only on the Armand pine, *P. armandii* Franchet, was also found in this region [13]. Both cases implied that host separation could be the driving force of the origin of these two *Tomicus* species.

The sampling range of the present study covered most of the distribution range of *T. yunnanensis* [3], therefore, the results can be used to infer the divergence time and origin area of this beetle. Our phylogenetic reconstruction showed that a haplotype from Mengzi (H19) directly connected to *T. piniperda* of the outgroups and then produced the shared H3 and a few related private

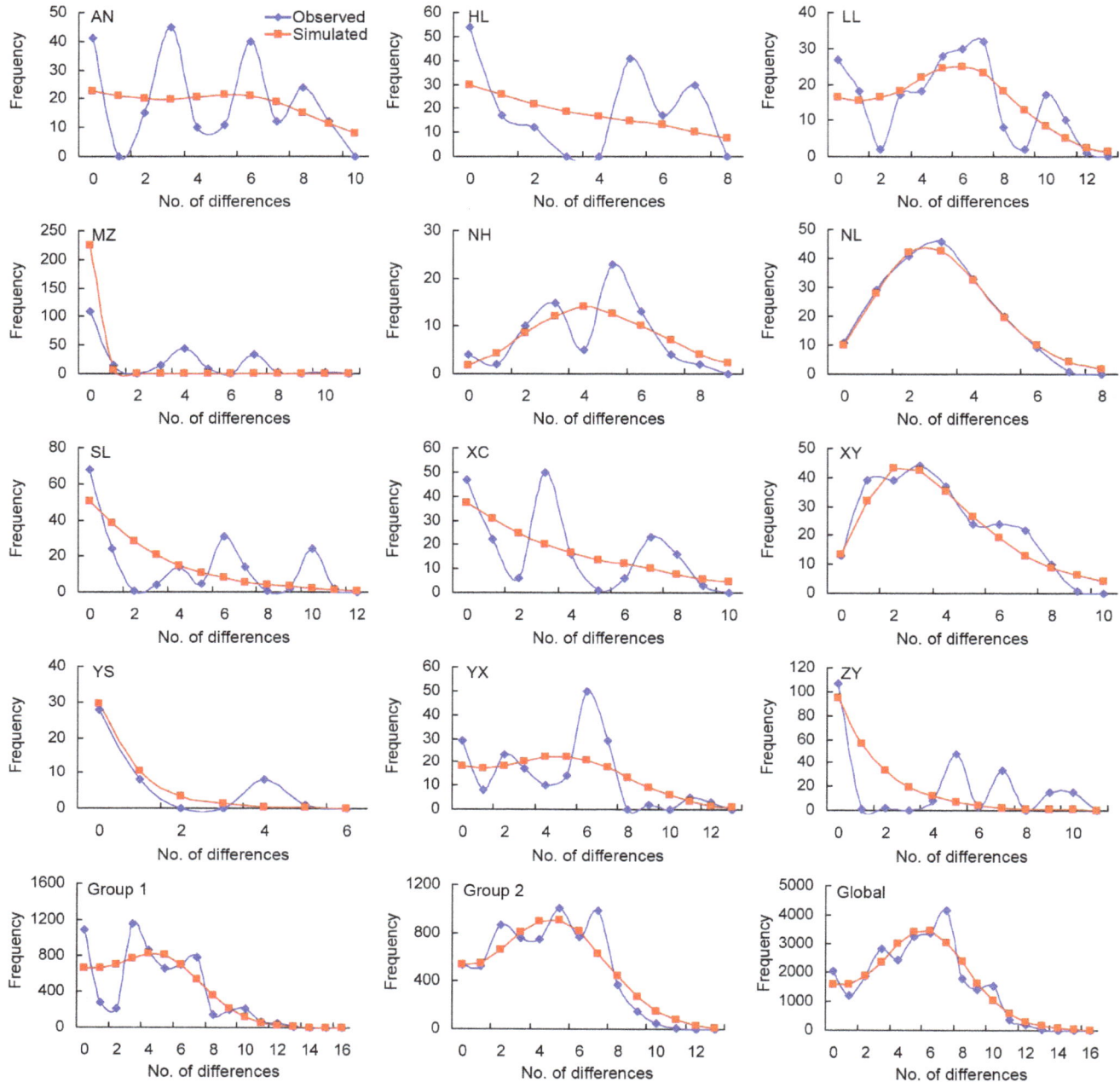

Figure 4. The mismatch distribution of the 12 populations of *T. yunnanensis*, the populations in Group 1, the populations in Group 2, and the global dataset. The blue lines indicated observed mismatch and the red lines indicated the simulation.

haplotypes which are all confined in Central and South Yunnan (Fig. 3A; Fig. S2), indicating the close relationship between these populations and the ancestor. Both the haplotype diversity (*H*) and nucleotide diversity (π) of these populations were greater; also Fu's *F_s* values and the mismatch distribution identified a constant growth pattern over time for these populations (Table 3; Table 5; Fig. 4) [39,41]. Hence, the authors of the present study tend to believe that Central-Yunnan was the likely area of origin for *T. yunnanensis*.

Population Divergence

Providing that there was no *Numts* in the dataset and our *Wolbachia* search did not show correlation between the infection rates and the molecular diversity indices of the 12 populations. The authors of the present research tend to attribute the following discussions to phylogenetic and geographic aspects.

Tomicus yunnanensis has expanded to different localities [3]. During the course of population expansion, genetic divergence occurred between populations due to isolation caused by geographical barriers and fragmented habitats. Our analyses divided the 12 populations into two groups with 32.16% of the

Table 6. Effective population size (θ) and migration rate (M) between the two population groups of $T.$ yunnanensis based on Bayesian inference and maximum likelihood methods.

Parameter	Bayesian Inference				Maximum Likelihood		
	2.5%	Mode	97.5%	Mean	2.5%	MLE	97.5%
θ Group1	0.003	0.006	0.011	0.007	0.004	0.006	0.009
θ Group2	0.014	0.023	0.043	0.027	0.017	0.026	0.034
M Group1→2	241.3	644.3	990.0	555.6	269.5	814.4	1101.8
M Group2→1	0.0	279.0	841.3	415.8	19.0	316.3	706.4

total variance in between ($F_{CT} = 0.322$, $P<0.01$) (Table 4; Fig. 1B), indicating obvious genetic differentiation. Haplotype distribution showed that the shared haplotypes in Group 2 mostly came from part of shared haplotypes in Group 1, and the ratios of private haplotypes (R_{prv}) of populations in Group 2 were evidently greater than that of Group 1 while the nucleotide diversity (π) were much poorer (Fig. 1; Table 3). Therefore, it can be inferred that populations of Group 2 were descendents of some individuals from populations of Group 1.

Haplotypes in Group 1 were directly connected to $T.$ piniperda and distributed near the root of the phylogenetic tree and the haplotype network (Fig. 3A, B), indicating the ancestral position of these populations. In comparison, haplotypes in Group 2 were situated in a more evolved position near the end of the tree and the network (Fig. 3A, B). The topology of phylogenetic tree and haplotype network also indicated that the six populations from western Yunnan, southwestern Sichuan, and Zhanyi came from populations from central Yunnan. The most likely process was that some individuals from central Yunnan populations 'moved' to novel localities and established their descendent populations there (see below).

The limited K2P distances and N_m values (Table S1) coupled with significant pairwise Fst values (Table S2) and significant Mantel test (Fig. 5) indicated obvious genetic differentiation and limited genetic exchange between the two groups. Adult $T.$ yunnanensis live in the truck phloem and the shoots of the host, and the limited flying ability only allows them to range from tree to tree for food and mate rather than long distance migration [49]. In Yunnan, the habitats of $P.$ yunnanensis between 1,500–2,800 m are fragmented and isolated due to mountainous terrain [6], and the fragmented patches of the pine trees added further barriers to the genetic exchange of the beetle. This is the biological and ecological contribution to obvious genetic differentiation between populations of $T.$ yunnanensis, especially between populations from the two groups.

Demographic History

Our phylogenetic analysis divided the haplotypes into two clades, the BI tree showed that most haplotypes of populations in Group 2 fell into clade b, which was evidently the descendant of clade a containing most haplotypes of populations in Group 1, and the haplotype divergence within populations from clade b was very limited (Fig. 3A). Meanwhile, the migration rate from Group 1 to Group 2 was obviously greater than the other way around (Table 6), indicating a history of population expansion of $T.$ yunnanensis after its origination in Central Yunnan. Spatial distribution of haplotypes demonstrated that five shared haplotypes (H2, H4, H7, H18, and H27) in the populations from Huili, Nanhua, Ninglang, Xichang, and Xiangyun all came from the populations in Group 1, while four haplotypes (H2, H7, H18, and H27) were completely shared by the five above-mentioned populations (Fig. 1; Table 2). The reduction trend of haplotypes from the ancestral populations to the descendant populations implied the trace of north- and westward expansion of $T.$ yunnanensis from Central Yunnan in the past.

It must be noted that the five above-mentioned populations in Group 2 contained a relatively higher ratio of private haplotypes compared to most populations in Group 1, while haplotypes such as H1, H3, and H6 which are extensively distributed in ancestral populations in Group 1 were completely absent (Fig. 1; Table 2). The median-joining network showed a missing haplotype (median vector) between the ancestral and descendant haplotypes (Fig. 3B), indicating that $T.$ yunnanensis underwent a certain kind of selection during its course of expansion, which likely resulted in the

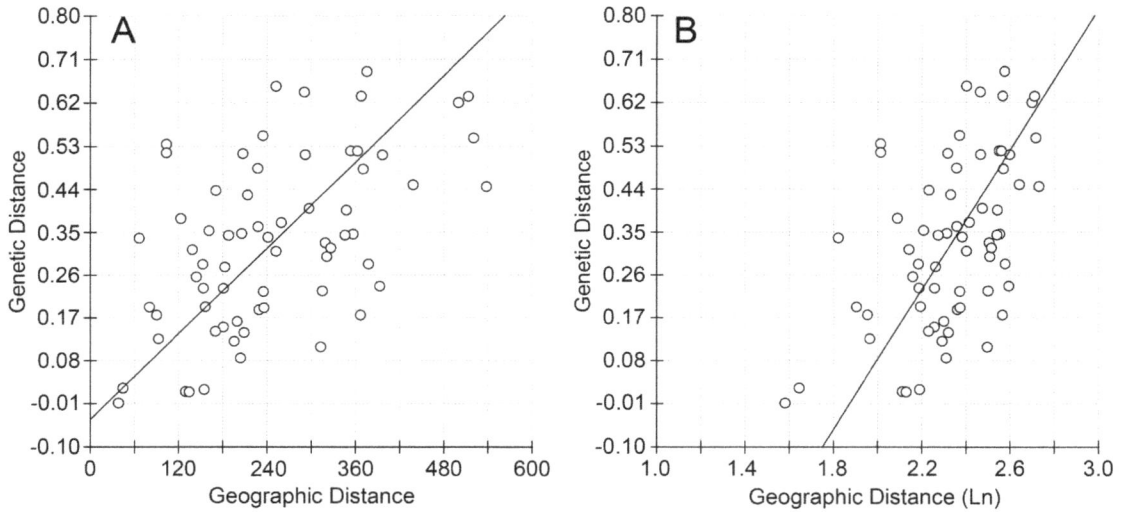

Figure 5. Plots of the isolation-by-distance (IBD) analysis of the 12 population of *T. yunnanensis,* **(A) regression between *F*st and geographic distance (*r*=0.5403, *P*<0.001, with 10,000 iterations), and (B) regression between *F*st and logarithm transformed geographic distance (*r*=0.5283, *P*<0.001, with 10,000 iterations).**

loss of haplotypes. It is generally accepted that the founder's effect, release of selective stress, and recent expansion are three major factors which cause negative deviation of evolutionary neutrality and produce a great number of private haplotypes with a star-like pattern in the network [50,51]. Our neutrality test and mismatch distribution showed that populations like Ninglang and Xiangyun exhibited traces of recent expansion (significant Fu's F_s values, small and insignificant SSD values, and typical unimodal curves of mismatch distribution) (Table 5; Fig. 4). The analyses further implied that recent expansion was not the only explanation of the higher ratio of private haplotypes in the five populations of Group 2.

The five populations are all located in the northern portion of the longitudinal range-gorge region (LRGR) of West Yunnan. Compared to the area of Central and East Yunnan, that particular area has experienced cycles of glacial and interglacial changes in approximately 0.70–0.13 Ma BP [52–56]. The cold climate of glacial periods could favour selections and bottlenecks to *T. yunnanensis*, which would likely reduce the genetic diversity, while the warmer climate of interglacial periods could release the stress and allow *T. yunnanensis* to expand quickly, which would likely produce a great number of private haplotypes due to the founder's effect. Our t_{MRCA} analysis inferred the divergence of Group 1 and Group 2 in approximately 0.60±0.10 Ma BP, when the northern portion of LRGR was experiencing an interglacial period and the climate was warmer, which allowed *T. yunnanensis* to expand into the area, however, *T. yunnanensis* underwent multiple cold and warm climate oscillations due to glacial and interglacial cycles in that area after the entry. Hence, the authors speculated that the five populations of Group 2 have experienced multiple selective processes as a result of regional climate oscillation, during which the loss of ancestral haplotypes and the generation of private haplotypes occurred.

The population from Zhanyi in Group 2 was different from the other five populations; despite its greater ratio of Group 2 component of the shared haplotypes, the ratio of private haplotypes was considerably lower (only one) (Fig. 1; Table 2). Moreover, Zhanyi is located in East Yunnan, where no obvious

climate oscillation occurred in the Pleistocene. Hence, the authors speculated that this particular population should have experienced a different demographic history compared to the other five populations in Group 2. Migration rate analysis showed a large value from Huili to Zhanyi (Table S3), which indicated possible introduction of non-local individuals from Huili to Zhanyi. However, Huili is located 200 km away from Zhanyi with the Wumeng Mountains and the Anning River in between, which constitute geographical barriers. Therefore it is logical to assume that the introduction of *T. yunnanensis* from Huili to Zhanyi should be facilitated by other factors, instead of natural dispersal via the flight of the beetles. Like other borers, *T. yunnanensis* lives in the trunk of its host, which could be easily relocated with untreated wood material [3]. The wood material from the Yunnan pine is commonly used in construction sites, and the transportation of such wood material across the research range could be regarded as a key factor in the relocation of *T. yunnanensis*.

Conclusion and Implications for Pest Management

The present research suggested that *T. yunnanensis* was originated in central Yunnan in more than 0.7 Ma BP, and expanded into northern, western Yunnan and Sichuan in approximately 0.6 Ma BP. Our analysis implied that regional climate oscillation in the Pleistocene might be responsible for the natural expansion of this beetle, it also the main driving force for shaping the different genetic profiles of these populations. However, the divergence time calculation was solely based on estimated molecular clock, thus the results should be applied with discretion. Moreover, our analysis also identified possible contemporary human-mediated relocation of the beetle from Sichuan to Yunnan.

Like many other pine borers, *T. yunnanensis* could be relocated to novel habitats and go through subsequent establishment. The present research inferred that the expansion of *T. yunnanensis* from central Yunnan to other localities in Yunnan and Sichuan occurred during an interglacial period took place and the regional climate was evidently warmer. Providing its capability of enduring and adapting to climate oscillation, *T. yunnanensis* may expand its range further driven by the warming of global climate change.

Fortunately, the occurrence of *T. yunnanensis* is currently confined to the distribution range of the Yunnan pine due to the restriction of host range [3]. Nevertheless, the Yunnan pine is also found in southern Tibet, western Sichuan, and western Guangxi [6], where *T. yunnanensis* has not been reported to date. Hence, quarantine precautions should be established in advance just to prevent the beetle from spreading into these areas.

Supporting Information

Figure S1　Gel images of *Wolbachia* search for 12 populations of *T. yunnanensis*. M: the TaKaRa DL2000 DNA marker, NC: negative control, PC: positive control.

Figure S2　BI phylogenetic tree of 57 haplotypes of *T. yunnanensis* and two outgroups.

Table S1　The Kimura two-parameter (K2P) distances (below diagonal) and the N_m values (above diagonal) between the 12 populations of *T. yunnanensis*.

Table S2　The pairwise F_{st} between populations (below diagonal), Nei's average number of differences within population (diagonal elements), and Nei's average number of differences between populations (above diagonal).

Table S3　Migration rates (migrate from row to column) between the 12 populations of *T. yunnanensis* via maximum likelihood method with 10 short chains consisted of 1,000,000 replications and three long chains consisted of 100,000,000 replications.

Acknowledgments

The authors wish to thank the following personnel for assistance during the course of sample collection, laboratory work, and data analysis: Di-yan Li and Jia-min Li (Yunnan Normal University), Tiao Ning (Laboratory for Animal Genetic Diversity and Evolution of Higher Education in Yunnan Province, Yunnan University), Xue-ying Wang, Li-min Dong, and Yan-ping Yu (Laboratory of Biological Invasion and Ecosecurity, Yunnan University); Andrew Bohonak (Department of Biology, San Diego State University) for interpreting the results of IBD; and Adam M. Cotton (Chiang Mai, Thailand) and two anonymous reviewers for improving the earlier drafts.

Author Contributions

Conceived and designed the experiments: SJH HY. Performed the experiments: JL SJH XYM. Analyzed the data: SJH JMC. Contributed reagents/materials/analysis tools: JL XYM QQL HY. Wrote the paper: SJH HY.

References

1. Ye H, Dang CL (1986) Studies on the feature of the bark beetle (*Tomicus piniperda* L.) injuring Yunnan pine. J Yunnan Univ (Nat Sci Ed) 8: 218–222.
2. Ye H (1992) Approach to the reasons of *Tomicus piniperda* (L.) population epidemics. J Yunnan Univ (Nat Sci Ed) 14: 211–215.
3. Ye H, Lü J, Chen P, Duan YQ, Liao ZY (2012) The Yunnan Shoot Borer (Yunnan Qieshao Xiaodu). Kunming: Yunnan Science and Technology Press.
4. Duan YQ, Kerdelhué C, Ye H, Lieutier F (2004) Genetic study of the forest pest *Tomicus piniperda* (Col., Scolytinae) in Yunnan province (China) compared to Europe: new insights for the systematics and evolution of the genus Tomicus. Heredity 93: 416–422.
5. Kirkendall LR, Faccoli M, Ye H (2008) Description of the Yunnan shoot borer, *Tomicus yunnanensis* Kirkendall & Faccoli sp. n. (Curculionidae, Scolytinae), an unusually aggressive pine shoot beetle from southern China, with a key to the species of *Tomicus*. Zootaxa 1819: 25–39.
6. Jin ZZ, Peng J (2004) The Yunnan Pine. Kunming: Yunnan Science and Technology Press.
7. Ye H (1991) On the bionomy of *Tomicus piniperda* (L.) (Col., Scolytidae) in the Kunming region of China. J Appl Entomol 112: 366–369.
8. Dai XX, Chen P, Huang C, Wu H (2006) Integrated management techniques against Tomicus piniperda in Yanshan County. For Pest Dis 25: 31–34.
9. Rui RJ (2007) Present status of harmful organisms to forestry in Yongsheng County and the corresponding countermeasures. J Southwest For Coll 27: 56–59.
10. He YH (2009) Preliminary report on causes and controls for *Myelophilus piniperda* in Lanping County. For Inventory Plann 34: 98–101.
11. Wang XD, Zhao Q (2005) Studies on the occurrence situation of *Tomicus piniperda* L. and its preventive measures in Liangshan Prefecture. Chinese Agr Sci Bull 21: 276–278.
12. Liu ZZ, Zheng HJ, Zeng XQ, Wan Y, Wu J (2007) General investigation on the harmful organisms in Bijie Area. Guizhou For Sci Technol 35: 57–64.
13. Li X, Zhang Z, Wang HB, Wu W, Cao P, et al. (2010) *Tomicus armandii* Li & Zhang (Curculionidae, Scolytinae), a new pine shoot borer from China. Zootaxa 2572: 57–64.
14. Hurst GDD, Jiggins FM (2005) Problems with mitochondrial DNA as a marker in population, phylogeographic and phylogenetic studies: the effects of inherited symbionts. Proc R Soc B: Biol Sci 272: 1525–1534.
15. Song H, Buhay JE, Whiting MF, Crandall KA (2008) Many species in one: DNA barcoding overestimates the number of species when nuclear mitochondrial pseudogenes are coamplified. PNAS 105: 13486–13491.
16. Werren JH, Baldo L, Clark ME (2008) Wolbachia: master manipulators of invertebrate biology. Nat Rev Micro 6: 741–751.
17. Bertheau C, Schuler H, Krumböck S, Arthofer W, Stauffer C (2011) Hit or miss in phylogenetic analyses: the case of the cryptic NUMTs. Mol Ecol Resour 11: 1056–1059.
18. Hu SJ, Ning T, Fu DY, Haack RA, Zhang Z, et al. (2013) Dispersal of the Japanese Pine Sawyer, *Monochamus alternatus* (Coleoptera: Cerambycidae), in Mainland China as Inferred from Molecular Data and Associations to Indices of Human Activity. PLoS ONE 8: e57568.
19. Simon C, Francesco F, Beckenbach A, Crespi B, Liu H, et al. (1994) Evolution, weighting, and phylogenetic utility of mitochondrial gene sequences and a compilation of conserved polymerase chain reaction primers. Ann Entomol Soc Am 87: 651–701.
20. Braig HR, Zhou WG, Dobson SL, O'Neill SL (1998) Cloning and characterization of a gene encoding the major surface protein of the bacterial endosymbiont *Wolbachia*. J Bacteriol 180: 2373–2378.
21. Thompson JD, Higgins DG, Gibson TJ (1994) CLUSTAL W: improving the sensitivity of progressive multiple sequence alignment through sequence weighting, position-specific gap penalties and weight matrix choice. Nucl Acid Res 22: 4673–4680.
22. Hall TA (1999) BioEdit: a user-friendly biological sequence alignment editor and analysis program for Windows 95/98/NT. Nucl Acid S 41: 95–98.
23. Tamura K, Peterson D, Peterson N, Stecher G, Nei M, et al. (2011) MEGA5: Molecular Evolutionary Genetics Analysis using maximum likelihood, evolutionary distance, and maximum parsimony methods. Mol Biol Evol 28: 2731–2739.
24. Librado P, Rozas J (2009) DnaSP v5: A software for comprehensive analysis of DNA polymorphism data. Bioinformatics 25: 1451–1452.
25. Kimura M (1980) A simple method for estimating evolutionary rates of base substitutions through comparative studies of nucleotide sequences. J Mol Evol 16: 111–120.
26. Nei M, Li WH (1979) Mathematical model for studying genetic variation in terms of restriction endonucleases. PNAS 76: 5269–5273.
27. Excoffier L, Laval G, Schneider S (2005) Arlequin (version 3.0): an integrated software package for population genetics data analysis. Evol Bioinform Online 1: 47–50.
28. Guindon S, Gascuel O (2003) A simple, fast and accurate method to estimate large phylogenies by maximum-likelihood. Syst Biol 53: 696–704.
29. Posada D (2008) jModelTest: phylogenetic model averaging. Mol Biol Evol 25: 1253–1256.
30. Lessa EP (1990) Multidimensional analysis of geographical genetic structure. Syst Zool 39: 242–252.
31. Dupanloup I, Schneider S, Excoffier L (2002) A simulated annealing approach to define the genetic structure of populations. Mol Ecol 11: 2571–2581.
32. Excoffier L, Smouse PE, Quattro JM (1992) Analysis of molecular variance inferred from metric distance among DNA haplotypes: application to human mitochondrial DNA restriction data. Genetics 131: 479–491.
33. Ronquist F, Huelsenbeck J, Teslenko M (2011) Draft MrBayes version 3.2 Manual: Tutorials and model summaries.

34. Rambaut A (2006) FigTree: tree figure drawing tool, version 1.0. Institute of Evolutionary Biology, University of Edinburgh.

35. Zuckerkandl E, Pauling L (1965) Evolutionary divergence and convergence in proteins. In: Bryson V, Vogel HJ, editors. Evolving Genes and Proteins. New York: Academic Press. pp. 97–166.

36. Drummond AJ, Rambaut A (2007) Bayesian evolutionary analysis by sampling trees. BMC Evol Biol 7: 214.

37. Pons J, Ribera I, Bertranpetit J, Balke M (2010) Nucleotide substitution rates for the full set of mitochondrial protein-coding genes in Coleoptera. Mol Phylogenet Evol 56: 798–807.

38. Slatkin M, Hudson RR (1991) Pairwise comparisons of mitochondrial DNA sequences in stable and exponentially growing populations. Genetics 129: 555–562.

39. Rogers AR, Harpending H (1992) Population growth makes waves in the distribution of pairwise genetic differences. Mol Biol Evol 9: 552–569.

40. Tajima F (1989) Statistical method for testing the neutral mutation hypothesis by DNA polymorphism. Genetics 123: 585–595.

41. Fu YX (1997) Statistical test of neutrality of mutations against population growth, hitchhiking and background selection. Genetics 147: 915–925.

42. Beerli P, Felsenstein J (1999) Maximum-likelihood estimation of migration rates and effective populations numbers in two populations using a coalescent approach. Genetics 152: 763–773.

43. Jensen JL, Bohonak AJ, Kelley ST (2005) Isolation by distance, web service. BMC Genetics 6: 13.

44. He DM, Wu SH, Peng H, Yang ZF, Ou XK, et al. (2005) A study of ecosystem changes in Longitudinal Range-Gorge Region and transboundary eco-security in Southwest China. Adv Earth Sci 20: 932–943.

45. Tang ZY, Wang ZH, Cheng CY, Fang JY (2006) Biodiversity in China's mountains. Front Ecol Environ 7: 347–352.

46. Eckert AJ, Hall BD (2006) Phylogeny, historical biogeography, and patterns of diversification for *Pinus* (Pinaceae): phylogenetic tests of fossil-based hypotheses. Mol Phylogenet Evol 40: 166–182.

47. Wood SL, Bright DE (1992) A Catalog of Scolytidae and Platypodidae (Coleoptera), Part 2: Taxonomic Index. Great Basin Nat Memo 13: 1–1553.

48. Långström B (1983) Life cycles and shoot feeding of the pine shoot beetles. Stud For Suecica 163: 1–29.

49. Liu H, Zhang Z, Ye H, Wang H, Clarke SR, et al. (2010) Response of *Tomicus yunnanensis* (Coleoptera: Scolytinae) to infested and uninfested *Pinus yunnanensis* bolts. J Econ Entomol 103: 95–100.

50. Lee CE (2002) Evolutionary genetics of invasive species. Trends Ecol Evol 17: 386–391.

51. Joy DA, Feng XR, Mu JB, Furuya T, Chotivanich K, et al. (2003) Early origin and recent expansion of *Plasmodium falciparum*. Science 300: 318–321.

52. Duan WT, Pu QY, Wu XH (1980) Climatic variations in China during the Quaternary. GeoJournal 4.6: 515–524.

53. Tong GB, Liu ZM, Wang SM, Yang XD, Wang SB (2002) Reconstruction of climatic sequence of the past 1 Ma in the Heqing Basin, Yunnan Province. Quatern Sci 22: 332–339.

54. Yang JQ, Zhang W, Cui ZJ, Yi CL, Liu KX, et al. (2006) Late Pleistocene glaciation of the Diancang and Gongwang Mountains, southeast margin of the Tibetan plateau. Quatern Int 154–155: 52–62.

55. Zhao XT, Zhang YS, Qu YX, Guo CB (2007) Pleistocene glaciations along the western root of the Yulong Mountains and their relationship with the formation and development of the Jinsha River. Quatern Sci 27: 35–44.

56. Yang JQ, Cui ZJ, Yi CL, Sun JM, Yang LR (2007) "Tali Glaciation" on Massif Diancang. Sci China Ser D 50: 1685–1692.

18

Genetic Differentiation, Niche Divergence, and the Origin and Maintenance of the Disjunct Distribution in the Blossomcrown *Anthocephala floriceps* (Trochilidae)

María Lozano-Jaramillo[1], Alejandro Rico-Guevara[2], Carlos Daniel Cadena[1]*

1 Laboratorio de Biología Evolutiva de Vertebrados, Departamento de Ciencias Biológicas, Universidad de los Andes, Bogotá, Colombia, **2** Department of Ecology & Evolutionary Biology, University of Connecticut, Storrs, Connecticut, United States of America

Abstract

Studies of the origin and maintenance of disjunct distributions are of special interest in biogeography. Disjunct distributions can arise following extinction of intermediate populations of a formerly continuous range and later maintained by climatic specialization. We tested hypotheses about how the currently disjunct distribution of the Blossomcrown (*Anthocephala floriceps*), a hummingbird species endemic to Colombia, arose and how is it maintained. By combining molecular data and models of potential historical distributions we evaluated: (1) the timing of separation between the two populations of the species, (2) whether the disjunct distribution could have arisen as a result of fragmentation of a formerly widespread range due to climatic changes, and (3) if the disjunct distribution might be currently maintained by specialization of each population to different climatic conditions. We found that the two populations are reciprocally monophyletic for mitochondrial and nuclear loci, and that their divergence occurred ca. 1.4 million years before present (95% credibility interval 0.7–2.1 mybp). Distribution models based on environmental data show that climate has likely not been suitable for a fully continuous range over the past 130,000 years, but the potential distribution 6,000 ybp was considerably larger than at present. Tests of climatic divergence suggest that significant niche divergence between populations is a likely explanation for the maintenance of their disjunct ranges. However, based on climate the current range of *A. floriceps* could potentially be much larger than it currently is, suggesting other ecological or historical factors have influenced it. Our results showing that the distribution of *A. floriceps* has been discontinous for a long period of time and that populations exhibit different climatic niches have taxonomic and conservation implications.

Editor: William J. Etges, University of Arkansas, United States of America

Funding: The study was financed by the Facultad de Ciencias at the Universidad de los Andes, Bogotá, Colombia. The funders had no role in study design, data collection and analysis, decision to publish, or preparation of the manuscript.

Competing Interests: The authors have declared that no competing interests exist.

* Email: ccadena@uniandes.edu.co

Introduction

The limits of the geographic ranges of populations and species reflect the interplay of a variety of ecological and evolutionary forces such as migration, extinction and speciation [1–3]. Understanding how such forces underlie the origin and maintenance of disjunct distributions, in which closely related taxa or members of the same species occur in widely separate areas, is of central interest in biogeography [4,5]. Hypotheses that may account for the disjunct distributions of species or close relatives include long-distance dispersal or the extinction of intermediate populations of a formerly continuous range, possibly as a result of geographic or climatic events, or human intervention. After disjunct distributions arise, the question becomes how are they maintained. Likely explanations for the maintenance of disjunct distributions are (1) environmental unsuitability of intervening areas and (2) adaptation to different environmental conditions in geographically separate areas [1,6–11].

When historical distributions cannot be studied directly (i.e., using the fossil record), testing hypotheses about the origin of disjunct distributions can be accomplished using molecular phylogenetic estimates of divergence times between populations,

which can be correlated with historical events [12–17]. This approach has provided insights into pervasive biogeographic patterns, such as the disjunct distribution of many organisms occurring in separate continents. For instance, based on the estimated time of lineage divergence, disjunct distributions of organisms occurring in America and Africa has been attributed to the split of Gondwana [17–20], transoceanic dispersal [21–23], human-mediated introductions [13], or various combinations of these processes [24].

Inferences about historical ranges and whether disjunct distributions might be the result of extinction of intermediate populations can also be made using ecological niche-modeling tools [25,26] to generate historical estimates of potential species distributions based on climatic data [27–30]. For example, such models have indicated that some species with currently disjunct distributions may have been widely distributed in the past [29,31]. If currently disjunct populations are relicts of more widespread lineages and one can construct models of the potential distributions at different times in the past, then one would expect to find a reduction in the connectivity between populations through time, with population separation matching the divergence dates estimated using molecular data. In addition, climatic data and statistical analyses based on null models can be used to evaluate the hypothesis that disjunct distributions are maintained at present time as a result of differentiation in climatic preferences between populations found in disjunct areas. Specifically, this hypothesis predicts that disjunct populations occur under different climatic environments as a result of niche divergence and that intervening areas are unsuitable for their occurrence [30].

The Blossomcrown (*Anthocephala floriceps* Gould, 1854), the single representative of a monotypic genus of hummingbird (Trochilidae) endemic to Colombia, is a good model in which to study disjunct distributions: two sedentary subspecies recognized based on plumage variation live in regions separated by more than 900 km (Fig. 1). *Anthocephala floriceps floriceps* is restricted to the foothills and mid elevations of the Sierra Nevada de Santa Marta in northern Colombia (500–1700 m), whereas *A. f. berlepschi* is found in the Andes (1200–2300 m) in Tolima and Huila departments [32–34]. In this study, we used DNA sequence data and niche modeling tools to (1) determine the timing of divergence between the two populations of *A. floriceps*, (2) assess whether the disjunct distribution of the species could have arisen as a result of fragmentation of a formerly widespread range owing to climate change over the Pleistocene, and (3) evaluate whether its disjunct distribution might be maintained by unsuitable intervening areas or specialization of each isolated population to different climatic conditions (niche divergence).

Materials and Methods

Molecular analyses

We used nuclear and mitochondrial DNA sequence data to examine genetic differentiation and to estimate the timing of divergence between populations of *A. floriceps*. These data allowed us to gain insight about factors potentially involved with the origin of their disjunct ranges. We extracted DNA from tissue samples of three museum specimens of *A. f. floriceps* and two of *A. f. berlepschi* (Table 1) using a phenol/chlorophorm protocol [35]. We then amplified and sequenced two mitochondrial (ND2 and ND4) and two nuclear genes (Bfib7 and ODC introns 6 and 7) for all individuals using published primers and protocols [36,37]. We did not estimate gametic phase for the nuclear loci; apparent heterozygosities were coded as ambiguities using IUPAC codes. We combined our data (GenBank accession numbers KJ826445–KJ826464) with sequences of the same genes from three individuals of *A. f. berlepschi* obtained from GenBank (GU167208.1, GU166876, GU167098.1, GU166955.1; Table 1; [38]). As outgroups, we used four of the closest living relatives of *Anthocephala* identified by phylogenetic analyses of the Trochilidae [37,39]. We obtained sequences for the ND2 and ND4 genes of the following outgroups from GenBank: *Campylopterus hemileucurus* (EU042534.1, EU042214.1), *Klais guimeti* (AY830495.1, EU042317.1), *Orthorhyncus cristatus* (AY830508.1, EU042328.1), and *Stephanoxis lalandi* (GU167250.1, GU166919.1).

To estimate the divergence time between the two populations of *A. floriceps*, we constructed a chronogram in BEAST 1.5.2 [40] based on a concatenated matrix including sequences of both mitochondrial genes for the two populations and outgroups. We conducted this analysis using the HKY+G substitution model, which was selected as the best fit to the data according to the Akaike Information Criterion (AIC) in ModelTest 3.7 [41]. To calibrate our tree based on analyses including ND4 data, we used the ND2 substitution rate of 2.5% divergence per million years [42], and related the corrected distances for ND2 with the

distances obtained combining ND2 and ND4 data using a linear regression. Because the slope of the regression was 1.11 ($r^2 = 0.99$), we multiplyed the ND2 per-lineage rate of 0.0125 by 1.11, and fixed the product (0.0139) as the mean rate for calibration. We fitted a relaxed molecular clock with lognormal rate-variation, and ran 50 million generations sampling every 1000 steps and discarding the first 10,000 as burn-in. We used TRACER v1.5 to check that effective sample sizes of parameter estimates were greater than 200. As an additional way to examine relationships among mtDNA and nucDNA sequences, we also constructed haplotype networks (for concatenated mitochondrial data and separately for each nuclear locus) using the median-joining algorithm in the software Network 4.5.1.6 [43].

Ecological niche modeling

We first used ecological niche modeling tools to (1) determine whether areas located in between the two disjunct distribution ranges of *A. floriceps* are unsuitable for its occurrence, and (2) to assess whether the distribution of *A. floriceps* could have been more widespread in the past (i.e., at different periods in the Pleistocene). For these analyses, we used 43 localities obtained from museum specimens ([44], Global Biodiversity Information Facility (GBIF: http://www.gbif.org)), field observations (N. Gutiérrez, pers. comm.), and published data [34]. We characterized each locality with 19 climatic variables at 1 km x 1 km resolution obtained from WorldClim [45]; these variables are commonly used in ecological niche modeling and indicate annual trends, seasonality, and extreme values in temperature and precipitation. We considered all of Colombia and western Venezuela and generated a model of the potential distribution of *A. floriceps* in this area at present using the maximum entropy algorithm implemented in Maxent 3.3.2 [27]. We used default settings to obtain a logistic model output with continuous values ranging from 0 to 100, with higher values indicating greater probabilities of occurrence. Following model-validation using the area under the receiver-operating-characteristic (ROC) curve and a binomial test of omission [27], we projected the model onto climate layers for 6,000 years before present (ybp), the Last Glacial Maximum (LGM; aprox. 21,000 ybp), and 130,000 ybp [45,46]. To distinguish climatically suitable from unsuitable sites, we applied the "fixed cumulative value 10" threshold rule in Maxent [47]. We visually assessed the extent of potential distributions at these different time periods.

We also used ecological niche modeling to evaluate whether the currently disjunct distribution of *A. floriceps* might be maintained by specialization of each population to different climatic conditions. To accomplish this, we first modeled the potential distribution at present of each population separately using the 19 climatic variables. We then projected models generated for each population onto the geographic region where the other population occurs to assess whether each model would classify the localities where the other population has been recorded as climatically suitable (i.e., model interprediction). Low model interprediction would support the hypothesis of climatic specialization maintaining disjunct ranges. However, because the two populations occur in geographically distinct areas where climate may differ considerably irrespective of the presence or absence of the study species, lack of interprediction of distribution models does not necessarily reflect intrinsic niche divergence between populations; populations may have equivalent fundamental niches yet occupy different environments (i.e., different realized niches) due solely to geographic differences in climate [10,48,49]. Thus, we sought to determine whether the environments where populations occurred were more or less similar that expected by chance based on

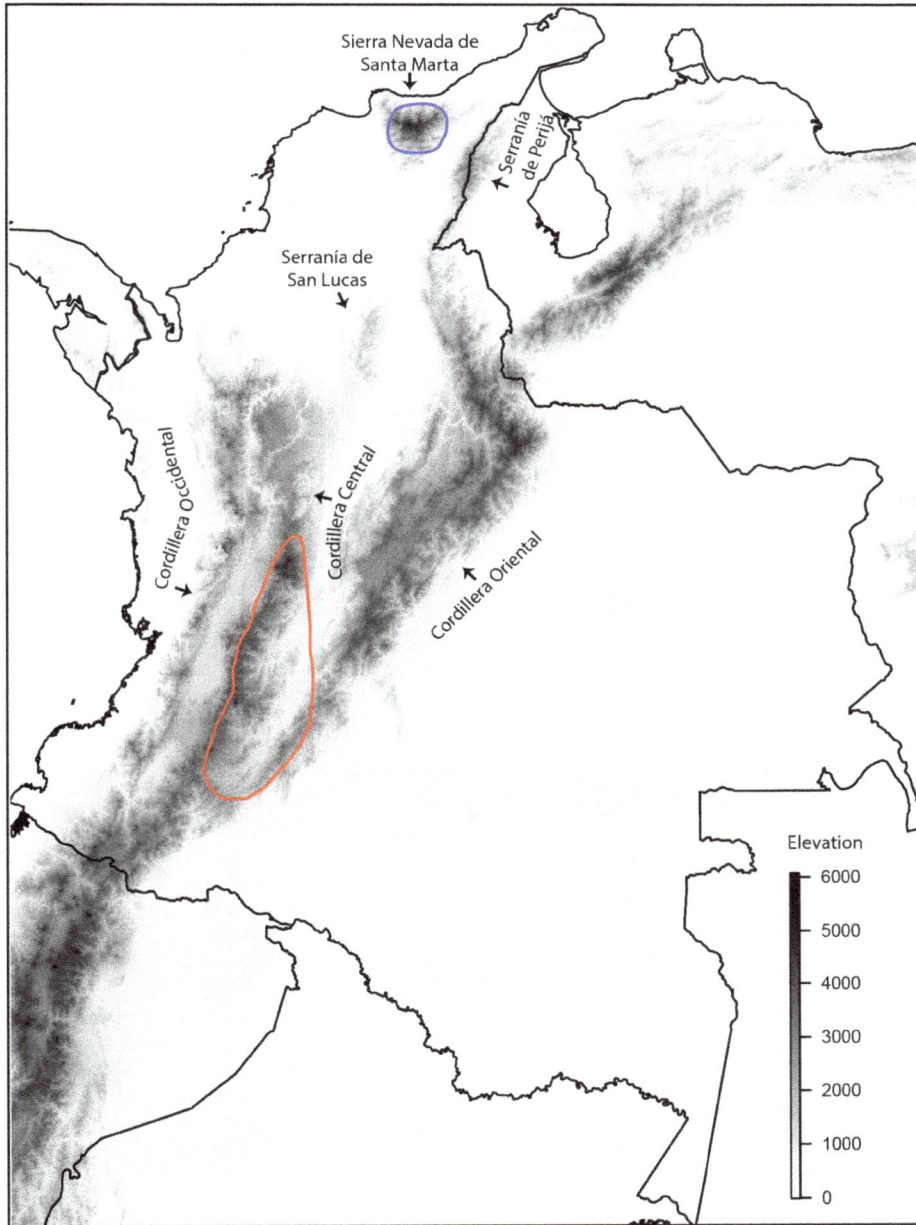

Figure 1. Current distribution of the Blossomcrown (*Anthocephala floriceps*). The blue area corresponds to *A. f. floriceps* from the Sierra Nevada de Santa Marta and the red to *A. f. berlepschi* from the Andes. The locations of different montane regions mentioned in the text are indicated.

differences in the climatic conditions of the regions within which the ranges of each population are embedded. To do so, we examined climatic divergence between populations relative to a null divergence model using the climatic background of the range of each population, an approach that allows for explicit testing of niche divergence vs. niche conservatism [10]. For this analysis we used the 19 WorldClim variables and also elevation; we reduced these 20 variables using a principal component analysis (PCA) and then employed the first four principal components (accounting for c. 97% of the variance, see below) as observed niche values. To establish background variation in climate, we extended polygons depicting the known distribution range of each population of *A. floriceps* [50] 20 km in all directions and randomly placed 1000

points within each expanded polygon. Values for elevation and the 19 climatic variables were extracted for all of these points. Niche divergence and conservatism were assessed by comparing the observed difference in mean niche values to the difference in mean background (i.e., null) values for each of the four principal components. Niche divergence, i.e., specialization to different climates, as a potential factor accounting for the maintenance of disjunct distributions would be supported if population niches were more divergent than expected based on background divergence [10]. Tests were conducted in R version 2.12.2 [51].

Table 1. Specimens of *A. floriceps* included in molecular phylogenetic analyses.

Taxon	Tissue number	Locality
A. f. floriceps	ICN 36492	Santa Marta, Cuchilla de San Lorenzo
A. f. floriceps	ICN 36491	Santa Marta, Cuchilla de San Lorenzo
A. f. floriceps	ICN 36467	Santa Marta, Cuchilla de San Lorenzo
A. f. berlepschi	ANDES-BT 1311	Huila, Algeciras, Vereda Las Brisas, Finca Bélgica
A. f. berlepschi	ANDES-BT 1315	Huila, Algeciras, Vereda Las Brisas, Finca Bélgica
A. f. berlepschi	IAvH 1253	Huila, Palestina, Parque Nacional Natural Cueva de los Guácharos
A. f. berlepschi	IAvH 1269	Huila, Palestina, Parque Nacional Natural Cueva de los Guácharos
A. f. berlepschi	IAvH 1255	Huila, Palestina, Parque Nacional Natural Cueva de los Guácharos

ICN: Instituto de Ciencias Naturales, Universidad Nacional de Colombia; ANDES-BT: Banco de Tejidos, Museo de Historia Natural de la Universidad de los Andes; IAvH: Instituto Alexander von Humboldt.

Results

Molecular analyses

Genealogies showed the same pattern for all genes: each subspecies of *A. floriceps* formed a monophyletic group comprising distinct haplotypes (Fig. 2). Although our sample sizes were low, this pattern suggested the two populations have been isolated for a considerable time span, long enough to have achieved reciprocal monophyly in both mitochondrial and nuclear loci. Furthermore, the chronogram based on mtDNA sequences indicated that the two subspecies were reciprocally monophyletic groups whose divergence dates to c. 1.4 million years before present (mybp; 95% credibility interval 0.7–2.1 mybp; Fig. 3).

Ecological niche modeling

The area under the ROC curve for the model predicting the potential distribution of *A. floriceps* at present was close to one (0.983), indicating it performed substantially better than chance. Additionally, the binomial test of omission was significant (p< 0.001), suggesting that the species' distribution was adequately predicted based on climate. This model suggested that environmental conditions suitable for the occurrence of *A. floriceps* existed well beyond the boundaries of its current range in the Andes (Fig. 4a). This indicates that, based on the climatic variables studied, at least part of the range disjunction cannot be attributed to climatic unsuitability of intervening areas.

Because the model based on climatic data adequately predicted the present-day distribution (i.e., point localities) of *A. floriceps*, assuming niche conservatism one can use such models to examine the potential distribution of the species in the past based on historical climate. None of the historical distribution ranges estimated by the model were sufficiently large suggesting there was potential for the species to be continuously distributed in the past (Fig. 4). However, the potential distribution for 6,000 ybp was considerably larger and more continuous than the potential distribution at present (Fig. 4b); at this time, the Sierra Nevada de Santa Marta appears to have been connected to the northern end of the Cordillera Oriental of the Andes (i.e., Serranía de Perijá) by areas suitable for the presence of *A. floriceps* across the intervening lowlands. Moreover, environments potentially suitable for the species appear to have been more extensively distributed in the northern sector of the Cordillera Central and in the Serranía de San Lucas and surrounding lowlands 6,000 ybp relative to the present. In contrast, much of the area now occupied by *A. floriceps* (including all of the range of *A. f. floriceps* in the Sierra Nevada de Santa Marta) appear to have been unsuitable for the species 21,000 ybp (Fig. 4c). Finally, for 130,000 ybp, the model identified continuous areas of potentially high climatic suitability along the eastern slope of the Cordillera Oriental and extending into lowland areas east of the Andes, but revealed no potential connections between the currently disjunct populations (Fig. 4d).

Potential distribution models constructed separately for each population based on present-day climate also had area under ROC curves close to one (*A. f. floriceps*: 0.981, *A. f. berlepschi*: 0.944). However, the distribution model constructed for each population did not predict the current distribution of the other (Fig. 5), implying that each population inhabits environments with different climatic conditions. This result was supported by tests of niche divergence and conservatism (Table 2). The axis explaining most of the variation (PC1; 40%) was largely associated with elevation and temperature and was the only one revealing significant niche conservatism. The other three axes (jointly accounting for c. 57% of environmental variation) revealed significant niche divergence between populations associated with precipitation and seasonality (Table 2; Table S1). The Andean

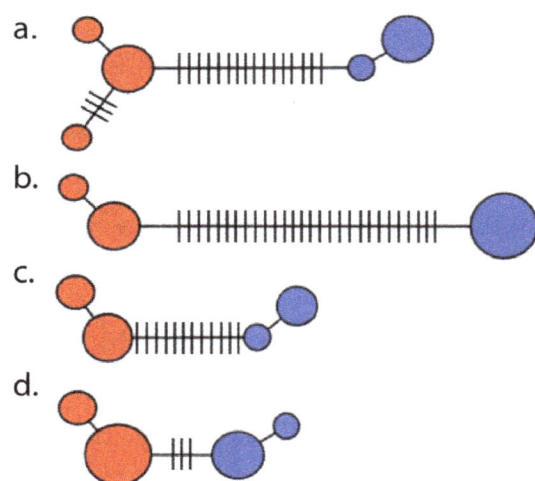

Figure 2. Haplotype networks showing that no alleles are shared between populations of *A. floriceps* in any of the genes analyzed. Blue corresponds to *A. f. floriceps* and red to *A. f. berlepschi*. Circle size is proportional to the number of individuals with each haplotype; hatches indicate mutational steps. (a) ND2, (b) ND4, (c) Bfib7 and (d) ODC.

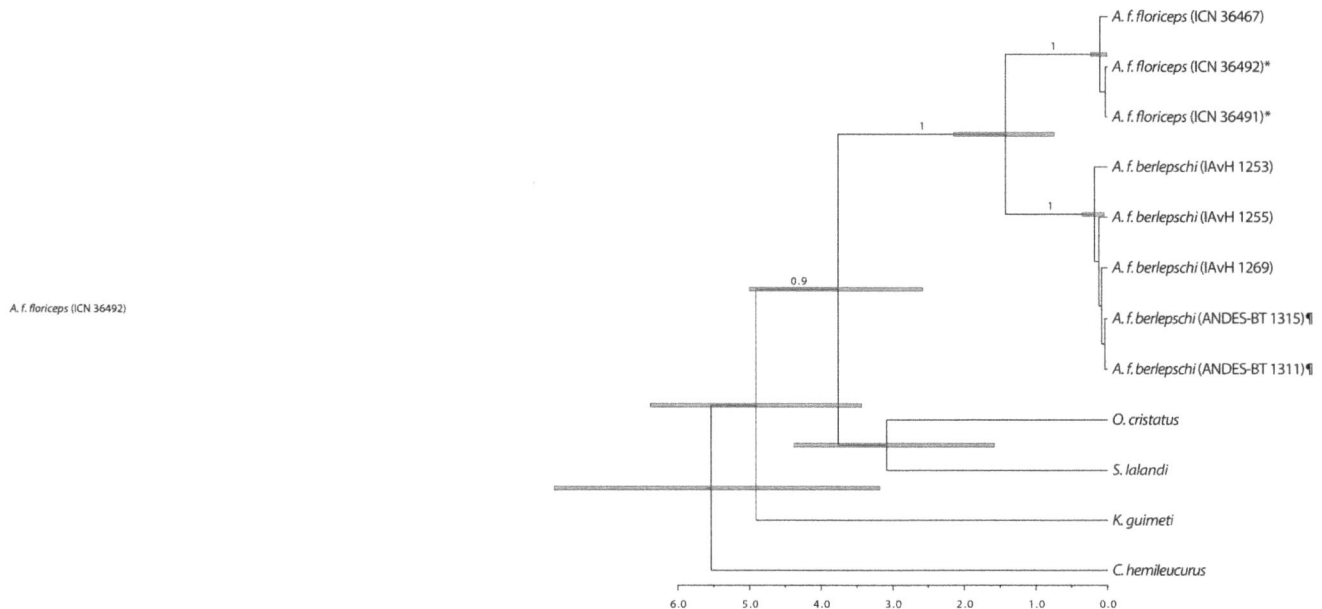

Figure 3. Divergence-time estimates (mya) between populations of *A. floriceps* **and outgroups, based on two mitochondrial genes using a Bayesian relaxed molecular-clock analysis.** Node bars indicate 95% credibility intervals on node ages; scale bar shows time in million years. Values on each clade indicate posterior probabilities when greater than 0.7. Symbols indicate individuals having identical sequences in *A. f. floriceps* (*) and *A. f. berlepschi* (¶).

population lives in less humid and less seasonal environments than the population from the Sierra Nevada de Santa Marta (Fig. S1).

Discussion

The origin and maintenance of the disjunct distribution in *A. floriceps*

Our estimates of potential distributions based on climatic data indicated that in four time periods over the last 130,000 ybp, including the present, climatic conditions have likely not been suitable for *A. floriceps* to have had a fully continuous distribution. The only possible exception to this pattern is the inferred connection between the Sierra Nevada de Santa Marta and the northern stretches of the Cordillera Oriental (Serranía de Perijá) suggested by the predicted potential distribution for 6,000 ybp (Fig. 4b). Also at 6,000 ypb, the species appears to have had a more extensive potential distribution along the Cordillera Central, which may have allowed for connectivity between this mountain range, the Serranía de Perijá and the Sierra Nevada de Santa Marta via the Serranía de San Lucas and surrounding areas, a region in which climatically suitable areas appeared to have been considerably more extensive than at present (Fig. 4). If either scenario is correct, then the species must have gone extinct not only from the lowland environments separating the Sierra Nevada de Santa Marta from the Perijá, but also from the full extent of the Perijá, the Serranía de San Lucas and the Cordillera Oriental, mountain systems where it does not presently exist.

We note, however, that estimates of potential historical distributions based on ecological niche modeling must be considered cautiously because the realized conditions under which species exist at present (i.e., those used to build ecological niche models) may not fully represent their fundamental niches and could lead to potentially misleading reconstructions of their geographic ranges at other times. Especially in scenarios where

combinations of climatic conditions that existed in the past are not equivalent to those existing in the present, i.e., non-analogous climates, models based only on present-day conditions may not accurately estimate historical distributions [52,53]. We suspect this likely applies to our estimate of potential distribution for *A. floriceps* at 21,000 ybp, when its potential range appeared to have been substantially reduced, with no suitable environments in the Sierra Nevada de Santa Marta, the region where one of its present-day populations is endemic (Fig. 4c). Based on patterns of genetic variation indicating marked distinctiveness of the Santa Marta population (see below), that the species was absent from this mountain range at this time and colonized it subsequently seems unlikely.

Because GIS layers depicting estimates of historical climate in our study region are unavailable for dates earlier than those we examined, we cannot address the possibility that the range of *A. floriceps* became disjunct at an earlier moment in history using ecological niche modeling. Can molecular data provide insights about the origin of its disjunct distribution? Our molecular-clock analysis suggests that the divergence between mtDNA clades dates to c 1.4 million mybp (credibility interval 0.7–2.1 mybp), suggesting that divergence occurred prior to the period for which historical climate data are available. However, we note that the inferred timing of divergence reflects gene divergence, which may be considerably older than population/taxon divergence [54,55]. This is a likely possibility considering that Neotropical montane birds often show strong population genetic differentiation even along continuous ranges [56,57].

Our results are consistent with the hypothesis that the currently disjunct distribution of *A. floriceps* may persist due to specialization of each isolated population to different climatic conditions. Ecological niche models suggest that populations of *A. floriceps* are divergent in their climatic niches beyond what one would expect given the climatic background where they exist, implying that a

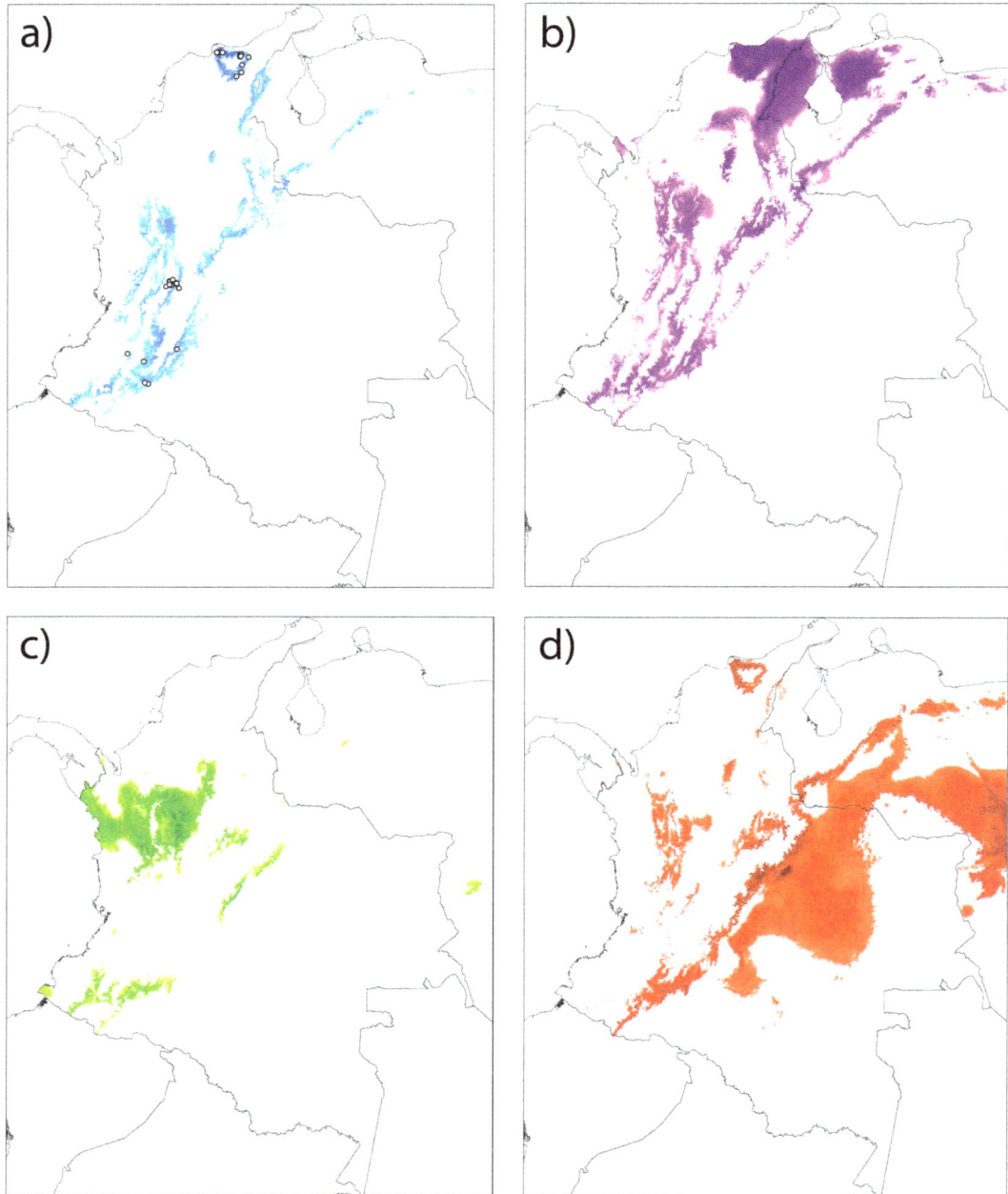

Figure 4. Potential distributions for _A. floriceps_ predicted using climatic data in Maxent. Models are shown for climatic conditions of (a) the present, (b) 6,000 ybp, (c) 21,000 ybp and (d) 130,000 ybp. Dots on the present distribution map indicate localities used to build the models. Darker colors denote areas of greater climatic suitability; areas in white are below the minimum suitability threshold and are therefore considered to be unsuitable.

plausible explanation for the maintenance of their disjunct ranges is climatic niche divergence. It makes sense that both populations exhibit a conserved niche axis related to elevation and temperature because their elevational ranges overlap broadly [34]. However, our analyses revealed significant niche divergence in relation to precipitation and seasonality, with the Andean population occupying less humid and less seasonal environments. If this reflects that each population is adapted to specific climatic conditions and not simply that realized climatic conditions differ between regions but fundamental climatic niches do not, then

climatic restrictions likely do not allow the species' geographic distribution to become fully continuous [10,48,49,58–60].

Although our models failed to reveal continuous potential distributions in the past and at present and populations showed significant climatic divergence, climatic unsuitability of intervening areas and niche divergence between populations are not sufficient explanations for the c. 900-km discontinuity in the present-day range of _A. floriceps_. The modeled potential distribution at present (Fig. 4a) indicates that environmental conditions suitable for its occurrence exist through much of the Cordillera Central of the

Table 2. Divergence on niche axes between populations of *A. floriceps*.

	Niche axes			
	PC1	**PC2**	**PC3**	**PC4**
Pairwise comparison				
A. f. floriceps vs *A. f. berlepshi*	**0.78C**	**1.23 D**	**0.70 D**	**1.37 D**
	(0.58, 1.26)	(0.30, 1.55)	(0.63, 0.89)	(1.13, 2.31)
Variance explained (%)	40%	24%	22%	11%
Top four variable loadings	elevation*	bio16	bio17*	bio3*
	bio6	bio13	bio14*	bio14
	bio11	bio12	bio12*	bio15
	bio10	bio18	bio18	bio17*

Instances of significant niche divergence (D) or conservatism (C) are shown in bold (t-test; $p < 0.05$). Values in parentheses represent the 95% confidence intervals of the null distributions based on background divergence between the geographic ranges of each population. For each niche axis, the top four environmental variables loading on it are shown (asterisks indicate opposite sign). bio3 = isothermality, bio6 = minimum temperature of coldest month, bio10 = mean temperature of warmest quarter, bio 11 = mean temperature of coldest quarter, bio12 = annual precipitation, bio13 = precipitation of wettest month, bio14 = precipitation of driest month, bio15 = precipitation seasonality, bio16 = precipitation of wettest quarter, bio17 = precipitation of driest quarter, bio18 = precipitation of warmest quarter. For full results of principal components analysis see Fig. S1.

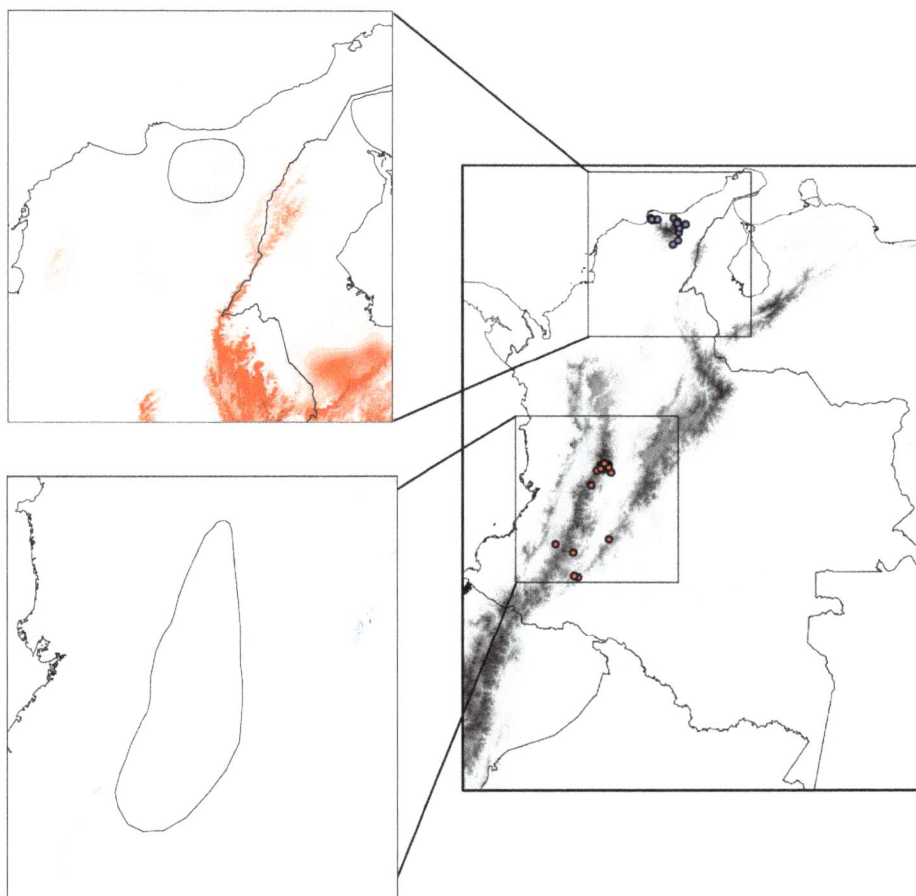

Figure 5. Model of potential distribution constructed based on localities of *A. f. berlepschi* **projected onto the region where** *A. f. floriceps* **occurs (indicated by a blue shape; (a)).** Model of potential distribution constructed based on localities of *A. f. floriceps* projected onto the region where *A. f. berlepschi* occurs (indicated by a red shape; (b)). Red and blue dots indicate localities used to build the models for *A. f. berlepschi* and *A. f. floriceps*, respectively. Darker colors denote areas of greater climatic suitability in a continuous scale (i.e., no cutoff threshold was established in Maxent). Note that localities of each population have low suitability according to the model constructed with data from the other population, indicating niche divergence.

Colombian Andes, a region lacking obvious environmental discontinuities [2]. Also, suitable conditions exist along the western slope of the Cordillera Oriental albeit with some notable environmental breaks (Fig. 4a; [2]). Thus, based on climatic conditions, the distribution range of *A. floriceps* could potentially be larger than it currently is, especially in the Cordillera Central. A similar result was obtained in a recent study examining disjunct populations of Painted Buntings (*Passerina ciris*) in North America, where areas not occupied by the species were found to be potentially suitable for its occurrence [30]. The restricted distribution range of the Andean form *A. f. berlepschi* likely reflects ecological factors not accounted for by climatic variation (e.g., biotic interactions) or historical factors limiting range expansion. The influence of historical factors is likely, considering *A. f. berlepschi* is one of several members of a distinctive assemblage of codistributed taxa restricted to an area of endemism in the departments of Tolima and Huila [8,61,62].

Taxonomic and conservation implications

Our divergence time estimates between populations of *A. floriceps* (1.4 mybp) suggest an older date than the reported divergence times for phylogroups within some Neotropical hummingbird species [60,63–65] and even between several lineages recognized as different species of hummingbirds [66]. Our analyses further showed that subspecies do not share haplotypes in four different genes including nuclear loci, with their four-fold higher coalescence times relative to mtDNA, indicating long-term isolation without gene flow. We realize our sample sizes are not large enough to provide a robust test of reciprocal monophyly, but given the strong divergence and geographic isolation, we suspect our conclusions would be robust to analyses with larger sample sizes.

In conclusion, our data suggest that the current distribution of *A. floriceps* has been disjunct for a relatively long time. Furthermore, each population occurs under distinct climatic conditions, which likely reflects evolved differences in their climatic niche. Our results revealing strong genetic and climatic divergence between populations of *A. floriceps*, together with morphological differences that led to their recognition as different subspecies, arguably have taxonomic implications. The evidence for marked divergence and reciprocal monophyly in mitochondrial and nuclear loci, in addition to climatic differentiation and morphological diagnosability, implies that each population could be considered a full species under several species concepts [67–71]. Applying the criterion of reproductive isolation central to the biological species concept is impossible owing to the allopatric distributions of the two populations, but divergence in several respects between them, relative to divergence between "good" species of hummingbirds [72], may suffice to consider them to be

reproductively isolated [73]. In any event, the likelihood that the two forms may eventually come into contact appears extremely unlikely, so their status as independently evolving units will most likely be maintained and should probably prevail in terms of establishing their taxonomic status [74]. At the very least, our work shows that these populations are divergent lineages meeting the criteria for recognition as evolutionarily significant units worthy of attention from a conservation standpoint and requiring independent management [75,76]. Their distinctiveness has likely been overlooked as a consequence of traditional taxonomy treating them as conspecific, a situation that may apply to several other populations of Neotropical birds with disjunct ranges [77].

Supporting Information

Figure S1   Bivariate plots showing climatic differences between localities occupied by *Anthocephala floriceps floriceps* in the Sierra Nevada de Santa Marta (blue) and *A. f. berlepschi* in the Andes (red). Note that *A. f. berlepschi* occurs in drier areas with more stable temperature and less seasonal precipitation than *A. f. floriceps*.

Table S1   Variables used to characterize the ecological niches of populations of *Anthocephala floriceps* and their loadings on the first four axes obtained following principal components analyses. These four axes accounted for 97% of the variation. The variables with the four highest loadings on each principal component are shown in bold.

Acknowledgments

We thank the Facultad de Ciencias at Universidad de Los Andes for funding and the Instituto de Genética at Universidad de Los Andes for allowing access to their facilities. We thank J. L. Parra for providing DNA sequences and F. G. Stiles for authorizing the use of tissue samples from the Instituto de Ciencias Naturales at Universidad Nacional de Colombia. We thank members of the Laboratorio de Biología Evolutiva de Vertebrados at Universidad de Los Andes, especially A. Morales, P. Pulgarín, N. Gutiérrez, and S. González, for sharing ideas and providing assistance throughout the study. E. Tenorio provided valuable help with analyses and figures. The manuscript was improved thanks to helpful comments by K. Hurme and two anonymous reviewers.

Author Contributions

Conceived and designed the experiments: MLJ CDC. Performed the experiments: MLJ CDC. Analyzed the data: MLJ ARG CDC. Contributed reagents/materials/analysis tools: MLJ CDC. Wrote the paper: MLJ ARG CDC.

References

1. Sexton JP, McIntyre PJ, Angert AL, Rice KJ (2009) Evolution and ecology of species range limits. Annual Review of Ecology, Evolution, and Systematics 40: 415–436.
2. Graham CH, Silva N, Velásquez-Tibatá J (2010) Evaluating the potential causes of range limits of birds of the Colombian Andes. Journal of Biogeography 37: 1863–1875.
3. Gatson JK (2003) The structure and dynamics of geographic ranges. Oxford, UK: Oxford University Press.
4. Donoghue MJ (2011) Bipolar biogeography. Proceedings of the National Academy of Sciences 108: 6341–6342.
5. Brown JH, Lomolino MV (1998) Biogeography. Sunderland, Massachusetts: Sinauer Associates. 305–333 p.
6. Schluter D (2001) Ecology and the origin of species. Trends in Ecology and Evolution 16: 372–380.
7. Wiens JJ (2004) Speciation and ecology revisited: phylogenetic niche conservatism and the origin of species. Evolution 58: 193–197.

8. Cavender-Bares J, Gonzalez-Rodriguez A, Pahlich A, Koehler K, Deacon N (2011) Phylogeography and climatic niche evolution in live oaks (*Quercus* series *Virentes*) from the tropics to the temperate zone. Journal of Biogeography 38: 962–981.
9. Wooten JA, Gibbs HL (2012) Niche divergence and lineage diversification among closely related *Sistrurus* rattlesnakes. Journal of Evolutionary Biology 25: 317–328.
10. McCormack JE, Zellmer AJ, Knowles LL (2010) Does niche divergence accompany allopatric divergence in *Aphelocoma* jays as predicted under ecological speciation?: Insights from tests with niche models. Evolution 64: 1231–1244.
11. Holt RD, Keitt TH (2005) Species' borders: a unifying theme in ecology. Oikos 108: 3–6.
12. Brown JH, Lomolino MV (2000) Concluding remarks: historical perspective and the future of island biogeography theory. Global Ecology and Biogeography 9: 87–92.

13. Carranza S, Arnold EN (2003) Investigating the origin of transoceanic distributions: mtDNA shows *Mabuya* lizards (Reptilia, Scincidae) crossed the Atlantic twice. Systematics and Biodiversity 1: 275–282.

14. Seutin G, Klein N, Ricklefs RE, Bermingham E (1994) Historical biogeography of the bananaquit (*Coereba flaveola*) in the Caribbean region: a mitochondrial DNA assessment. Evolution 48: 1041–1061.

15. Witt JDS, Zemlak RJ, Taylor EB (2011) Phylogeography and the origins of range disjunctions in a north temperate fish, the pygmy whitefish (*Prosopium coulterii*), inferred from mitochondrial and nuclear DNA sequence analysis. Journal of Biogeography 38: 1557–1569.

16. Popp M, Mirré V, Brochmann C (2011) A single Mid-Pleistocene long-distance dispersal by a bird can explain the extreme bipolar disjunction in crowberries (*Empetrum*). Proceedings of the National Academy of Sciences 108: 6520–6525.

17. Michalak I, Zhang LB, Renner SS (2010) Trans-Atlantic, trans-Pacific and trans-Indian Ocean dispersal in the small Gondwanan Laurales family Hernandiaceae. Journal of Biogeography 37: 1214–1226.

18. Conti E, Eriksson T, Schönenberger J, Sytsma KJ, Baum DA (2002) Early Tertiary out-of-India dispersal of Crypteroniaceae: Evidence from phylogeny and molecular dating. Evolution 56: 1931–1942.

19. Renner SS (2005) Relaxed molecular clocks for dating historical plant dispersal events. Trends in Plant Science 10: 550–558.

20. Barker NP, Weston PH, Rutschmann F, Sauquet H (2007) Molecular dating of the 'Gondwanan' plant family Proteaceae is only partially congruent with the timing of the break-up of Gondwana. Journal of Biogeography 34: 2012–2027.

21. de Queiroz A (2005) The resurrection of oceanic dispersal in historical biogeography. Trends in Ecology and Evolution 20: 68–73.

22. Poux C, Chevret P, Huchon D, de Jong WW, Douzery EJP (2006) Arrival and diversification of caviomorph rodents and platyrrhine primates in South America. Systematic Biology 55: 228–244.

23. Vidal N, Azvolinsky A, Cruaud C, Hedges SB (2008) Origin of tropical American burrowing reptiles by transatlantic rafting. Biology Letters 4: 115–118.

24. Gamble T, Bauer AM, Colli GR, Greenbaum E, Jackman TR, et al. (2011) Coming to America: multiple origins of New World geckos. Journal of Evolutionary Biology 24: 231–244.

25. Guisan A, Thuiller W (2005) Predicting species distribution: offering more than simple habitat models. Ecology Letters 8: 993–1009.

26. Peterson AT (2001) Predicting species geographic distributions based on ecological niche modeling. Condor 103: 599–605.

27. Phillips SJ, Anderson RP, Schapire RE (2006) Maximum entropy modeling of species geographic distributions. Ecological Modelling 190: 231–259.

28. Richards CL, Carstens BC, Knowles LL (2007) Distribution modelling and statistical phylogeography: an integrative framework for generating and testing alternative biogeographical hypotheses. Journal of Biogeography 34: 1833–1845.

29. Carstens BC, Richards CL, Crandall K (2009) Integrating coalescent and ecological niche modeling in comparative phylogeography. Evolution 61: 1439–1454.

30. Shipley JR, Contina A, Batbayar N, Bridge ES, Peterson AT, et al. (2013) Niche conservatism and disjunct populations: a case study with Painted Buntings (*Passerina ciris*). Auk 130: 476–486.

31. Powell M, Accad A, Shapcott A (2005) Geographic information system (GIS) predictions of past, present habitat distribution and areas for re-introduction of the endangered subtropical rainforest shrub *Triunia robusta* (Proteaceae) from south-east Queensland Australia. Biological Conservation 123: 165–175.

32. Hilty SL, Brown WL (1986) A guide to the birds of Colombia: Princeton, NJ: Princeton University Press.

33. Strewe R, Navarro C (2003) New distributional records and conservation importance of the San Salvador Valley, Sierra Nevada de Santa Marta, Northern Colombia. Ornitología Colombiana: 29–41.

34. Salaman PGW, Renjifo LM (2002) *Anthocephala floriceps*. in: Renjifo, L. M., A. M. Franco-Maya, J. D. Amaya-Espinel, G. Kattan y B. López-Lanús (eds.). 2002. Libro rojo de aves de Colombia. Serie Libros Rojos de Especies Amenazadas de Colombia. Instituto de Investigación de Recursos Biológicos Alexander von Humboldt y Ministerio del Medio Ambiente. Bogotá, Colombia. 257–259.

35. Sambrook J, Russell DW (2001) Molecular cloning: a laboratory manual. Cold Spring Harbor: Cold Spring Harbor Laboratory.

36. McGuire JA, Witt CC, Altshuler DL, Remsen JV (2007) Phylogenetic systematics and biogeography of hummingbirds: Bayesian and maximum likelihood analyses of partitioned data and selection of an appropriate partitioning strategy. Systematic Biology 56: 837–856.

37. Parra JL, Remsen JV Jr, Alvarez-Rebolledo M, McGuire JA (2009) Molecular phylogenetics of the hummingbird genus *Coeligena*. Molecular Phylogenetics and Evolution 53: 425–434.

38. Graham CH, Parra JL, Rahbek C, McGuire JA (2009) Phylogenetic structure in tropical hummingbird communities. Proceedings of the National Academy of Sciences 106: 19673–19678.

39. McGuire JA, Witt CC, Remsen J Jr, Corl A, Rabosky DL, et al. (2014) Molecular phylogenetics and the diversification of hummingbirds. Current Biology 24: 910–916.

40. Drummond AJ, Rambaut A (2007) BEAST: Bayesian evolutionary analysis by sampling trees. BMC Evolutionary Biology 7: 214.

41. Posada D, Crandall KA (1998) Model Test: testing the model of DNA substitution. Bioinformatics 14: 817–818.

42. Smith BT, Klicka J (2010) The profound influence of the Late Pliocene Panamanian uplift on the exchange, diversification, and distribution of New World birds. Ecography 33: 333–342.

43. Bandelt HJ, Forster P, Röhl A (1999) Median-joining networks for inferring intraspecific phylogenies. Molecular Biology and Evolution 16: 37–48.

44. Darwin Database (2014) Database: BioMap. Available: http://biomap.net/. Accessed 2014 Feb 17.

45. Hijmans RJ, Cameron SE, Parra JL, Jones PG, Javis A (2005) Very high resolution interpolated climate surfaces for global land areas. International Journal of Climatology 25: 1965–1978.

46. Otto-Bliesner BL, Marshall SJ, Overpeck JT, Miller GH, Hu A, et al. (2006) Simulating Arctic climate warmth and icefield retreat in the last interglaciation. Science 311: 1751–1753.

47. Phillips SJ, Dudík M (2008) Modeling of species distributions with Maxent: new extensions and a comprehensive evaluation. Ecography 31: 161–175.

48. Cadena CD, Loiselle BA (2007) Limits to elevational distributions in two species of emberizine finches: disentangling the role of interspecific competition, autoecology, and geographic variation in the environment. Ecography 30: 491–504.

49. Wellenreuther M, Larson KW, Svensson EI (2012) Climatic niche divergence or conservatism? Environmental niches and range limits in ecologically similar damselflies. Ecology 93: 1353–1366.

50. Natureserve (2014) NatureServe Web Service. Arlington, VA. U.S.A. Available http://services.natureserve.org. Accessed 2014 Feb 17.

51. R Development Core Team (2012) R: a language and environment for statistical computing. R Foundation for Statistical Computing, Vienna, Austria. Available: http://www.R-project.org.

52. Veloz SD, Williams JW, Blois JL, He F, Otto-Bliesner B, et al. (2012) No-analog climates and shifting realized niches during the late quaternary: implications for 21st-century predictions by species distribution models. Global Change Biology 18: 1698–1713.

53. Williams JW, Jackson ST (2007) Novel climates, no-analog communities, and ecological surprises. Frontiers in Ecology and the Environment 5: 475–482.

54. Edwards SV, Beerli P (2000) Perspective: gene divergence, population divergence, and the variance in coalescence time in phylogeographic studies. Evolution 54: 1839–1854.

55. Knowles LL (2004) The burgeoning field of statistical phylogeography. Journal of Evolutionary Biology 17: 1–10.

56. Gutiérrez-Pinto N, Cuervo AM, Miranda J, Pérez-Emán JL, Brumfield RT, et al. (2012) Non-monophyly and deep genetic differentiation across low-elevation barriers in a Neotropical montane bird (*Basileuterus tristriatus*; Aves: Parulidae). Molecular Phylogenetics and Evolution 64: 156–165.

57. Valderrama E, Pérez-Emán JL, Brumfield RT, Cuervo AM, Cadena CD (2014) The influence of the complex topography and dynamic history of the montane Neotropics on the evolutionary differentiation of a cloud forest bird (*Premnoplex brunnescens*, Furnariidae). Journal of Biogeography 41: 1533–1546.

58. Warren DL, Glor RE, Turelli M (2008) Environmental niche equivalency versus conservatism: quantitative approaches to niche evolution. Evolution 62: 2868–2883.

59. Ahmadzadeh F, Flecks M, Carretero MA, Böhme W, Ilgaz C, et al. (2013) Rapid lizard radiation lacking niche conservatism: ecological diversification within a complex landscape. Journal of Biogeography 40: 1807–1818.

60. Rodríguez-Gómez F, Gutiérrez-Rodríguez C, Ornelas JF (2013) Genetic, phenotypic and ecological divergence with gene flow at the Isthmus of Tehuantepec: the case of the azure-crowned hummingbird (*Amazilia cyanocephala*). Journal of Biogeography 40: 1360–1373.

61. Krabbe N, Salaman P, Cortés A, Quevedo A, Ortega LA, et al. (2005) A new species of *Scytalopus* tapaculo from the upper Magdalena Valley, Colombia. Bulletin of the British Ornithologists' Club 125: 93–108.

62. Stattersfield AJ, Crosby MJ, Long AJ, Wege DC (1998) Endemic bird areas of the world: priorities for bird conservation. Cambridge, U.K.: BirdLife International.

63. González C, Ornelas J, Gutiérrez-Rodríguez C (2011) Selection and geographic isolation influence hummingbird speciation: genetic, acoustic and morphological divergence in the wedge-tailed sabrewing (*Campylopterus curvipennis*). BMC Evolutionary Biology 11: 1–19.

64. Miller MJ, Lelevier MJ, Bermingham E, Klicka JT, Escalante P, et al. (2011) Phylogeography of the Rufous-tailed Hummingbird (*Amazilia tzacatl*). Condor 113: 806–816.

65. Ornelas JF, González C, los Monteros AE, Rodríguez-Gómez F, García-Feria LM (2014) In and out of Mesoamerica: temporal divergence of *Amazilia* hummingbirds pre-dates the orthodox account of the completion of the Isthmus of Panama. Journal of Biogeography 41: 168–181.

66. Roy MS, Torres-Mura JC, Hertel F (1998) Evolution and history of hummingbirds (Aves: Trochilidae) from the Juan Fernandez Islands, Chile. Ibis 140: 265–273.

67. Baum DA, Shaw KL (1995) Genealogical perspectives of the species problem. In: Hoch PC, Stephenson AG, editors. Experimental and molecular approaches to plant systematics. St. Louis, MO: Missouri Botanical Garden. 289–303.

68. Templeton AR (1989) The meaning of species and speciation: a genetic perspective. In: Otte D, Endler JA, editors. Speciation and its Consequences: Sinauer Associates, Sunderland, Massachusetts. 3–27.

69. De Queiroz K (2005) A unified concept of species and its consequences for the future of taxonomy. Proceedings of the California Academy of Sciences 56: 196–215.
70. De Queiroz K (2007) Species concepts and species delimitation. Systematic Biology 56: 879–886.
71. Cracraft J (1983) Species concepts and speciation analysis. Current Ornithology 1: 159–187.
72. Remsen J (2005) Pattern, process, and rigor meet classification. Auk 122: 403–413.
73. Collar NJ, Salaman P (2013) The taxonomic and conservation status of the *Oxypogon helmetcrests*. Conservación Colombiana 19: 31–38.

74. Cadena CD, Cuervo AM (2010) Molecules, ecology, morphology, and songs in concert: how many species is *Arremon torquatus* (Aves: Emberizidae)? Biological Journal of the Linnean Society 99: 152–176.
75. Moritz C (1994) Defining 'evolutionarily significant units' for conservation. Trends in Ecology and Evolution 9: 373–375.
76. Moritz C (1995) Uses of molecular phylogenetics for conservation. Philosophical Transactions of the Royal Society of London Series B: Biological Sciences 349: 113–118.
77. Laverde-RO, Cadena CD (2014) Taxonomy and conservation: a tale of two tinamou species groups (Tinamidae, *Crypturellus*). Journal of Avian Biology: In press.

19

Wing Shape of Four New Bee Fossils (Hymenoptera: Anthophila) Provides Insights to Bee Evolution

Manuel Dehon[1]*, **Denis Michez**[1], **André Nel**[2], **Michael S. Engel**[3], **Thibaut De Meulemeester**[1,4]

1 Laboratory of Zoology, Research Institute of Biosciences, University of Mons, Mons, Belgium, **2** Département d'entomologie, Muséum National d'Histoire Naturelle, Centre National de la Recherche Scientifique, Unité Mixte de Recherche, Paris, France, **3** Division of Invertebrate Zoology, American Museum of Natural History, New York, New York, United States of America, and Division of Entomology (Paleoentomology), Natural History Museum, and Department of Ecology and Evolutionary Biology, University of Kansas, Lawrence, Kansas, United States of America, **4** Naturalis Biodiversity Center, Leiden, the Netherlands

Abstract

Bees (Anthophila) are one of the major groups of angiosperm-pollinating insects and accordingly are widely studied in both basic and applied research, for which it is essential to have a clear understanding of their phylogeny, and evolutionary history. Direct evidence of bee evolutionary history has been hindered by a dearth of available fossils needed to determine the timing and tempo of their diversification, as well as episodes of extinction. Here we describe four new compression fossils of bees from three different deposits (Miocene of la Cerdanya, Spain; Oligocene of Céreste, France; and Eocene of the Green River Formation, U.S.A.). We assess the similarity of the forewing shape of the new fossils with extant and fossil taxa using geometric morphometrics analyses. Predictive discriminant analyses show that three fossils share similar forewing shapes with the Apidae [one of uncertain tribal placement and perhaps near Euglossini, one definitive bumble bee (Bombini), and one digger bee (Anthophorini)], while one fossil is more similar to the Andrenidae. The corbiculate fossils are described as *Euglossopteryx biesmeijeri* De Meulemeester, Michez, & Engel, gen. nov. sp. nov. (type species of *Euglossopteryx* Dehon & Engel, n. gen.) and *Bombus cerdanyensis* Dehon, De Meulemeester, & Engel, sp. nov. They provide new information on the distribution and timing of particular corbiculate groups, most notably the extension into North America of possible Eocene-Oligocene cooling-induced extinctions. *Protohabropoda pauli* De Meulemeester & Michez, gen. nov. sp. nov. (type species of *Protohabropoda* Dehon & Engel, n. gen.) reinforces previous hypotheses of anthophorine evolution in terms of ecological shifts by the Oligocene from tropical to mesic or xeric habitats. Lastly, a new fossil of the Andreninae, *Andrena antoinei* Michez & De Meulemeester, sp. nov., further documents the presence of the today widespread genus *Andrena* Fabricius in the Late Oligocene of France.

Editor: Guy Smagghe, Ghent University, Belgium

Funding: Funding provided by Naturalis Biodiversity Museum, Leiden, The Netherlands. Represented by Dr. Thibaut and De Meulemeester. The NBC allowed the authors to do a sampling of diverse bee tribes with three submarginal cells. Funding also provided by the Division of Entomology, University of Kansas, Kansas, United States of America. Represented by Prof. Engel. This co-funder provided us one of the four fossils described and analysed: Euglossopteryx biesmeijeri sp. nov.

Competing Interests: The authors have declared that no competing interests exist.

* Email: manuel.dehon@umons.ac.be

Introduction

Bees (Apoidea; Anthophila) constitute a monophyletic group of largely pollenivorous species that rely almost exclusively on flowers for their life cycle [1–3]. The lineage arose from among the carnivorous apoid wasps in the late Early Cretaceous and contemporaneous with the diversification of eudicots [1,2,4–9]. Indeed, the various families of bees all appear to have originated early in the history of the clade with the constituent subfamilies and tribes diversifying at different times and in response to a variety of intrinsic and extrinsic factors [1,9–11], ultimately producing a modern diversity of over 20,000 species [2]. Unfortunately, the paleontological record of bees is scant and 99% of all bee fossils are clustered in Tertiary, with most of the material spanning the Eocene and Miocene epochs [11]. Accordingly there is presently limited direct evidence of the earliest phases of bee evolution and the majority of available fossils

instead provide important insights into more recent phenomena among bee evolution, such as changing patterns of biogeography and responses to Paleogene and Neogene geological events.

Four of the most important deposits with bee fossils are the Eckfeld/Messel oil shales of Germany and extensive Baltic amber deposits from the middle Eocene (~47–44 Myr), the Florissant shale of Colorado from the Eocene-Oligocene boundary (~34 Myr), and Dominican amber from the Early Miocene (~19 Myr) [1,11–22]. These deposits have produced sizeable bee paleofaunas compared to other deposits that produced either single specimens only (e.g. [23–25]) or large series of a limited number of species (e.g. [26–28]). Because bee fossils are relatively scarce, additional records are of significant interest to evaluate the origin of particular groups (albeit younger groups given that most higher clades are of Cretaceous age), re-evaluate hypotheses of relationship, document patterns of extinction and distribution, ascertain paleobiological associations and ancient behaviors, estimate rates of diversification/extinction, and to calibrate accurately molecular phylogenies [1,3,10,11,29].

Herein we provide the description and analysis of four newly recognized compression fossils of bees from the Upper Oligocene of Céreste (Vaucluse, France), Miocene of la Cerdanya (Spain), and Eocene shale of the Green River Formation (Utah, U.S.A.). In order to assess their taxonomic affinity with extant and extinct bee taxa, we used geometric morphometric analysis of forewing shape of 639 specimens from 50 different bee tribes and representing 188 species. Admittedly, this method is a phenetic measure of similarity and cannot determine phylogenetic relationships owing to an inability to distinguish symplesiomorphies from synapomorphies. Nonetheless, such phenetic metrics permit a more accurate determination of gross affinities for such compression fossils whereby wings are often well preserved while other anatomical traits are either incomplete, obscured, or entirely absent (e.g. [18,27,30–32]). Thus, as a first approximation geometric morphometrics provides a robust phenetic estimate of taxonomic identity and is preferable to relying on often outdated species hypotheses (e.g. [21]). In fact, it has been emphasized that systematic melittology is in need of exploring new methods and character systems for both the circumscription of taxa and the recovery of genealogical relationships (e.g. [33,34]), and the methods employed herein are one step in that direction.

Materials and Methods

Description, terminology and repositories

The morphological terminology of the bee wings and bodies follows that of Engel (2001) [1], while the general classification follows that of Michener (2007) [2]. The two fossils from Cereste and the single specimen from Cerdanya are deposited in the Museum National d'Histoire Naturelle (Paris, France), while the fossil from Green River Formation is deposited in University of Kansas Natural History Museum (Lawrence, Kansas, USA). Representative bee specimens were sampled from the following collections: Laboratoire de Zoologie (University of Mons, Belgium); Département d'Entomologie Fonctionnelle et Evolutive (University of Liège, Belgium); Musée Royal de l'Afrique Centrale (Tervuren, Belgium); Musée de l'Institut Royal des Sciences Naturelles (Brussels, Belgium); Natural History Museum (London, United Kingdom); and Naturalis Biodiversity Center (Leiden, Netherlands).

No permit were required for the described study, which complied with all relevant regulations.

Nomenclatural Acts

The electronic edition of this article conforms to the requirements of the amended International Code of Zoological Nomenclature, and hence the new names contained herein are available under that Code from the electronic edition of this article. This published work and the nomenclatural acts it contains have been registered in ZooBank, the online registration system for the ICZN. The ZooBank LSIDs (Life Science Identifiers) can be resolved and the associated information viewed through any standard web browser by appending the LSID to the prefix "http://zoobank.org/". The LSID for this publication is urn:lsid:zoobank.org:pub:E4FE4837-1579-4BC7-BD43-9E08087B0351. The electronic edition of this work was published in a journal with an ISSN, and has been archived and is available from the following digital repositories: PubMed Central, LOCKSS.

Geological settings

The Green River Formation is exposed in a variety of sedimentary basins in northeastern Utah, southern Wyoming, and northwestern Colorado. One such basin is that of the Uinta, which spans between Utah and Colorado and represents a shallow paleolake nestled among mountains that persisted for ~20 Myr (latest Paleocene to Late Eocene). The Parachute Creek Member in this basin mainly formed during the middle Eocene (~47 Myr) [35,36], and is where one of the specimens analyzed herein originated. The outcrops contain many insect compressions in oil shale. The flora found in this section of the basin suggests a tropical to subtropical climate with a distinct dry season [37], and occurred at a paleoelevation of ~1500–2900 m [38].

The Oligocene lacustrine beds ('calcaire de Campagne-Calavon') of Céreste are exposed along the northern margin of the Lubéron mountain and represent a shallow paleolake. These beds were long considered as Rupelian (~30 Myr) [39], but the most recent study of Gregor (2002) [40] rather suggests a Late Oligocene age, the exact age of this formation remains controversial. The vertebrates fossils discovered would suggest a rather semi-arid palaeoenvironment [41], while the rich flora implies a mixed-mesophytic forest [40]. The high diversity and abundance of the entomofauna with thousands of different species belonging to most orders also supports a forested palaeoenvironment surrounding a lake. Moreover, observations from the Bibionidae (Diptera) imply a warmer episode than the latest Oligocene of Aix-en-Provence [42]. The rather reduced aquatic fauna reflects the presence of waters of poor quality rather than a dry climate (A.N. pers. obs.). Apoidea are not frequent in these beds most specimens belong to *Apis* [26].

The Late Miocene lacustrine beds (Vallesian – Turonian, ~10 Myr) of the Spanish Cerdanya are located around the small town of Bellver. They correspond to an association of fine-grained terrigenous sequences and lacustrine diatomites with levels of diagenetic phosphates [43]. They are supposed to correspond to a deep water palaeolake. The palaeoclimate of this mountain lake (~1100 m) was warmer than today [44]. The flora and entomofauna are very abundant and diverse, with lists of the fossil insects currently described given by Peñalver-Molla *et al.* (1999) [45] and Arillo (2001) [46]. Apoidea are rather frequent in these beds, although nearly all specimens belong to the genus *Apis* [26].

Dataset and morphometric analyses

We performed geometric morphometric analyses to assess the taxonomic affinity of the four new fossils. Geometric morphometrics is a core of procedures providing quantification of the global shape of a structure [47–49]. Geometric morphometrics can provide tools in paleontology for diagnosing fossil taxa at different levels, and for estimating their taxonomic affinities with extant taxa [31–32]. Phylogenetic relationships cannot be determined owing to an inability to distinguish symplesiomorphy from synapomorphy, but remains one of the most robust methods for ascertaining gross affinities of fragmentary or problematic fossils (e.g. [18,31,32]). Thus, while taxa can be robustly assigned to higher group any supraspecific taxa established based on geometric morphometrics run the risk of creating paraphyletic groups and therefore should still be tested in future studies by more complete material and cladistic analyses of additional character systems (e.g., non-wing traits).

Taxonomic affinities of the four new fossils discussed herein were based on their forewing shape. Wings have many methodological advantages: 2D structure, rigidity, species specificity, and typically excellent preservation in fossil specimens [50]. Moreover, wing veins and their intersections are unambiguously homologous among bees in taxa with three submarginal cells in their forewings [2,51]. Most importantly, for all of the specimens analyzed herein the wings were preserved flat (i.e., not folded or crumpled) and without postmortem tectonic distortion, thereby permitting meaningful comparison with extant species in the absence of retro-deformation (i.e., retrofitting a distorted wing as preserved back to its presumed living orientation of veins).

Given that all of the new fossil specimens exhibit three submarginal cells, we sampled broadly across tribes of bees and apoid wasps with the same number of submarginal cells, and with a maximum of 20 specimens per tribe. We attempted to maximize the morphological diversity of our dataset by selecting four species per tribe and five specimens per species, when possible. In addition, we included 14 extinct species with well-preserved forewings and robust taxonomic assignments, ideally based on a broad suite of available characters [Apidae: Apinae: *Paleohabropoda oudardi* Michez & Rasmont, 2009 (Anthophorini); *Bombus randeckensis* Wappler & Engel, 2012 (Bombini); *Electrapis meliponoides* (Buttel-Reepen, 1906), *E. krishnorum* (Engel, 2001), *Protobombus basilaris* Engel, 2001, *P. hirsutus* (Cockerell, 1908) and *Thaumastobombus andreniformis* Engel, 2001 (Electrapini); *Melikertes stilbonotus* (Engel, 1998), *Melissites trigona* Engel, 2001 and *Succinapis goeleti* Engel, 2001 (Melikertini); *Anthophorula persephone* Engel, 2012 (Exomalopsini); *Eufriesea melissiflora* (Poinar, 1998) (Euglossini); Halictidae: Halictinae: *Halictus petrefactus* Engel & Peñalver, 2006 (Halictini); *Electrolictus antiquus* Engel, 2001 (Thrinchostomatini)]. We assembled a global reference dataset of 632 female specimens representing 10 families, 21 sub-families, 39 tribes, 117 genera, and 188 species of Anthophila and apoid wasps (Table 1). As two of the four new fossils were attributed to the tribes Bombini and Anthophorini, we also assessed their generic attribution based on the dataset described in Wappler *et al.* (2012) [18] (n = 336 specimens) and Michez *et al.* (2009) [31] (n = 37 specimens) respectively (Tables 2 and 3 respectively).

Forewings were photographed using an Olympus SZH10 microscope coupled with a Nikon D200 camera. Photographs were input to tps-UTILS 1.56 [52]. The left forewing shape of the 632 specimens and of the four new fossils were captured from photographs by digitizing two-dimensional Cartesian coordinates of 18 landmarks placed on the wing veins (Fig. 1) with tps-DIG v2.17 [53]. The 636 landmark configurations of the reference dataset were scaled, translated and rotated against the consensus configuration using the GLS Procrustes superimposition method to remove all of the non-shape differences and to separate the size and shape components of the form [47,54]. The superimposition was performed using R functions of the package "geomorph" [55]. The aligned landmark configurations were projected into the Euclidean space tangent to the curved Kendall's shape space to aid further statistical analyses. The closeness of the tangent space to the curved shape space was tested by calculating the least-squares regression slope and the correlation coefficient between the Procrustes distances in the shape space with the Euclidean distances in the tangent space [56]. This variation amplitude of our dataset was calculated with tps-SMALL v1.25 [57].

As some cross-veins were difficult to define on the body fossil of *Euglossopteryx*, the landmarks positions were checked by three authors (MD, DM and TD) and by performing a superimposition of the right forewing on the left forewing to accentuate the veins. Landmarks coordinates resulting from digitalization of wing shape of Anthophorini [31], Bombini [18], four new fossils and 632 original specimens are available in Tables S1, S2, S3 and S4

respectively. The entomological collections from where the specimens were loaned are available in Table S4.

Validation of shape discrimination at different taxonomic levels

Prior to assignment of the four new fossils, shape variation within the reference dataset and discrimination of the different taxa was assessed by Linear Discriminant Analyses (LDA) of the projected aligned configuration of landmarks. These analyses were performed at suprafamily (i.e. bees *versus* apoid wasps), family, subfamily, and tribal levels as *a priori* grouping by using the software R version 3.0.2 [58]. The effectiveness of the LDA for discriminating taxon was assessed by the percentages of individuals correctly classified to their original taxon (hit-ratio, HR) in a leave-one-out cross-validation procedure based on the posterior probabilities of assignment. Given the observed scores of an "unknown", the posterior probability (pp) equals the probability of the unit to belong to one group compared to all others. The unit is consequently assigned to the group for which the posterior probability is the highest [59].

Assignment of the bee fossils

Taxonomic affinities of the new fossils were first assessed based on their score in the predictive discriminant space of shapes. After superimposition of the 636 landmark configurations (i.e. corresponding to the reference dataset and the four new fossils), aligned coordinates of the 632 specimens from the reference dataset were used to calculate the LDA. A unique superimposition of both the reference dataset and the assigned specimens is sometimes disregarded while it is of primary importance because GLS Procrustes superimposition is sampling dependent. We included a posteriori the four new fossil specimens in the computed LDA space as "unknown" specimens and calculated their score. Assignments of the new fossils were estimated by calculating the Mahalanobis Distance between "unknowns" and group mean of each taxa (Table S5). We also calculated posterior probabilities of assignment to confirm the assignment to one taxon (Table S5). Assignments of the new fossils were performed in four consecutive analyses corresponding to different taxonomic levels of *a priori* grouping: suprafamily (i.e. apoid wasps *versus* bees), family, subfamily, and tribe.

In order to further assess the generic taxonomic affinity of the two new fossils associated with the tribes Bombini and Anthophorini, the reference datasets assembled in Wappler *et al.* (2012) [18] and Michez *et al.* (2009) [31] were used. PCA were computed to visualize shape affinities between the fossils and the genera in question.

Results

Morphometric analysis

The regression coefficient between the Procrustes distances and the Euclidean distances is close to 1 (0.9999). This means that the linear tangent space closely approximates the shape space, thereby permitting us to be confident in the variation amplitude of our dataset.

Shape variation within the reference dataset. In the morphometrics space defined by the LDA based on suprafamily *a priori* grouping, the two groups are perfectly discriminated: all 602 bee specimens and the 30 apoid wasp specimens are assigned to their original group by the cross-validation procedure. In the LDA space with family *a priori* grouping, the 10 families are well isolated from each other and only 10 specimens are not assigned to their original group by the cross-validation procedure, accounting

Table 1. Reference data set for the geometric morphometric analysis.

Family	Subfamily	Tribe	N1	N2
Anthophila				
Andrenidae	Andreninae		4	20
	Oxaeinae		5	5
	Panurginae	Melitturgini	3	7
		Protandrenini	2	4
Apidae	Apinae	Ancylaini	4	16
		Anthophorini	8	20+1[†]
		Apini	4	20
		Bombini	5	20+1[†]
		Centridini	5	20
		Electrapini[†]	5	6
		Emphorini	8	20
		Ericrocidini	6	20
		Eucerini	4	20
		Euglossini	5	20+1[†]
		Exomalopsini	3	5+1[†]
		Isepeolini	1	1
		Melectini	6	20
		Melikertini[†]	3	3
		Osirini	3	7
		Protepeolini	2	2
		Tapinotaspidini	5	9
		Tetrapediini	3	7
	Nomadinae	Brachynomadini	2	2
		Epeolini	5	9
		Nomadini	4	20
	Xylocopinae	Ceratinini	5	20
		Xylocopini	4	20
Colletidae	Colletinae	Colletini	5	20
		Paracolletini	1	5
	Diphaglossinae	Caupolicanini	7	20
		Diphaglossini	4	2
		Dissoglottini	1	1
Halictidae	Halictinae	Augochlorini	4	20
		Caenohalictini	5	20
		Halictini	6	20+1[†]
		Sphecodini	5	20
		Thrinchostomatini	5	20+1[†]
	Nomiinae		7	20
	Nomioidinae		3	11
	Rophitinae		5	20
Megachilidae	Fideliinae	Fideliini	4	20
Melittidae	Meganomiinae		2	10
	Melittinae	Melittini	2	20
Stenotritidae			2	2
Sphecid Wasp				
Ampulicidae	Ampulicinae	Ampulicini	2	10
Crabronidae	Larrinae		1	5
	Philanthinae		1	5
Sphecidae	Ammophilinae		1	5

Table 1. Cont.

Family	Subfamily	Tribe	N1	N2
	Sphecinae		1	5

This sampling includes 632 specimens from 188 species, 117 genera, 39 tribes, 21 subfamilies, and 10 families of Apoidea. N1 = number of species. N2 = number of specimens.
† = extinct species.

for a global HR of 98.4%. The Ampulicidae, Crabronidae, Halictidae, Sphecidae, and Stenotritidae show a HR of 100%. The Andrenidae, Apidae, Colletidae, Megachilidae, and Melittidae have from a single to three specimens misclassified compared to the original classification, accounting for a minimal HR of 91.7% (Andrenidae). Among the 10 misclassified specimens, nine specimens show poorly supported assignment (i.e., with low posterior probability) and should be considered as dubiously classified. Moreover, the misclassified specimens belong to different species and groups, rejecting the hypothesis of poorly discriminated taxa that could affect the assignment. All of the 14 described fossils are attributed to their original family. As observed at the familial level, subfamily discrimination by LDA was effective, with a cross-validated HR of 97.5% (i.e., 16 misclassified specimens), and 14 of the 21 subfamilies account for a HR of 100% (Table S6). Five subfamilies have a HR between 95% and 99%, and two subfamilies have a HR lower than 95% (i.e., Diphaglossinae and Panurginae, Table S6). The low HR observed in Diphaglossinae and Panurginae are due to the large shape difference observed among tribes of these subfamilies where the intertribal shape differences are as large as subfamilial differences (e.g., among the Melitturgini and Protandrenini). It would be interesting in the future to include further tribes, genera, and species in these subfamilies so as to see if a denser sampling of taxa might alleviate some of these deficiencies. Among the 16 misclassified specimens, nine specimens show poorly supported assignment (i.e., with low posterior probability) and should be considered as dubiously classified. All of the 14 described fossils are

Table 2. Groups sampled for geometric morphometric analyses in Wappler *et al.* (2012).

Taxon	Number of species	Number of specimens
Bombini		
Genus *Bombus*		
Subgenus *Alpigenobombus*	4	13
Subgenus *Alpinobombus*	4	14
Subgenus *Bombias*	2	10
Subgenus *Bombus*	10	19
+*B. (Bombus) randeckensis* †	1	1
Subgenus *Cullumanobombus*	11	16
Subgenus *Kallobombus*	1	9
Subgenus *Megabombus*	16	36
Subgenus *Melanobombus*	7	13
Subgenus *Mendacibombus*	8	14
Subgenus *Orientalibombus*	2	6
Subgenus *Pyrobombus*	11	24
Subgenus *Sibiricobombus*	7	17
Subgenus *Subterraneobombus*	10	32
Subgenus *Thoracobombus*	29	93
Centridini		
Genus *Centris*	1	3
Anthophorini		
Genus *Anthophora*		
Subgenus *Anthophora*	1	3
Genus *Habropoda*	1	3
Genus *Pachymelus*		
Subgenus *Pachymelus*	1	3
Genus *Paleohabropoda* †	1	1

This sampling includes 328 bee specimens from three apine tribes [Bombini (n = 316), Centridini (n = 3), and Anthophorini (n = 9)].

Table 3. Groups sampled for geometric morphometric analyses in Michez *et al.* (2009).

Taxon	Number of sampled specimens
Anthophorini	
Amegilla albigena Lepeletier 1841	3
Am. quadrifasciata de Villers 1789	3
Anthophora aestivalis (Panzer 1801)	3
A. bimaculata (Panzer 1798)	3
A. plumipes (Pallas 1772)	3
A. quadrimaculata (Panzer 1798)	3
Deltoptila elefas (Friese 1917)	3
Elaphropoda moelleri Lieftinck 1966	1
E. percarinata (Cockerell 1930)	1
Habropoda tarsata (Spinola 1838)	3
H. zonatula Smith 1854	2
Habrophorula nubilipennis (Cockerell 1930)	1
Pachymelus ocularis Saussure 1890	3
P. radovae Saussure 1890	1
P. unicolor Saussure 1890	3
Paleohabropoda oudardi Michez & Rasmont 2009	1

This sampling includes 37 specimens representing 16 species and 8 genera of Apinae (Apidae).

attributed to their original subfamily. The 49 groups defined at the tribal level are well discriminated in the LDA, with 32 misclassified specimens, accounting for a global HR of 94.9%. Among the 49 groups, cross-validation cannot be performed on two groups because these each include a single specimen, 31 groups account for a HR of 100%, seven groups show HR between 90% and 99%, and nine groups have a HR lower than 90% (Table S7). Due to sampling size within groups, the HR drastically drops when one or two specimens are misclassified. This is the case for 14 groups among the 16 with HR lower than 100%. Two groups are poorly discriminated in the LDA: Ancylaini and Emphorini. Several described fossils are not attributed to their original tribe, particularly the Electrapini and Melikertini, but this is perhaps owing to the generally plesiomorphic nature of their wing venations. Moreover, some bee tribes (8 tribes with 3 submarginal cells out of 47 reported in [2]) are not included in the analyses and so it is perhaps not surprising that the analyses would have some difficulties at the tribal level for some groups. Nonetheless, the overall results demonstrate a remarkable fidelity in accurately placing particular taxa. Thus, the cross-validation assignments

(Tables S6, S7) allow us to be confident in the group discrimination at most higher taxonomical levels. The specimens' scores along the LDs (for the following taxonomic levels: suprafamily, family, subfamily and tribe) are available in the Table S8.

A posteriori assignment of the fossils. Taxonomic affinities are detailed for each four new fossils. The Green River specimen is assigned to Anthophila (MD = 1.09; pp = 1), to Apidae (MD = 7.43; pp = 0.999), to Apinae (MD = 13.01; pp = 0.999), and to Euglossini (MD = 20.02; pp = 0.999) in the respective LDA. The Cerdanya fossil is assigned to Anthophila (MD = 0.09; pp = 1), to Apidae (MD = 1.70; pp = 0.999), to Apinae (MD = 3.98; pp = 0.999), and to Bombini (MD = 4.82; pp = 1) in the respective LDA. The first fossil from Céreste is assigned to Anthophila (MD = 1.40; pp = 1), to Apidae (MD = 6.97; pp = 0.999), to Apinae (MD = 9.19; pp = 0.999), and to Anthophorini (MD = 12.73; pp = 0.998) in the respective LDA. Lastly, the second fossil from Céreste is assigned to Anthophila (MD = 1.25; pp = 1), to Andrenidae (MD = 5.12; pp = 0.926), and to Andreninae (MD = 6.74; pp = 0.999) in the respective LDA. Given that no

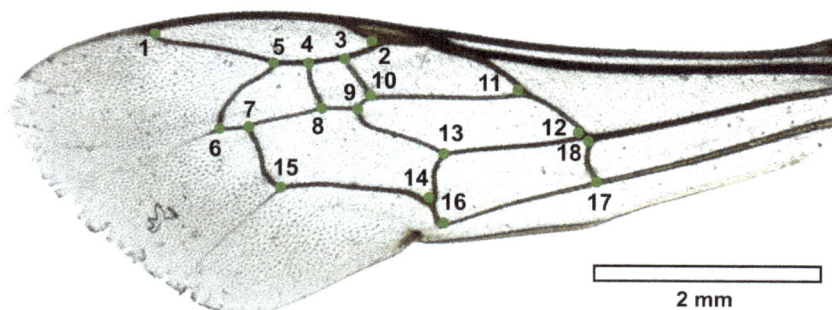

Figure 1. Left forewing of *Melitta leporina* with the 18 landmarks selected to describe the shape.

tribes were defined within the subfamily Andreninae, the "Andreninae" group was used for the tribal a priori grouping. The andrenine fossil was assigned to this group in the fourth LDA (MD = 9.30; pp = 1).

Taxonomic affinities of the bombine fossil from the lacustrine beds of Cerdanya were also assessed based on the reference dataset described in Wappler *et al.* (2012) [18] (Table 3). In the morphometric space defined by the PCA, this fossil is undoubtedly clustered with the genus *Bombus* (Fig. 2a). A-posteriori assignment of the fossil in the discriminant shape space based on the subgenera of *Bombus* does not allow a reliable subgeneric attribution.

Taxonomic affinities of the anthophorine fossil from Céreste were also assessed based on the reference dataset described in Michez *et al.* (2009) [31] (Table 3). In the morphometric space defined by the PCA, this fossil is undoubtedly clustered with the group of the *Habropoda sensu lato*, and more particularly with the genus *Habropoda* (Fig. 2b).

As mentioned previously, all of the specimens have three submarginal cells and therefore likely do not belong to the Xeromelissinae, Hylaeinae, or Euryglossinae (Colletidae), Dasypodainae (Melittidae), Lithurginae and Megachilinae (Megachilidae), and various tribes of Apidae (e.g., Allodapini, Ammobatini, Ammobatoidini, Biastini, Boreallodapini, Caenoprosopidini, Ctenoplectrini, Neolarrini, and Townsendiellini) which have 2 submarginal cells.

The positions of all 4 fossils in the shape space (PCA) and in the discriminant space (LDA) are available at http://beefossil.naturalis.nl/.

Systematic paleontology

Family: Apidae Latreille 1802
Subfamily: Apinae Latreille 1802
Clade: Corbiculata Engel, 1998
Tribe: incertae sedis
Genus *Euglossopteryx Dehon & Engel gen. nov.*

urn:lsid:zoobank.org:act:BC00394F-1FA5-4787-86E8-3D40E210BBB0

Type species. *Euglossopteryx biesmeijeri* De Meulemeester, Michez, & Engel, new species.

Diagnosis. Female of robust body form (body habitus is that of small *Bombus, Melipona, Protobombus,* or *Euglossella*), apparently without metallic coloration (although coloration is poorly pre-

served and may reflect only relative shades of areas as they were in life); metatibia corbiculate, with long fringe of setae along border, not dramatically expanded posteriorly (in this respect differing from non-parasitic Euglossini); forewing with first submarginal cell longest, shorter than combined lengths of second and third submarginal cells; pterostigma apparently relatively small (incompletely preserved but distinctly not the more triangular or elongate forms of some Meliponini; more similar to form observed in Bombini and Euglossini); marginal cell long, closed apically, apex acutely rounded (not open as in Meliponini nor greatly elongate as in Apini); 1m-cu meeting second submarginal cell, not strongly angulate (more distinctly angulate in many Euglossini and Bombini, more similar to some Electrapini or Melikertini in this respect); membrane not infuscate nor papillate (differing in this respect from most Bombini); first abscissa of Rs relatively straight (similar in this respect to many Euglossini, some Electrapini, and some Melikertini).

Etymology. The new genus-group name is a combination of *Euglossa*, type genus of the Euglossini and the tribe to which the fossil showed its greatest affinity, and *pteryx* (Greek, meaning, "wing"), and references to the *Euglossa*-like venation of the fossil. The name is feminine.

Euglossopteryx biesmeijeri De Meulemeester, Michez & Engel sp. nov.

urn:lsid:zoobank.org:act:2E0EF5AB-0B1B-4461-9A5F-7E3828D84B30

Holotype. Female; Division of Entomology (Paleoentomology), University of Kansas Natural History Museum, Lawrence, Kansas, U.S.A.

Type strata and locality. Eocene, oil shale deposit, Parachute Creek Member of the Green River Formation, Uinta Basin, Utah, U.S.A.

Diagnosis. As for the genus (*vide supra*).

Description. *Female* (Fig. 3). A female bee missing the head preserved with forewings outstretched (albeit slightly obliquely) with dorsoposterior oblique view of mesosoma, dorsal view of metasoma (slightly detached from mesosoma), and the hind and midlegs preserved alongside the body (left hind leg more outstretched and with posterior surface of metatibial corbicula evident. Wings preserved flat, with small portions of hind wings visible, leading edge of forewings not preserved or partially folded (thus pterostigma is foreshortened). Mesosomal length 4.00 mm as preserved (orientation is not a direct dorsal view), intertegular

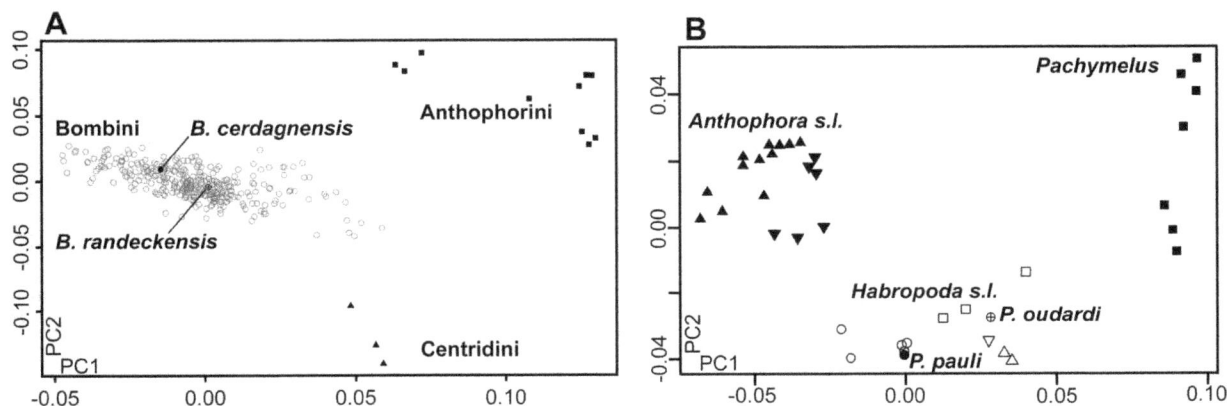

Figure 2. Generic taxonomic affinity of the two new fossils associated with the tribes Bombini and Anthophorini. A. Ordination of the Bombini and *Bombus cerdanyensis* sp. nov. along the first two axes of the PCA (PC1 = 32% and PC 2 = 19%). B. Ordination of the Anthophorini and *Protohabropoda pauli* gen. nov. sp. nov. along the first two axes of the PCA (PC1: 59%; PC2: 16%).

distance 2.00 mm. Tegulae preserved in oblique dorsal view, apparently ovoid with slightly angulate posterior inner border (much as in various corbiculate groups, including Euglossini: e.g., see tegular shape as depicted in Hinojosa-Díaz *et al.* (2011) [60]; posterior half of mesoscutum visible, 1.66 mm long as preserved; mesoscutellum 0.09 mm long; metanotum 0.32 mm long; propodeum 1.56 mm long as preserved across dorsal and posterior surfaces, dorsal surface apparently quite short, apparently no overhung by mesoscutellum or metanotum, propodeal pit present, narrow; pilosity of mesosoma generally scattered, short, only mesoscutum and mesoscutellum with setae over entire surfaces. Left mesofemur 2.49 mm long; mesotibia 1.75 mm long as preserved (incomplete); right mesofemur 2.57 mm long; mesotibia 2.14 mm long; mesotarsus 2.53 mm long; pretarsal claw 0.11 mm long (the state of preservation does not allow us to know if it is toothed or simple); left metafemur 1.30 mm long as preserved (incomplete); metatibia 4.23 mm long as preserved, slightly oblique, metatibia distinctly corbiculate, fringed by long setae; right metacoxa (incomplete) 0.31 mm long as preserved; metatrochanter 0.39 mm long as preserved (incomplete); left metacoxa 0.32 mm long as preserved (incomplete); metatrochanter 0.46 mm long as preserved (incomplete); right mesobasitarsus 1.91 mm long, 0.26 mm wide; right mesomediotarsus 0.68 mm long, 0.13 mm wide; right mesobasitarsus shorter than right mesotibia; black punctures mark base of setae on metatrochanters, mesofemora, mesotibiae, metafemora, and metatibiae. Left and right forewing each 8 mm long; three submarginal cells (i.e., 1rs-m present); first submarginal cell 1.23 mm long (as measured from origin of Rs+M to juncture of r-rs and Rs), 0.47 mm high (as measured from Rs+M to pterostigma); second submarginal cell 0.99 mm long (as measured from juncture of Rs+M and M to juncture of Rs and 1rs-m), 0.43 mm high (as measured from midpoint on M between 1m-cu and 1rs-m to juncture of r-rs and Rs); third submarginal cell 0.84 mm long (as measured from juncture of 1rs-m and M to juncture of M and 2rs-m), 0.41 mm high (as measured from juncture of M and 2m-cu to juncture of 2rs-m and Rs); first medial cell 2.47 mm long (measured from juncture of M+Cu and Cu to juncture of 1m-cu and M), 0.68 mm high (measured from juncture of M and Rs+M to midpoint on Cu between M+Cu and 1m-cu); pterostigma small, 0.2 mm long; marginal cell 2.04 mm long, apex acutely pointed; 2m-cu not entirely visible; 1m-cu joining second submarginal cell near its apex, not strongly angulate, relatively straight in basal half, then arched to meet M; first abscissa of Rs straight, 1.39 mm long. Metasomal length 5.43 mm as preserved (apical most segments slightly recessed into preceding segments; likely a postmortem factor), width 4.67 mm as preserved, with only scattered setae present apically.

Male. Unknown.

Etymology. The specific epithet is a patronym honoring Prof. Jacobus Biesmeijer, scientific director at Naturalis Biodiversity Centre and professor of Functional Biodiversity at the University of Amsterdam, a leading authority on pollinator-plant interactions and pollinator declines, and for his inspiration of many melittologists throughout the world. Prof. Biesmeijer is acknowledged for his career as a leading scientist who stimulates bee research.

Comments. As noted in the Results, *Euglossopteryx biesmeijeri* gen. nov. sp. nov. was confidently assigned to the Apinae. Moreover, aside from forewing shape, the specimen shows other traits of bees and apines, in particular, such as branched setae, an expanded tarsus, and even the posterior surface of a corbiculate metatibia (Fig. 3). Given that *E. biesmeijeri* exhibits pollen-collecting structures, it is clear that the species is not cleptopar-

asitic, thereby excluding the various cuckoo bee tribes and genera [2].

The most revealing structure as to the affinities of *E. biesmeijeri* is the presence of a metatibial corbicula, a unique synapomorphy of the corbiculate tribes in the Apinae. The clade Corbiculata comprises of four extant tribes (Apini, Bombini, Euglossini, and Meliponini) and three extinct groups (Melikertini, Electrapini, and Electrobombini) [1,14]. *E. biesmeijeri* can be excluded from the Meliponini and Apini as it does not have the dramatically reduced wing venation of the former, nor the uniquely modified and extended marginal cell (i.e., nearly four times as long as the distance from its apex to the wing tip) of the latter. The forewings of *E. biesmeijeri* do not possess alar papillae, suggesting that the species was not a bombine. The pterostigma is no longer that the prestigma, unlike the fossil tribes Melikertini, Electrapini, and Electrobombini, although admittedly there remains much to be learned about wing form diversity in these lineages. The 1m-cu vein is strongly angulate unlike Electrapini and Electrobombini and the body size was likely greater than the known melikertines, although body size is not necessarily a diagnostic feature of this group. The metatibia is not as grossly expanded as is true among the Euglossini. Overall the shape of the forewing is most similar to Euglossini (refer to Results, above), but this may very likely be a shared plesiomorphy given that the wing venation of Euglossini is very generalized relative to other corbiculate tribes and euglossines are frequently recovered as the basal lineage of the clade [14,61]. More importantly, the corbiculate clade is quite ancient, extending at least into the Late Cretaceous [62] and with significant levels of extinction [1,14], leaving only four, morphologically-isolated extant tribes. It is exactly for such a scenario that early fossils will be the most critical for resolving controversies over relationship, in this case meaning those taxa from the latter part of the Cretaceous, and there will likely be a larger number of Paleogene taxa that are difficult to assign as to tribe.

An affinity between *E. biesmeijeri* with euglossines is certainly plausible given the distribution and ecology of the latter. Moreover, orchid bees were certainly present in the Tertiary of North America as evidenced by fossils in Early Miocene amber of the region (e.g. [63,64]). We are confident that *E. biesmeijeri* is near to the Euglossini and may even belong to the crown group, but this will require further and more completely preserved material and it could be a stem-group euglossine. Future studies encompassing a dataset built around a maximal sampling of corbiculate bee genera would be most beneficial for further refining the potential affinities of this enigmatic fossil. In particularly, it would be very interesting to discover whether the jugal lobe of the hind wing was replaced by the distinctive jugal comb, a notable synapomorphy of Euglossini (e.g. [63]). Presently, there does not appear to be a jugal comb present but the hind wings are so poorly preserved that a definitive presence or absence cannot be stated. In the interim it is best to conservatively consider *E. biesmeijeri* as tribe *incertae sedis*.

Tribe: Bombini Latreille 1802

Genus: *Bombus* Latreille 1802

Bombus cerdanyensis *Dehon, De Meulemeester & Engel* sp. nov.

urn:lsid:zoobank.org:act:C61BA7C3-AD4E-4BA5-B277-4BA2-F7AD798E

Holotype. Sex unknown. Conserved in the Palaeontology department collection, Muséum National d'Histoire Naturelle, Paris, France. The fossil consists in 2 parts: the compression and the imprint of the compression.

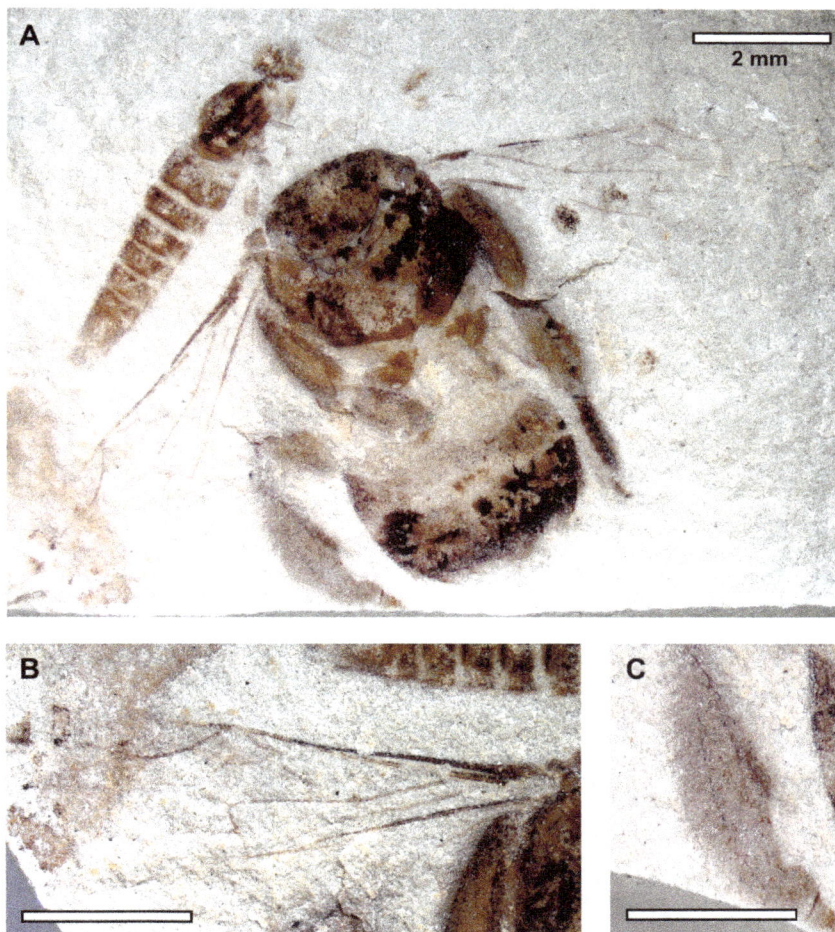

Figure 3. *Euglossopteryx biesmeijeri* **gen. nov sp. nov.** Dorsal view (photographs by N. J. Vereecken).

Etymology. The specific epithet is a reference to the lacustrine deposits of the la Cerdanya, Spain where the holotype was collected.

Type strata and locality. Late Miocene, lacustrine beds of Cerdanya, Spain.

Diagnosis. Forewing with alar papillae present on membrane beyond apical crossveins; membrane infuscated throughout, particularly in area beyond apical crossveins and along anterior borders of radial and marginal cells; pterostigma relatively small, trapezoidal, not greatly larger relative to prestigma and width not much shorter than length; marginal cell longer than distance from its apex to wing tip, tapering in width across its length, apex acutely rounded, not appendiculate, slightly offset from wing margin; three submarginal cells of relatively similar size (i.e., one not distinctly enlarged relative to the other two), anterior borders of second and third submarginal cells subequal; 1m-cu distinctly and strongly angulate anteriorly, meeting second submarginal cell near midpoint; 2m-cu weakly arched, meeting third submarginal cell in apical fifth; mesotibia five times longer than wide.

Description. Sex unknown (Fig. 4). A compressed individual in apparently dorsal oblique view with left forewing outstretched (right forewing not preserved); hind wing not preserved; head not preserved; mesosoma and metasoma largely incomplete and damaged; mid and hind legs preserved but somewhat jumbled and in some places partially overlapping forewing (e.g., obscuring

origin of basal vein relative to 1cu-a but orientation of veins implies 1cu-a was apical basal vein origin). Right profemur 1.43 long, 0.88 wide as preserved (incomplete). Left mesofemur 3.60 mm long, 0.93 mm wide; mesotibia 3 mm long, 0.59 mm wide; mesobasitarsus 3.22 mm long, 0.91 mm wide; remaining tarsomeres and pretarsal claws well preserved, with apical most tarsomere longer than individual lengths of preceding tarsomeres; pretarsal claws not toothed as preserved. Right mesofemur 3.45 long, 0.47 wide; mesotibia 2 mm long, 0.38 wide as preserved (incomplete). Left forewing 13.25 mm long, 4.56 mm wide; three submarginal cells of relatively equal sizes; first submarginal cell 1.50 mm long (as measured from origin of Rs+M to juncture of r-rs and Rs), 0.67 mm high (as measured from Rs+M to pterostigma); second submarginal cell 1.52 mm long (as measured from juncture of Rs+M and M to juncture of Rs and 1rs-m), 0.78 mm high (as measured from midpoint on M between 1m-cu and 1rs-m to juncture of r-rs and Rs); third submarginal cell 1.32 mm long (as measured from juncture of 1rs-m and M to juncture of M and 2rs-m), 1.08 mm high (as measured from juncture of M and 2m-cu to juncture of 2rs-m and Rs); first medial cell 3.41 mm long (measured from juncture of M+Cu and Cu to juncture of 1m-cu and M), 1.16 mm high (measured from juncture of M and Rs+M to midpoint on Cu between M+Cu and 1m-cu); pterostigma 0.93 mm long; marginal cell 3.44 mm long, width tapering gradually across its length, apex rounded, offset from anterior wing margin, not appendiculate; 1m-cu distinctly and strongly angulate, meeting second submarginal cell near midpoint; 2m-cu weakly arched, meeting third submarginal cell in apical fifth. Metasoma 5.80 mm wide as preserved; first two segments visible, first segment 1.76 mm long, second segment 1.23 mm long (incomplete as preserved).

Comments. The LDA place the specimen among the corbiculate tribes. Therein, the venation is plesiomorphic relative to the Apini and Meliponini and in the same characters as highlighted for *Euglossopteryx* gen. nov. (*vide supra*). The relative sizes of the prestigma and pterostigma exclude a placement in the Electrobombini (although the presence or absence of a jugal lobe in the hind wing cannot be determined in the holotype). The forewing is apically papillate (as in Bombini), and the marginal cell is not appendiculate and 1m-cu is strongly angulate together suggesting the species does not belong to the Electrapini or Melikertini (although some melikertines have 1m-cu more angulate, such as *M. trigona*, 1m-cu is always much shorter and not as long as in Bombini or Euglossini; a long 1m-cu is more plesiomorphic among Corbiculata). Indeed, the forewings of the present fossil are distinctly *Bombus*-like: presence of papillae, general infuscation of the membrane, three submarginal cells of relatively similar size (albeit the latter character is assuredly plesiomorphic). As noted above a more refined analysis with the present fossil and a diversity of species of *Bombus* further reinforced the great similarity of the

general wing shapes and supports inclusion within this genus, although attribution to a particular subgenus was not possible (refer to Results, above). Placement in *Bombus* is also in accordance with the geographic region and paleohabitat. Bumble bees are found throughout the Holarctic, Orient, and South America, but likely originated in the Palaearctic [65] and are principally associated with temperate and cold climates today [2,66,67]

Subfamily: Apinae Latreille 1802

Tribe: Anthophorini Dahlbom 1835

Genus Protohabropoda *Dehon & Engel gen. nov.*

urn:lsid:zoobank.org:act:809CFD9C-D3BD-4796-90E6-FEC-C65058AEC

Type species. *Protohabropoda pauli* De Meulemeester & Michez, new species.

Diagnosis. Female forewing with small pterostigma, pterostigma not tapering beyond r-rs, parallel-sided, length about equal to width; marginal cell broad, scarcely tapering apically, apex broadly rounded, slightly offset from anterior wing margin, not greatly extending beyond apical tangent of submarginal cells; three submarginal cells (i.e., 1rs-m present), first submarginal cell longer than individual lengths of second and third submarginal cells, but not longer than combined length of second and third submarginal cells; 1m-cu elongate, distinctly and strongly oblique, not angulate, relatively straight for most of its length, with weak arch in apical

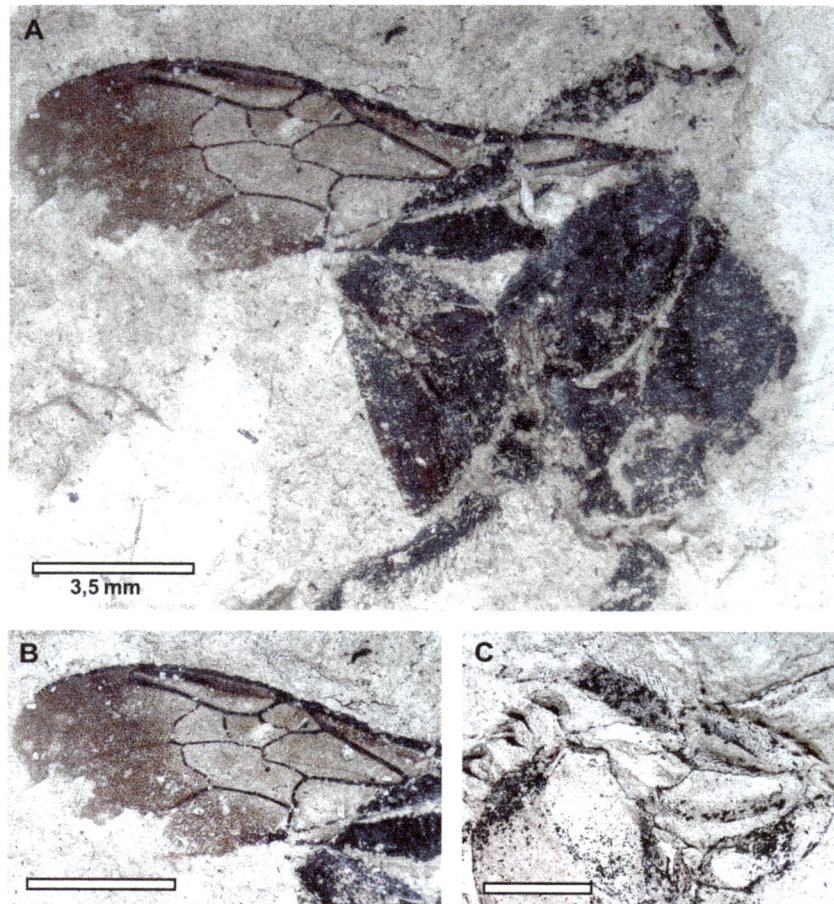

Figure 4. *Bombus cerdanyensis* **sp. nov.** A. General habitus. B. Detail of the left forewing. C. Detail of the middle leg on the imprint of the compression (photographs by T. De Meulemeester).

quarter (in this respect differing from *Habropoda*), meeting second submarginal cell near its apex (meeting cell only slightly basal to 1rs-m) (differing in this from *Anthophora* and *Amegilla*). Body of robust anthophoriform type, apparently densely setose (as in most modern Anthophorini); clypeus rounded.

Etymology. The new genus-group name is a combination of protos (Greek, meaning, "first") and and the generic name *Habropoda*, the most similar extant genus to the fossil. The name is feminine.

Protohabropoda pauli *De Meulemeester & Michez sp. nov.*

urn:lsid:zoobank.org:act:896FC513-0520-4EE0-A777-9CAEAFFBE5BC

Holotype. Female. Conserved in the Palaeontology department collection, Muséum National d'Histoire Naturelle, Paris, France. The fossil consists in 2 parts: the compression and the imprint of the compression.

Type strata and locality. Late Oligocene, lacustrine beds, Céreste, France.

Diagnosis. As for the genus (*vide supra*).

Description. *Female* (Fig. 5). Dorsal-obliquely compressed individual, with apparently right forewing outstretched at oblique angle to body (hind wings not preserved); head turned with left lateral surface visible and depicting compound eye, gena, and face. Head with clypeus rounded; compound eye length 1.15 mm; left antenna preserved but incomplete, details of individual flagellomeres not discernible. Mesosoma 5.88 mm long; individual segments not discernible as preserved; dense pilosity present, some setae branched. Legs only partially preserved overlayed with mesosoma and slightly beneath; individual podites not measureable as preserved but distinctly with dense setation. Forewing 7.79 mm long, 2.53 mm wide; pterostigma small, 0.78 mm long, not tapering beyond r-rs, parallel-sided; marginal cell 1.98 mm long, 0.46 mm wide, scarcely tapering apically, apex broadly rounded, slightly offset from anterior wing margin, not greatly extending beyond apical tangent of submarginal cells; three submarginal cells (i.e., 1rs-m present); first submarginal cell 1.21 mm long (as measured from origin of Rs+M to juncture of r-rs and Rs), 0.45 mm high (as measured from Rs+M to pterostigma); second submarginal cell 0.75 mm long (as measured from juncture of Rs+M and M to juncture of Rs and 1rs-m), 0.69 mm high (as measured from midpoint on M between 1m-cu and 1rs-m to juncture of r-rs and Rs); third submarginal cell 0.71 mm long (as measured from juncture of 1rs-m and M to juncture of M and 2rs-m), 0.75 mm high (as measured from juncture of M and 2rs-m to juncture of 2rs-m and Rs); first medial cell 2.86 mm long (measured from juncture of M+Cu and Cu to juncture of 1m-cu and M), 0.71 mm high (measured from juncture of M and Rs+M to midpoint on Cu between M+Cu and 1m-cu); 1m-cu elongate, distinctly and strongly oblique, not angulate, relatively straight for most of its length, with weak arch in apical quarter, meeting second submarginal cell only slightly basal 1rs-m; 2m-cu weakly arched. Metasoma in lateral aspect, 4.5 mm long as preserved, 6.59 mm wide; two segments discernible; with dense pilosity over integument.

Male. Unknown.

Etymology. The specific epithet is a patronym honoring Mr. Paul Léon Victor Vigot, young scientist. In acknowledgment of his interest dedicated to bee systematics.

Comments. The shape of pterostigma (small, not tapering beyond r-rs, and parallel-sided) in *P. pauli* is similar to those of both the Anthophorini and Centridini. However, the ratio of the first (larger) and second (smaller) submarginal cells and the global shape of the forewing (Fig. 5) are more indicative of an

anthophorine and the clypeus is rounded as in many Anthophorini. Seven extant genera are included in the Anthophorini: *Anthophora* (worldwide, 350 species), *Amegilla* (Old World, 250 species), *Deltoptila* (Central America, 10 species), *Elaphropoda* (Eastern Asia, 11 species), *Habrophorula* (Oriental, 3 species), *Habropoda* (Old and New World, 60 species), and *Pachymelus* (Southern Africa and Madagascar, 20 species) [68]. The new fossil specimen can easily be distinguished from *Anthophora* and *Amegilla* 1m-cu meeting the second submarginal cell near its apex, while the elongate and strongly oblique 1m-cu differs from that observed in *Habropoda*.

Family: Andrenidae Latreille, 1802
Subfamily: Andreninae Latreille, 1802
Genus: *Andrena* Fabricius, 1775

Andrena antoinei *Michez & De Meulemeester sp. nov.*

urn:lsid:zoobank.org:act:EE404B56-22F5-4FB9-9075-B1FC5BBA2F05

Holotype. Male. Conserved in the Palaeontology department collection, Muséum National d'Histoire Naturelle, Paris, France.

Type strata and locality. Late Oligocene, lacustrine beds, Céreste, France.

Diagnosis. Male with yellow clypeus, remainder of head black; mesosoma black; legs and antenna brown; metasoma largely brown although black on apical segments (as preserved); wing membranes hyaline, veins brown; inner margins of compound eyes relatively straight, parallel (i.e., not converging anteriorly); forewing with basal vein confluent with 1cu-a, the latter strongly oblique; basal vein weakly arched in basal half; pterostigma more than three times longer than wide, tapering inside of marginal cell, border inside of marginal cell slightly convex; marginal cell long, tapering gently along length to acutely rounded apex, not appendiculate; first submarginal cell elongate, slightly longer than combined lengths of second and third submarginal cells; third submarginal cell greatly projecting apically in posterior half (resulting from strongly arcuate 2rs-m); 1m-cu meeting second submarginal cell beyond cell midpoint, relatively straight over much of length, then strongly angled to meet M; 2m-cu meeting third submarginal cell in apical third, straight; 1rs-m straight; 2rs-m strongly arcuate such that posterior border of third submarginal cell is nearly twice length of anterior border.

Description. *Male* (Fig. 6). Dorsal compressed individual, with head thrust forward and showing facial view, antennae extending laterally from body, left antenna curling back under head; dorsal view of mesosoma; forewings outstretched orthogonal to long axis of body; legs preserved along side of body; metasoma preserving basal few segments. Head 2.56 mm long, 2.30 mm wide; compound eyes 1.62 mm long, 0.53 mm wide; inner margins parallel; clypeus off white (likely yellow in life), 0.41 mm long, 0.57 mm wide; labrum 0.13 mm long; mandibles simple; right antenna incomplete, 0.26 mm wide; scape 0.4 mm long, pedicel 0.27 mm long, basal seven flagellomeres preserved, each 0.38 mm long; left antenna with only six articles preserved; terminal part of left antenna curled under the head. Mesosoma 2.90 mm long, intertegular distance 2.83 mm; mesoscutum 1.43 mm long; mesoscutellum 0.57 mm long; metanotum 0.25 mm long; propodeum 0.81 mm long (measured as preserved across dorsal and posterior surfaces). Profemur only partially exposed; left mesofemur 1.26 mm long, 0.41 mm wide; mesotibia 1.39 mm long, 0.37 mm wide; mesobasitarsus 0.50 mm long, 0.15 mm wide (incomplete); right mesofemur 1.20 mm long, 0.41 mm wide; mesotibia 0.96 mm long, 0.37 mm wide (incomplete); left metafemur 1.65 mm long, 0.55 mm wide; metatibia 1.85 mm long, 0.45 mm wide (incomplete); right metafemur 1.67 mm long, 0.54 mm wide; metatibia 1.81 mm long, 0.47 mm wide; metaba-

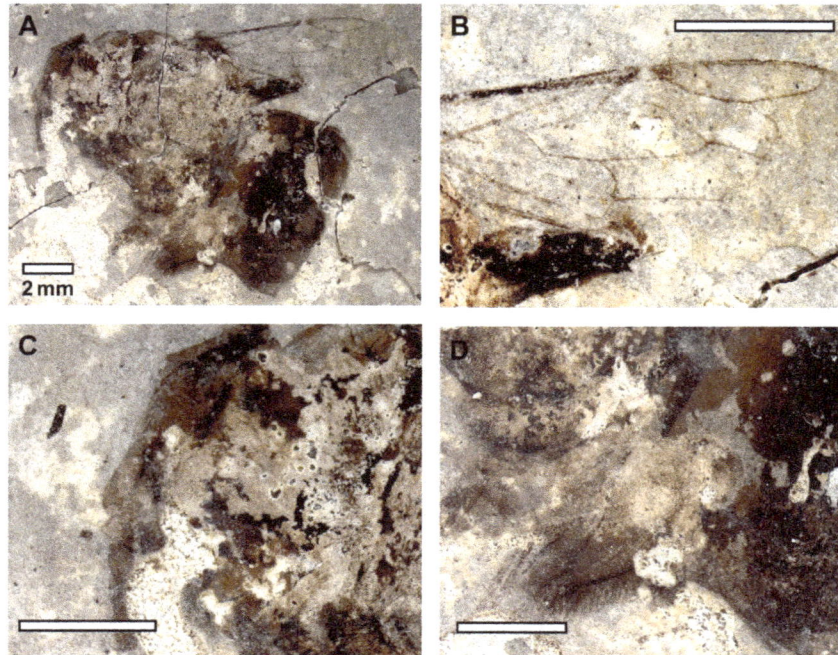

Figure 5. *Protohabropoda pauli* **gen. nov. sp. nov.** A. General habitus. B. Detail of wing. C. Detail of the head. D. Detail of the scopa (photographs by T. De Meulemeester).

sitarsus 1.34 mm long as preserved (incomplete). Forewing 5.94 mm long, 1.53 mm wide; pterostigma 0.72 mm long, 0.22 mm wide; marginal cell 1.62 mm long, tapering gently over length, apex acutely rounded; three submarginal cells, first submarginal cell 1.23 mm long (as measured from origin of Rs+M to juncture of r-rs and Rs), 0.28 mm high (as measured from Rs+M to pterostigma); second submarginal cell 0.50 mm long (as measured from juncture of Rs+M and M to juncture of Rs and 1rs-m), 0.26 mm high (as measured from midpoint on M between 1m-cu and 1rs-m to juncture of r-rs and Rs); third submarginal cell 0.66 mm long (as measured from juncture of 1rs-m and M to juncture of M and 2rs-m), 0.39 mm high (as measured from juncture of M and 2m-cu to juncture of 2rs-m and Rs); first medial cell 2.0 mm long (measured from juncture of M+Cu and Cu to juncture of 1m-cu and M), 0.29 mm high (measured from juncture of M and Rs+M to midpoint on Cu between M+Cu and 1m-cu); 1rs-m straight; 2rs-m strongly arcuate, thus projecting posterior half of third submarginal cell strongly apical with anterior border of second about one-half posterior border of cell; 1m-cu relatively straight for most of length; 2m-cu straight, 0.38 mm long. Metasoma 4.47 mm long as preserved, 3.10 mm wide; only basal few segments preserved, lighter in coloration than head and mesosoma, apparently with scattered minute setae.

Female. Unknown.

Etymology. The specific epithet is a patronym honoring Mr. Antoine Michez, in recognition of his dedication to melittology.

Comments. The morphometric analysis of forewing shape demonstrates that this species belongs to the Andrenidae (refer to Results, above). Unfortunately, diagnostic features of Andrenidae such as the two subantennal sulci below each antenna and the short to long-pointed glossa are not observable in the fossil. Nonetheless, the placement of the fossil in Andreninae is quite strong and the general habitus is certainly in accordance with such an assignment. The three submarginal cells and apically pointed marginal cell, with a slightly arched basal vein together exclude placement in any other subfamily of Andrenidae. The species does not exhibit any features which would permit its confident placement outside of the widespread and hyper diverse genus *Andrena* and so we have placed our species therein, pending the discovery of more complete material. The fossil further documents the presence of this lineage among the Late Oligocene fauna of France.

Discussion

Geometric morphometrics of wing shape to discriminate taxa

Michez *et al.* (2009) [31], De Meulemeester *et al.* (2012) [32], Wappler *et al.* (2012) [18], and Dewulf *et al.* (2014) [21] have shown that geometric morphometric analyses of forewing shape are a valuable tool for associating challenging bee fossils, particularly when other sources of character information are limited or lacking entirely. Naturally, it is most accurate and ideal to combine such geometric morphometric results with other anatomical traits for which cladistic polarity may be determined (e.g., leg morphology, mouthpart characters, etc.), but this is not always achievable with compression fossils such as the ones discussed herein. Thus, geometric morphometrics offers a way to open up a source of previously untapped data from such problematic specimens and thereby offer insights into bee evolution at different phases in their history. Indeed, these methods have proven useful in a variety of cases and lineages for discriminating taxa (e.g. [32,69–71]). Here we show that bees can be discriminated based on their forewing shape at a variety of taxonomic levels, ranging from subfamily and above. Discriminant analyses employed in our study discriminated Anthophila from the families of apoid wasps, and even among the currently recognized families and subfamilies with reasonable robustness.

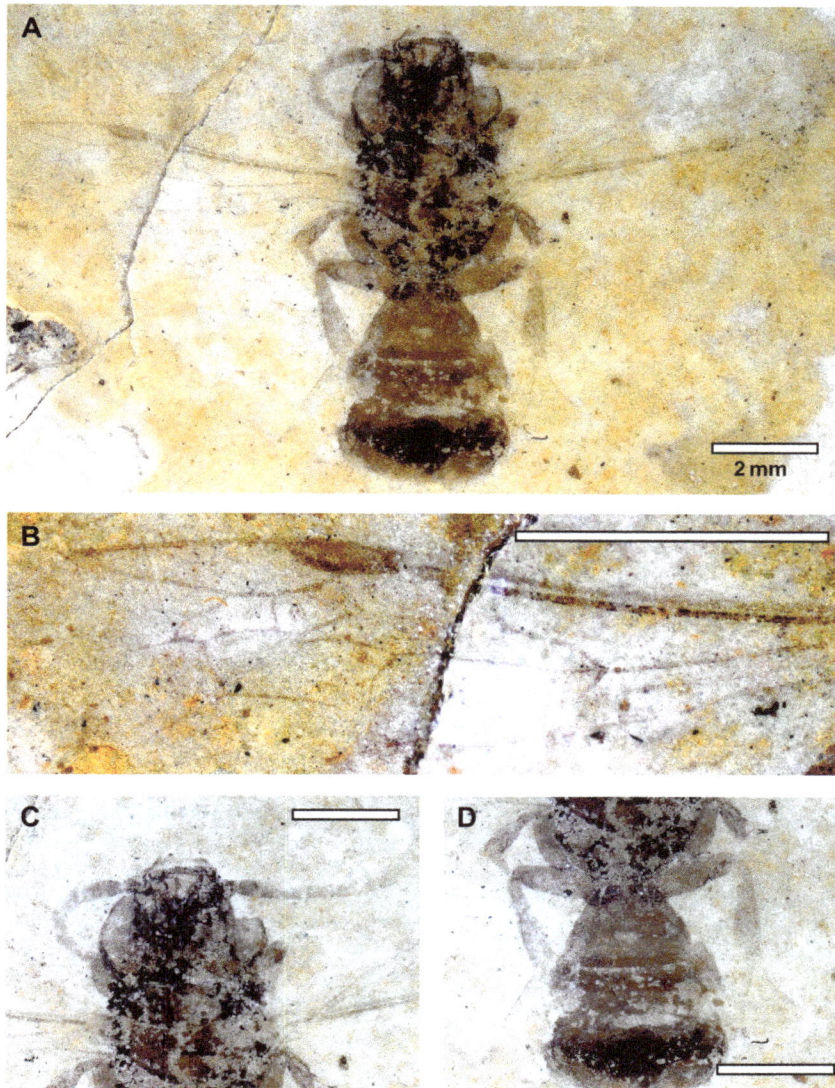

Figure 6. Male of *Andrena antoinei* **sp. nov.** A. General habitus. B. Detail of the wing. C. Detail of the head. D. Detail of the abdomen and the hind legs (photographs by T. De Meulemeester).

Morphological similarities at different taxonomical levels cannot be mistaken for phylogenetic relationships. They are merely similarities and this is highlighted by our own data. Exploration of our dataset representing significant parts of the bee forewing shape variation showed that phylogenetically close taxa are not always close in terms of overall similarity (e.g., Procrust distance is relatively high between Apidae and Megachilidae while these families are close phylogenetic relatives and among the families of Anthophila are actually sister) [1,3,72]. Despite these facts we observed a strong "taxonomic integrity" within forewing shape; i.e., all species were assigned to their corresponding tribe, all tribes were assigned to their corresponding subfamily, and all subfamilies were assigned to their corresponding family with high fidelity. We therefore argue that even though the morphological similarities do not represent phylogenetic relationships, there are patterns in forewing shape among members of individual clades. This statement is based on a supervised approach (i.e., LDA), so the assignment could be related to subtle common shape character-

istics. Therefore we cannot exclude that only a part of the forewing shape may be responsible for this "taxonomic integrity". Further fundamental studies on the phylogenetic signal of bee forewing shape, and morphometric integration and modularity in wing landmark configuration, remain to be undertaken and will certainly provide a considerable insight into apoid wing evolution. In fact, there are now well-established procedures for using morphometric data in cladistic analyses and this would be a fruitful line of investigation. In addition, such work might reveal those aspects of forewing shape tied to phylogeny and those more labile and perhaps tied to flight function and common to unrelated groups but whose flight mechanics are the same (e.g., correlates with body size, flight temperatures, etc.). This all highlights the broad potential for geometric morphometrics in evolutionary studies on bees.

Corbiculate apine history

Corbiculate bees are among the most common of bees found as fossils, particularly in a variety of amber deposits [11,73]. The clade is of Late Cretaceous age as evidenced by the occurrence of a crown-group meliponine in Maastrichtian-aged Raritan amber [61,74], and regardless of the preferred phylogenetic place of Meliponini (i.e., sister to Apini or Bombini: Cardinal & Packer, 2007) [60] demonstrates that the cladogenetic events among the tribes extend back to at least the latest Cretaceous. A Cretaceous crown-group meliponine means that stem- or crown-group members of the other tribes must have been present in the latest Cretaceous and all are, therefore, quite old. The morphological isolation of the respective living tribes, combined with their antiquity, also emphasizes that various extinct groups must have existed that intermingle subsets of their traits and, in fact, such taxa have been found persisting until the Eocene (i.e., the last global greenhouse epoch). Already from the Eocene are known three extinct tribes and some of these were assuredly advanced eusocial like the Apini and Meliponini [1]. The discovery of the corbiculate *Euglossopteryx biesmeijeri* gen. nov. sp. nov. provides new insights into corbiculate diversity in the middle Eocene of North America. This species also suggests that the global cooling and drying that marked the Eocene-Oligocene transition and that was likely responsible for the loss of corbiculate diversity in Europe [1,13,15] was probably a global phenomenon and impacted similar bee groups in the New World.

Among the corbiculate lineages, the Bombini is the only tribe with a higher extant diversity in temperate and cold habitats. Like most bee groups, the fossil record of bumble bees is sparse [11,18]. Twelve fossils have been documented from Oligocene through Miocene deposits as putative Bombini [18]; however, most of these fossils are poorly described and their placement within *Bombus* s.l. remains to be tested critically. Among those fossils, *B. cerdanyensis* and *B. randeckensis* have been the most thoroughly examined and in the case of the latter a confident subgeneric assignment has been demonstrated. These two specimens are of Late Miocene (~10 Ma: the Spanish fossil) and Early Miocene (16–18 Ma: the German species), respectively, and both have broad implications for the dating of bumble bee evolution. For example, *Bombus randeckensis* was shown to belong to *Bombus* s.str., regardless of whether it should be placed as a stem group or within the crown group, and therefore provided direct evidence regarding the age of divergence between *Bombus* s.str. and *Alpinobombus* (i.e., minimally at 16–18 Ma) and in stark contrast to molecular estimates which grossly under-dated the clade [63]. *Bombus randeckensis* was, therefore, a direct observational example of how many age estimates may be significantly off (with crown-group *Bombus* s.str. being almost twice as old as available estimates) [18]. The subgeneric placement of *B. cerdanyensis* is less obvious than that of *B. randeckensis* but the two fossils are certainly not representative of the same subgeneric clade. Together these species demonstrate that *Bombus* had well diversified by the Miocene and that the group as a whole is much older than molecular-only methods might imply. Continued paleontological exploration will only further refine our understanding, based on direct evidence, of bombine evolution and timing.

Anthophorine diversity and history

Multiple lines of evidence demonstrate the antiquity of the digger bees (Anthophorini) (e.g. [31,75]). Indeed, the tribe certainly extends well into the Cretaceous and the most basal among extant genera (i.e., *Habrophorula* and *Elaphropoda*) are found in tropical Asia while derived genera such as *Anthophora* are more widely distributed in mesic and xeric habitats [67]. The distribution of those few fossil records attributable to Anthophorini seem to loosely confirm the ancestral adaptation of these bees to tropical habitats with subsequent invasion and diversification within mesic and xeric environments. *Paleohabropoda oudardi* was recorded from the Paleocene deposits of Menat, France (~60 Myr) and which was at the time a distinctly tropical climate [31]. In contrast, the associated climate of the newly described *Protohabropoda pauli* gen. nov. sp. nov. was certainly mesic to xeric. It appears that the shift among anthophorines to non-tropical environments had already occurred during the Oligocene.

Supporting Information

Table S1 Landmarks coordinates of the Anthophorini data set (from Michez *et al.* 2009).

Table S2 Landmarks coordinates of the Bombini data set (from Wappler *et al.* 2012).

Table S3 Landmarks coordinates of the four new fossil specimens.

Table S4 Landmarks coordinates of the 632 original specimens and the entomological collections from where they are loaned.

Table S5 Mahalanobis distances (MD) between: (i) tribes centroids and the 632 specimens, (ii) the fossil *Euglossopteryx biesmeijeri* gen. nov. sp. n. and tribes centroids, (iii) the fossil *Bombus cerdanyensis* sp. nov. and tribes centroids, (iv) between the fossil *Protohabropoda pauli* gen. nov. sp. nov. and tribes centroids, (v) between the fossil *Andrena antoinei* sp. nov. and tribes centroids.

Table S6 Specimen assignment in subfamilies using the cross-validation procedure in the LDA of forewing shape. Original groups are along the rows, predicted groups are along the columns. The hit ratio (HR%) is given for each subfamily.

Table S7 Specimen assignment in tribes using the cross-validation procedure in the LDA of wing shape. Original groups are along the rows, predicted groups are along the columns. The hit ratio (HR%) is given for each tribe.

Table S8 Specimens scores along the LDs (for the following taxonomic levels: suprafamily, family, subfamily and tribe).

Acknowledgments

We are very grateful for collection access to the curators of the following museum: David Notton, National History Museum (London, UK); Frederique Bakker, Naturalis Biodiversity Center (Leiden, NL); Jeannine Bortels, University of Liège (Gembloux, BE); Eliane De Coninck, Royal Museum of Central Africa (Tervuren, BE); and Wouter Dekoninck, Royal Belgian Institute of Natural Sciences (Bruxelles, BE). We also would like to thank Maryse Vanderplanck and Thomas Lecocq (University of Mons, BE) for their precious help, and Ahmet Murat Aytekin (Hacettepe University, TR) for providing data on bumble bee wings.

Author Contributions

Conceived and designed the experiments: MD DM TD. Performed the experiments: MD TD. Analyzed the data: MD DM AN MSE TD. Contributed reagents/materials/analysis tools: MD DM AN MSE TD. Wrote the paper: MD DM MSE TD.

References

1. Engel MS (2001) A monograph of the Baltic amber bees and evolution of the apoidea (Hymenoptera). Bull Am Museum Nat Hist 259: 1–192.
2. Michener CD (2007) The bees of the world, 2nd edition. The Johns Hopkins University Press, Baltimore, 953 p.
3. Danforth BN, Cardinal S, Praz C, Almeida EAB, Michez D (2013) The impact of molecular data on our understanding of bee phylogeny and evolution. Annu Rev Entomol 58: 57–78.
4. Michener CD (1944) Comparative external morphology, phylogeny, and a classification of the bees. Bull American Museum Nat Hist 82: 151–326.
5. Michener CD (1979) Biogeography of the bees. Ann Mo Bot Gard 66: 277–347.
6. Alexander BA (1992) An exploratory analysis of cladistic relationships within the superfamily Apoidea, with special reference to sphecid wasps (Hymenoptera). J Hymenopt Res 1: 25–61.
7. Melo GAR (1999) Phylogenetic relationships and classification of the major lineages of Apoidea (Hymenoptera), with emphasis on the crabronid wasps. Scientific Papers, Natural History Museum, University of Kansas 14: 1–5.
8. Engel MS (1996) New augochlorine bees (Hymenoptera: Halictidae) in Dominican amber, with a brief review of fossil halictidae. J Kansas Entomol Soc 69: 334–345.
9. Cardinal S, Danforth BN (2013) Bees diversified in the age of eudicots. Proc R Soc B Biol Sci 280: 1–9.
10. Ohl M, Engel MS (2007) Fossil history of the bees and their relatives (Hymenoptera: Apoidea). Denisia 20: 687–700.
11. Michez D, Vanderplanck M, Engel MS (2012) Fossil bees and their plant associates. In: Patiny S, ed. Evolution of plant-pollinator relationships. Cambridge University Press, Cambridge 103–164.
12. Zeuner FE, Manning FJ (1976) A monograph on fossil bees (Hymenoptera: Apoidea). Bulletin of the British Museum (Natural History), Geology 27(3): 149–268.
13. Engel MS (2001) The first large carpenter bee from the tertiary of North America, with a consideration of the geological history of Xylocopinae (Hymenoptera: Apidae). Trans Am Entomol Soc 127: 245–254.
14. Engel MS (2001) Monophyly and extensive extinction of advanced eusocial bees: insights from an unexpected Eocene diversity. Proc Natl Acad Sci USA 98: 1661–1664.
15. Engel MS (2002) Halictine bees from the Eocene-Oligocene boundary of Florissant, Colorado (Hymenoptera: Halictidae). Neues Jahrb fur Geol und Palaontologie - Abhandlungen 225: 251–273.
16. Engel MS (2008) A new species of Ctenoplectrella in Baltic amber (Hymenoptera: Megachilidae). Acta Zool Acad Sci Hungaricae 54: 319–324.
17. Gonzalez VH, Engel MS (2011) A new species of the bee genus Ctenoplectrella in middle Eocene Baltic amber (Hymenoptera, Megachilidae). Zookeys 111: 41–49.
18. Wappler T, De Meulemeester T, Murat Aytekin A, Michez D, Engel MS (2012) Geometric morphometric analysis of a new Miocene bumble bee from the Randeck Maar of southwestern Germany (Hymenoptera: Apidae). Syst Entomol 37: 784–792.
19. Engel MS, Grimaldi DA, Gonzalez VH, Hinojosa-Díaz IA, Michener CD (2012) An exomalopsine bee in early Miocene Amber from the Dominican Republic (Hymenoptera: Apidae). Am Museum Novit 3758: 1–16.
20. Engel MS, Breitkreuz LCV (2013) A male of the bee genus Agapostemon in Dominican amber (Hymenoptera: Halictidae). Journal of Melittology 16: 1–9.
21. Dewulf A, De Meulemeester T, Dehon M, Engel MS, Michez D (2014) A new interpretation of the bee fossil Melitta willardi Cockerell (Hymenoptera, Melittidae) based on geometric morphometrics of the wing. Zookeys 389: 35–48.
22. Engel MS, Breitkreuz LCV, Ohl M (2014) The first male of the extinct bee tribe Melikertini (Hymenoptera: Apidae). Journal of Melittology 30: 1–18.
23. Michez D, Nel A, Menier J-J, Rasmont P (2007) The oldest fossil of a melittid bee (Hymenoptera: Apiformes) from the early Eocene of Oise (France). Zool J Linn Soc 150: 701–709.
24. Engel MS (2006) A giant honey bee from the Middle Miocene of Japan (Hymenoptera: Apidae). Am Museum Novit 3504: 1–12.
25. Engel MS, Hinojosa-Díaz IA, Rasnitsyn AP (2009) A honey bee from the Miocene of Nevada and the biogeography of Apis (Hymenoptera: Apidae: Apini). Proc Calif Acad Sci Series 4 60: 23–38.
26. Nel A, Martínez-Delclòs X, Arillo A, Peñalver E (1999) The fossil Apis Linnaeus, 1758 (Hymenoptera, Apidae). Palaeontology 42: 243–285.
27. Kotthoff U, Wappler T, Engel MS (2011) Miocene honey bees from the Randeck Maar of southwestern Germany (Hymenoptera: Apidae). Zookeys 96: 11–37.
28. Engel MS (2004) Notes on a megachiline bee (Hymenoptera: Megachilidae) from the Miocene of Idaho. Trans Kans Acad Sci 107: 97–100.
29. Engel MS (2004) Geological history of the bees (Hymenoptera: Apoidea). Revista de Tecnologia e Ambiente 10(2): 9–33.
30. Roberts DW (2008) Statistical analysis of multidimensional fuzzy set ordinations. Ecology 89:1246–1260.
31. Michez D, De Meulemeester T, Rasmont P, Nel A, Patiny S (2009) New fossil evidence of the early diversification of bees: Paleohabropoda oudardi from the French Paleocene (Hymenoptera, Apidae, Anthophorini). Zool Scr 38: 171–181.
32. De Meulemeester T, Michez D, Aytekin AM, Danforth BN (2012) Taxonomic affinity of halictid bee fossils (Hymenoptera: Anthophila) based on geometric morphometrics analyses of wing shape. J Syst Palaeontol 10: 755–764.
33. Engel MS (2011) Systematic melittology: where to from here? Syst Entomol 36: 2–15.
34. Gonzalez VH, Griswold T, Engel MS (2013) Obtaining a better taxonomic understanding of native bees: where do we start? Syst Entomol 38: 645–653.
35. Franczyk KJ, Fouch TD, Johnson RC, Molenaar CM, Cobban WA (1992) Cretaceous and Tertiary paleogeographic reconstructions for the Uinta-Piceance Basin study area, Colorado and Utah. US Geol Surv Bull 1787-Q, 37 p.
36. Hail WJ Jr, Smith MC (1997) Geology map of the southern part of the Piceance Creek Basin, northwestern Colorado. United States Geological Survey. IMAP 2529.
37. MacGinitie HD (1969) The Eocene Green River flora of northwestern Colorado and eastern Utah. University of California Publications in Geological Sciences 83: 1–140.
38. Forest CE, Wolfe JA, Molnar P, Emanuel KA (1999) Paleoaltimetry incorporating atmospheric physics and botanical estimates of paleoclimate. Bull Geol Soc Am 111: 497–511.
39. Mayr G (2000) Charadriiform birds from the early Oligocene of Céreste (France) and the Middle Eocene of Messel (Hessen, Germany). Geobios 33(5): 625–636.
40. Gregor H-J von (2002) Die fossile Megaflora von Cereste in der Provence – I (Coll. Lutz). Flora Tertiaria Mediterranea 4: 1–55.
41. Schmidt-Kittler N, Storch G (1985) Ein vollständiges Theridomiden-skelett (Mammalia: Rodentia) mit Renmaus-Anpassungen aus dem Oligozän von Céreste, S-Frankreich. Senckenbergiana Lethaea 66: 89–109.
42. Collomb FM, Nel A, Fleck G, Waller A (2008) March flies and European Cenozoic palaeoclimates (Diptera: Bibionidae). Ann Soc Entomol Fr 44: 161–179.
43. Diéguez C, Nieves-Aldrey J, Barrón E (1996) Fossil galls (zoocecids) from the Upper Miocene of La Cerdaña (Lérida, Spain). Rev Palaeobot Palynol 94: 329–343.
44. Jiménez-Moreno G, Fauquette S, Suc J-P (2010) Miocene to Pliocene vegetation reconstruction and climate estimates in the Iberian Peninsula from pollen data. Rev Palaeobot Palynol 162: 403–415.
45. Peñalver-Molla E, Martínez-Delclòs X, Arillo A (1999) Yacimientos con insectos fósiles en España. Revista Española de Paleontologia 14: 231–245.
46. Arillo A (2001) Presencia de la familia Pompilidae (Insecta, Hymenoptera) en el Mioceno superior de la Cuenca de la Cerdaña (Lleida, NE de España). Coloquios de Paleontología 52: 79–83.
47. Bookstein FL (1991) Morphometric tools for landmark data: geometry and biology. Cambridge University Press, Cambridge, 435 p.
48. Rohlf FJ, Marcus LF (1993) A revolution in morphometrics. Trends Ecol Evol 8: 129–132.
49. Adams DC, Rohlf FJ, Slice DE (2004) Geometric morphometrics: Ten years of progress following the "revolution". Ital J Zool 71: 5–16.
50. Pavlinov IY (2001) Geometric morphometrics, a new analytical approach to comparison of digitized images. Saint Petersburg: Information Technology in Biodiversity Resarch, Abstract of the 2nd International Symposium, pp. 41–90.
51. Ross HH (1936) The ancestry and wing venation of the Hymenoptera. Ann Entomol Soc Am 29: 99–111.
52. Rohlf FJ (2013) tpsUTIL Version 1.56. Department of Ecology and Evolution, State University of New York at Stony Brook, New-York.
53. Rohlf FJ (2013) tpsDIG Version 2.17. Department of Ecology and Evolution, State University of New York at Stony Brook, New-York.
54. Rohlf FJ, Slice D (1990) Extensions of the Procrustes method for the optimal superimposition of landmarks. Syst Zool 39: 40–59.
55. Adams DC, Otárola-Castillo E (2013) Geomorph: An R package for the collection and analysis of geometric morphometric shape data. Methods Ecol Evol 4(4): 393–399.
56. Rohlf FJ (1999) Shape statistics: Procrustes superimpositions and tangent spaces. J Classif 16: 197–223.
57. Rohlf FJ (2013) tpsSMALL Version 1.25. Department of Ecology and Evolution, State University of New York at Stony Brook, New-York.
58. R Development Core Team (2013) A language and environment for statistical computing, version 3.0.2, ISBN 3-900051-07-0, R Foundation for Statistical Computing. Vienna, Austria.
59. Huberty CJ, Olejnik S (2006) Applied MANOVA and discriminant analysis. Second Edition. New Jersey, 488 p.
60. Hinojosa-Díaz IA, Melo GAR, Engel MS (2011) Euglossa obrima, a new species of orchid bee from Mesoamerica, with notes on the subgenus Dasystilbe Dressler (Hymenoptera, Apidae). Zookeys 97: 11–29.

61. Cardinal S, Packer L (2007) Phylogenetic analysis of the corbiculate Apinae based on morphology of the sting apparatus (Hymenoptera: Apidae). Cladistics 23: 99–118.
62. Engel MS (2000) A new interpretation of the oldest fossil bee (Hymenoptera: Apidae). Am Museum Novit 3296: 1–11.
63. Engel MS (1999) The first fossil *Euglossa* and phylogeny of the orchid bees (Hymenoptera: Apidae; Euglossini). Am Museum Novit 3272, 14 pp.
64. Engel MS (2014) An orchid bee of the genus *Eulaema* in Early Miocene Mexican amber (Hymenoptera: Apidae). Novit Pal 7: 1–15.
65. Hines HM (2008) Historical biogeography, divergence times, and diversification patterns of bumble bees (Hymenoptera: Apidae: Bombus). Syst Biol 57: 58–75.
66. Rasmont P (1988) Monographie écologique et zoogéographique des bourdons de France et de Belgique (Hymenoptera, Apidae, Bombinae). Faculté des Sciences agronomiques de L'Etat, Gembloux, 310+LXII p.
67. Williams PH, Cameron SA, Hines HM, Cederberg B, Rasmont P (2008) A simplified subgeneric classification of the bumblebees (genus *Bombus*). Apidologie 39: 46–74.
68. Dubitzky A (2007) Phylogeny of the world Anthophorini (Hymenoptera: Apoidea: Apidae). Syst Entomol 32: 585–600.
69. Kandemir İ, Özkan A, Fuchs S (2011) Reevaluation of honeybee (*Apis mellifera*) microtaxonomy: a geometric morphometric approach. Apidologie 42: 618–627.
70. Friedman M (2010) Explosive morphological diversification of spiny-finned teleost fishes in the aftermath of the end-Cretaceous extinction. Proc Biol Sci 277: 1675–1683.
71. Tabatabaei Yazdi F, Adriaens D (2013) Cranial variation in *Meriones tristrami* (Rodentia: Muridae: Gerbillinae) and its morphological comparison with *Meriones persicus*, *Meriones vinogradovi* and *Meriones libycus*: a geometric morphometric study. J Zool Syst Evol Res 51: 239–251.
72. Roig-Alsina A, Michener CD (1993) Studies of the phylogeny and classification of long-tongued bees (Hymenoptera: Apoidea). University of Kansas Science Bulletin 55(4): 123–162.
73. Engel MS, Ortega-Blanco J, Nascimbene PC, Singh H (2013) The bees of early Eocene Cambay amber (Hymenoptera: Apidae). Journal of Melittology 25: 1–12.
74. Michener CD, Grimaldi DA (1988) A *Trigona* from Late Cretaceous amber of New Jersey (Hymenoptera: Apidae: Meliponinae). Am Museum Novit 2917: 1–10.
75. Cardinal S, Straka J, Danforth BN (2010) Comprehensive phylogeny of apid bees reveals the evolutionary origins and antiquity of cleptoparasitism. Proc Natl Acad Sci USA 107: 16207–16211.

20

Field Evidence of Colonisation by Holm Oak, at the Northern Margin of Its Distribution Range, during the Anthropocene Period

Sylvain Delzon[1,2]*, Morgane Urli[1,2], Jean-Charles Samalens[3], Jean-Baptiste Lamy[1,2], Heike Lischke[4], Fabrice Sin[5], Niklaus E. Zimmermann[4], Annabel J. Porté[1,2]

1 INRA, UMR 1202 BIOGECO, Cestas, France, 2 Université de Bordeaux, UMR 1202 BIOGECO, Cestas, France, 3 INRA, UR 1263 EPHYSE, Villenave d'Ornon, France, 4 Swiss Federal Research Institute WSL, Birmensdorf, Switzerland, 5 ONF, Agence de Bordeaux, Bruges, France

Abstract

A major unknown in the context of current climate change is the extent to which populations of slowly migrating species, such as trees, will track shifting climates. Niche modelling generally predicts substantial northward shifts of suitable habitats. There is therefore an urgent need for field-based forest observations to corroborate these extensive model simulations. We used forest inventory data providing presence/absence information from just over a century (1880–2010) for a Mediterranean species (*Quercus ilex*) in forests located at the northern edge of its distribution. The main goals of the study were (i) to investigate whether this species has actually spread into new areas during the Anthropocene period and (ii) to provide a direct estimation of tree migration rate. We show that *Q. ilex* has colonised substantial new areas over the last century. However, the maximum rate of colonisation by this species (22 to 57 m/year) was much slower than predicted by the models and necessary to follow changes in habitat suitability since 1880. Our results suggest that the rates of tree dispersion and establishment may also be too low to track shifts in bioclimatic envelopes in the future. The inclusion of contemporary, rather than historical, migration rates into models should improve our understanding of the response of species to climate change.

Editor: Han Y.H. Chen, Lakehead University, Canada

Funding: This study was supported by a grant from the Aquitaine region and the BACCARA and MOTIVE projects, which were funded by the European Community's Seventh Framework Programme (FP7/2007-2013) under grant agreements nos. 226299 and 226544, respectively. The funders had no role in study design, data collection and analysis, decision to publish, or preparation of the manuscript.

Competing Interests: The authors have declared that no competing interests exist.

* E-mail: sylvain.delzon@u-bordeaux1.fr

Introduction

Global temperature rises, increasing atmospheric CO_2 concentration, nitrogen deposition, changes in land use and forest management have altered the production and biodiversity of the terrestrial biosphere [1,2]. Concerns have been raised about the responses to ongoing climatic change in tree species, particularly given the rapid rate of environmental change and the long life-span of these species [3]. Trees may adopt various strategies (migration, adaptation) to cope with rapid environmental changes and landscape fragmentation, but adaptation alone (through genetic adaptation and phenotypic plasticity) is unlikely to be rapid enough to maintain standing populations [4]. A combination of migration and adaptation is, therefore, likely to be the only solution permitting the sustainable survival of tree populations. As a consequence, the boundaries of species distribution ranges may shift. Recent changes in plant distributions have, indeed, been reported in recent decades [1,5]. These migrations are however evident in mountain areas only, in which plant populations are mostly shifting upwards [6–12]. Studies of tree responses to rapid changes in climate are thus essential, to provide us with insight into the possible future distributions of these species and biodiversity.

A better understanding of the link between climate change and distribution range dynamics can be obtained from studies of past changes. Phylogeographic, palynologic and anthracological studies have quantified shifts in tree distribution ranges due to past climate change (alternations of glacial and post-glacial warming periods), providing essential information for the determination of migration directions and rates and for forecasting future shifts of tree species distribution ranges [13,14]. Fossil pollen data have suggested that northward migration is typical for many tree species responding to post-glacial climate warming [15,16]. The ranges of many temperate tree species were estimated to have expanded at rates of $500–1000 \text{ m year}^{-1}$ during the early Holocene [17], but these palynological studies showed that migration rates slow considerably after an initial phase of rapid invasion [17]. One well documented example of range shift is that for Holm oak (*Quercus ilex*), which displayed westward colonisation in the Mediterranean Basin during the Tertiary, demonstrated by studies of chloroplast DNA variation [18]. It remains unclear whether past migrations of trees were limited by rates of climate change or by their dispersal abilities [19], but past rates of migration suggest that even relatively rapid changes in range limits would be insufficient to keep pace with predicted future climatic changes [20,21].

Vegetation distribution models provide additional insight, improving our understanding of the impact of climate change on species distribution and biodiversity. Niche-based models accounting for observed geographical distribution ranges on the basis of climatic and other environmental variables are frequently used to forecast the potential impact of ongoing climatic changes on the distribution and size of plant and animal ranges [22–26]. Many studies based on the use of such models have suggested that species distributions may shift in response to climate change (mostly towards higher latitudes and altitudes; e.g. [27,28]). In the future, climatic conditions may be such that the magnitude of potential range changes is large even for trees [29]. Iverson and Prasad [30] evaluated potential changes in suitable habitats for tree species in the eastern United States and showed a substantial northward shift for 60 of the 80 species studied. Using eight models, ranging from niche-based models to process-based models including ecophysiological processes but no structural dynamics or seed dispersal, Cheaib et al. [31] predicted a significant range expansion, by 2055, for Mediterranean broad-leaved evergreen species (mostly Holm oak) over the western-most two thirds of France. All models have predicted massive range expansions for this species, but they yield very different predictions concerning the magnitude and location of the changes. Thus, although there is sufficient space and suitable habitats for the northward expansion of the distribution ranges of many species, the degree to which species are actually able to achieve such rapid migration remains poorly understood (see [32] for review): model uncertainties concerning the rate and direction of migration create broad differences in projections of the impact of climate change on tree species ranges [31,33]. Thus, in addition to more explicit modelling of the processes underlying migration [34], there is a need for field-based forest observations to confirm these extensive model simulations. Appropriate field observation data for the derivation of migration rates are scarce but of great importance, if we are to understand the response of tree species to global changes [32].

We investigated changes in the distribution of a Mediterranean tree species (Quercus ilex) at the northern edge of its range. This species is expected to be strongly affected by ongoing climate change [31]. Our main purpose was to check the veracity of the recent forecasts of changes in distribution area for this species. We focused on the situation in western France, an area corresponding to the northern limit of the distribution range for this Mediterranean oak. We used field-based forest inventories from the French National Forestry Office, providing presence/absence information for a period of just over a century (1880–2010) for Quercus ilex in four state-owned forests. The two specific goals of the study were (i) to check whether this Mediterranean species is currently colonising new sites at its leading edge, with a magnitude similar to that predicted by models and (ii) to provide a direct estimate of the rate of colonisation from field observations. By quantifying contemporary changes in tree distribution based on reliable information, this study will improve our understanding of the extent to which species are likely to move in the future.

Materials and Methods

Study area and climate

This study was conducted at the northern edge of the distribution range of Holm oak (Quercus ilex L.) in Europe, along the French Atlantic coast, from Bordeaux to Nantes (figure 1). This evergreen tree of the Fagaceae family is a typical component of the flora in Mediterranean climates, its largest populations extending throughout the Iberian Peninsula. The Holm oak

populations of the Atlantic coast dunes in France constitute the northern-most occurrences of Mediterranean evergreen trees. Q. ilex is considered to be drought-tolerant [35,36]. It displays tight stomatal control over transpiration, but may nevertheless display embolism and a loss of hydraulic conductance under severe drought stress [37,38]. The major limitations to the persistence of Holm oak at its northern margin are damage to photosynthetic capacity [39] and cold-related damage to young plants [40]. Holm oak acorns are dispersed during fall mainly by rodents and European jays (Garrulus glandarius). They usually transport acorns far from adult oaks [41,42]. The average distance of acorn dispersal is around 250 m, with dispersal events occurring up to 1 km from source Oak trees. Usually age at maturity of resprouter individuals is 10 years and acorn size can vary strongly between years and sites, with the biggest acorns preferably consumed by predators (wild boars), which might hamper regeneration.

We carried out a retrospective analysis of four state-owned forests covering a broad range of Atlantic dune forests; these forests have been managed by the National Forestry Office (ONF) since 1880 (figure 2). They are mixed pine-oak forests composed principally of a maritime pine (Pinus pinaster) overstorey and a codominant or dominant Holm oak (Quercus ilex) canopy, interspersed with scarce, sparse patches of pedunculate oak (Quercus robur) at the base of the dunes. These Atlantic coastal dune forests (maximum canopy height of 60 m) are subject to a maritime climate, with a mean annual temperature of 12.5°C and precipitation of 850 mm (see table 1). The soil is an arenosol with a filtering siliceous sandy texture and low organic matter content, varying little between sites. The climate warming recorded in France in the 20th century is about 30% greater than mean global warming levels. It has been particularly marked in south-western France, where temperatures have risen by 1.1°C [43] mostly during the second half of the 20th century (1.5°C from 1960 to 2000 in the studied areas, figure S1). This temperature increase has been accompanied by an increase in precipitation during the autumn and winter and lower levels of precipitation in the summer, exacerbating the summer drought period [43].

Retrospective survey data

For more than 120 years, the French National Forestry Office (ONF) has managed the state-owned forests along the Atlantic coast, for which it has carried out a comprehensive inventory of the species present. Four to seven surveys have been carried out during this extensive period, depending on the forest considered. However, in all the forests studied, the first survey was conducted between 1880 and 1891, and the last survey was carried out around the start of the 21st century. All inventories were carried out according to the same methodology in all plots (15 to 90) depending on the studied forest). Briefly, forest managers investigated forest plots, scoring various types of tree species (broadleaved oaks, evergreen oak, pines) as present or absent (including seedlings, saplings and adult trees without distinction). As Holm oak is the only evergreen oak growing in these forests, its presence/absence within all plots was known with certainty. Forest-covered plots were explored at least once during each of the survey periods, such that geographic coverage could be considered complete and identical for all surveys. The observed presence of this species in these forest plots was then used for the construction of maps. The distribution maps obtained during different periods therefore provide estimates of changes in the area occupied by Holm oak.

The range of a colonising species spreads to become irregular in shape over geographic space, due to stochastic processes, geographic boundaries and multiple colonisation fronts. We

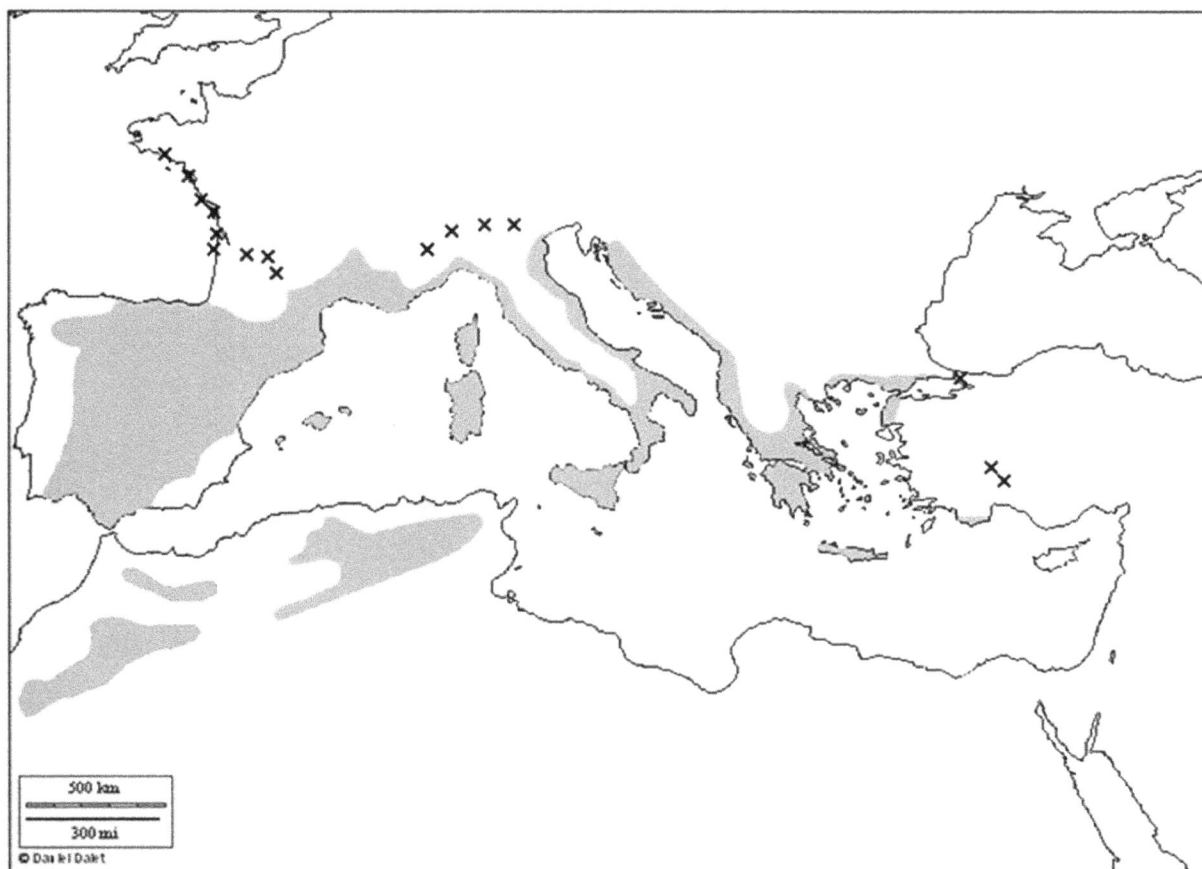

Figure 1. Map of the current distribution range of Holm oak (*Quercus ilex*) in the Mediterranean basin. The distribution range of *Quercus ilex* is shown in grey and crosses represent geographically isolated populations. The map was established according to the findings of Barbero *et al.* [74], Michaud *et al.* [75] and raw data from the French National Forest Inventory (http://inventaire-forestier.ign.fr/edb/query/show-query-form#consultation_panel). The blank base map was provided free of charge and use by Daniel Dalet/Académie Aix-Marseille (http://www.ac-aix-marseille.fr/pedagogie/jcms/c_67064/fr/cartotheque).

therefore estimated the colonisation rate based on the colonized area rather than using a unidirectional distance, which might result in overestimation. For each inventory date, the colonised area, a, was measured by summing the areas of plots in which *Q. ilex* was observed. Assuming an omnidirectional spread of this species within each studied forest, the radius of a circle of the same area was calculated. The rate of colonisation between two inventories (t_{+1}, t) was then estimated as follows:

$$rate = \frac{\sqrt{\frac{a_{(t+1)}}{\pi}} - \sqrt{\frac{a_{(t)}}{\pi}}}{t_{+1} - t}$$

The maximum colonisation rate for each forest studied (highest values throughout the entire period) was preferably used in order to (i) determine the capacity of tree populations to migrate, and (ii) avoid bias due to spatial constraints related to the forest boundaries. Thus we did not compare mean values of colonization rate during the studied period because recent rates were constrained by the limited size of the forests.

Results and Discussion

We found that the distribution of *Q. ilex* had changed markedly over the last century, with massive colonisation occurring in all the studied forests along the Atlantic coast (figure 3). From relict individuals in the studied forests, this species began to colonise new areas in the early 1900s, spreading considerably in the middle of the 20th century. Most inventories now indicate the presence of Holm oak throughout the entire area of the forests investigated. There have been only a few detailed case studies of range expansion in trees during the Anthropocene period, although such events may be very common. Indeed, long-term presence/absence data based on field observations are very scarce. Walther *et al.* [44] demonstrated range expansion for *Ilex aquifolium* in northern Europe in association with an increase in winter temperatures. Forest inventories have been used to estimate tree migration, but always in an indirect manner, by comparing the latitude for seedling and tree biomass [45], or by comparing latitudinal distributions of abundance and occupancy [46], for example. Such studies have concluded that most tree species are currently colonising new areas at their northern margins in the eastern USA. However, these studies have tended to analyse dynamics at the core of the distribution, rather than at its limits. Their analyses

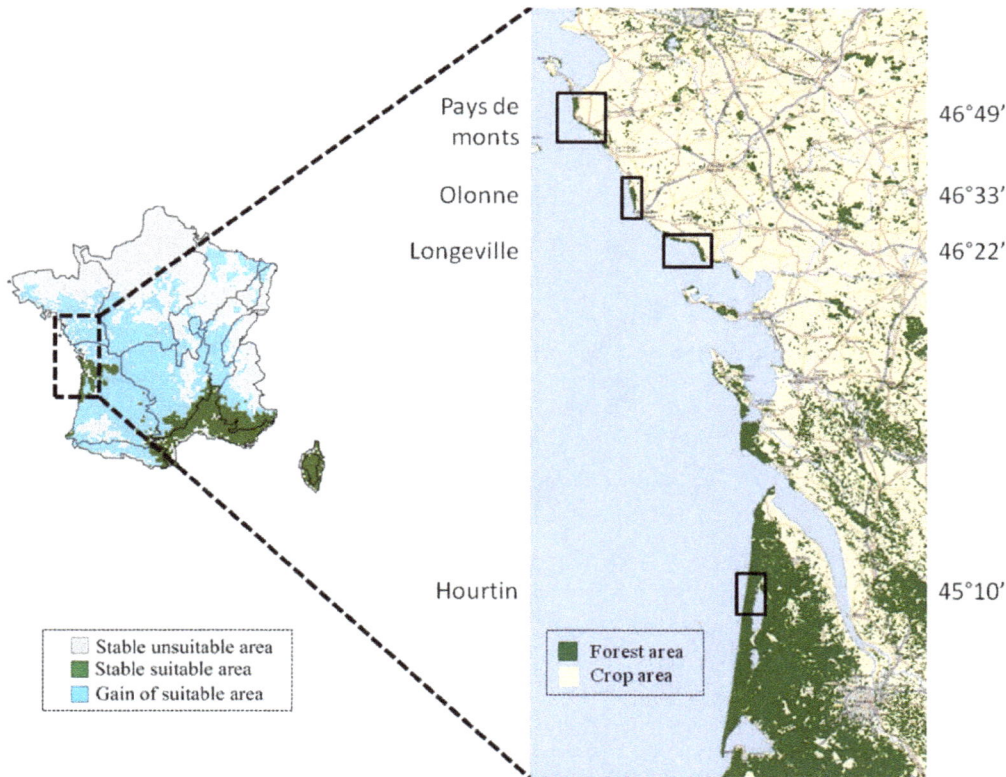

Figure 2. Map of the study area, with locations and mean latitudes of the four state-owned forests used in the retrospective analysis (Pays de Monts, Olonne, Longeville, and Hourtin; right panel). These studied forests represent the northern limit of Holm oak distribution in Europe. All forest covers are represented in green colour. The map in the left panel shows the changes in Holm oak suitable habitat by 2055 in France as predicted by the BIOMOD niche-based model [31]. The base map featuring the forested areas was extracted free of charge and use from the IGN National Forest Inventory cartography site ((© IGN 2013 – BD Forêt V1 des départements 33,17, 85 and FranceRaster® V.1 (2007 ©IGN/CARTOSPHERE))).

therefore fail to portray the dynamics at the northern margin of the species concerned [45]. Zhu *et al.* [47] recently reported that there was little evidence of climate-induced tree migration, with only a few species (20%) presenting a pattern consistent with a northward shift of their northern distribution margin. In a similar study, Gehrig-Fasel *et al.* [48] found that only a small proportion (5%) of observed shift close to or at the upper tree line in Switzerland could readily be attributed to climate-induced migration, the remaining observed changes instead reflecting changes in land use. In the same way, the treeline of *Abies alba* has

Table 1. Location and climatic characteristics of the four forests studied.

Forest	Area ha	Latitude	Longitude	MAP mm	MAT °C
Pays de Monts	2009	46°49′ N	2°08′ W	897	12.11
Olonne	1106	46°33′ N	1°49′ W	821	12.20
Longeville	1520	46°22′ N	1°28′ W	772	12.53
Hourtin	3621	45°10′ N	1°09′ W	864	12.93

Climate data were obtained from Meteo France. MAP is the mean annual precipitation over the period 1961-1990; MAT is the mean annual temperature over the period 1961-1990.

shifted upward in the Alps during the second half of the 20th century, mainly in response to land use change, even though the role of climatic warming cannot be ruled out [49].

The data presented in table 2 are derived from the colonisation maps generated over periods of 112 to 130 years, for similar local spatial scales in each case. The highest values of colonisation rates for the forests studied were between 22 and 57 m yr^{-1}. Colonisation rates were fastest in the southern-most forest of the study area (Hourtin) and decreased with increasing latitude (figure 4), but the changes observed were of similar magnitude in all forests over the period studied. The estimated colonisation rates were much lower than those estimated for post-glacial migrations of trees (> 500 m year^{-1} [1,50,51]), probably because colonising individuals do not encounter established forests during the Holocene [17]. Competition with other species at a local scale may thus have slowed Holm oak colonisation during the Anthropocene [34]. The potential of a species to colonise a new area inevitably depends on its performance relative to those of the resident species, and interactions with other species may clearly slow establishment and maturation, thereby decreasing the rate of migration of a tree species, thereby decreasing the capacity of that species to track the changing climate [52]. The difference in colonisation rates between the Holocene and Anthropocene periods may also reflect habitat fragmentation due to human activity [53,54], forest management or, simply, an overestimation of past migration rates in cases in which outlier populations invisible to the pollen record nucleate range expansions [55].

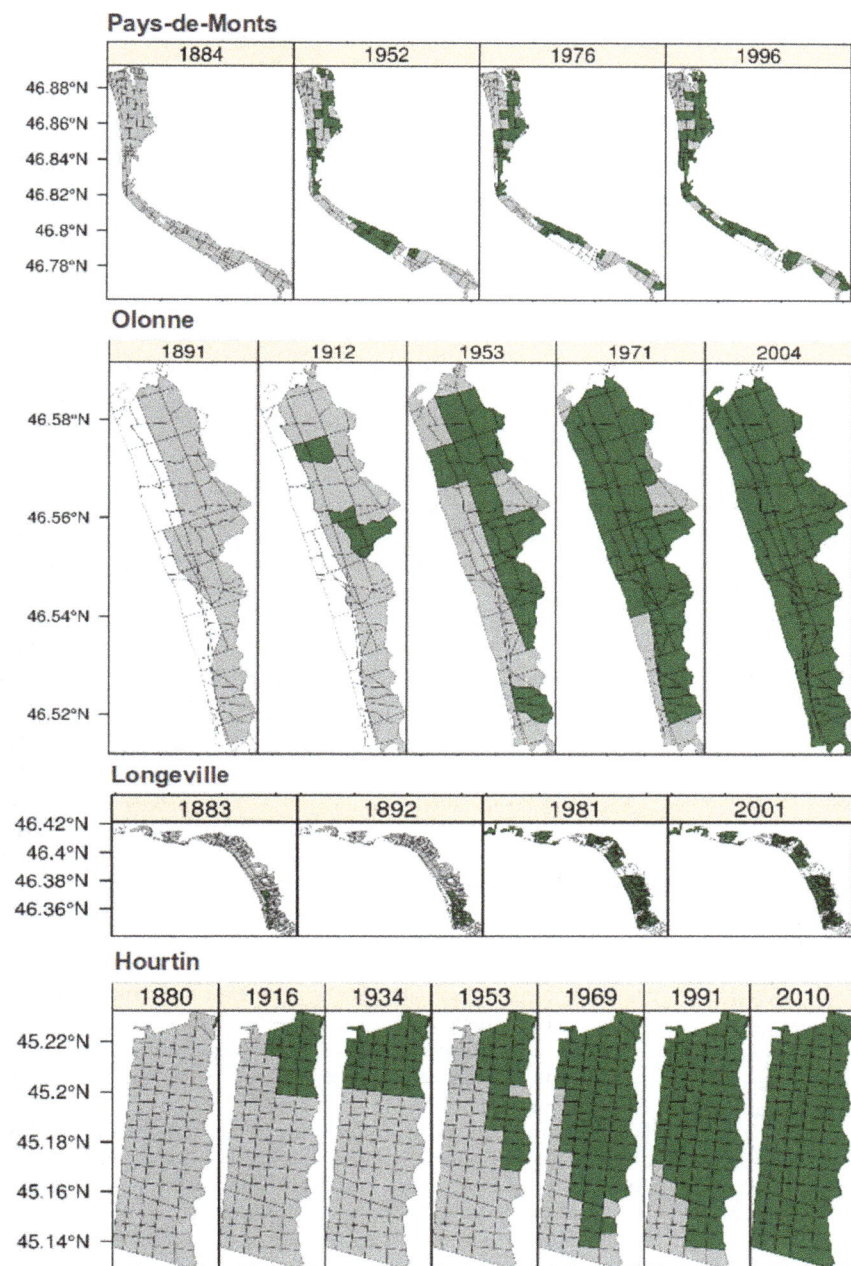

Figure 3. Changes in the area of Holm oak over the 20th century in each of the forests studied. The presence of Holm oak is indicated in green, the forest area is indicated in grey and the non-forest area is indicated in white. Forests are ordered by latitude, from bottom to top: Hourtin, Longeville, Olonne and Pays de Monts. Please note the different scales used.

Indeed, some recent studies have estimated rates of tree species migration during the Holocene of only one fifth to one tenth (<100 m year^{-1}) previously reported values. Such discrepancies may be due to the presence of late glacial refugia close to the northern limits, and a lack of knowledge of these refugia may have biased earlier estimates [56,57]. Our estimated rates (<60 m year^{-1}) are consistent with model predictions based on population dynamics, competition and dispersal [58,59], seed dispersal data (12.5 m year^{-1} for *Q. robur* and *Q. petraea* [60]) and cell-based migration simulation models (probability of colonisation within a zone of 10–20 km around the area currently occupied [61]).

By implicitly including biotic interactions between species, our migration rate estimates represent the actual capacity of the species to disperse into existing forests. They are, therefore, more representative when assessing the response of trees to current climate change than migration rate estimates from paleoecological studies of the last glaciation–deglaciation cycle. However, our field observations were made at a local spatial scale (<4000 ha), so an underestimation of migration rates cannot be excluded. Indeed, the importance of long-distance dispersal for migration has been demonstrated theoretically [62,63]. Seed dispersal is usually considered to be long-distance if it occurs over a distance of more

Table 2. Colonisation rates (m year^{-1}) for Holm oak, estimated by the analysis of its presence distribution maps in the four studied forests along the latitudinal gradient.

Hourtin		Longeville		Olonne		Pays de Monts	
year	m yr^{-1}	year	m yr^{-1}	year	m yr^{-1}	year	m yr^{-1}
1880–1916	36.5	1883–1892	37.3	1891–1912	28.9	1884–1952	21.8
1916–1934	21.0	1892–1981	10.5	1912–1953	17.5	1952–1976	3.9
1934–1953	5.0			1953–1971	20.4	1976–1996	17.3
1953–1969	56.7			1971–2004	5.2		
1969–1991	6.8						
1991–2010	23.1						
maximum rates	56.7		37.3		28.9		21.8

than 100 m. Our study therefore excluded only rare, very long-distance dispersal events (>5 km). However, Iverson *et al.* [61] demonstrated that a species has to have a reasonable abundance close to the limit of its range for colonisation to be likely. These authors concluded that very long-distance migration events alone might not be sufficient to rescue migration. The decreasing colonisation rate with increasing latitude found here confirms the importance of population abundance in the process of migration, with much lower rates of colonisation occurring in the northern-most populations (Pays de Mont/Olonne exhibit the same temperature change over the period studied but dramatically lower abundance of Holm oak).

Our findings confirm that this Mediterranean species is colonising new areas at the northern limits of its distribution range and that it therefore has the potential to shift the edge of its range northwards under projected climate change, as predicted by both niche-based and process-based models [31]. Indeed, all the various types of model indicate that the climate will probably become favourable for this species well to the north of its current bioclimatic range. However, we show here that despite the clear occurrence of migration in this tree species, colonisation rates are much lower than predicted by models. The estimated migration rates obtained here are two orders of magnitude below those

required to track climate change. Indeed, estimates of the migration rates required for plant species to keep pace with climate change over the coming century are frequently >1000 m year^{-1} [64,65], exceeding even the fastest migration rates observed during early Holocene colonisation [66]. However, none of these models takes dispersal capacity into account [31], and their projections of increases in range size should therefore be interpreted with caution. Only a few models have included specific migration rates, estimated from data obtained from paleoecological studies of the last glaciation–deglaciation cycle. The rates used in these models were thus between 1000 and 10000 m year^{-1}, depending on the species considered [67]. Sensitivity analyses of these models have generally identified migration rate as a crucial issue, and their results have indicated that colonisation remains strongly limited by migration rates, even if a rate of 1000 m year^{-1} is used [65]. Finally, our results clearly confirm that the migration of the species studied here will be limited by its ability to disperse and to colonise habitats that have recently become suitable. The time lag for species establishment in new areas is unknown, and similar field estimates of migration rates should be used and combined with modelling approaches, to improve the accuracy of projections and inferences concerning the future distributions of tree species.

This range expansion observed at the northern margin of the Holm oak distribution is likely explained by both climate and recent management changes but their relative contribution cannot be disentangled. Across the region studied, mean annual temperatures have rapidly increased, by 1.5°C during the 20th century, leading to more arid conditions, due to higher evapo-transpiration rates and a change in the seasonal pattern of annual precipitation [68]. This climate warming is probably one of the key drivers of the observed range expansion of Holm oak over the last century, as suggested by the relationship between colonisation rates and latitude or mean annual temperature (MAT; table 1, figure 4). Northern range limitation by temperature has commonly been reported for various plant species [44,69,70]. On the other hand, during the first half of the 20th century (slight change in temperature; see figure S1), forest management probably initially moderated colonisation by Holm oak, limiting the expansion of the range of this species, because there was a general policy of cutting down Holm oaks to favour monospecific maritime pine stands [71,72]. However, this systematic felling policy ceased after the 1950s, when forest managers began to support natural forest dynamics, including Holm oak colonisation. This change in forestry practices has strongly influenced the recruitment of *Quercus ilex* and therefore its colonization rate. However, in all the forests

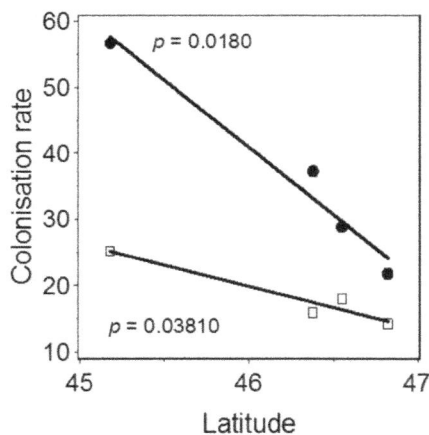

Figure 4. Colonisation rates (m year^{-1}) as a function of the latitude (°) of the studied forests. Maximum (closed circle) and mean (open square) values were estimated for each forest. The *P*-values indicated are those for linear regression analysis.

studied, no land abandonment or Holm oak plantation has been observed; thus the pattern of colonization reported here cannot be attributed to land-use change while in contrary it played a major role in mountainous areas [49]. Both these successive changes in management and the sharp increase in temperature during the second half of the 20th century may have led to an increase in the rate of expansion of the distribution of Holm oak, as this species is known to recover rapidly from disturbances, such as felling and drought, through highly dynamic resprouting [73]. These changes were indeed reflected in estimated colonisation rates, which were higher in the second half of the 20th century (table 2).

Conclusions

The area along the Atlantic coast of France colonised by Holm oak increased steadily throughout most of the 20th century due to both management and climate changes, but the rate of range expansion by migration is much lower than would be required to track future climate change according to the predictions of niche modelling and phylogeographic studies. At its current rate of expansion, this species will not be able to colonise all the climatically suitable habitats that are likely to appear in France and elsewhere over the next 50 years or so. Even if a population disperses to a new region with a favourable climate, interactions with other species may prevent its establishment and further spread. Thus, predictions of substantial range shifts should probably be tempered, at least for plant species, and particularly for long-lived species, such as trees. The prediction of shifts in species ranges as a result of climate change could be improved by

taking migration processes into account explicitly in models. Measurements of contemporary colonisation rates are invaluable for providing realistic estimates of population range dynamics, and should be obtained and used more frequently in assessments of the potential of a species to track climate change.

Supporting Information

Figure S1 Temperature trends in the four studied areas over the twentieth century (left) and over the four last decades (right). Linear regressions fitted to the annual means are also depicted and showed an increase of 1.5°C during the last four decades. Temperature changes were statistically significant at $P<0.001$ level and no significant differences of temperature changes have been found between areas. Data sources: Météo France (stations n° 33009001, 33236002, 85216001 and 85234001).

Acknowledgments

We thank Arndt Hampe and Antoine Kremer for providing stimulating advice, comments and discussion.

Author Contributions

Conceived and designed the experiments: SD NEZ AJP. Performed the experiments: MU JBL FS. Analyzed the data: SD MU JCS HL. Contributed reagents/materials/analysis tools: SD MU JCS HL. Wrote the paper: SD NEZ AJP.

References

1. Parmesan C, Yohe G (2003) A globally coherent fingerprint of climate change impacts across natural systems. Nature 421: 37–42.
2. Vitousek PM, Mooney HA, Lubchenco J, Melillo JM (1997) Human domination of Earth's ecosystems. Science 277: 494–499.
3. Aitken SN, Yeaman S, Holliday JA, Wang T, Curtis-McLane S (2008) Adaptation, migration or extirpation: climate change outcomes for tree populations. Evol Applic 1: 95–111. 10.1111/j.1752-4571.2007.00013.x.
4. Lindner M, Maroschek M, Netherer S, Kremer A, Barbati A et al. (2010) Climate change impacts, adaptive capacity, and vulnerability of European forest ecosystems. For Ecol Manage 259: 698–709.
5. Bertin RI (2008) Plant phenology and distribution in relation to recent climate change. J Torrey Bot Soc 135: 126–146. http://dx.doi.org/10.3159/07-RP-035R.1.
6. Grabherr G, Gottfried M, Pauli H (2009) Climate effects on mountain plants. Nature 369: 448. 10.1038/369448a0.
7. Kullman L (2002) Rapid recent range-margin rise of tree and shrub species in the Swedish Scandes. J Ecol 90: 68–77. 10.1046/j.0022-0477.2001.00630.x.
8. Lenoir J, Gegout JC, Marquet PA, De Ruffray P, Brisse H (2008) A significant upward shift in plant species optimum elevation during the 20th century. Science 320: 1768–1771. 10.1126/science.1156831.
9. Peñuelas J, Boada M (2003) A global change-induced biome shift in the Montseny mountains (NE Spain). Global Change Biol 9: 131–140.
10. Shiyatov SG, Terent'ev MM, Fomin VV, Zimmermann NE (2007) Altitudinal and horizontal shifts of the upper boundaries of open and closed forests in the Polar Urals in the 20th century. Russian J Ecol 38: 223–227.
11. Urli M, Delzon S, Eyermann A, Couallier V, García-Valdés R et al. (2013) Inferring shifts in tree species distribution using asymmetric distribution curves: a case study in the Iberian mountains. J Veg Sci in press.
12. Wardle P, Coleman MC (1992) Evidence for rising upper limits of four native New Zealand forest trees. NZ J Bot 30: 303–314.
13. Petit RJ, Hu FS, Dick CW (2008) Forests of the past: a window to future changes. Science 320: 1450–1452.
14. Bush MB, Silman MR, Urrego DH (2004) 48,000 years of climate and forest change in a biodiversity hot spot. Science 303: 827–829. 10.1126/science.1090795.
15. Davis MB, Shaw RG (2001) Range shifts and adaptive responses to Quaternary climate change. Science 292: 673–679. 10.1126/science.292.5517.673.
16. King GA, Herstrom AA (1996) Holocene tree migration rates objectively determined from fossil pollen data. In: Huntley B, Cramer W, Marsa AV, Prentice IC, Allen JRM, editors. Past and future environmental changes: the spatial and evolutionary responses of terrestrial biota. New York, USA: Springer Verlag. pp. 91–102.
17. Birks HJB (1989) Holocene isochrone maps and patterns of tree-spreading in the British Isles. J Biogeogr 16: 503–540.
18. Lumaret R, Mir C, Michaud H, Raynal V (2008) Phylogeographical variation of chloroplast DNA in holm oak (Quercus ilex L.). Mol Ecol 11: 2327–2336.
19. Clark JS, Fastie C, Hurtt G, Jackson ST, Johnson C et al. (1998) Reid's paradox of rapid plant migration. BioScience 48: 13–24.
20. Gear AJ, Huntley B (1991) Rapid changes in the range limits of Scots pine 4000 years ago. Science 251: 544. 10.1126/science.251.4993.544.
21. Pearson RG (2006) Climate change and the migration capacity of species. Trends Ecol Evol 21: 111–113.
22. Araújo MB, Thuiller W, Pearson RG (2006) Climate warming and the decline of amphibians and reptiles in Europe. J Biogeogr 33: 1712–1728. 10.1111/j.1365-2699.2006.01482.x.
23. Huntley B, Collingham YC, Willis SG, Green RE (2008) Potential impacts of climatic change on European breeding birds. PLoS One 3: e1439. 1371/journal.pone.0001439.
24. Jetz W, Wilcove DS, Dobson AP (2007) Projected impacts of climate and land-use change on the global diversity of birds. PLoS Biology 5: e157. 10.1371/journal.pbio.0050157.
25. Thomas CD, Cameron A, Green RE, Bakkenes M, Beaumont LJ et al. (2004) Extinction risk from climate change. Nature 427: 145–148.
26. Thuiller W, Lavorel S, Araújo MB, Sykes MT, Prentice IC (2005) Climate change threats to plant diversity in Europe. PNAS 102: 8245–8250.
27. Guisan A, Thuiller W (2005) Predicting species distribution: offering more than simple habitat models. Ecol Lett 8: 993–1009. 10.1111/j.1461-0248.2005.00792.x.
28. Thuiller W, Albert C, Araújo MB, Berry PM, Cabeza M et al. (2008) Predicting global change impacts on plant species' distributions: future challenges. Perspect Plant Ecol 9: 137–152.
29. Shafer SL, Bartlein PJ, Thompson RS (2001) Potential changes in the distributions of western North America tree and shrub taxa under future climate scenarios. Ecosystems 4: 200–215.
30. Iverson LR, Prasad AM (2001) Potential changes in tree species richness and forest community types following climate change. Ecosystems 4: 186–199. 10.1007/s10021-001-0003-6.
31. Cheaib A, Badeau V, Boe J, Chuine I, Delire C et al. (2012) Climate change impacts on tree ranges: model intercomparison facilitates understanding and quantification of uncertainty. Ecol Lett. 10.1111/j.1461-0248.2012.01764.x.
32. Corlett RT, Westcott DA (2013) Will plant movements keep up with climate change? Trends Ecol Evol 28: 482–488.
33. Higgins SI, Clark JS, Nathan R, Hovestadt T, Schurr FM et al. (2003) Forecasting plant migration rates: managing uncertainty for risk assessment. J Ecol 91: 341–347. 10.1046/j.1365-2745.2003.00781.x.

34. Neilson RP, Pitelka LF, Solomon AM, Nathan R, Midgley GF et al. (2005) Forecasting regional to global plant migration in response to climate change. BioScience 55: 749–759.

35. Damesin C, Rambal S, Joffre R (1998) Seasonal and annual changes in leaf delta C-13 in two co-occurring Mediterranean oaks: relation to leaf growth and drought progression. Funct Ecol 12: 778–785. 10.1046/j.1365-2435.1998.00259.x.

36. Tyree MT, Cochard H (1996) Summer and winter embolism in oak: impact on water relations. Ann Sci For 53: 173–180.

37. Limousin JM, Rambal S, Ourcival JM, Rocheteau A, Joffre R et al. (2009) Long-term transpiration change with rainfall decline in a Mediterranean Quercus ilex forest. Global Change Biol 15: 2163–2175. 10.1111/j.1365-2486.2009.01852.x.

38. Martínez-Vilalta J, Prat E, Oliveras I, Piñol J (2002) Xylem hydraulic properties of roots and stems of nine Mediterranean woody species. Oecologia 133: 19–29.

39. Larcher W, Tisi F (1990) Bioclima invernale e rendimento carbonico di Quercus ilex al limite settentrionale delle leccete prealpine. Atti della Accademia Nazionale dei Lincei 387: 3–22.

40. Larcher W, Mair B (1969) Die Temperaturresistenz als ökophysiologisches Konstitutionsmerkmal. 1. Quercus ilex und andere Eichenarten des Mittel-meergebietes. Oecologia Plantarum 4: 347–376. 10.1111/j.1752-4571.2007.00013.x.

41. Gómez JM, Puerta-Piñero C, Schupp EW (2008) Effectiveness of rodents as local seed dispersers of Holm oaks. Oecologia 155: 529–537.

42. Gómez JM (2003) Spatial patterns in long-distance dispersal of Quercus ilex acorns by jays in a heterogeneous landscape. Ecography 26: 573–584.

43. Moisselin JM, Schneider M, Canellas C (2002) Les changements climatiques en France au XXème siècle. Etude des longues séries homogénéisées de données de température et de précipitations. La météorologie 38: 45–56.

44. Walther GR, Berger S, Sykes MT (2005) An ecological footprint of climate change. P Roy Soc B-Biol Sci 272: 1427–1432.

45. Woodall CW, Oswalt CM, Westfall JA, Perry CH, Nelson MD et al. (2009) An indicator of tree migration in forests of the eastern United States. For Ecol Manage 257: 1434–1444.

46. Murphy HT, VanDerWal J, Lovett-Doust J (2010) Signatures of range expansion and erosion in eastern North American trees. Ecol Lett 13: 1233–1244.

47. Zhu K, Woodall CW, Clark JS (2012) Failure to migrate: lack of tree range expansion in response to climate change. Global Change Biol 18: 1042–1052.

48. Gehrig-Fasel J, Guisan A, Zimmermann NE (2007) Treeline shifts in the Swiss Alps: climate change or land abandonment? J Veg Sci 18: 571–582. 10.1111/j.1654-1103.2007.tb02571.x.

49. Chauchard S, Beilhe F, Denis N, Carcaillet C (2010) An increase in the upper tree-limit of silver fir (Abies alba Mill.) in the Alps since the mid-20th century: A land-use change phenomenon. For Ecol Manage 259: 1406–1415.

50. Noss RF (2002) Beyond Kyoto: forest management in a time of rapid climate change. Conservation Biology 15: 578–590.

51. Schwartz MW (1993) Modelling effects of habitat fragmentation on the ability of trees to respond to climatic warming. Biodiv Conserv 2: 51–61.

52. Ibañez I, Clark JS, Dietze MC (2008) Estimating colonization potential of migrant tree species. Global Change Biol 15: 1173–1188. 10.1111/j.1365-2486.2008.01777.x.

53. Higgins SI, Lavorel S, Revilla E (2003) Estimating plant migration rates under habitat loss and fragmentation. Oikos 101: 354–366. 10.1034/j.1600-0706.2003.12141.x.

54. Honnay O, Verheyen K, Butaye J, Jacquemyn H, Bossuyt B et al. (2002) Possible effects of habitat fragmentation and climate change on the range of forest plant species. Ecol Lett 5: 525–530. 10.1046/j.1461-0248.2002.00346.x.

55. McLachlan JS, Clark JS (2004) Reconstructing historical ranges with fossil data at continental scales. For Ecol Manage 197: 139–147.

56. Stewart JR, Lister AM (2001) Cryptic northern refugia and the origins of the modern biota. Trends Ecol Evol 16: 608–613.

57. McLachlan JS, Clark JS, Manos PS (2005) Molecular indicators of tree migration capacity under rapid climate change. Ecology 86: 2088–2098.

58. Meier ES, Lischke H, Schmatz DR, Zimmermann NE (2011) Climate, competition and connectivity affect future migration and ranges of European trees. Global Ecology and Biogeography 21: 164–178.

59. Lischke H (2005) Modeling tree species migration in the Alps during the Holocene: What creates complexity? Ecol Complex 2: 159–174.

60. Gerber S, Latouche-Hallé C, Lourmas M, Morand-Prieur ME, Oddou-Muratorio S et al. (2003) Mesure directe des flux de gènes en forêt. Les actes du BRG 4: 349–368.

61. Iverson LR, Schwartz MW, Prasad AM (2004) How fast and far might tree species migrate in the eastern United States due to climate change? Global Ecology and Biogeography 13: 209–219. 10.1111/j.1466-822X.2004.00093.x.

62. Cain ML, Milligan BG, Strand AE (2000) Long-distance seed dispersal in plant populations. Am J Bot 87: 1217–1227.

63. Higgins SI, Richardson DM (1999) Predicting plant migration rates in a changing world: the role of long-distance dispersal. Am Nat 153: 464–475. 10.1086/303193.

64. Malcolm JR, Markham A, Neilson RP, Garaci M (2002) Estimated migration rates under scenarios of global climate change. J Biogeogr 29: 835–849.

65. Morin X, Thuiller W (2009) Comparing niche-and process-based models to reduce prediction uncertainty in species range shifts under climate change. Ecology 90: 1301–1313.

66. Huntley Brian and Birks, H. J B. (1983) An atlas of Past and Present Pollen Maps for Europe: 0 - 13000 years ago. Cambridge, UK: Cambridge University Press.

67. Morin X, Viner D, Chuine I (2008) Tree species range shifts at a continental scale: new predictive insights from a process-based model. J Ecol 96: 784–794.

68. IPCC 2007 (2007) Climate change 2007: the physical science basis. In: Solomon S, Qin D, Manning M, Chen Z, Marquis M et al., editors. Contribution of working group I to the fourth assessment report of the intergovernmental panel on climate change. United Kingdom and New York, NY, USA: Cambridge University Press. pp. -996.

69. Morin X, Augspurger C, Chuine I (2007) Process-based modeling of species distributions: what limits temperate tree species' range boundaries? Ecology 88: 2280–2291.

70. Woodward FI, Lomas MR, Kelly CK (2004) Global climate and the distribution of plant biomes. Phil Trans R Soc B 359: 1465–1476.

71. ONF (1936) Plan d'aménagement à Hourtin. Agence interdépartementale de Bordeaux, France: ONF Direction territoriales du Sud Ouest.

72. ONF (1951) Plan d'aménagement à Hourtin. Agence interdépartementale de Bordeaux, France: ONF Direction territoriales du Sud Ouest.

73. Espelta JM, Retana J, Habrouk A (2003) Resprouting patterns after fire and response to stool cleaning of two coexisting Mediterranean oaks with contrasting leaf habits on two different sites. For Ecol Manage 179: 401–414. 10.1016/S0378-1127(02)00541-8.

74. Barbero M, Loisel R, Quézel P (1992) Biogeography, ecology and history of Mediterranean Quercus ilex ecosystems. Plant Ecol 99: 19–34. 10.1007/BF00118207.

75. Michaud H, Toumi L, Lumaret R, Li TX, Romane F et al. (1995) Effect of geographical discontinuity on genetic variation in Quercus ilex L.(holm oak). Evidence from enzyme polymorphism. Heredity 74: 590–606.

Permissions

All chapters in this book were first published in PLOS ONE, by The Public Library of Science; hereby published with permission under the Creative Commons Attribution License or equivalent. Every chapter published in this book has been scrutinized by our experts. Their significance has been extensively debated. The topics covered herein carry significant findings which will fuel the growth of the discipline. They may even be implemented as practical applications or may be referred to as a beginning point for another development.

The contributors of this book come from diverse backgrounds, making this book a truly international effort. This book will bring forth new frontiers with its revolutionizing research information and detailed analysis of the nascent developments around the world.

We would like to thank all the contributing authors for lending their expertise to make the book truly unique. They have played a crucial role in the development of this book. Without their invaluable contributions this book wouldn't have been possible. They have made vital efforts to compile up to date information on the varied aspects of this subject to make this book a valuable addition to the collection of many professionals and students.

This book was conceptualized with the vision of imparting up-to-date information and advanced data in this field. To ensure the same, a matchless editorial board was set up. Every individual on the board went through rigorous rounds of assessment to prove their worth. After which they invested a large part of their time researching and compiling the most relevant data for our readers.

The editorial board has been involved in producing this book since its inception. They have spent rigorous hours researching and exploring the diverse topics which have resulted in the successful publishing of this book. They have passed on their knowledge of decades through this book. To expedite this challenging task, the publisher supported the team at every step. A small team of assistant editors was also appointed to further simplify the editing procedure and attain best results for the readers.

Apart from the editorial board, the designing team has also invested a significant amount of their time in understanding the subject and creating the most relevant covers. They scrutinized every image to scout for the most suitable representation of the subject and create an appropriate cover for the book.

The publishing team has been an ardent support to the editorial, designing and production team. Their endless efforts to recruit the best for this project, has resulted in the accomplishment of this book. They are a veteran in the field of academics and their pool of knowledge is as vast as their experience in printing. Their expertise and guidance has proved useful at every step. Their uncompromising quality standards have made this book an exceptional effort. Their encouragement from time to time has been an inspiration for everyone.

The publisher and the editorial board hope that this book will prove to be a valuable piece of knowledge for researchers, students, practitioners and scholars across the globe.

List of Contributors

Fengjiang Li, Naiqin Wu, Yajie Dong and Dan Zhang
Key Laboratory of Cenozoic Geology and Environment, Institute of Geology and Geophysics, Chinese Academy of Sciences, Beijing, China

Denis-Didier Rousseau
Laboratoire de Meteorologie Dynamique, UMR INSU-CNRS 8539 & CERES-ERTI, Ecole Normale Superieure, Paris, France
Lamont-Doherty Earth Observatory of Columbia University, Palisades, New York, United States of America

Yunpeng Pei
School of the Earth Sciences and Resources, China University of Geosciences, Beijing, China

Juan Viruel
Departamento de Agricultura y Economía Agraria, Escuela Politécnica Superior de Huesca, Universidad de Zaragoza, Huesca, Spain

Pilar Catalán
Departamento de Agricultura y Economía Agraria, Escuela Politécnica Superior de Huesca, Universidad de Zaragoza, Huesca, Spain
Department of Botany, Institute of Biology, Tomsk State University, Tomsk, Russia

José Gabriel Segarra-Moragues
Departamento de Ecología, Centro de Investigaciones sobre Desertificación (CIDE), Consejo Superior de Investigaciones Científicas (CSIC), Moncada, Valencia, Spain

Jaelyn J. Eberle
University of Colorado Museum of Natural History and Department of Geological Sciences, University of Colorado at Boulder, Boulder, Colorado, United States of America

Michael D. Gottfried
Department of Geological Sciences and Museum, Michigan State University, East Lansing, Michigan, United States of America

J. Howard Hutchison
University of California Museum of Paleontology, Berkeley, California, United States of America

Christopher A. Brochu
Department of Earth and Environmental Sciences, University of Iowa, Iowa City, Iowa, United States of America

Hugo I. Martínez-Cabrera
Estación Regional del Noroeste, Instituto de Geología, Universidad Nacional Autónoma de México, Hermosillo, México

Emilio Estrada-Ruiz
Laboratorio de Ecología, Departamento de Zoología, Escuela Nacional de Ciencias Biológicas – Instituto Politécnico Nacional, Ciudad de México, México

Trisha L. Spanbauer and Sherilyn C. Fritz
Department of Earth and Atmospheric Sciences and School of Biological Sciences, University of Nebraska–Lincoln, Lincoln, Nebraska, United States of America

Craig R. Allen
U.S. Geological Survey, Nebraska Cooperative Fish and Wildlife Research Unit, School of Natural Resources, University of Nebraska–Lincoln, Lincoln, Nebraska, United States of America

David G. Angeler
Department of Aquatic Sciences and Assessment, Swedish University of Agricultural Sciences, Uppsala, Sweden

Tarsha Eason and Ahjond S. Garmestani
Office of Research and Development, National Risk Management Research Laboratory, U.S. Environmental Protection Agency, Cincinnati, Ohio, United States of America

Kirsty L. Nash
Australian Research Council Centre of Excellence for Coral Reef Studies, James Cook University, Townsville, Queensland, Australia

Jeffery R. Stone
Department of Earth and Environmental Systems, Indiana State University, Terre Haute, Indiana, United States of America

Bjorn M. M. Boysen
Department of Environmental and Primary Resources, State of Victoria, East Melbourne, Victoria, Australia

Michael N. Evans
Department of Geology and Earth System Science Interdisciplinary Center, University of Maryland, College Park, Maryland, United States of America

Patrick J. Baker
Department of Forest and Ecosystem Science, Melbourne School ofLand and Environment, University of Melbourne, Victoria, Australia

Debapriyo Chakraborty
Nature Conservation Foundation, Gokulam Park, Mysore, India
National Centre for Biological Sciences, GKVK Campus, Bangalore, India

Anindya Sinha
Nature Conservation Foundation, Gokulam Park, Mysore, India
National Centre for Biological Sciences, GKVK Campus, Bangalore, India
National Institute of Advanced Studies, Indian Institute of Science Campus, Bangalore, India

Uma Ramakrishnan
National Centre for Biological Sciences, GKVK Campus, Bangalore, India

Luciano Bosso
Wildlife Research Unit, Dipartimento di Agraria, Universita` degli Studi di Napoli Federico II, Portici, Napoli, Italy

Danilo Russo
Wildlife Research Unit, Dipartimento di Agraria, Università degli Studi di Napoli Federico II, Portici, Napoli, Italy
School of Biological Sciences, University of Bristol, Bristol, United Kingdom

Hugo Rebelo
School of Biological Sciences, University of Bristol, Bristol, United Kingdom
CIBIO, Centro de Investigação em Biodiversidade e Recursos Genéticos da Universidade do Porto, University of Porto, Vairã, Portuga

Mirko Di Febbraro
EnvixLab, Dipartimento Bioscienze e Territorio, Universitá del Molise, Pesche, Italy

Mauro Mucedda
Centro per lo studio e la protezione dei pipistrelli in Sardegna, Sassari, Italy

Luca Cistrone
Forestry and Conservation, Cassino, Frosinone, Italy

Paolo Agnelli
Museo di Storia Naturale dell'Universitá di Firenze, Sezione di Zoologiá La Specolá, Firenze, Italy

Pier Paolo De Pasquale
Wildlife Consulting, Palo del Colle, Bari, Italy

Adriano Martinoli
Unitá di Analisi e Gestione delle Risorse Ambientali, Guido Tosi Research Group, Dipartimento di Scienze Teoriche e Applicate, Universitá degli Studi dell'Insubria, Varese, Italy

Dino Scaravelli
Dipartimento di Scienze Mediche Veterinarie, Universitá degli Studi di Bologna, Ozzano dell'Emilia, Bologna, Italy

Cristiano Spilinga
Studio Naturalistico Hyla snc, Tuoro sul Trasimeno, Perugia, Italy

Fernando Rodrigues da Silva and Mariana Victorino Nicolosi Arena
Departamento de Ciências Ambientais, Universidade Federal de São Carlos, Sorocaba, São Paulo, Brazil

Mário Almeida-Neto
Departamento de Ecologia, Universidade Federal de Goiás, Goiânia, Goiás, Brazil

Timme H. Donders and Friederike Wagner-Cremer
Palaeoecology, Department of Physical Geography, Faculty of Geosciences, Utrecht University, Laboratory of Palaeobotany and Palynology, Utrecht, The Netherlands

Kimberley Hagemans
Palaeoecology, Department of Physical Geography, Faculty of Geosciences, Utrecht University, Laboratory of Palaeobotany and Palynology, Utrecht, The Netherlands
Department of Environmental Sciences, Copernicus Institute, Faculty of Geosciences, Utrecht University, Utrecht, The Netherlands

Stefan C. Dekker
Department of Environmental Sciences, Copernicus Institute, Faculty of Geosciences, Utrecht University, Utrecht, The Netherlands

Letty A. de Weger
Department of Pulmonology, Leiden University Medical Centre, Leiden, The Netherlands

Pim de Klerk
Botany section, Staatliches Museum für Naturkunde Karlsruhe, Karlsruhe, Germany

Tony Chang, Andrew J. Hansen and Nathan Piekielek
Department of Ecology, Montana State University, Bozeman, Montana, United States of America

Danielle Fraser and Natalia Rybczynski
Department of Biology, Carleton University, Ottawa, Ontario, Canada
Palaeobiology, Canadian Museum of Nature, Ottawa, Ontario, Canada

Christopher Hassall
Department of Biology, Carleton University, Ottawa, Ontario, Canada
School of Biology, University of Leeds, Leeds, United Kingdom

Root Gorelick
Department of Biology, Carleton University, Ottawa, Ontario, Canada
Department of Mathematics and Statistics, Carleton University, Ottawa, Ontario, Canada Institute of Interdisciplinary Studies, Carleton University, Ottawa, Ontario Canada

Yiling Wang, Guiqin Yan
College of Life Sciences, Shanxi Normal University, Linfen, China

Mark O. Cuthbert
Connected Waters Initiative Research Centre, UNSW Australia, Sydney, NSW, Australia,

School of Geography, Earth and Environmental Sciences, University of Birmingham, Birmingham, United Kingdom

Gail M. Ashley
Department of Earth and Planetary Sciences, Rutgers University, Piscataway, New Jersey, United States of America

Ron Pinhasi
Earth Institute and School of Archaeology, University College Dublin, Dublin, Ireland

Tengiz Meshveliani, David Lordkipanidze and Nino Jakeli
Georgian State Museum, Department of Prehistory, Tbilisi, Georgia

Zinovi Matskevich
Israel Antiquities Authority, Jerusalem, Israel

Guy Bar-Oz and Lior Weissbrod
Zinman Institute of Archaeology, University of Haifa, Haifa, Israel

Christopher E. Miller
Institute for Archaeological Sciences, and Senckenberg Centre for Human Evolution and Paleoenvironment, University of Tübingen, Tübingen, Germany

Keith Wilkinson
Department of Archaeology, University of Winchester, Winchester, United Kingdom

Eliso Kvavadze
Institute of Paleobiology, National Museum of Georgia, Tbilisi, Georgia

Thomas F. G. Higham
Oxford Radiocarbon Accelerator Unit, Research Laboratory for Archaeology & the History of Art, University of Oxford, Oxford, United Kingdom

Anna Belfer-Cohen
Institute of Archaeology, Hebrew University, Jerusalem, Israel

Hedvig K. Nenzén
Département des sciences biologiques, Université du Québec à Montréal, Montréal, Québec, Canada

Daniel Montoya
School of Biological Sciences, Life Sciences Building, University of Bristol, Bristol, United Kingdom

Sara Varela
Department of Ecology, Faculty of Science, Charles University, Prague, Czech Republic

Jun Lü, Shao-ji Hu and Hui Ye
Laboratory of Biological Invasion and Ecosecurity, Yunnan University, Kunming, 650091, China
Yunnan Key Laboratory of International Rivers and Transboundary Ecosecurity, Yunnan University, Kunming, 650091, China

Xue-yu Ma
Laboratory of Biological Invasion and Ecosecurity, Yunnan University, Kunming, 650091, China
School of Mathematics and Computer Science, Yunnan University of Nationalities, Kunming, 650031, China

Jin-min Chen
Laboratory for Conservation and Utilization of Bio-resources, Yunnan University, Kunming, 650091, China

Qing-qing Li
Life Science College, Yunnan Normal University, Kunming, 650092, China

María Lozano-Jaramillo and Carlos Daniel Cadena
Laboratorio de Biología Evolutiva de Vertebrados, Departamento de Ciencias Biológicas, Universidad de los Andes, Bogotá, Colombia

Alejandro Rico-Guevara
Department of Ecology & Evolutionary Biology, University of Connecticut, Storrs, Connecticut, United States of America

Manuel Dehon and Denis Michez
Laboratory of Zoology, Research Institute of Biosciences, University of Mons, Mons, Belgium

André Nel
Département d'entomologie, Muséum National d'Histoire Naturelle, Centre National de la Recherche Scientifique, Unité Mixte de Recherche, Paris, France

Michael S. Engel
Division of Invertebrate Zoology, American Museum of Natural History, New York, New York, United States of America, and Division of Entomology (Paleoentomology), Natural History Museum, and Department of Ecology and Evolutionary Biology, University of Kansas, Lawrence, Kansas, United States of America

Thibaut De Meulemeester
Laboratory of Zoology, Research Institute of Biosciences, University of Mons, Mons, Belgium
Naturalis Biodiversity Center, Leiden, the Netherlands

Sylvain Delzon, Morgane Urli, Jean-Baptiste Lamy and Annabel J. Porté
INRA, UMR 1202 BIOGECO, Cestas, France
Université de Bordeaux, UMR 1202 BIOGECO, Cestas, France

Jean-Charles Samalens
INRA, UR 1263 EPHYSE, Villenave d'Ornon, France

Niklaus E. Zimmermann and Heike Lischke
Swiss Federal Research Institute WSL, Birmensdorf, Switzerland

Fabrice Sin
ONF, Agence de Bordeaux, Bruges, France

Index

www.ingramcontent.com/pod-product-compliance
Lightning Source LLC
Chambersburg PA
CBHW061247190326
41458CB00011B/3598